High-Entropy Materials for Energy Storage Devices

High-Entropy Materials for Energy Storage Devices

Edited by Chien-Te Hsieh, Pradeep Kumar Panda, and Arpan Kumar Nayak

WILEY-VCH

Editors

Chien-Te Hsieh
Department of Chemical Engineering
and Materials Science
Yuan Ze University
Taoyuan 32003
Taiwan

Pradeep Kumar Panda
Department of Chemical Engineering
and Materials Science
Yuan Ze University
Taoyuan 32003
Taiwan

Arpan Kumar Nayak
Department of Education in Science and
Mathematics (Physics Section)
Regional Institute of Education
(National Council of Educational
Research and Training)
Mysore Karnataka
Karnataka 570006
India

Cover Design: Wiley
Cover Image: © YiuCheung/stock.adobe.com

Typesetting: Straive, Chennai, India
Printing and Binding: CPI Group (UK) Ltd, Croydon, CR0 4YY

Library of Congress Card No.: applied for

British Library Cataloguing-in-Publication Data A catalogue record for this book is available from the British Library.

Bibliographic information published by the Deutsche Nationalbibliothek
The Deutsche Nationalbibliothek lists this publication in the Deutsche Nationalbibliografie; detailed bibliographic data are available on the Internet at <http://dnb.d-nb.de>.

Print ISBN: 978-3-527-35558-7
ePDF ISBN: 978-3-527-85366-3
ePub ISBN: 978-3-527-85365-6
oBook ISBN: 978-3-527-85367-0

C9783527355587_270226

Contents

Preface

The development of advanced materials for energy storage devices, such as batteries, supercapacitors, and dielectric capacitors is becoming increasingly important as the world's energy demands continue to rise. The broad adoption of renewable energy technologies is constantly hampered by the inherent performance, durability, and cost-effectiveness limitations of traditional materials, despite their extensive research and optimization. With the introduction of high-entropy materials (HEMs) in recent decades, a paradigm shift in materials science has started to address these issues. HEMs adopt a multi-elemental philosophy, usually incorporating five or more elements in equiatomic or near-equiatomic ratios, in contrast to the traditional method of designing materials around one or two principal elements. Increased strength, outstanding thermal stability, superior corrosion and wear resistance, and intriguing electrochemical behavior are just a few of the remarkable properties that result from this unusual compositional complexity, which also creates special synergistic effects. These properties are frequently better than those of their conventional counterparts.

This book, *High-Entropy Materials for Energy Storage Devices*, aims to provide a comprehensive and timely exploration of this exciting and rapidly evolving field. Herein, we highlights the basic concept of various high-entropy materials (especially alloy and oxides-based materials), advanced synthesis techniques, characterization, mechanism, and potential energy storage applications for the battery, dielectric capacitor, and supercapacitor. Our goal is to bridge the gap between fundamental materials science and practical energy device engineering. We seek to equip researchers, engineers, and students with a thorough understanding of how the unique attributes of high-entropy materials can be harnessed to overcome current limitations in energy storage technologies. By bringing together diverse perspectives on synthesis, characterization, theoretical modeling, and device integration, we hope to inspire further innovation and accelerate the development of next-generation energy storage solutions. We sincerely hope that this book will be a vital resource and a driving force behind future innovations, making a substantial contribution to a world that is more energy-secure and sustainable.

Dr. Chien-Te Hsieh
Dr. Pradeep Kumar Panda
Dr. Arpan Kumar Nayak

List of Abbreviations

Abbreviation	Definition
AC	Activated carbon
AI	Artificial intelligence
BCC	Body-centered cubic
CVD	Chemical vapor deposition
CV	Cyclic voltammetry
DFT	Density functional theory
EDLC	Electrical double-layer capacitance
FCC	Face-centered cubic
FSP	Flame spray pyrolysis
GITT	Galvanostatic intermittent titration technique
HEA	High-entropy alloy
HEM	High-entropy material
HEO	High-entropy oxide
HEPO	High-entropy perovskite oxide
HEPBA	High-entropy Prussian blue analogue
HE-SC-N88	$LiNi_{0.88}Mn_{0.03}Mg_{0.02}Fe_{0.02}Ti_{0.02}Mo_{0.02}Nb_{0.01}O_2$
LIB	Lithium-ion battery
ML	Machine learning
M-HEM	Metal-doped high-entropy material
MD	Molecular dynamics
PBA	Prussian blue analogue
SSB	Solid-state battery
SC	Supercapacitor
SC-NCM88	SC $LiNi_{0.88}Co_{0.09}Mn_{0.03}O_2$
SIB	Sodium-ion battery

1

Overview of High-Entropy Materials for Energy Storage

Surface Chemistry and Its Functionality

Mukarram Ali[1,2], *Mohsin Saleem*[3], *Tahir Sattar*[4], *Muhammad Zubair Khan*[5], *Yoon-Cheol Ha*[1], *and Jung Hyuk Koh*[6]

[1] *Next-Generation Battery Research Center, Korea Electrotechnology Institute (KERI), Changwon 51543, Republic of Korea,*
[2] *2D Materials and Electrochemical Devices Lab, Department of Chemical Engineering, University of Waterloo, Waterloo N2L 3G1, Canada*
[3] *School of Chemical and Materials Engineering (SCME), National University of Sciences and Technology (NUST), Islamabad 44000, Pakistan*
[4] *Department of Energy Engineering, Korea Institute of Energy Technology (KENTECH), Naju 58330, Republic of Korea*
[5] *Department of Materials Science and Engineering, Pak-Austria Fachhochschule: Institute of Applied Sciences and Technology, Haripur 22620, Pakistan*
[6] *School of Electrical and Electronics Engineering, Chung-Ang University, Seoul 06974, Republic of Korea*

1.1 Introduction

High-entropy materials (HEMs) have emerged as a revolutionary class of materials that promises unparalleled possibilities to develop energy storage and energy conversion devices [1–3]. This class of materials is defined by the presence of multiple-principal elements in equimolar or near-equimolar compositions. HEMs are characterized by entropy-stabilized composition, which dictates the structural, phase, and electrochemical stability [4–7]. The configurational entropies of these materials drive single-phase solid solution forms, suppressing phase segregation and improving stability under aggressive electrochemical conditions [8] (see Scheme 1.1). These characteristics make HEMs of special interest to next-generation lithium-ion batteries (LIBs), sodium-ion batteries (SIBs), and all-solid-state batteries (ASSBs) where the conventional materials are prone to limitations in performance encompassing structural degradation, capacity decay, and kinetic inefficiency.

One of the key aspects of HEMs that enables them to exhibit superior electrochemical performance is their modified compositional chemistry. Unlike energy storage materials of the past, where there are limited elements that control performance, HEMs take advantage of multi-principal element combinations to develop tunable electrochemical properties, such as increased ionic conductivity, decreased charge transfer resistance, and phase stability [9–11]. Through the strategic choice

High-Entropy Materials for Energy Storage Devices, First Edition.
Edited by Chien-Te Hsieh, Pradeep Kumar Panda, and Arpan Kumar Nayak.

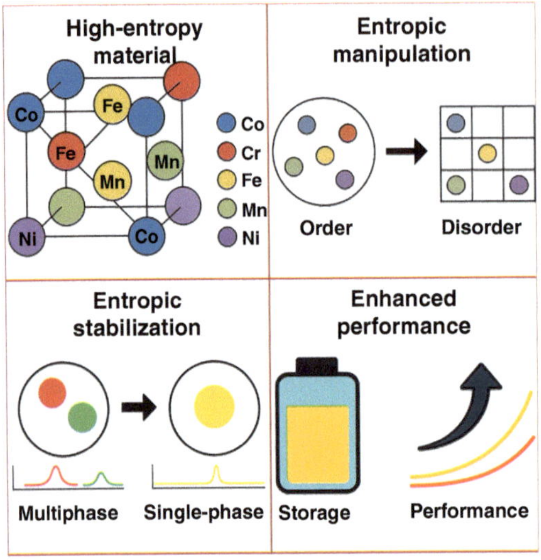

Scheme 1.1 Schematic illustration of HEMs enabling entropic disorder, phase stabilization, and enhanced electrochemical performance in energy storage for next-generation energy storage devices.

and optimization of elemental compositions, it is possible to design these materials to have low activation energies (E_a), high-power densities, and extended cycling life, making them potential materials of choice for electrodes (anode/cathode) and electrolyte (liquid/solid) in high-performance battery systems [12–15].

In this chapter, the fundamental role of compositional chemistry in determining the electrochemical properties of HEM-based energy devices is discussed. This includes the kinetic and thermodynamic fundamentals of entropy stabilization and how these affect phase transformations, structural stability, and electrochemical robustness [16–18]. The relation between multiple-elemental correlations, redox activity, and defect chemistry is discussed, detailing how each of these influences the functional characteristics of HEMs in batteries. Another key focus of this chapter is the introduction of analysis methods used to characterize phase and compositional stability within high-entropy systems. Advanced in situ methods, including X-ray diffraction (XRD), X-ray photoelectron spectroscopy (XPS), X-ray absorption spectroscopy (XAS), and electrochemical impedance spectroscopy (EIS), yield rich information on dynamic phase transformation within HEMs under operating conditions [19–25]. These techniques are used to monitor phase transformation, defect formations, and interfacial behavior in real time, allowing the optimization of HEMs to achieve ideal electrochemical properties. The chapter also discusses case studies that demonstrate compositional engineering application to HEM-based electrolytes and electrodes. Strategic control of compositions is shown which lead to significantly improved cycling stability, rate capacity, and energy density, highlighting the real benefits of high-entropy systems in next-generation battery systems [26–28]. By combining the fundamental concepts of compositional entropy, defect thermodynamics, and surface chemistry, the chapter presents a holistic view of the rationale-based design of HEMs for energy storage devices. The information discussed here is anticipated to chart the direction of future research in high-entropy battery materials to allow the development of high-performance, and sustainable energy storage devices.

1.2 Fundamental Principles in HEMs

1.2.1 Definition and Fundamental Principles

HEMs are a new class of materials characterized by multi-principal elements. Unlike materials based on a binary, ternary, or quaternary compositions, HEMs have five or more elements in near-equimolar or equimolar compositions, resulting in high configurational entropy [29]. The fundamental design principle of an HEM lies in the fine-tuning of configurational entropy, which measures the disorder brought into a system of a single phase by multiple elements [30, 31]. Through the entropic pathway, stable solid solutions are favored, and phase segregation is reduced, enhancing the phase stability of the material under harsh conditions.

Configurational entropy of a multicomponent system is quantified by the Boltzmann equation [3]:

$$S^{Ideal}_{Config} = -k_B \sum_{i=1}^{N} x_i \ln x_i \tag{1.1}$$

where x_i is the mole fraction of the ith component, and k_b is the Boltzmann constant. For a system of N number of elements in equimolar concentration, Eq. 1.1 simplifies to:

$$S^{Ideal}_{Config} = -R \ln (N) \tag{1.2}$$

where R is the general gas constant. Empirically, a system is classified as an HEM if its configurational entropy exceeds 1.5 k_B per site. For instance, when $N = 5$, the entropy reaches approximately 1.6 k_B, which aligns with the threshold for high-entropy classification [3, 32].

Thus, the threshold criteria are as follows [3]:

- High-entropy phase: $S^{Ideal}_{Config} \geq 1.5\, k_B$
- Medium-entropy phase: $1.0\, k_B \leq S^{Ideal}_{Config} < 1.5\, k_B$
- Low-entropy phase: $S^{Ideal}_{Config} < 1.0\, k_B$

This entropic effect promotes the formation of thermodynamically stable, single-phase structures that can withstand electrochemical cycling without significant phase decomposition.

1.2.2 Classification and Types of Entropy in HEMs

While configurational entropy is a defining feature of HEMs, other entropy contributions also play critical roles in material stability and performance, including conformational entropy $\left(S^{Ideal}_{Config}\right)$ which is related to the spatial arrangement of atoms and their degrees of freedom, vibrational entropy $\left(S_{Vib}^{Ideal}\right)$ arising from atomic lattice vibrations, influencing phase stability at different temperatures, rotational and translational entropy significantly considered in liquid-state or polymer-based multicomponent systems and cavity entropy which is associated with voids and defects in crystalline or amorphous structures.

One needs to understand that just by virtue of having a number of elements within the material does not mean that it is automatically a material of high-entropy. How the elements are distributed, what constitutes the phases, what the role of entropy is all play a cumulative role in deciding whether or not the system itself is of a high-entropy nature. Terms such as "entropy stabilization" and "compositionally complex materials (CCMs)" are often used interchangeably with HEMs but possess distinct technical meanings.

1.2.3 High-Entropy Materials in Energy Storage Systems

Incorporation of HEMs with electrolyte and electrode materials has opened new avenues for the development of energy storage technologies. They consist of severely disordered lattice structures with many internal defects, increasing the ion migration and decreasing the activation energies. These characteristics have placed the HEMs at the forefront for future battery technologies, such as high-entropy oxides (HEOs) as anode materials for LIBs, exhibiting high cycling stability and exceptional rate capabilities [33]; high-entropy sulfides (HESs) and high-entropy phosphides with fast ionic conductivity, making them promise for solid-state electrolyte (SSE) applications [34, 35]; and high-entropy alloys (HEAs) and ceramics for structural components in SIBs and Li-S batteries [3, 8].

Recent innovations in zero/low-strain cathodes and multicomponent anodes have demonstrated remarkable improvements in electrochemical stability, making HEMs highly attractive for high-rate and high-voltage applications. Additionally, their ability to replace scarce or toxic elements with earth-abundant alternatives promotes sustainability in energy storage technologies.

1.2.4 Ion Transport Mechanisms in HEMs

Liquid- and solid-state HEM preparations generally entail combinations of several salts and additives in inorganic or organic solvents [2, 3]. Productive solute–solvent interactions alter order–disorder features in favor of increasing the stability of the electrolyte and ionic transport functionality. In high-entropy liquid electrolytes (HELEs) and high-entropy solid-state electrolytes (HESEs), the presence of multiple ionic species introduces a complex network, leading to uneven energy distribution among ions and modification of Li^+ hopping site energy, enhancing ion mobility in SSEs.

The diffusion coefficient of Li^+ in such systems follows the relation:

$$D_i = ae^{\left(\frac{bS_{Config}^{Ideal}}{k_B}\right)} \tag{1.3}$$

where D_i is the ion diffusion coefficient, and a and b are the fitting parameters. Higher entropy results in enhanced ion transport efficiency, leading to lower polarization and improved battery performance. This property is particularly valuable in solid-state LIBs and SIBs, where interface resistance is a key challenge.

1.2.5 Configurational Entropy in Ceramic HEMs

Among HEMs for energy storage, ceramic HEMs have gained prominence as electrode materials. Their configurational entropy is typically resolved into subspecies contributions, considering cation, anion, and oxygen-site entropy:

$$S_{Config}^{Ideal} = -k_B \left(\sum_{a=1}^{M} x_a \ln x_a + \sum_{b=1}^{N} y_b \ln y_b + \sum_{V=1}^{O} V_O \ln V_O \right) \tag{1.4}$$

where x_a, y_b, and V_o denote the mole fractions of cations, anions, and oxygen-site elements, respectively. The presence of multiple elements in near-equimolar ratios enhances electronic conductivity, defect chemistry, and redox stability, making HEOs highly competitive as battery electrode materials.

1.3 Design and Synthesis of High-Entropy Materials for Energy Storage

With the introduction of multicomponent designs, battery compositions have notably enhanced their performance. Specifically, it boosts energy of carrier ions (Li+, Na+, K+) of SSEs and decreases liquid electrolyte solvation of ions to enhance ion motion. Moreover, it induces a frustrated energy landscape helping to suppress volumetric changes of the anode/cathode materials and decreases the dependency upon critical metals, i.e. cobalt (Co) and manganese (Mn). This tunable performance–chemistry interface renders the HEMs specially prominent [36–38]. HEMs employ several key components to enhance stability, ionic conductivity, and general electrochemical performance for rechargeable batteries. Ball milling with high energy and solid-state sintering are widely utilized for the synthesis of solid HEMs, but these have phase segregation and inhomogeneous problems. Moreover, these processes are very energy-intensive, increasing the material cost. The structural complexity due to a number of components also requires fast and sharp conditions for synthesis to maintain only a single phase, which is not easily achievable for classical techniques. Ultra-fast methods of preparing these materials have been recently conceptualized where processing time and cost are reduced drastically with ultrahigh purity. Table 1.1 outlines various synthesis strategies used for HEMs, highlighting their processes.

Methods of spark plasma sintering (SPS), solution schemes of sol-gel and co-precipitation, and plasma approaches enable highly adjustable atomic structuring and particle sizes, which are considerably significant for battery material advancements [31, 36, 39–42]. Furthermore, novel methods of carbothermal shock synthesis and microwave-assisted methodologies enable quick and large-quantity

Table 1.1 Various synthesis strategies used for HEMs, highlighting their processes.

Sr. No	Approach	Method	Description	References
1	Top-down	High-energy ball milling	Mechanical alloying method for synthesizing HEM powders	[1, 8]
2	Top-down	Cryomilling	Refinement of grain structure by milling at cryogenic temperatures	[8]
3	Top-down	Spark plasma sintering (SPS)	Rapid sintering with limited grain growth	[3]
4	Top-down	Attrition milling	Intensive mixing of powders to obtain fine and homogeneous HEMs	[6]
5	Top-down	Laser ablation	Physical breakdown of bulk to nanoform using laser pulses	[10]
6	Bottom-up	Sol–gel	Chemical solution method allowing fine control of composition	[2, 37]
7	Bottom-up	Co-precipitation	Simultaneous precipitation of multiple precursors	[3]
8	Bottom-up	Spray pyrolysis	Formation of particles from aerosol precursor in heated reactor	[5]
9	Bottom-up	Combustion synthesis	Exothermic reaction yielding phase-pure powders	[4]
10	Bottom-up	Microwave-assisted synthesis	Efficient and rapid heating for uniform nucleation	[31]

productions along with maintaining favorable microstructures. In anode and cathode materials, the design stage is of relatively more importance than the synthesis process. In most cases, by the time a novel system of cathode material appears as a need, the necessary schemes for synthesis are already present and not likely to change in a foreseeable time span.

However, the addition of more-than-one dopant toward the design step requires much more profound experimentations and theories of understanding. In a situation of HELEs, though the component mix does not alter, the designs can be highly distinct and involve Single-salt/Multi-solvent (SSMS), Multi-salt/Single-solvent (MSSS), or suspension methodologies [43, 44]. In the SSMS method, one of the salts is soluble in a group of more-than-five solvents, enhancing solvation entropy by adding molecular diversity, which decreases Li+–solvent aggregation, reduces Li–metal anode concentration polarization, suppresses dead-Li formation, and increases rate capability. This disparity of MSSS design for single-salt solubilities is compensated for by involving a set of four Li-salts of equal amounts in a single solvent. This increases anionic species concentrations, lowers solvation strength, prefers anion–Li+ interactions, increases ionic conductivity, and constructs a stable solid electrolyte interphase (SEI) on anode electrodes for a high-rate performance. Both of these schemes aim at the maximization of the solvation entropy. Li+ coordination by the solvents is dependent on solvent polarity and by donor number through which higher Li+–anion coordinations form inorganic-rich SEIs through Li–anode inorganically and thus increase the cycling stability. A suspension electrolyte is a distinct alternative, wherein inorganic nano-particles are suspended in liquid-type electrolytes. This suspension acts passively as a pseudo-in situ SEI catalyst, forming an inorganic-rich, anion-based SEI coating on the Li–metal anode for improved electrochemical performance and preservation. In the case of solid-state HEEs, other than design, the synthesis route is of significance. Common methods are solid-state reactions and high-energy milling. However, time-consuming high-temperature procedures like solid-state sintering can cause secondary phases and element volatility, which are detrimental to electrochemical performance. Ultrafast synthesis of a novel class of materials owes many interests due to their ability of synthesizing a variety of materials in seconds [42, 45]. These are most productive for synthesizing metastable materials, such as HEAs and HEOs.

Cathode materials mainly dictate battery cost and performance. State-of-the-art cathode chemistries have poor stability and capacity at more than 4 V, and can mechanically degrade due to strain buildup and the formation of cracks, which weakens mechanical integrity in ASSBs. High-entropy cathodes strive to improve LIB and ASSB stability and performance by placing several main elements in a single-phase (ordered or disordered) structure. This optimizes both structural preservation of the shape of the crystal and electrochemical durability, including cycle lifespan, rate capability, and thermal durability [46]. Element selection is essential; typically, transition metals like Ni, Co, Mn, with addition of Ti, Cr, Zr, Fe, and Al are combined to create an HEM [47, 48]. Elements are chosen for structural stability and electrochemical functionality. High-entropy anode materials are also vital for future-generation batteries. By improving structural integrity, conductivity, dendrite inhibition, and affording design freedom, high-entropy anodes improve shortcomings of standard anodes. They are at the heart of more secure, efficient, and durable energy storage devices. Research on High-Entropy Battery Material (HEBMs) had a beginning with HEOs being anode promises [49, 50]. Such HEOs provide good structural stability due to their multicomponent nature, which prevents structural degradation during charge–discharge operations and prevents mechanical breakdown or phase segregation. Their evenly dispersed nature of metal cation helps in having a smoother surface, which prevents Li or Na dendrite formation, a major safety problem in batteries. Thus, HEOs offer a safer option for LIBs and SIBs.

1.4 High-Entropy Phase Stabilization and Structural Integrity

The thermodynamic stability of any material is governed by the Gibbs free energy equation:

$$\Delta G = \Delta H - T\Delta S \tag{1.5}$$

For conventional materials, stability is largely dictated by enthalpy (ΔH), which depends on the strength of atomic bonds. However, in HEMs, the entropy (ΔS) becomes dominant at elevated temperatures, significantly lowering the free energy (ΔG) and favoring the formation of stable single-phase structures. For a large ΔS, entropy-driven stabilization occurs, which reduces multi-phase segregation. Conversely, for a dominant ΔH, phase decomposition takes place into multiple phases. It is the balance between ΔS and ΔH that determines if a multi-principal element system will be a single-phase system or will decompose into multi-phase system.

HEMs are particularly stable owing to their distinctive multi-principal element compositions, which are responsible for the phase stabilization by entropic stabilization. This stabilization mechanism plays a pivotal role in mitigating phase segregation, ensuring homogeneous elemental distribution, and enhancing the overall structural robustness under extreme electrochemical conditions [49, 51]. The intrinsic lattice disorder in the structure of HEMs derived from multiple-elemental random distribution as part of a single-phase solid solutions significantly affects ionic conductivity and electrochemical and mechanical stability [2]. With the help of stacking faults, distortions of the crystal structure, and the control of defects, scientists have maximized the structure of the high-entropy regimes for electrode and electrolyte-driven applications in next-generation energy storage devices (see Figure 1.1).

The incorporation of five or more principal elements in near-equimolar ratios generates high configurational entropy, which thermodynamically favors the formation of single-phase solid solutions over phase-segregated states as can be seen Figure 1.1. This entropy-driven stabilization enhances the material's ability to withstand prolonged electrochemical cycling and suppresses undesirable phase transformations. In SSB applications, this effect is particularly crucial for ensuring structural retention during repeated lithium or sodium intercalation and deintercalation cycles.

The high configurational entropy in HEMs disrupts the periodicity of the crystal lattice, inducing lattice distortions and local disorder. While this disorder might seem detrimental in conventional materials, in HEMs, it plays a beneficial role by generating diverse migration pathways for mobile ions. The presence of multiple cations with varying atomic radii and electronegativities results in heterogeneous electrostatic potentials, which, in turn, reduce activation barriers for ion transport. This effect is particularly pronounced in HEOs and high-entropy halides, where the disorder facilitates a more isotropic Li-ion or Na-ion conduction network. One notable example is the high-entropy lithium halide solid electrolyte

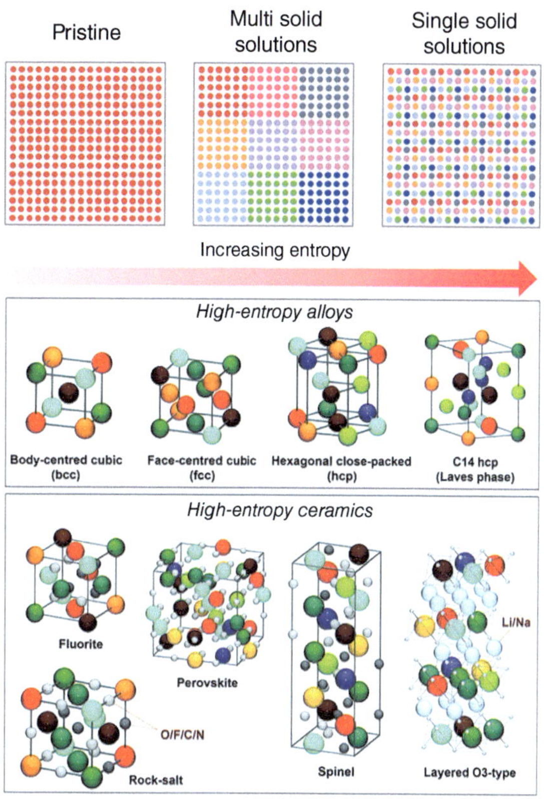

Figure 1.1 Evolution of high-entropy materials from pristine to single-phase solid solutions, highlighting structural diversity in alloys and ceramics. *Source:* Reproduced from Refs. [6, 52] / with permission of Royal Society of Chemistry.

($Li_{2.8}In_{0.2}Sc_{0.2}Yb_{0.2}Lu_{0.2}Zr_{0.2}Cl_6$) [6], which was designed using ionic potential descriptors. This approach allowed researchers to predict stable crystal structures based on charge-to-radius ratios of cations [32]. By optimizing the ionic interactions between halide anions and the multi-element cation sublattice, the high-entropy lithium halide exhibited enhanced lattice cohesion, leading to superior thermal stability and mechanical robustness. Subsequently, $Li_{2.8}In_{0.2}Sc_{0.2}Yb_{0.2}Lu_{0.2}Zr_{0.2}Cl_6$ achieved a high ionic conductivity of 2.13 mS cm^{-1} at room temperature, surpassing its conventional single-component architecture.

1.4.1 Lattice Disorder and Its Impact on Ion Transport

The high configurational disorder in HEMs imposes large lattice distortions, breaking long-range order and developing favorable short-range disorder. This disorder alters the localized electrostatic potential in the vicinity of the ions, lowering ion migration activation barriers and increasing ionic conductivity [53, 54]. In SSEs like high-entropy lithium halides, this effect results in a more isotropic network for Li-ion conduction. In lithium-halide solid electrolytes, for instance, the presence of multiple cationic species contributes to faster Li-ion diffusivity because of increased defect density and dynamic fluctuations in the lattice [6, 32]. Structural studies, such as the use of neutron powder diffraction (NPD) and XRD, have

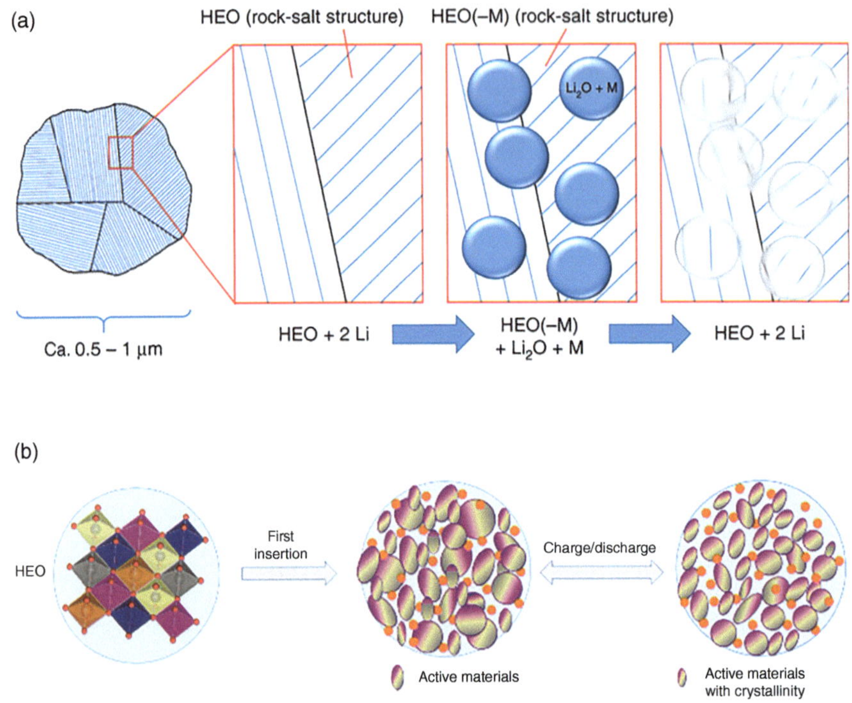

Figure 1.2 (a,b) Illustration of lattice disorder in HEOs during lithiation, highlighting structural evolution, phase transformations, and its impact on ion transport.
Source: Reproduced from Refs. [33, 55] / with permission of Elsevier.

established the broadened diffraction peaks of the high-entropy halides as evidence for enhanced stacking faults and cation disorder [32]. These defects contribute to the formation of highly conductive grain boundaries, which further improve overall ionic conductivity. Similarly, in HEOs such as (MgCoNiCuZn)O, disorder in the cation sublattice led to a reduced Li-ion migration barrier, increasing diffusion rates by 30% compared to conventional oxide electrolytes as can be seen in Figure 1.2. Density functional theory (DFT) calculations confirmed that ionic conductivity in HEOs is enhanced due to the presence of variable local environments for Li-ion diffusion.

1.4.2 Role of Stacking Faults and Crystallographic Defects

Stacking faults, disordered atomic arrangements along specific crystallographic planes, are common in HEMs and have been identified as key factors in optimizing electrochemical performance. In layered halide solid electrolytes, the presence of stacking faults has been directly correlated with increased Li-ion transport rates, as they create interconnected diffusion channels. Moreover, cation disorder enhances interstitial site occupancy variations, effectively increasing the number of available migration pathways. In SSB electrolytes, such as high-entropy lithium halides, controlled introduction of stacking faults has been shown to modulate

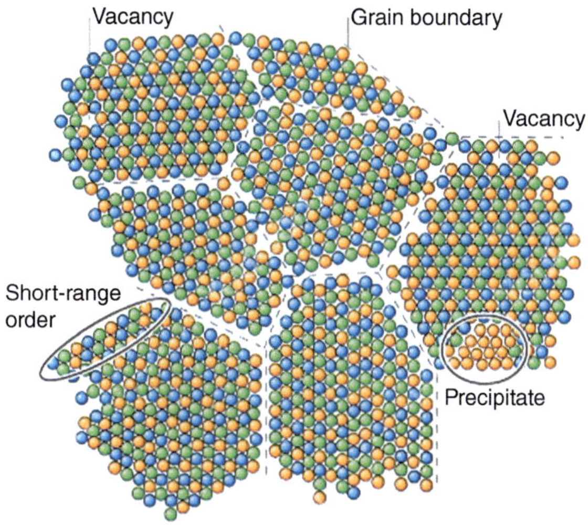

Figure 1.3 Schematic representation of stacking faults and crystallographic defects in high-entropy materials, highlighting vacancies, grain boundaries, short-range order, and precipitate formation. *Source:* Reproduced from Ref. [56] / with permission of Springer Nature.

ionic transport properties. By tailoring the ratio of metal cations to halide anions, researchers have been able to engineer materials with high ionic conductivities ($>2.0\,mS\,cm^{-1}$) while maintaining structural stability (Figure 1.3).

Stacking faults and twinning mechanisms significantly influence mechanical stability and phase transformation as well. Studies on FCC-based HEAs (CoCrFeNiMn) revealed that low stacking fault energies (SFEs) enhance twinning-induced plasticity (TWIP) and transformation-induced plasticity (TRIP) mechanisms, leading to increased toughness [57].

1.4.3 Morphology and Microstructural Considerations

The microstructural characteristics of HEMs, including grain size, morphology, and particle packing density, significantly influence their electrochemical stability. For instance, recent studies have demonstrated that high-entropy ceramic materials tend to form isotropic spherical particles rather than lamellar morphologies, leading to improved compaction densities and reduced grain boundary resistance. The spherical morphology not only facilitates better densification during fabrication but also enhances interfacial contact between electrode and electrolyte layers in ASSBs (Figure 1.4a).

High-entropy lithium halide electrolytes ($Li_{2.8}In_{0.2}Sc_{0.2}Yb_{0.2}Lu_{0.2}Zr_{0.2}Cl_6$) were observed to have a more uniform particle morphology, improving densification and minimizing interfacial resistance in ASSBs [6]. The transition from a layered anisotropic structure (as seen in Li_3InCl_6) to an isotropic spherical morphology was attributed to enhanced configurational entropy and a higher concentration of stacking faults, which reduced preferred crystal growth directions (Figure 1.4b).

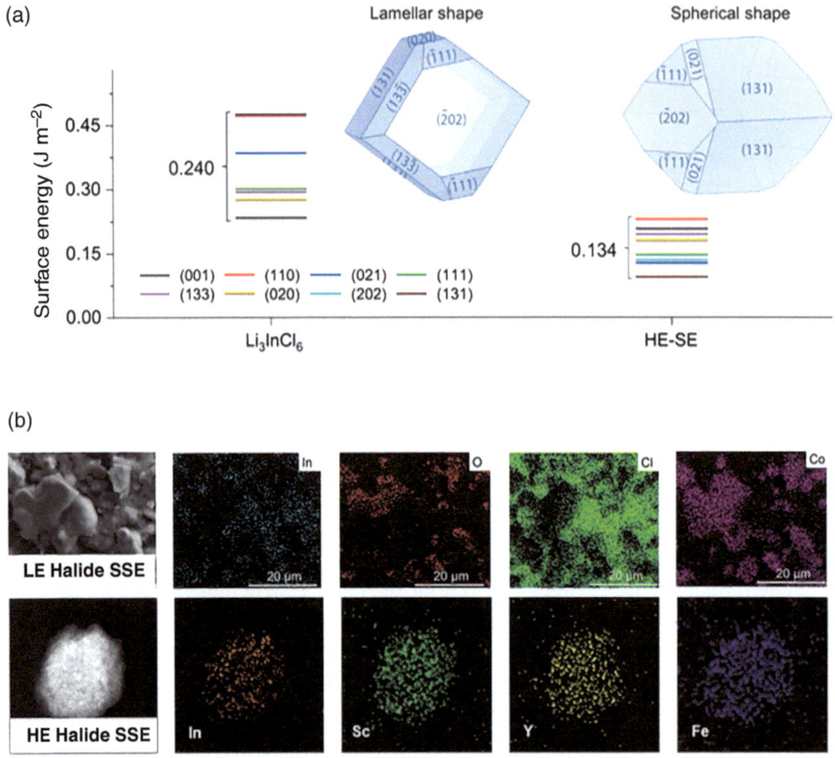

Figure 1.4 (a) Surface energy analysis of HEO illustrating structural differences between lamellar and spherical morphologies. *Source:* Adapted from Ref. [33]. (b) SEM and elemental mapping of low-entropy (LE) and high-entropy (HE) halide solid-state electrolyte, highlighting morphological transformation from layered to spherical morphology. *Source:* Reproduced with permission. Copyright 2019, Royal Society of Chemistry. Copyright 2019, American Chemical Society Refs. [6, 58].

1.5 Compositional Engineering in High-Entropy Materials

The ability to tailor the elemental composition of materials plays a fundamental role in determining their structural, thermal, and electrochemical properties (see Figure 1.5a–c). The rational design of HEOs, sulfides, halides, and polyanionic compounds has paved the way for the development of next-generation electrodes and electrolytes with superior stability and transport properties. The influence of the compositional design in HEMs is reviewed in this section with a focus on their application in energy storage devices.

1.5.1 HEOs and Transition Metal Stabilization

HEOs have attracted much interest because of their capability to include multiple cationic (normally transition metals) species in a single-phase lattice, which

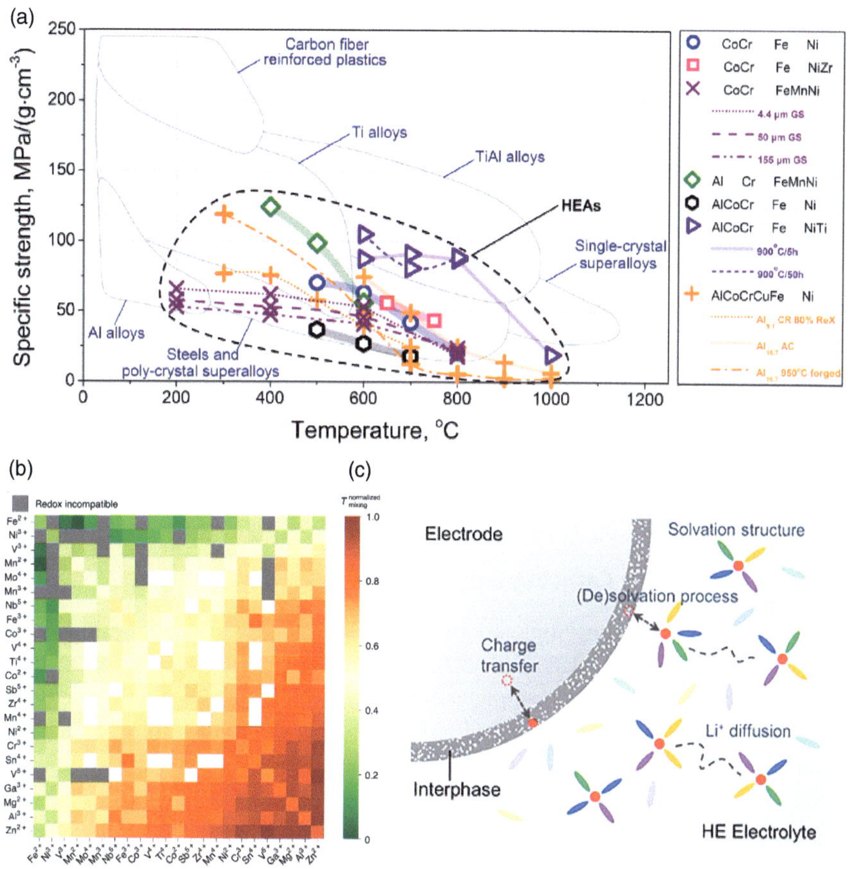

Figure 1.5 (a) Specific strength vs. temperature comparison of HEAs and conventional high-temperature materials. *Source:* Reproduced from Ref. [1] / with permission of Elsevier. (b) Normalized mixing temperature of transition metal species in disordered rock-salt (DRX) compounds, highlighting redox compatibility. *Source:* Reproduced from Ref. [5] / with permission of Springer Nature. (c) Schematic representation of charge transfer and Li^+ diffusion in high-entropy electrolytes. *Source:* Reference [2] / Springer Nature / CC BY 4.0.

contributes to phase stability as well as electrochemical reversibility [33, 55, 59]. The choice of transition metals in HEOs plays a crucial role in ionic diffusion pathways, mechanisms for charge storage, and redox dynamics, and therefore, have excellent prospects in LIB as well as SIB electrodes. Recent phase-property studies have shown that it is possible to anticipate as well as tailor phase stability of multicomponent oxide systems. For example, the addition of Fe and Cu in MnCoNi-alloy as a high-entropy anode component of LIBs has revealed enhanced cycling performance and rate capability [53]. The synergistic interaction between multiple transition metals mitigates phase transitions and facilitates a more uniform charge distribution during lithiation and delithiation cycles. Furthermore, dopant selection plays a crucial role in modulating the electronic band structure, thereby influencing the conductivity and electrochemical activity of HEOs.

1.5.2 High-Entropy Solid Electrolytes: Enhancing Ionic Transport

Solid electrolytes with high-entropy compositions have exhibited significant advantages over their conventional counterparts, primarily due to their ability to stabilize fast-ion conduction pathways. In high-entropy lithium argyrodite (HE-Li_6PS_5X) electrolytes, compositional engineering has been employed to optimize anion disorder, thereby reducing the activation energy for lithium-ion transport [9, 60]. The introduction of halide-rich anions (Cl^-, Br^-, I^-) in Li-argyrodite electrolytes has led to an increase in Li-ion diffusivity, resulting in an exceptionally high ionic conductivity of 22.7 mS cm^{-1} at room temperature, surpassing conventional sulfide-based electrolytes [7]. Similarly, high-entropy polyanionic lithium conductors, such as $Li_6PS_5(Cl_{0.33}Br_{0.33}I_{0.33})$, have demonstrated that compositional complexity can lead to lower activation barriers for ion migration. By incorporating a diverse set of polyanions, researchers have successfully developed high-entropy lithium superionic conductors with reduced grain boundary resistance and improved electrochemical stability.

1.5.3 High-Entropy Cathodes and Multi-Electron Reactions

One of the most promising applications of HEMs lies in the design of multi-electron reaction cathodes for LIBs and SIBs [54, 59]. The integration of multiple redox-active cations into a single-cathode structure enables the participation of multiple charge carriers, thereby increasing the overall energy storage capacity. A notable example is the development of a high-entropy polyanionic phosphate cathode, $Na_3VFe_{0.5}(TiMnZrCuAl)_{0.5}(PO_4)_3$, which facilitates the activation of six different redox couples ($V^{2+/3+}$, $Ti^{3+/4+}$, $Fe^{2+/3+}$, $V^{3+/4+}$, $Mn^{2+/3+}$, and $V^{4+/5+}$) [61]. This compositional engineering strategy has enabled a high specific capacity of 158.8 mA h g^{-1}, with 95% capacity retention after 170 cycles, highlighting the potential of high-entropy cathodes in next-generation energy storage systems.

1.5.4 High-Entropy Doping for Cathode Stability

To further improve the structural stability of cathode materials, researchers have introduced high-entropy doping strategies that incorporate low-electronegativity transition metal elements. In high-Ni-layered oxide cathodes ($LiNi_{0.5}Mn_{0.43}Ti_{0.02}Mg_{0.02}Nb_{0.01}Mo_{0.02}O_2$), the introduction of multiple dopants has been shown to stabilize the anionic redox process, thereby preventing irreversible oxygen release during cycling [62]. The cationic disorder introduced by high-entropy doping effectively anchors oxidized lattice oxygen species, suppressing the formation of detrimental O_2 gas and extending the lifespan of the cathode. Electrochemical performance tests indicate that this high-entropy-doped cathode retains 95% of its capacity after 1,000 cycles at 1C, significantly outperforming conventional NCM-811 cathodes.

1.5.5 Vacancy Engineering in High-Entropy Sulfide and Polyanionic Conductors

Another promising avenue for compositional engineering in HEMs involves vacancy-driven ionic conduction. In Ca-doped Na_3PS_4, aliovalent substitution creates a high concentration of Na-vacancies, leading to a substantial increase in Na-ion conductivity, reaching ~1 mS cm^{-1} at room temperature [63]. Similarly, in $Li_{3.25}[Si_{0.25}P_{0.75}]S_4$, the introduction of aliovalent Si^{4+} cations induces site disorder, resulting in a three-order magnitude increase in ionic conductivity compared to conventional β-Li_3PS_4. As compositional engineering strategies continue to evolve, the integration of high-entropy materials in energy storage technologies is expected to drive advancements toward high-performance, sustainable, and commercially viable battery systems.

1.6 High-Entropy Electrodes for Energy Storage

The multi-element composition of high-entropy electrodes enables the activation of multiple redox centers, increasing charge storage capacity and electrochemical reversibility. Current developments in High-Entropy Phosphates (HEOs), HESs, high-entropy phosphates (HEPs), and HEAs have shown enhanced cycling performance as well as rate capability, making them a candidate for next-generation electrode material for high-energy and high-power battery applications [1, 4, 57]. The structural design, electrochemical aspects, as well as commercialization prospects of the use of high-entropy electrodes for energy application are briefly discussed in this section [64]. Table 1.2 details a summary of existing methods to synthesize HEMs for energy storage materials.

1.6.1 HEOs as Anodes and Cathodes

1.6.1.1 HEOs as Anodes

HEOs have been widely investigated as an anode material because of their multi-electron redox reactions and phase stability under repeated charging–discharging cycling. The presence of multiple transition metal elements in a single-phase oxide structure inhibits unwanted phase transitions, thus providing enhanced cycling and rate capability. The entropy-driven stabilization in these oxides prevents lithium trapping, increases ionic conductivity, and mitigates structural degradation during lithiation and delithiation cycles. Moreover, the presence of oxygen vacancies and cationic disorder in the HEOs enhances charge transportation characteristics as well, further increasing their electrochemical activities.

1.6.1.2 HEOs as Cathodes

The use of HEOs as LIB and SIB cathodes is of particular interest because such cathodes can accommodate multiple redox species with enhanced structural stability.

Table 1.2 Summary of existing methods to synthesize HEMs for energy storage materials.

Synthesis of high-entropy electrolytes				
Material	**Type**	**Synthesis method**	**Ionic conductivity**	**Activation energy**
0.1 m LiFSI/0.1 m LiTFSI/0.1 m LiDFOB/0.1 m $LiNO_3$/1.0 m $LiPF_6$ in EC/DMC [65]	Liquid	Multi-salt Single-solvent	18 mS cm^{-1}	—
0.15 mol/l (M) lithium bis(fluorosulfonyl)imide (LiFSI), 0.15 M lithium bis(trifluoromethanesulfonyl)imide (LiTFSI), 0.15 M lithium difluoro(oxalato)borate (LiDFOB) and 0.15 M lithium nitrate ($LiNO_3$) in a single solvent of dimethoxyethane (DME) [2]	Liquid	Single-salt Multi-solvent	12.1 mS cm^{-1}	—
$Li_{5.5}PS_{4.5}Cl_{0.8}Br_{0.7}$ [7]	Solid	Polyanionic	22.7 mS cm^{-1}	0.2 eV
$Li_{2.2}In_{0.2}Sc_{0.2}Zr_{0.2}Hf_{0.2}Ta_{0.2}Cl_6$ [6]	Solid	Multicationic	4.7 mS cm^{-1}	0.3 eV

Synthesis of high-entropy cathode				
Material	**Type**	**Synthesis method**	**Electrochemical performance**	**Working potential**
$Li_{1.3}Mn^{2+}{}_{0.2}Mn^{3+}{}_{0.2}Ti_{0.1}Nb_{0.2}O_{1.7}F_{0.3}$ [5]	DRX	Solid state	270 mA h g^{-1}	4.7 V
$LiNi_{0.5}Mn_{0.43}Ti_{0.02}Mg_{0.02}Nb_{0.01}Mo_{0.02}O_2$ [62]	Layered	Co-precipitation	185 mA h g^{-1}	4.3 V

Synthesis of high-eEntropy anode				
Material	**Type**	**Synthesis method**	**Electrochemical performance**	**Working potential**
$(FeCoNiCuRu)S_2$ [66]	Sulfide	Solid state	900 mA h g^{-1}	3.0 V
$(CoMnZnNiMg)_2CrO_4$ [67]	Oxide	Solid state	608 mA h g^{-1}	3.0 V
SnSbMnBiTe [49]	Alloy	Ball milling	1,000 mA h g^{-1}	3.0 V

Multi-transition metal-based HEO cathodes promote cationic and anionic redox activity with a strong increase in energy density and a longer cycle life. The presence of multiple cations stabilized the anionic redox process by inhibiting the reversible oxygen loss and providing a very high level of structural reversibility. Additionally, the entropy stabilization effect in these cathodes suppressed phase transitions and reduced capacity fading under high-voltage cycling conditions.

In SIBs, HEO cathodes such as $Na_3(VFeMnCoNi)O_4$ have exhibited superior cycling stability and rate performance, owing to their structural robustness and tunable redox potentials. These materials benefit from entropy-induced stabilization, which minimizes the volume changes associated with sodium intercalation and enhances electrode durability [61].

1.6.2 High-Entropy Sulfides and Phosphides for Conversion Reactions

HESs and HEPs have been explored as conversion-type electrodes, offering high theoretical capacities through multi-step electrochemical reactions. Unlike conventional intercalation-based electrodes, which rely on reversible ion insertion, conversion electrodes store charge through chemical transformation reactions, often involving the formation and dissolution of intermediate phases.

1.6.2.1 High-Entropy Sulfide Anodes

HESs, such as $(CoCuZnSnFe)S_2$, have demonstrated superior cycling stability due to entropy-stabilized phase formation [66]. The incorporation of multiple cations in the sulfide matrix suppresses structural collapse and ensures long-term electrochemical reversibility. These anodes exhibit minimal volume expansion, reducing mechanical stress and enhancing battery lifespan.

1.6.2.2 High-Entropy Phosphate Cathodes

High-entropy phosphate cathodes offer structural robustness and tunable voltage profiles, making them suitable for LIBs and SIBs. One such example, a NASICON-type $Na_3VFe_{0.5}(TiMnZrCuAl)_{0.5}(PO_4)_3$ cathode, demonstrated a high specific capacity of $158.8\ mA\ h\ g^{-1}$ and retained 95% of its capacity after 170 cycles, making it one of the most stable sodium-ion cathodes reported [10].

HEAs have gained attention as cathode materials due to their exceptional electronic conductivity and phase stability. Unlike traditional cathodes, which often suffer from phase segregation and degradation over extended cycling, HEAs provide a homogeneous reaction environment that supports multi-electron redox processes.

For example, CoCrFeMnNi-based HEAs have demonstrated excellent cycling stability and mechanical robustness, making them attractive for high-power energy storage applications. Additionally, in Li-S batteries, high-entropy metallic alloys have been used as cathode hosts, effectively suppressing the polysulfide shuttle effect and extending cycle life.

High-entropy electrodes represent a new materials design strategy for energy storage applications, offering enhanced electrochemical performance, superior structural stability, and prolonged cycle life. HEOs, HESs, high-entropy phosphates, and HEAs have been successfully employed in LIBs, SIBs, and ASSBs, demonstrating their potential as next-generation energy storage materials. Future advancements in computational modeling, in situ characterization, and scalable synthesis techniques will be critical for the commercialization of high-entropy electrodes. By addressing current challenges and optimizing their properties, these materials are expected to redefine energy storage technologies, paving the way for high-performance and sustainable battery systems.

1.7 High-Entropy Electrolytes and Interface Engineering

Aside from the bulk attributes of high-entropy electrolytes (HEEs), interface engineering is also vital for maximizing battery performance. Engineering stable electrolyte–electrode interfaces is critical in minimizing the interfacial resistance, enhancing ion transfer, and suppressing undesirable side reactions that lower battery durability [68, 69]. The design rules, structural aspects, and electrochemical attributes of the HEEs are covered under this section in addition to the contribution of interface engineering to enhance their incorporation within energy storage devices.

1.7.1 High-Entropy Solid-State Electrolytes

SSEs are an essential constituent of ASSBs because of their enhanced safety, thermal stability, and mechanical robustness. High-entropy solid-state electrolytes (HESEs) extends the above advantages with the inclusion of multiple cationic and anionic species to further enhance structural stability and ionic conductivity.

1.7.1.1 Composition and Structural Characteristics

HESEs typically consist of a combination of oxides, sulfides, halides, or polyanionic frameworks. These materials exhibit disordered lattice structures, where entropy-driven stabilization suppresses phase decomposition and promotes the formation of single-phase solid solutions. Notable examples of HESEs include high-entropy lithium argyrodite (HE-Li_6PS_5X) electrolytes, which incorporate Cl^-, Br^-, and I^- anions to enhance Li-ion mobility, polyanionic sulfide-based HESEs, such as $Li_6PS_5(Cl_{0.33}Br_{0.33}I_{0.33})$, which exhibit an exceptionally high ionic conductivity of 22.7 mS cm^{-1} at room temperature and $Li_{3.25}[Si_{0.25}P_{0.75}]S_4$, an entropically stabilized fast-ion conductor that achieves 1.22 mS cm^{-1} ionic conductivity, significantly outperforming conventional β-Li_3PS_4.

1.7.1.2 Ionic Transport Mechanisms in HESEs

The presence of multiple cations and anions in HEEs creates a complex energy landscape for ion migration. This compositional disorder due to the presence of multiple ions results in reduced activation energy for ion diffusion, facilitating fast ionic transport, increased number of percolation pathways, that leads efficient ion conduction even in polycrystalline phases and suppression of grain boundary resistance which improves overall electrolyte performance in SSBs.

Computational studies and experimental observations have confirmed that high-entropy-driven site disorder leads to an increase in lithium and sodium diffusion coefficients, making HESEs highly competitive with conventional SSEs.

1.7.1.3 Interfacial Engineering in High-Entropy Solid Electrolytes

One of the primary challenges in HEEs is achieving stable electrode–electrolyte interfaces that minimize interfacial resistance and prevent side reactions. Interface engineering strategies for HESEs include surface coatings (e.g. $LiNbO_3$, $Li_4Ti_5O_{12}$) to stabilize SEIs in lithium-metal batteries and interfacial engineering with high-entropy fluorides (HEFEs), which form stable passivation layers that suppress dendrite formation.

1.7.1.4 Mitigation of Interfacial Decomposition

Interfacial decomposition remains a major concern in HEE systems. Strategies to mitigate these effects is focused on grain boundary modifications to reduce contact resistance between polycrystalline phases in HESEs and pre-lithiation and artificial SEI formation to prevent electrolyte degradation upon cycling.

Advanced characterization techniques, such as in situ XPS and EIS, have been employed to monitor and optimize interface formation in HESE systems. HEEs represent a transformative approach to electrolyte design, offering enhanced ionic conductivity, superior phase stability, and improved interfacial compatibility for next-generation energy storage systems. By harnessing configurational entropy and multicomponent interactions, HESEs, HELEs, and high-entropy polymer electrolytes (HEPEs) exhibit outstanding electrochemical performance. With increasing advancements, the future of energy storage will be revolutionized by the increasing prominence of HEEs, which will facilitate safer, more efficient, and enhanced performance battery technology.

1.7.2 High-Entropy Liquid and Gel Electrolytes

Although SSEs have gained significant attention, liquid and gel-based HEEs are also being developed for applications in hybrid batteries and flexible energy storage devices. These electrolytes take advantage of solute–solvent interactions in multicomponent ionic systems, modifying transport properties and electrochemical stability.

1.7.2.1 Multi-Salt HELEs

HELEs incorporate multiple lithium or sodium salts to optimize solvation dynamics and enhance ionic conductivity. Studies on multi-salt electrolyte formulations containing LiTFSI, LiFSI, and $LiClO_4$ have demonstrated that the disordered solvation environment reduces ion association, promoting free-ion conductivity.

In SIBs, HEEs have been designed using a combination of NaFSI, $NaClO_4$, and NaTFSI, achieving improved electrochemical window stability and increased Na-ion diffusion rates compared to conventional single-salt electrolytes.

1.7.2.2 High-Entropy Polymer Electrolytes

The development of HEPEs has enabled the fabrication of flexible and highly conductive gel electrolytes for next-generation energy storage devices. These materials integrate multiple polymeric matrices and ionic liquid additives to improve mechanical flexibility, ionic transport, and thermal stability. For instance, HEPEs are composed of PEO, PVDF, and PAN matrices, which exhibit enhanced compatibility with lithium and sodium metal anodes and multicomponent ionic gels incorporating LiTFSI, NaFSI, and deep eutectic solvents, achieving stable performance in flexible batteries.

1.8 Advanced Characterization Techniques for HEMs

It is vital to utilize a combination of structural, spectroscopic, and electrochemical techniques to unravel the intrinsic characteristics of HEMs. A discussion in this section focuses on the most critical X-ray, electron microscopy, electrochemical, and in situ/operando techniques used to characterize HEMs and their usage specifically in electrodes, electrolytes, and interfaces for energy storage applications.

1.8.1 Structural and Phase Analysis Techniques

1.8.1.1 X-ray Diffraction (XRD) and Rietveld Refinement

Given that HEMs often form single-phase solid solutions due to entropy stabilization, XRD is instrumental in distinguishing between homogeneous solid solutions and multiphase composites. Rietveld refinement of XRD data enables precise lattice parameter determination, site occupancy analysis, and phase fraction quantification. Similarly, synchrotron XRD provides high-resolution diffraction patterns, allowing for the study of short-range disorder, stacking faults, and lattice distortions in HEMs and pair distribution function (PDF) analysis extends XRD capabilities by providing insights into local atomic arrangements, particularly in materials exhibiting short-range ordering despite long-range crystallinity.

1.8.1.2 Neutron Diffraction and Total Scattering Techniques

Neutron diffraction offers advantages over XRD for HEMs containing low-Z elements (e.g. Li, Na, O, F, S), which have weak X-ray scattering contrast. Time-of-flight neutron diffraction (TOF-ND) enables high-resolution structure determination for complex multi-element HEMs. In the same manner, total scattering and atomic PDF analysis reveal local atomic environments, enabling the study of lattice distortions and cation disorder. Moreover, *in situ* neutron diffraction is particularly useful for tracking phase transformations and defect evolution in HEM electrodes and electrolytes during charge/discharge cycles.

1.8.2 Chemical State and Electronic Structure Characterization

HEMs exhibit complex multi-element interactions, variable oxidation states, and defect-driven electronic properties, necessitating advanced spectroscopic techniques.

1.8.2.1 X-ray Photoelectron Spectroscopy (XPS) and Auger Electron Spectroscopy (AES)

XPS provides elemental composition, oxidation states, and surface chemistry information. Due to the diverse chemistry of HEMs, XPS is crucial for deconvoluting multiple oxidation states in multi-cationic HEMs (e.g. $Fe^{2+/3+/4+}$, $Mn^{2+/3+/4+}$), identifying cation migration and surface reconstructions in battery electrodes and tracking electrochemical degradation in HEMs through post-cycling XPS analysis. In addition, Auger Electron Spectroscopy (AES) is often used alongside XPS to study electronic structure modifications and defect chemistry at electrode–electrolyte interfaces.

1.8.2.2 X-ray Absorption Spectroscopy

XAS is particularly powerful for probing local electronic structures and coordination environments in high-entropy systems. X-ray absorption near-edge structure (XANES) provides oxidation state information, which is critical for understanding redox activity and charge compensation mechanisms in HEM-based electrodes [21]. Similarly, extended X-ray absorption fine structure (EXAFS) analysis permits the exact measurement of cation–anion bonding geometries and coordination distortions, which determine the manner by which structural disorder is influenced by entropy in HEMs.

1.8.3 Electron Microscopy for Microstructural and Compositional Analysis

High-resolution electron microscopy is an essential method used to examine the morphology, grain boundary structure, and atomic-scale defects of HEMs.

1.8.3.1 Scanning Electron Microscopy and Energy-Dispersive X-ray

Spectroscopy Scanning electron microscopy (SEM) enables microstructural characterization for the detection of the grain size, surface morphology, and phase distribution of the HEM and energy-dispersive X-ray spectroscopy (EDS) mapping attests to the uniform distribution of various main elements, one of the main characteristics of well-synthesized HEMs.

1.8.3.2 Transmission Electron Microscopy and Electron Energy Loss Spectroscopy (EELS)

High-resolution transmission electron microscopy (HRTEM) facilitates imaging of lattice structures, dislocation defects, and grain boundary characteristics at the atomic level [70]. Selected area electron diffraction (SAED) enables phase identification of nanostructured HEMs and information from the local bonding state, defect state, and electronic state can be obtained from electron energy loss spectroscopy (EELS) analysis, which is a complementary tool to XAS [23].

1.8.4 In Situ and Operando Characterization Techniques

To understand the electrochemical and structural development of HEMs in real time, in situ and operando methodologies are vital. In situ XRD records phase transitions and lattice parameters in real time during charge–discharge cycling [71]. Operando XAS monitors dynamic changes of oxidation states in battery electrodes. In situ TEM facilitates the observation of the process of lithiation/delithiation-induced defect formation, providing valuable information regarding structural degradation mechanisms in high-entropy electrodes.

The in-depth characterization of HEMs involves a multidimensional approach involving X-ray, neutron, and electron microscopy, spectroscopic, and electrochemical studies. All these methodologies yield useful information about the phase stability, defect chemistry, ion transport, and interface evolution so the rational design of high-performance HEMs for application in energy storage is facilitated. Future studies need to concentrate on combining sophisticated in situ and operando methodologies to further define the basic mechanisms that control electrochemical performance, durability, and degradation in HEMs. All this will open the way for the large-scale application of HEMs for next-generation LIBs, SIBs, and ASSBs.

1.9 Challenges, Prospects, and Commercialization Pathways

The development of HEMs has created a new paradigm in the field of material science, especially in applications for energy storage. The application of configurational entropy helps materials attain increased phase stability, tunable electrochemical performance, and improved ionic conductivity. Though HEMs offer the said attributes, their large-scale usage in commercial battery systems is hindered by a number of scientific as well as technical roadblocks (see Figure 1.5). The complexity factor at the synthesis step, electrochemical stability, financial feasibility, and large-scale production are the main challenges hindering the use of HEMs. The following section discusses the major barriers to the commercialization of HEM, presents prospective research directions for their exploitation, and details how the large-scale production of HEMs can be achieved.

One of the main difficulties in the construction of high-entropy battery material is the complexity and scalability of synthesis. For maintaining phase purity and uniform distribution of the multiple elements, which is essential for consistent electrochemical activity, synthesis of the HEM is usually performed at a high temperature using the solid-state sintering method, the sol–gel process, SPS, and mechanochemical synthesis. All such syntheses are power-hungry and hard to scale. Batch-to-batch differences in composition further complicate their industrial production. Another key challenge is the electrochemical and structural stability of the long-term cycling of the HEMs. Although configurational disorder alleviates phase segregation, a level of gradual compositional ordering is still present in certain high-entropy systems, affecting their capacity retention and overall electrochemical performance. Interfacial stability is also an issue with the high-entropy electrodes and HEEs specifically in ASSBs since such materials can show large interfacial resistance and reactivity, which can eventually cause degradation. Interface engineering strategies involving surface coatings and artificial SEI films will be essential for long-term electrochemical stability. The availability and the cost of raw materials are also vital factors that determine the commercialization of HEMs. The inclusion of costly components like cobalt (Co), nickel (Ni), and vanadium (V) can dramatically raise the cost of production. For further economic feasibility, researchers are investigating the application of earth-abundant and low-cost elements like iron (Fe), manganese (Mn), and titanium (Ti) as replacements. Moreover, the availability of resources and geopolitical considerations for the extraction of rare metals might also affect the future supply chain of HEM.

Despite these limitations, HEMs have tremendous potential for the next wave of energy storage. One of the most promising research avenues is the application of high-throughput computational simulation and artificial intelligence (AI) for material discovery and to speed up the development of advanced high-performance HEMs [72, 73]. Machine learning (ML) techniques can examine large datasets to identify the best combination of elements that maximize electrochemical performance and reduce material expenses. Furthermore, density functional theory (DFT) calculations and molecular dynamics (MD) simulations can gain insight into ion diffusion mechanisms, defect energies, and phase stability, which can be used for rational material design with the aid of high-throughput screening strategies. Another area of focus is the interfacial engineering to enhance the stability of electrolytes and high-entropy electrodes. Thin-film coatings, artificial SEI films, and grain boundary modifications can all be used to lower interfacial resistance and enhance charge transfer. Multi-layer electrode designs coupling high-entropy phases with highly conductive scaffolds could further be beneficial. Another vital area for future studies is the development of low-cost and sustainable materials. To make high-entropy battery materials economically viable, researchers are also looking at the application of naturally prevalent elements. Recycling and upcycling methodologies for spent battery materials can also decrease reliance on critical resources. The application of green process technologies like low-sintering temperature and solvent-free synthesis can also reduce the overall impact of the development of high-entropy materials.

To gain universal industrial applications, scalable and economic manufacturing methods have to be developed for high-entropy compounds. Dry roll-to-roll processing and chemical vapor deposition (CVD) would be used to mass-produce thin-film high-entropy electrodes [74–76]. 3D printing and additive manufacturing also open possibilities for the creation of tailored electrode geometries with tailored porosity and mechanics [77]. Sustainable synthesis strategies like aqueous processing can further enhance cost-effectiveness. Another essential aspect of commercialization is that the HEMs should be integrated well with conventional battery technology. The HEEs and electrodes must be made compatible with conventional LIB and SIB designs with straightforward adoption by battery manufacturers. Hybrid designs with conventional battery material combined with the high-entropy phase can be an intermediate step toward the use of entirely entropy-maximized energy storage devices. For the development of HEMs closer to commercialization, cooperation between government agencies, academic institutions, and industry will be essential. Technology transfer can be accelerated by public–private collaborations, enabling research ideas to go from lab-scale demonstrations to pilot-scale production.

Standardized testing procedures should also be put in place to guarantee that HEM-based battery materials fulfill industry and regulatory standards for durability, safety, and dependability. Governments and funding organizations ought to encourage research projects aimed at the scalable manufacture and recycling of HEMs, as well as offer incentives for the development of sustainable batteries.

1.10 Summary and Outlook

A novel approach of developing next-generation energy storage systems has been brought about by the development of HEMs. These materials are intriguing options for solid electrolytes, battery electrodes, and interface engineering because they use combinations of multiple primary elements to increase ionic transport, phase stability, and electrochemical characteristics. The basic ideas, synthesis methods, compositional tweaking, electrochemical performance, and characterization methodologies of HEMs in energy storage applications have all been thoroughly covered in this book chapter. An introduction to HEMs and the function of configurational entropy in stabilizing single-phase solid solutions kicked off the conversation. In order to prevent phase segregation and guarantee uniform ionic and electronic conduction, the idea of entropy-driven phase stabilization was examined. The chapter went on to discuss compositional engineering techniques, emphasizing the ways in which multi-element interactions affect the redox activity, defect chemistry, and electronic structure of electrodes and electrolytes. High-entropy electrodes, including HEOs, HESs, HEPs, and HEAs, accounted for a large amount of the chapter. These materials are appealing for LIBs, SIBs, Li-S, and ASSBs due to their better cycling stability, multi-electron redox capabilities, and programmable electrochemical potentials. The function of HEEs was also explored, including how compositional disorder and interface engineering can improve electrochemical stability, inhibit dendritic development, and increase ionic conductivity. The chapter additionally covers advanced characterization methods that are necessary for studying HEMs, such as in situ/operando analysis, transmission electron microscopy (TEM), XRD, neutron scattering, XPS, and EIS. These methods allow for the logical design of high-entropy battery components by offering crucial insights into phase evolution, defect chemistry, interfacial phenomena, and long-term cycling performance. HEMs have several advantages, but a number of issues prevent them from being commercialized. These issues include cost considerations, interfacial compatibility, electrochemical stability, and synthesis scalability. To overcome these obstacles and hasten the industrial deployment of HEM-based battery materials, techniques such as sustainable synthesis methods, ML-guided material discovery, and enhanced interface engineering have been put forth.

1.11 Outlook

In energy storage research, HEMs mark a paradigm shift by providing new avenues for creating battery materials that are efficient, long-lasting, and reasonably priced.

To fully realize the promise of HEMs and make their commercialization easier, several important research avenues need to be explored in the future.

1.11.1 Accelerating High-Throughput Computational Discovery

Advancements in ML and AI-assisted materials design will play a significant role in adjusting the compositions of HEMs. DFT simulations, MD simulations, and computational high-throughput screening can be employed to accelerate the discovery of novel HEMs with predesigned electronic, ionic, and mechanical characteristics. Phase stability, defect energy of formation, and electrochemical performance can be estimated through AI-assisted models, and hence trial-and-error experimentations can be reduced significantly.

1.11.2 Interface Engineering for Next-Generation Batteries

For HEMs to be successfully integrated into commercial batteries, interface compatibility with electrolytes and current collectors must be optimized. Strategies such as thin-film coatings, artificial SEI layers, and grain boundary engineering will be crucial in reducing interfacial resistance and enhancing charge transfer kinetics. In SSBs, improving electrolyte–electrode contact through engineered interfaces will be key to achieving high-energy density and prolonged cycle life.

1.11.3 Sustainable and Cost-Effective Material Development

The economic viability of HEMs depends on their scalability and cost-effectiveness. Future research should focus on replacing expensive transition metals with earth-abundant elements while maintaining high electrochemical performance. Additionally, low-energy synthesis techniques, such as aqueous processing, solvent-free synthesis, and roll-to-roll manufacturing, should be explored to minimize production costs. Developing recycling and upcycling strategies for HEM-based electrodes and electrolytes will further contribute to sustainability and circular economy initiatives.

1.11.4 Integration with Emerging Energy Storage Technologies

Beyond LIBs and SIBs, HEMs hold promise for next-generation energy storage systems, including multivalent batteries (Mg, Zn, Al-based systems), supercapacitors, and hybrid energy storage devices. Their intrinsic phase stability, multi-electron charge storage capabilities, and defect-tolerant structures make them attractive candidates for high-energy density and high-power applications. Further research should explore the potential of HEMs in advanced battery chemistries, enabling breakthroughs in energy storage beyond current lithium-ion technologies.

1.11.5 Industry–Academia Collaborations and Commercialization Pathways

To accelerate the transition of HEMs from research laboratories to commercial-scale production, strong collaborations between academic institutions, industry leaders, and government agencies will be essential. Establishing standardized testing protocols and performance benchmarks will facilitate the comparison of HEMs with conventional battery components. Additionally, government policies, funding initiatives, and industry investments will play a crucial role in driving the commercialization of HEM-based energy storage solutions.

HEMs have shown great potential as electrolytes for next-generation batteries because they are more stable, have customizable electrochemical characteristics, and perform better than traditional materials. Energy storage issues including capacity fading, interfacial degradation, and safety can be resolved by these materials by utilizing entropy-driven stabilization and compositional flexibility. However, before HEMs can be widely adopted in the commercial sector, much research is still needed to overcome scalability, interfacial, and cost obstacles. Optimizing high-entropy compositions for useful energy storage applications will need the combination of computational modeling, sophisticated characterization methods, and sustainable material design. HEMs are expected to transform energy storage as developments proceed, opening the door to the creation of long-lasting, high-performing, and ecologically friendly battery technologies. Their successful commercialization could pave the way for the next generation of high-efficiency energy storage devices by bringing about innovations in portable electronics, grid-scale storage, and electric cars. HEMs could influence battery technology in the future and help create a more sustainable, energy-efficient, and electrified society by resolving outstanding issues and utilizing new scientific discoveries.

Acknowledgment

This work was supported by the National Research Program of Universities (NRPU) and Higher Education Commission (HEC), Pakistan research fund "Development of lab/pilot scale facilities for the production of piezoelectric material for multilayer energy devices" 20-15673/NRPU/R&D/HEC/2021.

References

1 George, E.P., Curtin, W.A., and Tasan, C.C. (2020). High entropy alloys: a focused review of mechanical properties and deformation mechanisms. *Acta Mater* 188: 435–474.

2 Wang, Q., Zhao, C., Wang, J. et al. (2023). High entropy liquid electrolytes for lithium batteries. *Nat Commun* 14 (1): 440.

3 Ali, M., Saleem, M., Sattar, T. et al. (2025). High-entropy battery materials: revolutionizing energy storage with structural complexity and entropy-driven stabilization. *Mater Sci Eng R Rep* 163: 100921.

4 Pei, Z., Zhao, S., Detrois, M. et al. (2023). Theory-guided design of high-entropy alloys with enhanced strength-ductility synergy. *Nat Commun* 14 (1): 2519.

5 Lun, Z., Ouyang, B., Kwon, D.H. et al. (2021). Cation-disordered rocksalt-type high-entropy cathodes for Li-ion batteries. *Nat Mater* 20 (2): 214–221.

6 Ye, Y., Gu, Z., Geng, J. et al. (2025). Advanced high-entropy halide solid electrolytes enabling high-voltage, long-cycling all-solid-state batteries. *Nano Lett* 25: 3747–3755.

7 Li, S., Lin, J., Schaller, M. et al. (2023). High-entropy lithium argyrodite solid electrolytes enabling stable all-solid-state batteries. *Angew Chem Int Ed Eng* 62 (50): e202314155.

8 Fu, M., Ma, X., Zhao, K. et al. (2021). High-entropy materials for energy-related applications. *iScience* 24 (3): 102177.

9 Strauss, F., Lin, J., Duffiet, M. et al. (2022). High-entropy polyanionic lithium superionic conductors. *ACS Mater Lett* 4 (2): 418–423.

10 Zeng, Y., Ouyang, B., Liu, J. et al. (2022). High-entropy mechanism to boost ionic conductivity. *Science* 378 (6626): 1320–1324.

11 Sun, G., Lin, H., Yao, S. et al. (2023). High-entropy solid-state Na-ion conductor for stable sodium-metal batteries. *Chem Eur J* 29 (28): e202300413.

12 Nam, K.H., Wang, Z., Luo, J. et al. (2024). High-entropy spinel oxide ferrites for battery applications. *Chem Mater* 36 (9): 4481–4494.

13 Cui, Y., Sukkurji, P.A., Wang, K. et al. (2022). High entropy fluorides as conversion cathodes with tailorable electrochemical performance. *J Energy Chem* 72: 342–351.

14 Li, H., Xu, M., Long, H. et al. (2022). Stabilization of multicationic redox chemistry in polyanionic cathode by increasing entropy. *Adv Sci* 9 (25): 2202082.

15 Patra, J., Nguyen, T.X., Tsai, C.C. et al. (2022). Effects of elemental modulation on phase purity and electrochemical properties of Co-free high-entropy spinel oxide anodes for lithium-ion batteries. *Adv Funct Mater* 32 (17): 2110992.

16 Zhou, S., Sun, Y., Gao, T. et al. (2023). Enhanced Li+ diffusion and lattice oxygen stability by the high entropy effect in disordered-rocksalt cathodes. *Angew Chem Int Ed Eng* 62 (42): e202311930.

17 Du, X.Y., Meng, Y., Yuan, H., and Xiao, D. (2023). High-entropy substitution: a strategy for advanced sodium-ion cathodes with high structural stability and superior mechanical properties. *Energy Storage Mater* 56: 132–140.

18 Spiridigliozzi, L., Di Feo, A.F., Accardo, G. et al. (2024). Li-doped transition metal high-entropy oxides (Li/TM-HEOs) as Li-ion batteries cathodes: a first report on capacity fading and cycling stability. *Mater Res Express* 11 (4).

19 Chen, R., Leung, C.L.A., and Huang, C. (2024). Exploring the properties of disordered rocksalt battery cathode materials by advanced characterization. *Adv Funct Mater* 34: 2308165.

20 Tanibata, N., Noi, K., Hayashi, A., and Tatsumisago, M. (2014). Preparation and characterization of highly sodium ion conducting Na3PS4–Na4SiS4 solid electrolytes. *RSC Adv* 4 (33): 17120–17123.

21 Luo, X., Patra, J., Chuang, W. et al. (2022). Charge-discharge mechanism of high-entropy Co-free spinel oxide toward Li+ storage examined using operando quick-scanning X-ray absorption spectroscopy. *Adv Sci* 9 (21): 2201219.

22 Chen, T.Y., Wang, S.Y., Kuo, C.H. et al. (2020). In operando synchrotron X-ray studies of a novel spinel $(Ni_{0.2}Co_{0.2}Mn_{0.2}Fe_{0.2}Ti_{0.2})3O_4$ high-entropy oxide for energy storage applications. *J Mater Chem A* 8 (41): 21756–21770.

23 Ghigna, P., Airoldi, L., Fracchia, M. et al. (2020). Lithiation mechanism in high-entropy oxides as anode materials for Li-ion batteries: an operando XAS study. *ACS Appl Mater Interfaces* 12 (45): 50344–50354.

24 Han, S.C., Ali, M., Kim, Y.J. et al. (2024). Unraveling electrochemo-mechanical aspects of core-shell composite cathode for sulfide-based all-solid-state batteries. *J Mater Chem A* 12: 24896–24905.

25 Sattar, T., Ali, M., Doo, S.G. et al. (2025). Investigating the impact of calcination temperature to improve the discharge capacity of $LiNi_{0.95}Co_{0.03}Mn_{0.02}O_2$ cathode material synthesized by Taylor-Couette reactor. *J Power Sources* 641: 236853.

26 Kang, R., Zhang, D., Du, Y. et al. (2024). Configurational entropy strategy enhanced structure stability achieves robust cathode for aluminum batteries. *Small* 20 (5): 4369–4381.

27 Kawaguchi, T., Bian, X., Hatakeyama, T. et al. (2022). Influences of enhanced entropy in layered rocksalt oxide cathodes for lithium-ion batteries. *ACS Appl Energy Mater* 5 (4): 4369–4381.

28 Liu, X., Yu, Y., Li, K. et al. (2024). Integrating hollow multishelled structure and high entropy engineering toward enhanced mechano-electrochemical properties in lithium battery. *Adv Mater* 36 (19): 2312583.

29 Ouyang, B. and Zeng, Y. (2024). The rise of high-entropy battery materials. *Nat Commun* 15 (1): 973.

30 Gu, S., Huang, H.Y., Lin, Y.C. et al. (2025). High-entropy alloys with low platinum content: novel catalysts for oxygen reduction and evolution reactions. *Emerg Mater* 8: 3649–3662.

31 Bib Khan, J., Kumar Panda, P., Dash, P., and Hsieh, Te, C. (2025). Microwave-assisted synthesis of platinum-free high-entropy alloy catalysts on reduced graphene oxide sheets for enhanced oxygen reduction and evolution reactions. *Chem Eur J* 31: e202403863.

32 Wang, Q., Zhou, Y., Wang, X. et al. (2024). Designing lithium halide solid electrolytes. *Nat Commun* 15 (1): 1050.

33 Sarkar, A., Velasco, L., Wang, D. et al. (2018). High entropy oxides for reversible energy storage. *Nat Commun* 9 (1): 3400.

34 Mohili, R., Hemanth, N.R., Jin, H. et al. (2023). Emerging high entropy metal sulphides and phosphides for electrochemical water splitting. *J Mater Chem A* 11: 10463–10472.

35 Lin, L., Wang, K., Sarkar, A. et al. (2022). High-entropy sulfides as electrode materials for Li-ion batteries. *Adv Energy Mater* 12 (8): 2103090.

36 Panda, P.K., Huang, H.Y., Dash, P. et al. (2025). Liquid-phase microwave synthesis of platinum-based high-entropy alloy catalysts on carbon supports for electrochemical hydrogen adsorption/desorption and oxygen evolution/reduction reactions. *Int J Hydrog Energy* 111: 536–545.

37 Liu, K.C., Kumar Panda, P., Chandra Mallick, B. et al. (2024). Solid-phase microwave synthesis of high-entropy graphene quantum dots as metal-free electrochemical catalysts. *Appl Surf Sci* 648: 159061.

38 Khan, J.B., Panda, P.K., Yang, P.C. et al. (2024). Microwave synthesis of high-entropy alloy catalysts on graphene oxide sheets for oxygen reduction and evolution reactions. *Int J Hydrog Energy* 53: 999–1008.

39 Ali, M., Kim, S., Kim, S. et al. (2025). Lithiation-driven cascade dissolution coprecipitation of sulfide superionic conductors. *Energy Storage Mater* 74.

40 Ali, M., Han, S.C., Park, H. et al. (2022). Solvent exchange-induced facile recrystallisation and particle size control of sulphide solid electrolytes for all-solid-state Li-ion batteries. *J Mater Chem A* 10 (48): 25471–25480.

41 Wang, Q., Velasco, L., Breitung, B., and Presser, V. (2021). High-entropy energy materials in the age of big data: a critical guide to next-generation synthesis and applications. *Adv Energy Mater* 11 (47): 2102355.

42 Feng, Y., Yang, L., Yan, Z. et al. (2023). Discovery of high entropy garnet solid-state electrolytes via ultrafast synthesis. *Energy Storage Mater* 63: 103053.

43 Kim, S.C., Wang, J., Xu, R. et al. (2023). High-entropy electrolytes for practical lithium metal batteries. *Nat Energy* 8 (8): 814–826.

44 Kim, M.S., Zhang, Z., Rudnicki, P.E. et al. (2022). Suspension electrolyte with modified Li+ solvation environment for lithium metal batteries. *Nat Mater* 21 (4): 445–454.

45 Zuo, D., Yang, L., Zou, Z. et al. (2023). Ultrafast synthesis of NASICON solid electrolytes for sodium-metal batteries. *Adv Energy Mater* 13 (37): 2301540.

46 Ali, M., Park, S., Raza, A. et al. (2024). Thermal stability analysis of nitrile additives in LiFSI for lithium-ion batteries: an accelerating rate calorimetry study. *Heliyon* 10 (9): e29397.

47 Li, S., Wu, L., Fu, H. et al. (2023). Entropy-tuned layered oxide cathodes for potassium-ion batteries. Small. *Methods* 7 (11): 2300893.

48 Zhang, R., Wang, C., Zou, P. et al. (2022). Compositionally complex doping for zero-strain zero-cobalt layered cathodes. *Nature* 610 (7930): 67–73.

49 Zhao, J., Wei, Z., Chen, N. et al. (2024). High-entropy alloy anodes for low-strain and high-volumetric lithium-ion storage at ambient and subzero temperatures. *Energy Storage Mater* 65: 103127.

50 Xiao, B., Wu, G., Wang, T. et al. (2022). High-entropy oxides as advanced anode materials for long-life lithium-ion batteries. *Nano Energy* 95: 106962.

51 Yang, C., Liu, X., Lin, Y. et al. (2023). Entropy-driven solvation toward low-temperature sodium-ion batteries with temperature-adaptive feature. *Adv Mater* 35 (28): 2301817.

52 Ma, Y., Ma, Y., Wang, Q. et al. (2021). High-entropy energy materials: challenges and new opportunities. *Energy Environ Sci* 14: 2883–2905.

53 Wang, F., Zou, P., Zhang, Y. et al. (2023). Activating lattice oxygen in high-entropy LDH for robust and durable water oxidation. *Nat Commun* 14 (1): 6019.

54 Chen, Y., Liao, X., Xie, M. et al. (2024). An advanced high-entropy cathode achieves a multi-electron reaction via the activation of multicationic redox in polyanionic phosphates for sodium-ion batteries. *ACS Sustain Chem Eng* 12 (36): 13568–13577.

55 Qiu, N., Chen, H., Yang, Z. et al. (2019). A high entropy oxide ($Mg_{0.2}Co_{0.2}Ni_{0.2}Cu_{0.2}Zn_{0.2}O$) with superior lithium storage performance. *J Alloys Compd* 777: 767–774.

56 George, E.P., Raabe, D., and Ritchie, R.O. (2019). High-entropy alloys. *Nat Rev Mater* 4: 515–534.

57 Shahmir, H., Mehranpour, M.S., Arsalan Shams, S.A., and Langdon, T.G. (2023). Twenty years of the CoCrFeNiMn high-entropy alloy: achieving exceptional mechanical properties through microstructure engineering. *J Mater Res Technol* 23: 3362–3423.

58 Li, X., Liang, J., Luo, J. et al. (2019). Air-stable Li_3InCl_6 electrolyte with high voltage compatibility for all-solid-state batteries. *Energy Environ Sci* 12 (9): 2665–2671.

59 Velasco, L., Castillo, J.S., Kante, M.V. et al. (2021). Phase–property diagrams for multicomponent oxide systems toward materials libraries. *Adv Mater* 33 (43): e2102301.

60 Zhou, L., Assoud, A., Shyamsunder, A. et al. (2019). An entropically stabilized fast-ion conductor: $Li_{3.25}[Si_{0.25}P_{0.75}]S_4$. *Chem Mater* 31 (19): 7801–7811.

61 Xu, C., Zhao, J., Wang, E. et al. (2021). A novel NASICON-typed $Na_4VMn_{0.5}Fe_{0.5}(PO_4)_3$ cathode for high-performance Na-ion batteries. *Adv Energy Mater* 11 (22): 2100729.

62 Zhang, R., Wang, C., Zou, P. et al. (2023). Long-life lithium-ion batteries realized by low-Ni, Co-free cathode chemistry. *Nat Energy* 8 (7): 695–702.

63 Moon, C.K., Lee, H.J., Park, K.H. et al. (2018). Vacancy-driven Na+ superionic conduction in new Ca-doped Na_3PS_4 for all-solid-state Na-ion batteries. *ACS Energy Lett* 3 (10): 2504–2512.

64 Liang, P., Qi, K., Chen, S. et al. (2024). Low-electronegativity cationic high-entropy doping to trigger stable anion redox activity for high-Ni Co-free layered cathodes in Li-ion batteries. *Angew Chem Int Ed Eng* 63 (10): e202318186.

65 Wang, Q., Zhao, C., Yao, Z. et al. (2023). Entropy-driven liquid electrolytes for lithium batteries. *Adv Mater* 35 (17): 2210677.

66 Cheng, W., Liu, J., Hu, J. et al. (2023). Pressure-stabilized high-entropy $(FeCoNiCuRu)S_2$ sulfide anode toward simultaneously fast and durable lithium/sodium ion storage. *Small* 19 (29): e2301915.

67 An, Q., Li, S., Zhou, J. et al. (2023). Novel spinel multicomponent high-entropy oxide as anode for lithium-ion batteries with excellent electrochemical performance. *Adv Eng Mater* 25 (20): 2300585.

68 Chen, G., Ji, H., Fang, H. et al. (2023). Dual modification of P3-type layered cathodes to achieve high capacity and long cyclability for sodium-ion batteries. *ACS Appl Mater Interfaces* 15 (28): 33682–33692.

69 Zhao, J., Yang, X., Huang, Y. et al. (2021). Entropy stabilization effect and oxygen vacancies enabling spinel oxide highly reversible lithium-ion storage. *ACS Appl Mater Interfaces* 13 (49): 58674–58681.

70 Yeh, Y., Huang, C., Hou, A. et al. (2024). In situ TEM observation of (Cr, Mn, Fe, Co, and $Ni)_3O_4$ high-entropy spinel oxide formation during calcination at atomic scale. *Small* 20 (15): 159–171.

71 Lin, C.C., Liu, H.Y., Kang, J.W. et al. (2022). In-situ X-ray studies of high-entropy layered oxide cathode for sodium-ion batteries. *Energy Storage Mater* 51: 159–171.

72 Park, S., Lee, H., Choi, S. et al. (2024). Microstructure-based digital twin thermo-electrochemical modeling of LIBs at the cell-to-module scale. *eTransportation* 22: 100370.

73 Ali, M., Doh, C.H., Lee, Y.J. et al. (2021). Current trends in nanoscale interfacial electrode engineering for sulfide-based all-solid-state Li-ion batteries. *Energy Technol* 9: 2001096.

74 Ali, H.G., Khan, K., Hanif, M.B. et al. (2023). Advancements in two-dimensional materials as anodes for lithium-ion batteries: exploring composition-structure-property relationships, emerging trends, and future perspective. *J Energy Storage* 73: 108980.

75 Gohar, O., Khan, M.Z., Bibi, I. et al. (2024). Nanomaterials for advanced energy applications: recent advancements and future trends. *Mater Des* 241: 112930.

76 Gohar, O., Ishfaq, H.A., Iqbal, M.A. et al. (2025). Recent advancements in high-performance and durable electrode materials for magnesium-ion batteries. *Coord Chem Rev* 538: 216702.

77 Zhu, G., Hou, Y., Lu, J. et al. (2023). MXene decorated 3D-printed carbon black-based electrodes for solid-state micro-supercapacitors. *J Mater Chem A* 11: 25422–25428.

2

Perovskite-Based High-Entropy Materials for Energy Storage Applications

Asfaq Ali[1], *Sanjeev Verma*[2], *Pradeep Kumar Panda*[3], *and Tapas Das*[4]

[1] *NIT, Department of Chemical Engineering, Rourkela, Odisha 769008, India*
[2] *NIT Patna, Department of Chemical Engineering & Technology, Patna, Bihar 800005, India*
[3] *Yuan Ze University, Department of Chemical Engineering and Materials Science, Taoyuan 32003, Taiwan*
[4] *NIT, Department of Chemical Engineering, Rourkela, Odisha 769008, India*

2.1 Introduction

Entropy is a thermodynamic variable that measures disorder or randomness of the system. High-entropy alloys (HEAs) were the first to offer alloy formation as a novel idea in 2004. HEAs consist of five principal elements, with each element comprising between 5% and 35% of the overall composition. These alloys' high configurational entropy causes the presence of different chemicals to produce simple single-phase structures. HEAs have performance characteristics distinct from conventional alloys. For example, they have slow diffusion, significant lattice distortion, and "cocktail effect" in their performance. Additionally, HEAs come in an extensive variety of compositions, allowing for acquisition of specific features. In recent years, inorganic non-metallic materials have been subjected to the concept of high-entropy [1–3].

There are several types of perovskite oxides, including $Pb(Mg_{1/3}Nb_{2/3})O_3$ (an electrostrictive material), Pb (Ti, Zr)O_3 (piezoelectric material), and (La, Ca)MnO_3 (a magnetoresistive material). High-entropy perovskite oxides (HEPOs) were the subject of the initial investigation. Improved ferroelectric relaxation, decreased thermal conductivity, increased catalytic activity, improved thermal and chemical stability, increased proton conductivity, improved oxygen permeability, and a unique luminescence sequence are just a few of the intriguing properties that have been reported to distinguish HEPOs from traditional perovskite oxides. High-Entropy Ceramics (HECs) have been the subject of numerous evaluations, while HEPOs have not been the subject of any of them. This study focuses on reviewing the latest findings and advancements in the field of HEPOs. The focus will be on structural descriptors, synthesis approaches, compositional plan techniques, and the application of various energy materials [4–7].

High-Entropy Materials for Energy Storage Devices, First Edition.
Edited by Chien-Te Hsieh, Pradeep Kumar Panda, and Arpan Kumar Nayak.

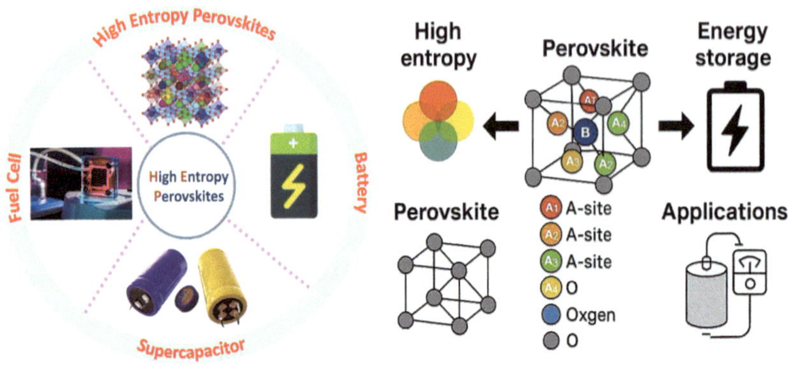

Figure 2.1 Overview of perovskite material application.

Figure 2.1 depicts the crystal structure of a HEPO, known as ABO_3. In this structure, the A-site locations are taken by different types of cations (A_1–A_4), represented as colored spheres (e.g. La^{3+}, Sr^{2+}, Ba^{2+}, Ca^{2+}), placed at the corners of the cubic unit cell. The deliberate mixing of these cations in equimolar or near-equimolar ratios results in high configurational entropy, which is critical for stabilizing single-phase solid solutions and improving thermal and structural stability. A_1 (red), A_2 (orange), A_3 (yellow), and A_4 (green) represent different A-site metal ions in the high-entropy perovskite. In high-entropy materials (HEMs), multiple types of A-site cations are mixed in near-equimolar ratios to enhance configurational entropy.

The B-site, depicted by a center blue sphere, usually contains transition metal (TM) cations as Ti^{4+}, Mn^{3+}, Fe^{3+}, or Co^{3+}. These B-site ions are octahedrally coordinated by oxygen anions (O^{2-}), represented as gray spheres situated at the face centers of the cube to form BO_6 octahedra. This octahedral coordination is required to enable a variety of physical processes such as electron conduction, ionic transport, and redox activity all of which are vital to the operation of energy storage systems. The high-entropy configuration provides for an extensive control over the material's electrical structure, oxygen vacancy formation energy, and ionic diffusion channels. As a result, HEPOs perform better in terms of capacity retention, rate performance, and thermal robustness when used in lithium-ion batteries (LIBs), sodium-ion batteries (SIBs), and supercapacitors. The schematic emphasizes the structural versatility and functional adaptability of perovskite-based HEMs as potential options for next-generation energy storage systems.

2.2 Design Strategies and Entropy Engineering in HEPOs

Systems exhibiting configurational entropy more than 1.5R are categorized as high-entropy systems.

$$\Delta S_{config} = -R\left(\sum_{i=1}^{N} A_i \ln A_i + \sum_{j=1}^{M} B_j \ln B_j + 3\sum_{q=1}^{P} O_q \ln O_q\right)$$

where

- A_i, B_j, O_q are the mole fractions of different cations at A-site, B-site, and anionic O-site, respectively.
- M N, P are the number of species at each site.
- R is molar gas constant.

Currently, a few HEPOs documented, that use the oxygen lattice's involvement to entropy to attain elevated configurational entropy. Therefore, removing the anionic lattice's impact on the molecule's configurational entropy, Eq. (2.1) may be condensed to $\Delta S_{config} = \Delta SA_{config} + \Delta SB_{config}$. Consequently, there stand primarily two main approaches for developing HEPOs: Route 1 entails several cations inhabiting either the A-site or the B-site, whereas Route 2 entails different cations inhabiting both sites [8].

One sublattice for Route 1 will be swapped out, resulting in the formula $\Delta S_{config} = \sum_{i=1}^{N} x_i \ln x_i$, where x_i denotes the mole fraction of the ith ion at the specified location of perovskite. Upon conducting equimolar substitution, the formula may be further refined to $\Delta S_{config} = \ln(N)\ R$, where N represents the quantity of species on static sublattice which is the most common design methodology. Using this technique, Jiang and associates developed 13 Ba/Sr-based HEPOs. Six products having structures of pure perovskite are produced when tetravalent ions (Zr, Sn, Ti, and Hf) are substituted for the B-site along with one variable ion. There are two distinct cation sublattices in perovskite structures. The A-site often contains larger alkali or rare-earth cations, whereas the B-site generally accommodates smaller TM cations. Sarkar et al. [8] conducted a pivotal investigation in which they synthesized 11 high-entropy rare-earth–TM (RE–TM) perovskite oxides by methodically altering the substitutional composition at both A- and B-sites. The A-site is modified by including rare-earth elements, such as La, Nd, Sm, Gd, and Y. We modified the B-site by including TMs, such as Fe, Co, Cr, Mn, and Ni. This method produced stable single-phase perovskite structures characterized by significant configurational entropy. This suggests that multi-component design may be used to alter the properties of buildings, magnets, and electrochemicals. In Route 2 of HEPO design, configurational entropy is raised by replacing many cations at both the A-site and the B-site of the perovskite lattice. The sum of the configurational entropies at the A-site (ΔSA_{config}) and the B-site (ΔSB_{config}) must be more than 1.50R, where R is the universal gas constant, for a system to be called "high-entropy." The ΔSB_{config} values for the two composition $Ba_{0.4}Sr_{0.4}Bi_{0.2}(Zr_{0.3}Hf_{0.3}Ti_{0.2}Fe_{0.2})O_3$ and $Ba_{0.5}Sr_{0.5}(Zr_{0.4}Hf_{0.3}Ti_{0.3})O_3$ are 1.05R and 0.69R, respectively. The ΔSB_{config} values are 1.37R and 1.09R. This suggests that the overall configurational entropy values (ΔS_{config}) for both systems are greater than 1.50R, which is the lowest threshold required to be called HEMs [9].

The compositions, ion-substitution locations, and development techniques of HEPOs reported in literature are listed in Table 2.1. Ba, Bi, Ca, and La are often discovered at the A-site, whereas Ti, Fe, Zr, Ni, and Co are primarily identified at the B-site. The "A/B-site" in Figure 2.1 denotes the area inside the crystal structure of HEPOs that may hold multiple ions, leading to elevated configurational entropy.

Table 2.1 Representative high-entropy perovskite oxides for energy storage applications.

HEPO composition	Synthesis method	Application area	Key property/ advantage	References
BaSrBi $(ZrHfTiFe)O_3$	Sonochemical	Supercapacitors	High surface area and mesoporosity	[8]
$(Y_{0.2}Nd_{0.2}Sm_{0.2}Eu_{0.2}Er_{0.2})AlO_3$	Co-precipitation	Dielectric storage	High-energy density and dielectric constant	[10]
$Pb(Ni_{0.177}Sc_{0.015}In_{0.06}Ti_{0.32}Nb_{0.428})O_3$	Solid-state reaction	Piezoelectric energy devices	Adjustable polarization and phase stability	[11]
$Li_x(LiLaCaSrBa)Ti_{1-x}Al_xO_3$	Combustion synthesis	Lithium storage	Low density and porous architecture	[12]
$(1-x)Pb(Mg_{0.2}Zn_{0.2}Nb_{0.2}Ta_{0.2}W_{0.2})O_3-xPbTiO_3$	Valence state strategy	Ferroelectric capacitors	MPB design enhances ferroelectric and piezoelectric response	[13]

The periodic table indicates the type of replacement site and the rate of recurrence of substitution by height and column color, respectively, illustrating the wide range of ions utilized in the synthesis of HEPOs. Because of their large radii, alkali and rare-earth ions are found at the A-site, while TM ions are found at the B-site. For example, a rare-earth perovskite oxide with ultrahigh entropy and 16 ions at the A-site.

Figure 2.2a shows the relationship between domain structure and performance of energy storage as entropy function. Nanoscale domains are quite big at low-entropy levels, which leads to low-energy storage efficiency and density. Nanoscale domains get smaller as entropy rises, and the density and efficiency of energy storage are significantly increased when polar nanodomains are refined. This finding suggests that high-entropy tactics enhance the material's capacity for energy storage.

Lattice distortion from high-entropy methods can disrupt symmetry and long-range order in crystal structures with high symmetry and long-range order, resulting in dielectric ceramics with extremely dynamic and flexible polarization. Figure 2.2b illustrates the high-entropy technique used in the design of $Bi_{0.2}Na_{0.2}Ba_{0.2}Sr_{0.2}Ca_{0.2}TiO_3$ with high-energy storage performance. Plasmonic elements doped at the A-site produce polar nano-regions (PNRs), destroy long-range order, and increase cationic disorder and random fields. Furthermore, the inclusion of Li_2CO_3 improves overall performance and narrows the P–E loop from a saturated state by achieving high density and fine grains in addition to lowering the sintering temperature. Li^+ inclusion also balances grain size and density, increases breakdown field strength, and inhibits vacancy formation brought on by element volatilization during high-temperature sintering. Figure 2.2c displays the design of the piezoresponse force microscopy (PFM) probe domain, where narrow fringe domains (i) and (ii) emerge at a voltage of 2 V AC. 180° domain flipping is induced

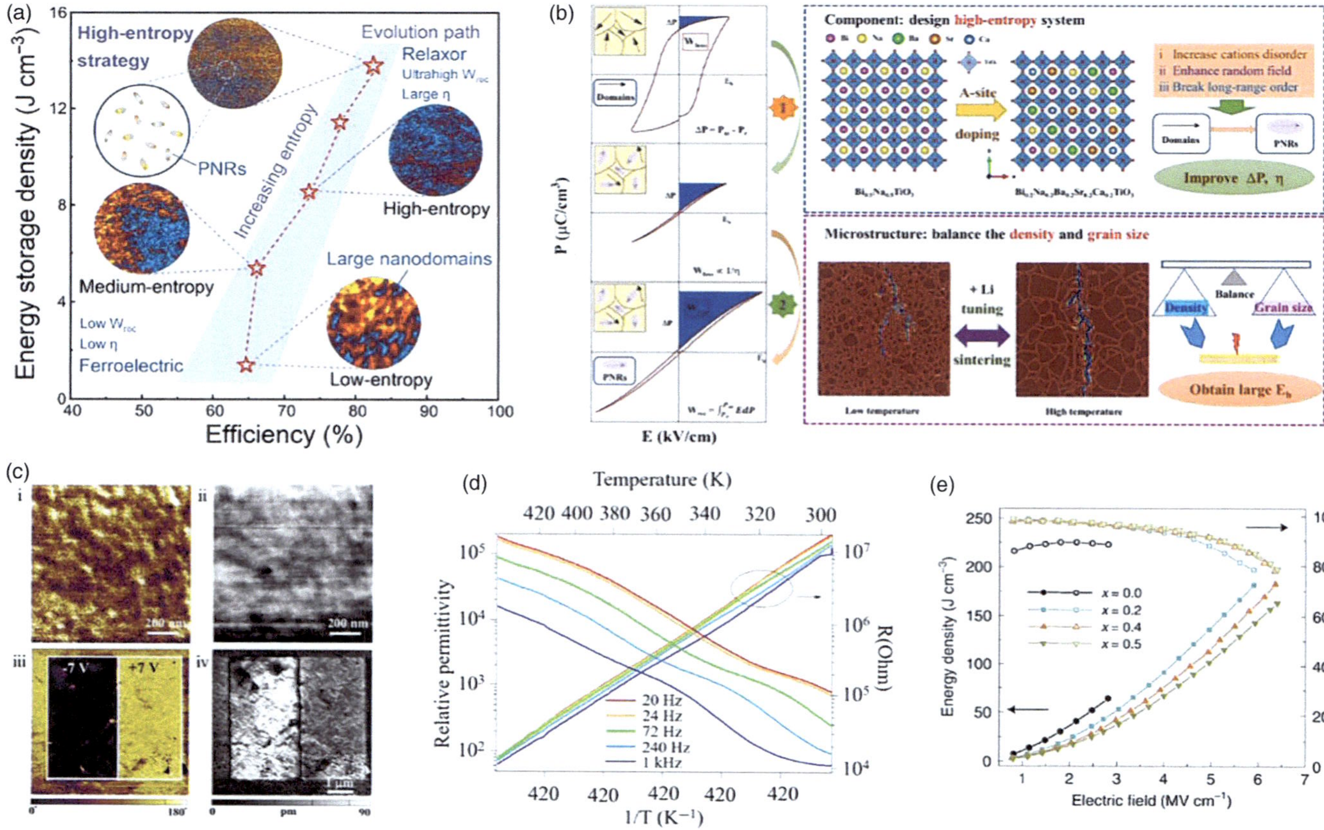

Figure 2.2 (a) An illustration of the domain structure and energy storage performance as entropy increases. (b) Diagrammatic representation of the high-entropy method used to achieve high-energy storage qualities. (c) PFM phase and amplitude images at different voltages. (d) Resistance and relative permittivity of HEO_x-Li_5 as a function of temperature at low frequencies. (e)The energy density and efficiency of $(Bi_{3.25}La_{0.75})(Ti_{3-3x}Zr_xHf_xSn_x)O_{12}$ as a function of the electric field are as high as E_b. *Source:* Reference [14] / with permission of Springer Nature.

by applying a 7 V DC voltage; domain transitions (iii and iv) are associated with the massive $P_m - P_r$ and low-hysteresis polarization shifts in the P–E loop, and distinct domain boundaries are apparent.

As seen in Figure 2.2d, the relative dielectric constant of HEO_x-Li_5, as determined by an LCR bridge at 20 Hz, approached 2×105 at 440 K. Their research offers fresh concepts for modifying HEM characteristics and demonstrates the viability of high-entropy tactics in dielectric ceramics. Dielectric ceramics with high-energy storage performance are essential to capacitors, which polarize an electrolyte to store energy. Dielectric capacitors are attractive energy storage components because of their high-energy density, high efficiency, and quick charge/discharge rates.

HEMs are perfect for producing energy storage devices like capacitors because of their high dielectric constant, breakdown strength, and low dielectric losses, which are all attained using high-entropy techniques. In order to reduce dielectric losses and enhance overall energy storage performance, Sun et al. [15] used a hydrothermal process to create a HEC $(Bi_{0.2}Ca_{0.2}Sr_{0.2}Na_{0.2}Ba_{0.2})TiO_3$ with a single tetragonal phase. Dielectric relaxation and improved energy storage ability in HECs are the outcomes of substantial lattice distortions caused by the large variances in cation radius and electronegativity, which break octahedral symmetry. By adding equimolar Zr, Hf, and Sn (x) to the Ti site and La to the Bi site, Yang et al. [16] created a $Bi_2Ti_2O_7$-based high-entropy dielectric film with a pyrochlorite phase, producing $(Bi_{3.25}La_{0.75})(Ti_{3-3x}Zr_xHf_xSn_x)$O12. Figure 2.2e illustrates how, as entropy increases, the energy density Ue rises from 63 J cm^{-3} at $x = 0.0$ to a maximum of 182 J cm^{-3} at $x = 0.4$. A 5.0 MV cm^{-1} electric field causes the energy storage density to be 90%. Lattice distortion, grain refinement, and amorphous phases brought on by increased entropy are credited with improving energy storage performance by lowering dielectric losses and raising resistivity and breakdown strength. Excessive time spent searching for HEPOs that are not equimolar may result in missed opportunities. Ma et al. [17] developed a method to address this issue by using Ba-based HEPOs via a valence-state combination strategy. Numerous compounds with average valences ranging from +2 to +5 were discovered. Their study indicates that systems with high-entropy may remain stable even with changes in the ratios of components as little as 1%. This provides designers of non-equimolar HEPOs with more options.

As HEPOs are developed, design methodologies are anticipated to transition from a focus on composition to an emphasis on functionality. The morphotropic phase boundary (MPB) in ferroelectric systems has facilitated structural optimization; however, the primary objective for energy storage is to enhance lattice distortion, compositional fluctuations, and random internal fields, which collectively facilitate dielectric energy storage. From a thermodynamic perspective, the high-entropy effect contributes to increased stability. The synergistic interactions of multivalent ions also enhance performance in electrochemical and catalytic applications. Consequently, HEPOs are promising candidates for multifunctional energy storage devices [18–21].

2.3 Synthesis Techniques

So far, many synthesis processes have been used to produce HEPOs in diverse dimensions, including powders, films, and bulk solids. Table 2.1 enumerates the various synthesis techniques of HEPOs. This section succinctly presents many prevalent techniques for synthesizing HEPOs.

2.3.1 Advanced Synthesis Strategies for High-Entropy Perovskites

Metal alloys are the most commonly synthesized and researched HEMs as colloidal nanoparticles, owing to their desired catalytic characteristics. Other types of high-entropy colloidal nanoparticles have been identified, including intermetallic complexes, metal sulfides, metal oxides, and metal oxysulfides [22]. Additionally, novel colloidal high-entropy nanoparticle forms are becoming more accessible. Figure 2.3 highlights some of the crystal structures that have been synthesized so far and shows a high-entropy outline of some that are known to be synthetically available as colloidal nanoparticles that have been modified to present them as high-entropy phases.

Enhancing the effectiveness of catalysts and electrodes in electrochemical energy storage devices such as batteries and capacitors requires the acquisition of materials characterized. The structural characteristics are closely linked to the choice and design of synthesis pathways for HEPOs. A mechanochemically

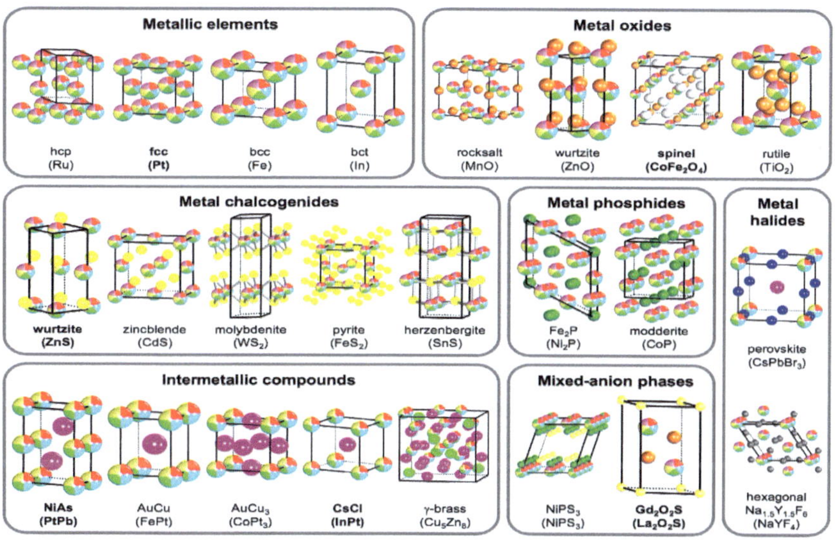

Figure 2.3 Schematic representation of crystal structures of some of the most frequently synthesized materials as colloidal nanocrystals, along with sample compositions. The randomized site or sites are marked by multicolored atoms, and all are displayed as high-entropy variants. *Source:* Reproduced from Ref. [23] / with permission of American Chemical Society.

facilitated redox approach using $KMnO_4$ as an oxidizing agent during the milling of precursor materials resulted in the in situ creation of an intermediate phase, $La_5(Co_{0.2}Cu_{0.2}Fe_{0.2}Mn_{0.2}Ni_{0.2})(OH)_{30-\delta}$. This intermediate stabilized the system at markedly lower synthesis temperatures owing to decreased reaction enthalpy. HEPOs nanopowders were successfully manufactured utilizing the redox-assisted mechanochemical approach (Route A) at temperatures below 500 °C. In contrast, since no intermediate phase developed in Route B, temperatures exceeding 800 °C were required for phase stability [24, 25].

The process of particle size reduction is enhanced, synthesis durations are minimized, and calcination temperatures are decreased through the application of mechanochemical synthesis, notably via high-energy ball milling. The compound $La(Co_{0.2}Cr_{0.2}Fe_{0.2}Mn_{0.2}Ni_{0.2})O_{3-\delta}$ achieved complete crystallization within seconds, facilitated by an expedited synthesis method employing Ni–Pt Joule heating, thereby enhancing both energy and temporal efficiency markedly. The co-precipitation approach is favored for synthesizing HEPOs due to its simplicity and scalability. It utilizes substances such as sodium hydroxide or ammonia to facilitate the precipitation of metal nitrates from solution. The solution is then dried and calcined. This methodology enables the production of multicomponent oxides, such as $(Y_{0.2}Nd_{0.2}Sm_{0.2}Eu_{0.2}Er_{0.2})AlO_3$ and $La(Co_{0.2}Mn_{0.2}Cr_{0.2}Fe_{0.2}Ni_{0.2})O_3$, characterized by uniform composition and stable phases [11, 13, 26].

In the solvothermal synthesis process, the solution of precursor undergoes elevated pressures and temperatures within a sealed reactor. This is done to assist recrystallization and ensure that the composition is homogeneous. A hybrid co-precipitation–hydrothermal technique has been employed for synthesizing $(Gd_{0.2}Nd_{0.2}La_{0.2}Sm_{0.2}Y_{0.2})CoO_3$ with improved compositional homogeneity. Utilizing acoustic cavitation in ultrasonication enhances the reactivity of precursors in sonochemical synthesis. This indicates that HEPOs may form without undergoing calcination. Systems, such as $BaSr(ZrHfTi)O_3$ and $Ru/BaSrBi(ZrHfTiFe)O_3$, exhibit substantial specific surface areas as a result of cavitation-induced precipitation. This is an excellent method for energy storage [27–30].

The sol–gel method provides an alternative approach for synthesizing fine and uniformly dispersed HEPO powders at comparatively low temperatures. Citric acid, ethylene glycol, or Ethylenediaminetetraacetic Acid (EDTA) are used as complexing agents to dissolve metal nitrates in water. Heating and evaporation are employed to achieve gelation, which is subsequently followed by calcination and maturation of the resulting xerogel. Several HEPO compositions have been synthesized successfully using this method [31, 32]. The fabrication of distinctive morphologies, such as yolk-shell structures, can further improve electrochemical performance. The perovskite $La_{0.7}Bi_{0.3}Mn_{0.4}Fe_{0.3}Cu_{0.3}O_3$, synthesized using ethylene glycol and isopropanol, demonstrates optimized bond angles and decreased bond lengths, resulting in an improved electronic configuration and increased specific capacity. During calcination at 600 °C, oxygen transport gradients between the core and the shell create the yolk-shell structure [33].

A hollow spherical multilayer framework for HEPOs may be fabricated using acidic carbon spheres (ACSs) as sacrificial models. This method, using ultrasonication-assisted metal precursor deposition and calcination at 700 °C, significantly enhances the quantity of oxygen vacancies (up to 37.71%) compared to traditional structures (15.9%). This increases the number of active sites, enhancing its catalytic efficiency.

2.3.2 Advanced Thin-Film Deposition Techniques for High-Entropy Perovskite Oxides

Pulsed laser deposition (PLD) is a widely utilized method for the fabrication of high-quality thin films of HEPOs, due to its ability to precisely control stoichiometry and crystallinity. In another studies, deposited $Ba(Zr_{0.2}Sn_{0.2}Ti_{0.2}Hf_{0.2}Nb_{0.2})O_3$ single-crystal thin films on $SrTiO_3$ and MgO substrates were aligned along the (001) crystallographic direction. This represents the initial documentation of HEPO thin films fabricated through PLD. Subsequently, epitaxial films including $La_{1-x}Sr_x(Cr_{0.2}Fe_{0.2}Co_{0.2}Mn_{0.2}Ni_{0.2})O_3$ and $Tb(Fe_{0.2}Mn_{0.2}Cr_{0.2}Co_{0.2}Ni_{0.2})O_3$ have been successfully deposited on $SrTiO_3$ substrates, demonstrating the capability of PLD to support a variety of multicomponent compositions [34, 35]. Additionally, various alternative substrates have been utilized to broaden the synthesis landscape. $Pb(Hf_{0.2}Zr_{0.2}Ti_{0.2}Nb_{0.2}Mn_{0.2})O_3$ and $Pb(Hf_{0.2}Zr_{0.2}Ti_{0.2}Nb_{0.2}Al_{0.2})O_3$ films were produced utilizing a $Pb(Zr,Ti)O_3$ (PZT) seed layer on platinum-coated silicon dioxide/silicon wafers. The deposition of $(La_{0.2}Pr_{0.2}Nd_{0.2}Sm_{0.2}Eu_{0.2})NiO_{3-\delta}$ thin films on $NbGaO_3$ substrates illustrate the influence of substrate characteristics on the modification of film properties [36].

In addition to PLD, alternative deposition techniques, including spin coating, have shown effectiveness in the fabrication of HEPO films. The spin coating technique to fabricate single-crystal $Ba(Zr_{0.2}Sn_{0.2}Ti_{0.2}Hf_{0.2}Nb_{0.2})O_3$ thin films on magnesium oxide substrates. In a separate investigation, Chen et al. utilized a method that integrated spin coating with hydrothermal synthesis on fluorine-doped SnO_2 substrates to produce $Ba(Ti_{0.2}Hf_{0.2}Zr_{0.2}Fe_{0.2}Sn_{0.2})O_3$ thin films. The outlined approaches highlight the growing range of methodologies available for engineering HEPO films with customized microstructures and functionalities, essential for the progression of next-generation energy storage applications [37–42].

2.3.3 Solid-State and Advanced Sintering Techniques

The conventional solid-state reaction (SSR) method is a widely utilized technique for the synthesis of bulk-type oxide ceramics, including HEPOs. This approach generally includes high-temperature calcination, followed by extended annealing to guarantee adequate ionic diffusion and uniformity within the multicomponent lattice.

Due to the inherent sluggish diffusion kinetics associated with high-entropy systems, the process requires prolonged thermal treatment and mechanical milling durations, resulting in significant time and energy requirements. Additionally, the high-temperature sintering process linked with SSR frequently results in structural imperfections, including porosity and the volatilization of elements. The combined effects undermine the densification and structural integrity of the final ceramic product. Alternative advanced sintering techniques have been explored to address the limitations of conventional SSR. For example, utilized spark plasma sintering (SPS) to produce $Sr[(Zr_{0.94}Y_{0.06})_{0.2}Sn_{0.2}Ti_{0.2}Hf_{0.2}Mn_{0.2}]O_{3-\delta}$, resulting in enhanced microstructural densification and decreased porosity relative to the conventionally sintered version, even with the use of considerably lower temperatures and shorter processing times. The microstructural evaluation indicates a more compact and homogeneous grain distribution, demonstrating the effectiveness of SPS in improving densification kinetics [43].

Flash sintering presents a promising alternative, involving the thermal input to accelerate the densification process. This method has shown significant capability in the swift production of HEPOs. For instance, $Sr(Ti_{0.2}Y_{0.2}Zr_{0.2}Sn_{0.2}Hf_{0.2})O_{3-\delta}$ and $(Bi_{0.2}Na_{0.2}K_{0.2}Ba_{0.2}Ca_{0.2})TiO_3$ were synthesized through flash sintering by Wang et al. [44] and Liu et al. [45] separately. This represents a significant enhancement compared to conventional sintering methods, which generally require several hours at temperatures above 1,000 °C. Similar HEPOs compositions synthesized through SSR have necessitated sintering conditions of 1,500 °C for 10 hours [9] or 1,020 °C for 2 hours [46], frequently resulting in the formation of secondary phases.

The findings collectively emphasize the constraints of conventional SSR methods in the synthesis of HEPOs. They also illustrate the significant advantages of advanced sintering techniques, such as SPS and flash sintering, in improving the efficiency, structural integrity, and applicability of high-entropy perovskite materials within energy storage technologies.

2.4 Energy Storage Applications

The inherent characteristics of HEMs, such as elevated configurational entropy, substantial lattice distortion, sluggish atomic diffusion, and the synergistic "cocktail effect" have led to the growing application of high-entropy design strategy in the advancement of energy materials due to these distinct physicochemical properties.

2.4.1 Lithium-Ion Batteries

LIBs are a rechargeable energy source that includes a lithium-based material as one of its electrodes. During usage, lithium ions travel between the anode and the cathode. The cathode, anode, separator, and electrolyte are all essential components of the battery. Bismuth, Sodium, and a composition of La, Li, Ce, K, Ca, and Sr in the formula TiO_3 have exceptional cyclic performance, maintaining performance after lengthy charge–discharge cycles. Bi^{3+} continuously remains stable in this environment, whereas Ce^{4+} and Ti^{4+} are reduced, suggesting that Bi^{3+} improves the structural integrity of high-entropy molecules during redox reactions. The reduction of Ce^{4+} occurs before the redox process, accredited to the doping of isovalent ions. During the first discharge cycle, Ti^{4+} reduction leads to the development of a solid electrolyte interphase (SEI), which results in a 50%

coulombic efficiency (CE). A coulombic efficiency of 100% is seen beginning with the second discharge cycle and stays constant as the count of discharge cycles rises, representing exceptional reversible lithium storage capability. Additionally, the elimination of the testing current (100–3,000 mA g^{-1}) restores the typical capacitance of the compound to its pre-test value, indicating significant structural stability. The features of HEPOs are attributed to the charge compensation process enabled by isovalent mixing and the steady structure resulting from the high-entropy effect. $Li_x(LiLaCaSrBa)Ti_{1-x}Al_xO_3$ is an effective lithium storage material due to its lightweight composition and porous structure. $La(Co_{0.2}Fe_{0.2}Cr_{0.2}Mn_{0.2}Ni_{0.2})O_3$ exhibits superior specific capacity, cyclic stability, and rate performance compared to conventional $LaCoO_3$ [47]. The mesoporous structure enhances the quantity of active sites, and the collaboration of several main sections facilitates improved Li^+ diffusion and structural stability, enabling these benefits. The high-entropy effect employs charge compensation mechanisms to enhance structural stability. This suggests that using this to produce the upcoming generation of LIBs would be wise.

In order to investigate the effects of Mn^{2+} doping and thermal field treatment on electrochemical performance at a current density of 70 mA g^{-1} (using lithium metal as a reference electrode), a number of Mn^{2+}-doped samples were synthesized as anode constituents for LIBs, as illustrated in Figure 2.4. The structural stabilizing effect of manganese, or the stable $[MnX_6]$ octahedral framework created when manganese is doped into the anode material, is responsible for the highest performance gain observed in 1% Mn-doped samples. In addition to stabilizing the structure, this framework aids in preventing volume expansion and structural

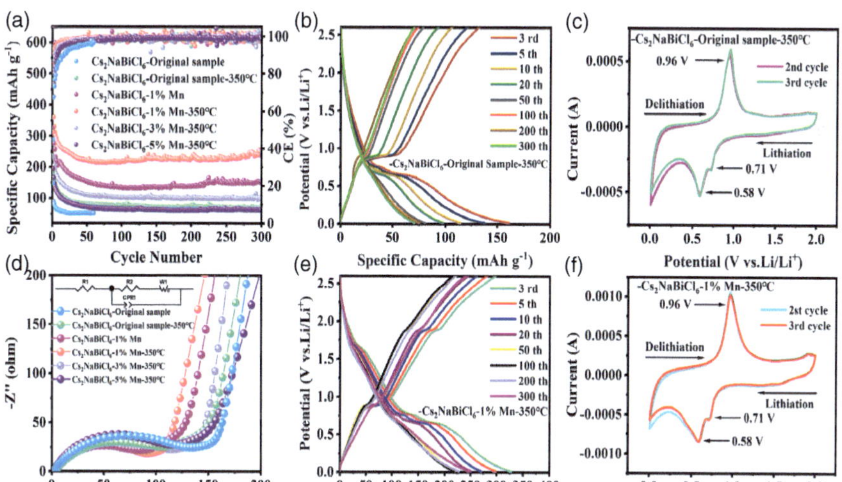

Figure 2.4 (a) Comparison of cycling specific capacity in different samples. (b, c) Charge–discharge curves and CV curves of the pristine sample Cs_2NaBiC_{l6}. (d) Comparison of EIS in different samples. (e, f) Charge–discharge curves and CV curves of the 1% Mn^{2+}-doped sample with thermal field treatment. *Source:* Reproduced from Ref. [48] / with permission of American Chemical Society.

phase transition during charge/discharge cycle. Additionally, the multivalent Mn may affect the material's structural and electrochemical stability.

However, additional thermal field treatment can enhance the samples' electrochemical performance, which may be linked to their increased crystallinity. At the same time, high-temperature treatment can optimize the ion migration paths by promoting the uniform arrangement of Mn^{2+} and reducing lattice defects. The strong self-discharge property of the battery, which causes a faster buildup of Li^+ and the development of SEI during the initial charging and discharging of the battery, is likely the cause of the rapid decrease in specific capacity of the pristine Cs_2NaBiC_{16} in the first 20 cycles. However, this also creates a protective mechanism for the lithium anode to some extent, which gradually stabilizes the discharge capacity in the subsequent period and then stabilizes at 71 mA h g^{-1} after 300 cycles. The performance of the materials doped with 1% and 3% Mn is shown to be considerably better in Figure 2.4a. At the 50th cycle, the discharge specific capacity of the 1% Mn material reaches 152 mA h g^{-1}, whereas the original undoped material only has 81 mA h g^{-1}. The heat-treated 1% Mn samples still show self-discharge during the first 20–30 cycles after the modification, but after 300 cycles, the specific capacity will stabilize and remain at 240 mA h g^{-1} with almost 100% coulombic efficiency [48].

As a result, it suggests that Mn^{2+} doping and thermal field treatment can both enhance the Cs_2NaBiC_{16} anode's cycling stability. They will also work in concert to greatly raise the cycle specific capacity. After a certain quantity of Mn^{2+} doping (between 3% and 5%), the discharge specific capacity is less than that of the original substrate material. This could be because of excessive ionic doping, which alters the material's local field or causes more distortions in the material's lattice structure, which alters the migration and intercalation of Li^+ and lowers performance. Figure 2.4b,e displays the comparable charge–discharge curves. After tens of cycles, we may observe that the relevant materials' discharge specific capacity and coulombic efficiency tend to be more stable.

To confirm the impact of Mn^{2+} doping and heat field treatment on the Li^+ intercalation–deintercalation reaction, the electrochemical response was reflected in the current–voltage curve (I–V curve). The intercalation (Li–Bi and Li_3Bi alloying reactions) and deintercalation (Li–Bi dealloying reaction) processes of Li^+ are represented by the oxidation peak at 0.96 V and the two reduction peaks at 0.71 and 0.58 V, respectively. Additionally, the figure's oxidation and reduction peaks have a tiny peak center spacing difference (ΔE_p), indicating improved battery reversibility. The area of the CV curve of the heat-treated 1% Mn-doped material is also noticeably larger than that of the pristine material when comparing Figure 2.4c, f. This suggests that the heat-treated 1% Mn sample has a higher discharge specific capacity and confirms the synergistic enhancement effect of Mn^{2+} doping and thermal field, which is also consistent with the results in Figure 2.4a.

However, the migration and intercalation of Li^+ are also directly impacted by the material's charge transfer resistance. For this reason, the electrochemical impedance spectroscopy (EIS) spectra of the samples with varying Mn^{2+} doping concentrations and thermal field treatment were examined, as illustrated in Figure 2.4d. The 1% Mn-doped material's R_{ct} in the high-frequency region is found to be substantially smaller than that of the 3% and 5% Mn-doped materials. This indicates that the material has a high ionic conductivity and a low charge transfer resistance, which can be ascribed to the formation of $[MnCl_6]$ to mix the lattice. Mn^{2+} doping, on the other hand, can increase the material's porosity to enhance its electrochemical qualities while encouraging particle boundary fusion to lower the associated resistance [49, 50]. High-concentration Mn^{2+} doping, on the other hand, will cause some of the Mn ions to dissolve into the electrolyte and stick to the SEI, raising the material's interface resistance and polarization and ultimately lowering Li^+ conductivity [51].

2.4.2 Thermoelectric Properties of HEPOs

The efficiency of thermoelectric materials is quantified by the dimensionless figure of merit, $ZT = S^2\sigma T/\kappa$, where S is the Seebeck coefficient, κ kappa is the thermal conductivity, σ sigma is the electrical conductivity, and T is the absolute temperature. Since κ comprises both electronic κ_e and κ_l lattice components, reducing κ_l through phonon scattering is a preferred strategy, as it enhances ZT with minimal impact on σ\sigma$\sigma$, thereby improving thermoelectric performance. HEPOs demonstrate intrinsically low lattice thermal conductivity due to their severe lattice distortion, substantial compositional variations, and arbitrarily distributed internal stresses and electric fields resulting from many cations with valence states and fluctuating ionic radii. This makes them outstanding candidates for thermoelectric applications. For example, $SrTiO_3$-based ceramics exhibiting diverse degrees of configurational entropy have significantly reduced thermal conductivities. Similar decreases in thermal conductivity are seen in HEPO systems, including $La(Fe_{0.2}Co_{0.2}Ni_{0.2}Cr_{0.2}Mn_{0.2})O_3$ and $(Y_{0.2}Nd_{0.2}Sm_{0.2}Eu_{0.2}Er_{0.2})AlO_3$, when compared to their conventional equivalents, $LaCoO_3$ and $YAlO_3$. The HEPO $Sr(Ti_{0.2}Fe_{0.2}Mo_{0.2}Nb_{0.2}Cr_{0.2})O_3$ has a value of 0.7 $W\,m^{-1}\,K^{-1}$ at 1,100 K as thermal conductivity. The performance is qualified to the creation of oxygen vacancies, cationic defects, and intricate defect states caused by fluctuating B-site oxidation states, all of which augment phonon scattering. Moreover, high-order phonon interactions, stemming from lattice anharmonicity caused by multivalent B-site TMs, significantly diminish κ_e [52]. Thermal conductivity in HEPOs may be adjusted with post-synthetic thermal treatments. $(Ca_{0.2}Sr_{0.2}Ba_{0.2}La_{0.2}Pb_{0.2})TiO_3$ exhibits varying thermal conductivity and ZT values contingent upon the calinating procedure used. Zheng et al. [53] indicate that samples subjected to calcination at 1,000 °C for one hour in a 5% H_2, 95% air environment, with an electric current of 0.4 $A\,mm^{-2}$, exhibit a thermal conductivity value of 1.17 $W\,m^{-1}\,K^{-1}$ at 923 K and a ZT value of 0.2 at 873 K. Zhang et al. [54] illustrate that samples annealed at 1,150–1,300 °C for eight hours in argon and carbon atmospheres have enhanced thermal conductivity. The enhanced performance of the former is attributable to the reducing environment, which facilitates the conversion of Ti^{4+} to Ti^{3+}, hence augmenting oxygen vacancy concentration and charge carrier density to maintain charge neutrality [55]. Improved properties result from lattice distortions, compositional complexities from A-site cations, and production of Ti^{3+} ions and oxygen vacancies. The synergistic interplay of increasing carrier concentration and electrical conductivity boosts ZT values [56].

In contrast, $(Pb_{1/6}Ba_{1/6}Sr_{1/6}Ca_{1/6}Na_{1/6}Bi_{1/6})TiO_3$ demonstrates a thermal conductivity of 1.15 $W\,m^{-1}\,K^{-1}$ at ambient temperature [57]. The dielectric behavior, characterized by relaxation events, is associated with PNRs created by the inclusion of Bi^{3+} and Na^{+} ions. Additionally, a p-type perovskite oxide, $(La_{0.2}Nd_{0.2}Pr_{0.2}Sm_{0.2}Eu_{0.2})_{0.95}Sr_{0.05}CoO_3$, attains a ZT value of 0.23 at 350 K, while $Sr_{0.9}La_{0.1}(Zr_{0.25}Sn_{0.25}Ti_{0.25}Hf_{0.25})O_3$ has a thermal conductivity of 1.89 $W\,m^{-1}\,K^{-1}$ at 873 K, accompanied by a Seebeck coefficient of 393 $\mu V\,K^{-1}$, which is among the highest documented for HEPOs at this temperature [58–62].

The presence of several cations with differing ionic radii and valence states in the lattice structure of HEPOs causes significant lattice distortions, unpredictable stress and electric fields, and the creation of point defects such as oxygen vacancies. The structural complexity collectively augments phonon scattering, therefore diminishing thermal conductivity. The ensuing defect chemistry and diversity in valence states simultaneously enhance electrical conductivity. Thus, the high-entropy approach presents a feasible framework for the design and advancement of sophisticated thermoelectric materials derived from perovskite structures.

In the NLTO-NF/Li half cell, the electrochemical characteristics of NLTO-NF were examined. The CV profiles of NLTO-NF were obtained for the first three cycles at 0.1 $mV\,s^{-1}$ within an operating window of 0.01–3.0 V vs Li^+/Li (Figure 2.5a). During the first cycle, there was an irreversible cathodic peak at 0.77 V due to electrolyte breakdown and the formation of the SEI. After the first cycle, the NLTO-NF CV curves show significant reversibility. The observation of two sets of reversible anodic/cathodic peaks linked to Ti redox is noteworthy. Ti^{4+}/Ti^{3+} redox couples are responsible for one pair of them at 2.15/1.49 V, while Ti^{3+}/Ti^{2+} redox couples are responsible for the other pair at 1.29/0.76 V.

The NLTO-NF anode has an initial CE of 58.6%, which is in line with the CV curve (Figure 2.5a), as shown in the initial cycle curve at 0.05 $A\,g^{-1}$. Due to the combined effects of a large specific surface area, low discharge cutoff voltage (0.01 V vs Li^+/Li), and metal oxide-type anodes, the carbonate electrolyte (EC/DEC) breaks down at low potentials (typically below 0.8 V versus Li^+/Li) to produce an SEI in the initial discharging process, which is the primary cause of such a low ICE. Figure 2.5b shows the galvanostatic charge/discharge curves of NLTO-NF at various rates. The NLTO-NF reaches a reasonable reversible capacity of 265 $mA\,h\,g^{-1}$ at 0.1 $A\,g^{-1}$. Furthermore, at 0.1, 0.2, 0.5, 1.0, 2.0, 3.0, and 4.0 $A\,g^{-1}$, NLTO-NF's reversible specific capacity reaches 265, 219, 183, 157, 135, 124, and 118 $mA\,h\,g^{-1}$ (Figure 2.5c). Significantly better rate performance and reversible specific capacity are demonstrated by the NLTO-NF anode when compared to other perovskite-type or Ti-based materials (Figure 2.5e) [63].

Additionally, NLTO-NF uses galvanostatic cycling charge/discharge measurement to provide ultrastable cycling performance at a range of current densities. After 100 cycles, NLTO-NF maintains its high reversible capacity of 217 $mA\,h\,g^{-1}$ at a low current density of 0.2 $A\,g^{-1}$ without experiencing capacity decline (Figure 2.5d).

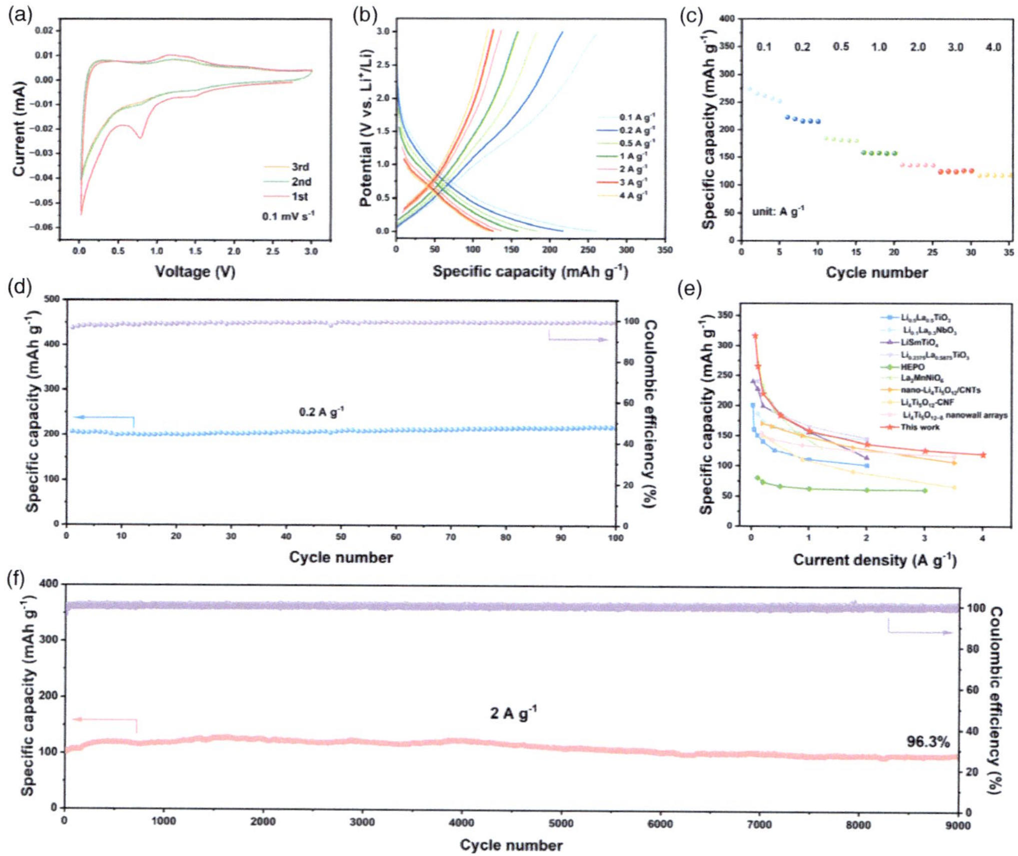

Figure 2.5 (a) Three CV profiles at 0.1 mV s^{-1} at the beginning. (b) Profiles of charge and discharge from 0.1 to 4.0 A g^{-1}. (c) NLTO-NF's rate capabilities. (d) The capacity to cycle at 0.2 A g^{-1}. (e) The NLTO-NF and reported anodes' specific capacity and current density are compared. (f) Cyclability at 2 A g^{-1}. *Source:* Reproduced from Ref. [63] / with permission of American Chemical Society.

Furthermore, the NLTO-NF exhibits a reversible capacity of $166\,mA\,h\,g^{-1}$ with an average CE of about 100% after 1,000 cycles at $1\,A\,g^{-1}$. Furthermore, after up to 9,000 cycles at $2\,A\,g^{-1}$, a 96.3% capacity retention rate with minor deterioration and almost 100% CE are attained (Figure 2.5f).

2.4.3 Electrocatalytic Applications of Perovskite-Based High-Entropy Materials

Catalysis is very important for chemical changes in many different industrial processes. One of the key goals of catalysis research is still to make catalysts that work well and are strong. HEPOs have lately become intriguing prospects because they can be made with different compositions, and their surface chemistry may be changed, making them useful as both catalysts and catalytic supports. Researchers have looked at using HEPO-based materials for a number of electrocatalytic processes, such as CO oxidation, water splitting, and oxygen reduction to make hydrogen [53, 64–66].

HEPOs are a different kind of catalyst than traditional noble metals like Ru and Ir. They may be used in water electrolysis, which includes the oxygen evolution reaction (OER) at the anode and the hydrogen evolution reaction (HER) at the cathode [42]. These noble metals work well, but they are expensive and hard to get. HEPOs are a good choice since they are cheap and easy to get on Earth. The specific ways that OER and HER work change with pH, but this chapter doesn't have room for a full explanation. Some important factors to look at when judging electrocatalytic performance include overpotential, Tafel slope, and turnover frequency. Overpotential is the extra voltage needed to drive a reaction at a certain current density beyond the thermodynamic potential. The Tafel slope shows the overpotential needed for a 10-fold increase in current density. Lower values of both show better catalytic performance. Turnover frequency tells you how many catalytic cycles happen at each active site in a given amount of time. This gives you information about the intrinsic catalytic activity.

HEPOs have lower overpotentials and Tafel slopes than regular perovskite oxide catalysts. This means that they are better at converting energy and have quicker kinetics. For example, $La(Cr_{0.2}Mn_{0.2}Fe_{0.2}Co_{0.2}Ni_{0.2})O_3$ films have shown greater turnover frequencies, which means they have more intrinsic activity. This better performance comes from the way that the different TM cations work together. In $La(5TM)O_3$, Cr may be a good place to start the oxidation process, while Ni and Co help it along. This cooperative impact between nearby TM sites lowers the energy barrier for the reaction as a whole, which boosts OER activity. Also, HEPOs like $La(Cr_{1/6}Mn_{1/6}Fe_{1/6}Co_{2/6}Ni_{1/6})O_3$ and $(La_{0.6}Sr_{0.4})(Co_{0.2})[FeMnNiMg]_{0.8})O_3$ are very stable in electrochemical reactions. This is because the high-entropy effect gives it a lot of thermodynamic stability, and the cations don't move around very quickly and the lattice distorts a lot. These kinds of structural characteristics make it harder for phase separation and degradation to happen while the system is working electrochemically. Because HEPOs may be tuned, their composition can also be optimized for specific catalytic performance.

2.4.4 Perovskite-Based Electrodes for Supercapacitors

Supercapacitors have garnered significant interest as cutting-edge energy storage solutions, owing to their impressive power density, rapid charge–discharge capabilities, and extended cycle longevity. The utilization of perovskite oxides in supercapacitors, especially as anode materials, has been thoroughly examined due to their distinctive energy storage mechanisms that involve oxygen-anion intercalation. Recent advancements indicate HEPOs, including $La(Co_{0.2}Cu_{0.2}Fe_{0.2}Mn_{0.2}Ni_{0.2})O_3$ and $La_{0.7}Bi_{0.3}Mn_{0.4}Fe_{0.3}Cu_{0.3}O_3$, are emerging as viable options for supercapacitor electrodes [67].

The cycle stability of $La(CoCrFeMnNiAl_x)_{1/(5+x)}O_3$ is markedly affected by the entropy-stabilized crystal structure. The electrode devoid of Al^{3+} doping ($x = 0$) maintained 79.8% of its specific capacitance, while the HEPO electrodes with Al^{3+} doping ($x = 0.4$, 0.5, 0.6, and 0.7) exhibited improved retention rates of 85.01, 88.61, 86.37, and 91.25%, respectively. This enhancement is mainly due to the increased configurational entropy resulting from Al^{3+} inclusion, which stabilizes the lattice structure and reduces deterioration during the cycle.

For example, $La(Co_{0.2}Cu_{0.2}Fe_{0.2}Mn_{0.2}Ni_{0.2})O_3$, with a hollow spherical multilayer structure retained 88% of its specific capacity after 10,000 cycles at a high current density of $15\,A\,g^{-1}$. The hierarchical structure enhances durability by effectively accommodating volumetric changes and minimizing mechanical stress during electrochemical cycling. The degradation of HEPOs in high-power battery-supercapacitor hybrid (BSH) systems has been clearly elucidated. The main challenges identified include irreversible phase transitions and volume expansion, which arise from discrepancies between electrochemical kinetics and the charge–discharge processes.

The accumulation and movement of cations throughout the cycling process exacerbate the decline in performance, attributable to the deterioration induced by segregation. To address these challenges, it has been proposed that doping the A-site with larger-radius cations could improve oxygen-ion transport, thereby bolstering electrochemical stability. Striking a harmonious equilibrium between high specific capacitance and improved oxygen reduction reaction (ORR) activity within a singular perovskite composition presents a notable challenge, due to the intrinsic tension between ion intercalation and ORR mechanisms. Compositional tuning in systems such as $La_{0.7}Bi_{0.3}Mn_{0.7-x}Fe_{0.3}Cu_xO_3$ ($x = 0$–0.4) has demonstrated the efficacy of simultaneously enhancing both characteristics.

2.4.5 Sodium-Ion Batteries

The high-entropy layered oxide $NaMn_{0.2}Fe_{0.2}Co_{0.2}Ni_{0.2}Ti_{0.2}O_2$ is a notable example. It shows a high reversible capacity of approximately $160\,mA\,h\,g^{-1}$ at a current density of $0.1\,A\,g^{-1}$. This performance is ascribed to the synergetic effect of numerous TM cations, which facilitate multiple redox reactions and enhance the kinetics of sodium-ion diffusion.

Moreover, the elevated configurational entropy contributes to improved structural stability throughout cycling, resulting in better capacity retention over prolonged cycles. Another example is the multiprincipal component P2-type $Na_{0.6}(Ti_{0.2}Mn_{0.2}Co_{0.2}Ni_{0.2}Ru_{0.2})O_2$ cathode, which demonstrates a notable discharge capacity of 68 mA h g^{-1} at a high rate of 86 C and achieves a discharge capacity of 164 mA h g^{-1} at 0.1 C. The uniform distribution of various cations contributes to the notable rate capability, enhancing electronic conductivity and structural integrity throughout swift charge–discharge cycles. The findings highlight the promise of HEPOs in enhancing SIB technology, especially for uses that require sustained cycling stability and elevated energy density.

SIBs' intrinsic features provide them with distinct competitive advantages as shown in Figure 2.6. First, sodium salts' higher electrostatic energies result in greater thermal stability than their lithium counterparts (e.g. $NaPF_6$ versus $LiPF_6$), indicating promising safety and superior Na-storage behavior in high-temperature settings [69]. Second, Na has a lower initial ionization energy than Li (495.8 vs 520.2 kJ mol^{-1}), which contributes to enhanced electrochemical activity and stability, resulting in improved electrochemical processes and a decreased risk of dendrite development at low temperatures. Third, because Na has a larger atomic radius (compared to Li), its atomic nucleus is more weakly attached to the outermost electrons, allowing it to lose an electron and form Na^+ ions. Na^+ has a

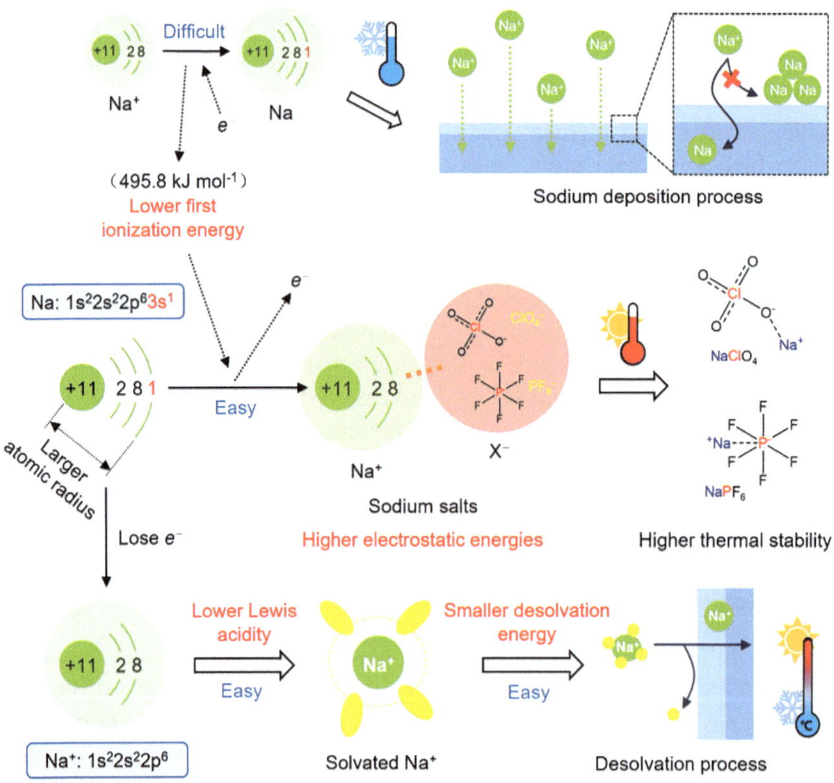

Figure 2.6 Intrinsic advantages of SIBs for acceptable wide-temperature performance. *Source:* Reproduced from Ref. [68] / with permission of American Chemical Society.

lower desolvation energy in most organic solvents due to its lower Lewis acidity, allowing for a faster desolvation of active ions than LIBs. The decreased charge transfer resistance greatly enhances electrode kinetics, laying the groundwork for Na^+ to maintain significant transport properties across a wide-temperature range.

SIBs based on intercalation chemistry will fail at high temperatures because Na^+ has a greater ionic radius than Li^+. SIBs' possible operating temperature range cannot extend beyond the real ambient temperature range, indicating that there is much of room for these cathode materials to advance. Furthermore, the majority of cathode materials are still in the early stages of development for the half battery system, rather than the complete battery system. This necessitates in-depth knowledge of the foundations and developments in cathode materials. Furthermore, due to the interfacial compatibility and adaptability of these cathodes and Na-free anodes, achieving satisfactory wide-temperature performance is more challenging in a whole battery than in a half cell [68]. As major components of SIBs, LOEs and anode materials must be optimized for harsh-temperature applications.

Figure 2.7a displays voltage profiles for charging and discharging R-FeHCF in a Na half-cell with a current of 10 mA g^{-1}. The voltages differ for the Fe^{III}/Fe^{II} redox energies of the HS-FeN_6 and LS-FeC_6 octahedral over $0.08 < x < 1$ and $1 < x < 2$. The initial high Na-concentration results in a capability of 160 mA h g^{-1} approximately in the first cycle, corresponding to a value of around 490 Wh kg^{-1} as energy density. In contrast to the previously conveyed cubic FeHCF [71], the R-FeHCF shows substantially sharper peaks in the chronoamperograms and flatter two-phase voltage profiles, as seen in the inset of Figure 2.7a, indicating first-order transitions among both structures during charge and discharge. The two plateaus have voltages of 3.11 and 3.30 V on charge, and 3.00 and 3.29 V on discharge. As illustrated in Figure 2.7b, redox reactions are highly reversible, with initial capacity of 80%, remaining after Na half-cell's 750 cycles. R-FeHCF has a higher CE from first cycle, even with negligible interstitial water present.

In a Na half-cell, Figures 2.7c,d depicts Na removal rate and reinsertion following R-FeHCF. As the current increased from 0.1 C (15 mA g^{-1}) to 10 C (1,500 mA g^{-1}), the discharge capacity of the R-FeHCF electrode, which contains 86 weight percent active material, 7 weight percent Ketjen black, and 7 weight percent polytetrafluoroethylene (PTFE) binder, dropped from 157 to 145 mA h g^{-1}. This high-rate capacity and capabilities are significantly higher than those found in earlier cubic experiments of Na_{2-x}FeHCF ($x > 1$) electrodes. Remarkably, the discharge capacity declines sharply at 15 C (Figure 2.7c) due to a larger capacity loss at the low-voltage plateau (Figure 2.7d), indicating a slow rate of reduction of HS-$Fe^{III}N_6$ octahedra than of LS-$Fe^{III}C_6$ octahedra.

2.4.6 HEPOs for Enhanced Dielectric Energy Storage

In HEPOs, conventional cation sublattice arrangements are substituted with multiple cations of varying ionic radii and valence states. This compositional complexity induces local compositional fluctuations, random internal

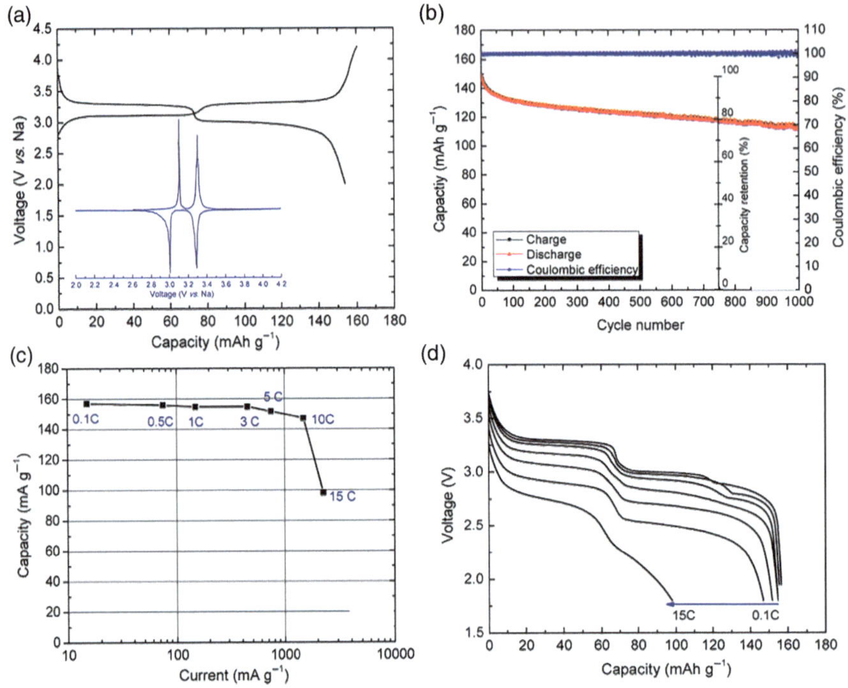

Figure 2.7 An R-FeHCF electrode with 86 weight percent active material, 7 weight percent Ketjen black, and 7 weight percent PTFE were tested for electrochemical performance. (a) Electroamperograms was placed in the galvanostatic charge and discharge curves at a current of 10 mA g^{-1} during the first cycle. (b) A R-FeHCF/Na half-cell's capacity retention across 1,000 cycles (beginning with the first cycle). It was charged at 75 mA g^{-1} at 0.5 C and discharged at 300 mA g^{-1} at 2 C. (c) The R-FeHCF/Na half-cell's rate capability. It was discharged at various currents after being charged at 0.1 C (15 mA g^{-1}). (d) R-FeHCF/Na half-cell discharge curves at various currents. *Source:* Reproduced from Ref. [70] / with permission of American Chemical Society.

stresses, and electric fields, significantly affecting the electronic structure and defect dynamics, including the formation and behavior of point and complex defects. These effects are further intensified under the influence of external electric fields.

Chemical disorder-induced lattice distortions have a profound effect on the ferroelectric properties of HEPOs. For instance, the γ value increases in high-entropy (Ba, Bi, Ni)x(Ca, Sr){(1–3x)/2}TiO_3 systems, although no consistent correlation between configurational entropy and ferroelectricity has been established. In $(Bi_{0.2}Na_{0.2}K_{0.2}La_{0.2}Sr_{0.2})TiO_3$, phase transitions between ferroelectric, antiferroelectric, and paraelectric states are significantly influenced by A-site lattice distortions. Similarly, polar units with local distortions in $Ba(Ti_{0.2}Zr_{0.2}Sn_{0.2}Hf_{0.2}Ce_{0.2})O_3$ enhance system relaxation. In (5A)BO_3 (A = Ag, K, Sr, Ca, Na, Ba, Li, Pb) systems, lattice distortion due to isovalent-ion substitution suppresses grain growth, reduces symmetry, and alters the Landau free energy landscape, thereby modulating ferroelectricity [46, 72].

The formation of high-entropy solid solutions effectively enhances dielectric energy-storage properties by disrupting long-range ferroelectric ordering via random stresses and electric fields. This disruption promotes the formation of PNRs, leading to increased relaxation. Furthermore, defects resulting from lattice alteration, sluggish dispersion, and charge compensation at grain borders suppress the growth of grain, thereby enhancing Broadband Dielectric Spectroscopy (BDS), albeit at the cost of reduced dielectric constant and polarization. Therefore, a balanced enhancement of both BDS and relaxation is essential for optimal energy storage performance in HEPOs [73].

Relying solely on compositional complexity to induce relaxation without grain size control yields suboptimal performance. For example, in $(1-x)(0.9Ba(Ti_{0.97}Ca_{0.03})O_3-0.1Bi_{0.55}Na_{0.45}TiO_3)$-$x$HEPO, the sample with $x = 0.09$ displays superior energy storage behavior. Likewise, in $(1-x)(Bi_{0.4}Na_{0.2}K_{0.2}Ba_{0.2})TiO_3-xSr(Mg_1/_3Nb_2/_3)O_3$, γ decreases from 1.84 to 1.69 as x increases, while grain size increases; the best performance occurs at $x = 0.12$. Thus, both microstructural and domain engineering mediated by HEPO formation are critical for optimizing dielectric performance [74].

PbO doping further illustrates the synergistic tuning of microstructure and relaxation. For instance, $(Bi_{0.2}Na_{0.2}Ba_{0.2}Sr_{0.2}Ca_{0.2})TiO_3$ with 2 mol% PbO demonstrates enhanced dielectric energy storage, attributed to reduced Pr, increased BDS, and enhanced Pm. Pb^{2+} incorporation induces lattice distortion, enhances relaxation, and decreases grain size via reduced diffusion and mitigation of A-site and oxygen vacancies during sintering, thereby improving overall energy storage performance [75].

It has underscored the importance of synergistically tuning BDS and relaxation. Their 2022 study on $(K_{0.2}Na_{0.8})_{0.8}Li_{0.08}Ba_{0.02}Bi_{0.1}O_3$ revealed ultrahigh-energy density and BDS [76], while their 2023 report on $(Bi_{0.5}Na_{0.5})(Ti_{1-2x}Fe_xNb_x)O_3$ demonstrated a high W_{rec} of 13.8 J cm^{-3}. These findings affirm that arbitrary electric fields as well as stress fields induced by high-entropy design substantially improve BDS and energy storage efficiency. Although configurational entropy contributes to these effects, superior performance cannot be solely attributed to entropy magnitude. Instead, the suppression of grain growth, increase in density, induction of common lattice distortions, and disturbance of ferroelectric long-range ordering arising from multicomponent A/B-site substitutions collectively enhance both BDS and relaxation. Thus, high-entropy engineering emerges as a robust strategy for advancing dielectric energy storage materials.

2.4.7 Enhanced Piezoelectric Properties in High-Entropy Perovskite Oxides

The way perovskite oxides are manufactured and what chemicals they are comprised of affect their crystal formations. It is remarkable that only the non-cubic perovskite oxides exhibit piezoelectric and ferroelectric characteristics. The cations on the A-site, present in the gaps between the oxygen octahedra, change the way the A—O—Ti bonds work and make the octahedra tilt. This change in structure has a direct influence on the ferroelectric characteristics of perovskite materials. For instance, polarization rotation processes have been connected to the higher piezoelectric response found near the morphotropic phase boundary (MPB) in piezoelectric systems made of $Pb(Zr,Ti)O_3$ (PZT). Adding cations with varied ionic radii along with valence states makes the Landau free energy profile flatter, which is the simplest explanation for this effect. These alterations make the lattice bend in some places, which makes the crystal less symmetrical. Because of this, the Landau free energy landscape of high-entropy perovskite ferroelectrics is not the same as that of normal ferroelectrics. Because of this, high-entropy perovskite ferroelectrics may operate better as piezoelectrics. Several ideas have been put up on how to make the piezoelectric response better in perovskite systems. These include producing MPBs, stabilizing monoclinic phases, making the structure different in different places, and helping phases coexist.

We mixed $PbTiO_3$ into the high-entropy system $Pb(Mg_{0.2}Zn_{0.2}Nb_{0.2}Ta_{0.2}W_{0.2})O_3$ and discovered that the MPB composition range was between 0.275 and 0.35 molar fraction of $PbTiO_3$. The materials that came out of it, on the other hand, had extremely low piezoelectric coefficients (<100 pC/N), which is far lower than those of ordinary Pb-based piezoelectrics. The fundamental explanation for this reduced performance was that atoms moved slowly, which made the material porous, and the presence of antiferroelectric phases like $Pb(Mg_{0.5}W_{0.5})O_3$. On the other hand, Liu et al.'s recent work found that certain HEPOs may have extremely high piezoelectric coefficients. The composition $Pb(Ni_{0.173}In_{0.07}Zr_{0.034}Ti_{0.306}Nb_{0.417})O_3$ has an incredible 1,200 pC/N value. They named this change a "flexible polarization configuration" since they deliberately added B-site cations with varying ionic radii and ferroelectric activity. This made it easy to produce a broad zone of polarization change. This flexible setup is not the same as the uniform rhombohedral–tetragonal (R–T) polarization distribution that is seen in normal ferroelectrics. There are more kinds of elements at the B-site, which means that it has less long-range polarization order. The piezoelectricity in this system is enhanced because of continuous polarization rotation, particularly in places where there are a lot of monoclinic phases at the same time. The angle range of 0°–45° reveals that the landscape has a number of various phases. When the rhombohedral (R), monoclinic (M), and tetragonal (T) phases are all present at the same time, it is easier for polarization to rotate, which makes the piezoelectric response greater.

2.4.8 HEPOs for Electrochemical Energy Devices

HEPOs are good materials for answering the rising requirement for reliable and efficient energy storage devices because they offer great electrochemical performance and structural stability. More and more studies are suggesting that HEPOs might be employed in LIBs and solid oxide fuel cells (SOFCs). A lot of research has proven that they help electronics work better and survive longer. $La_{1-x}Sr_x(Co,Cr,Fe,Mn,Ni)O_{3-\delta}$ compounds are great for SOFC air-electrode materials because their structures are so stable. For instance, $La_{0.7}Sr_{0.3}(Co,Ni,Fe,Cr,Mn)O_{3-\delta}$ has a linear thermal expansion coefficient value of $16.0 \times 10^{-6}\ K^{-1}$ and a very low chemical expansion throughout a wide range of temperatures, from ambient temperature to 1,000 °C.

This cathode is not like the conventional $La_{0.8}Sr_{0.2}Co_{0.2}Fe_{0.8}O_{3-\delta}$. $La_{0.7}Sr_{0.3}(Co,Cr,Fe,Mn,Ni)O_{3-\delta}$'s structure remains robust even after being heated at 1,100 °C for two hours. Electrochemical measurements reveal that this material has a value of 553 mW cm^{-2} as power density, an electrode polarization resistance value of 0.126 Ω cm^{-2} at 900 °C, and an electronic conductivity value of 16.03 S cm^{-1} at 950 °C. This means that it is a great candidate for a SOFC air electrode. One common difficulty with Sr-doped cathode materials is that Sr ions do not stay together. This makes the materials less stable over time and slows down the rate of oxygen reduction. HEPOs may be able to block Sr from separating because they have a distorted lattice and interact with several components. This improves the performance of the electrochemical system. The $(La_{0.2}Sr_{0.2}Pr_{0.2}Y_{0.2}Ba_{0.2})Co_{0.2}Fe_{0.8}O_{3-\delta}$ electrodes revealed no evidence of Sr segregation during Cr-tolerance test of 41-day. Adding Sm, Pr, and Nd to $La_{0.8}Sr_{0.2}MnO_{3-\delta}$ (LSM) also generated $(La_{0.2}Pr_{0.2}Nd_{0.2}Sm_{0.2}Sr_{0.2})MnO_{3-\delta}$ (HE-LSM), which made it difficult for Sr to migrate. The stress field that forms when the lattice deforms because of the evenly spaced multivalent ions around Sr^{2+} makes it harder for it to migrate.

The $\delta(r)$ value of HE-LSM was 5.30%, which is a lot higher than the value of LSM, which was 2.33%. But large $\delta(r)$ values do not necessarily suggest that Sr segregation is lower. The highest $\delta(r)$ value for sample, which was 8.49%, was the only one that showed Sr segregation. There was no segregation in the samples with $\delta(r)$ values of 7.09% and 5.30%. This proves that lattice strain is not the only thing that controls Sr segregation in HEPOs. For instance, the sluggish diffusion kinetics that are common in high-entropy systems are also a factor. The next stage is to put HEPOs inside ceramic fuel cells that conduct protons.

For instance, $Ba(Sn_{0.16}Ce_{0.35}Y_{0.1}Dy_{0.05}Yb_{0.1}Zr_{0.24})O_{3-\delta}$ and $Ba(Co_{0.2}Fe_{0.2}Zr_{0.2}Sn_{0.2}Pr_{0.2})O_3$, as well as $(Pr_{1/6}La_{1/6}Nd_{1/6}Ba_{1/6}Sr_{1/6}Ca_{1/6})CoO_{3-\delta}$, have been employed successfully as air electrodes in reversible proton ceramic cells. The establishment of the entropy-stabilized phase makes these HEPO-based electrodes very stable in both structure and cycling. Adding multivalent cations to similar sublattices makes it easier for protons and electrons to move about and retain water, which makes the electrochemical properties better. The high-entropy effect keeps the structure of HEPO electrodes constant by preventing phase shifts from developing when they are utilized for a long period. Also, the combination of sluggish ion diffusion and built-in lattice distortion keeps ions from separating, which keeps the intense oxygen-reduction activity going and helps the electrodes last longer. Also, having more than one cation with distinct valence states makes the electrode material far more conductive as a whole.

2.5 Challenges and Future Prospects

HEPOs have gone a long way, but there are still a lot of things that need to be corrected. First, we do not fully understand how entropy affects performance, which makes it challenging to create things in a rational way. We need to learn more about how configurational entropy directly influences electrochemical or catalytic properties, even if it renders phases more stable. Second, scalability is still an issue; many synthesis methods work well for generating things in the lab but not yet for making things on a wide scale. Third, not much research has been done on non-equimolar design techniques, even though they provide you a lot of options for making things perform better. In the future, the discipline should cease using design paradigms that concentrate on composition and start using ones that focus on function. This may be done with the aid of computer modeling, machine learning, and high-throughput testing. Adding sophisticated in situ characterization tools might also help us understand more about how flaws and phase stability vary over time.

Also, it will be vitally necessary to develop methods to make the process cheaper, cooler, and better for the environment so that the public may use it. In the end, HEPOs are a game-changing platform for energy storage devices of the future, but they will not realize their full potential until researchers from other areas work together to come up with new concepts.

2.6 Conclusions

HEPOs enable the storage of energy through innovative methods by utilizing compositional complexity to demonstrate physicochemical properties that have not been previously observed. The distinct arrangement of multiple cations within a single-phase lattice enhances structural stability, enables alterations in electrical properties, and reduces the likelihood of breakdown during cycling. The features stated above are particularly essential for electrochemical systems like LIBs and SIBs, dielectric capacitors, and supercapacitors. These systems need to work well, endure a long period, and store a lot of energy. This chapter talks about how HEPO works better when lattice distortion, charge compensation mechanisms, defect engineering, and targeted synthesis approaches all work together. There has been a lot of growth, but the area is still in the early phases of development. There are still a number of difficulties that need to be fixed, such as making the system more scalable, discovering the optimum non-equimolar compositions, and working out how entropy impacts performance. In the future, it will be vitally necessary to stop using design methodologies that focus on composition and start adopting ones that focus on function. Using both computational modeling and experimental validation together makes it easier to build HEPOs with certain functionality. Also, it will be highly crucial to find cheap, scalable ways to produce these materials that can be employed in commercial energy storage systems. HEPOs are generally a very promising next-generation platform for better energy storage technologies. They are a versatile and long-lasting approach to meet the world's energy demands.

References

1 Ding, Q., Zhang, Y., Chen, X. et al. (2019). Tuning element distribution, structure and properties by composition in high-entropy alloys. *Nature* 574: 223–227. https://doi.org/10.1038/s41586-019-1617-1.

2 Ren, J., Kumkale, V.Y., Hou, H. et al. (2025). A review of high-entropy materials with their unique applications. *Adv Compos Hybrid Mater* 8: 195. https://doi.org/10.1007/s42114-025-01275-4.

3 Huang, R., Zhao, H., and Chen, Z. (2024). High-entropy materials for photocatalysis. *Nano Mater Sci.* https://doi.org/10.1016/j.nanoms.2024.09.002.

4 Ma, J., Liu, T., Ye, W. et al. (2024). High-entropy perovskite oxides for energy materials: a review. *J Energy Storage* 90: 111890. https://doi.org/10.1016/j.est.2024.111890.

5 Kumar, A. and Gangawane, K.M. (2022). Synthesis and effect on the surface morphology & magnetic properties of ferrimagnetic nanoparticles by different wet chemical synthesis methods. *Powder Technol* 410: 117867. https://doi.org/10.1016/j.powtec.2022.117867.

6 Kumar, A. and Gangawane, K.M. (2022). Effect of precipitating agents on the magnetic and structural properties of the synthesized ferrimagnetic nanoparticles by co-precipitation method. *Powder Technol* 401: 117298. https://doi.org/10.1016/j.powtec.2022.117298.

7 Kumar, A., Gangawane, K.M., and Ramanjaneyulu, B. (2022). Ferrofluids for waste-water treatment. In: *International Conference on Chemical, Bio and Environmental Engineering*, 723–744. Springer. https://doi.org/10.1007/978-3-030-96554-9_48.

8 Sarkar, A., Djenadic, R., Wang, D. et al. (2018). Rare earth and transition metal based entropy stabilised perovskite type oxides. *J Eur Ceram Soc* 38: 2318–2327. https://doi.org/10.1016/j.jeurceramsoc.2017.12.058.

9 Jiang, S., Hu, T., Gild, J. et al. (2018). A new class of high-entropy perovskite oxides. *Scr Mater* 142: 116–120. https://doi.org/10.1016/j.scriptamat.2017.08.040.

10 Zhao, G., Xu, X., Zhu, G. et al. (2020). Flexible nitrogen-doped carbon heteroarchitecture derived from ZIF-8/ZIF-67 hybrid coating on cotton biomass waste with high supercapacitive properties. *Microporous Mesoporous Mater* 303: 110257. https://doi.org/10.1016/j.micromeso.2020.110257.

11 Guo, M., Zhang, F., Miao, Y. et al. (2021). Preparation and electrical properties of high entropy $La(Co_{0.2}Cr_{0.2}Fe_{0.2}Mn_{0.2}Ni_{0.2})O_3$ perovskite ceramics powder. *J Inorg Mater* 36: 431. https://doi.org/10.15541/jim20200380.

12 Natesan, K. and (2022). Recent trends in lithium-ion battery – a critical review. *Int J Renew Energy Res.* https://doi.org/10.20508/ijrer.v12i3.13015.g8536.

13 Guo, M., Liu, Y., Zhang, F. et al. (2022). Inactive Al_{3+}-doped $La(CoCrFeMnNiAl_x)_{1/(5+x)}O_3$ high-entropy perovskite oxides as high performance supercapacitor electrodes. *J Adv Ceram* 11: 742–753. https://doi.org/10.1007/s40145-022-0568-4.

14 Gu, X., Guo, X.-B., Li, W.-H. et al. (2024). High-entropy materials for application: electricity, magnetism, and optics. *ACS Appl Mater Interfaces* 16: 53372–53392. https://doi.org/10.1021/acsami.4c11898.

15 Sun, W., Zhang, F., Zhang, X. et al. (2022). Enhanced electrical properties of $(Bi_{0.2}Na_{0.2}Ba_{0.2}Ca_{0.2}Sr_{0.2})TiO_3$ high-entropy ceramics prepared by hydrothermal method. *Ceram Int* 48: 19492–19500. https://doi.org/10.1016/j.ceramint.2022.04.139.

16 Yang, B., Zhang, Y., Pan, H. et al. (2022). High-entropy enhanced capacitive energy storage. *Nat Mater* 21: 1074–1080. https://doi.org/10.1038/s41563-022-01274-6.

17 Ma, J., Chen, K., Li, C. et al. (2021). High-entropy stoichiometric perovskite oxides based on valence combinations. *Ceram Int* 47: 24348–24352. https://doi.org/10.1016/j.ceramint.2021.05.148.

18 Ahart, M., Somayazulu, M., Cohen, R.E. et al. (2008). Origin of morphotropic phase boundaries in ferroelectrics. *Nature* 451: 545–548. https://doi.org/10.1038/nature06459.

19 Kumar, A., Gangawane, K.M., and Jena, H.M. (2025). Variations in magnetic behaviour of ferrimagnetic nanoparticles with temperature. *Indian Chem Eng* 1–10. https://doi.org/10.1080/00194506.2025.2471412.

20 Chaturvedi, K.R., Kumar, A., and Chaubey, A.R. (2025). Functional groups and functionalized nanoparticles in EOR techniques. In: *Nanotechnology in Enhanced Oil Recovery*, 1e (ed. A.K. Krunal and M. Gangawane), 92–103. Boca Raton: CRC Press. https://doi.org/10.1201/9781032650685-6.

21 Chaturvedi, K.R., Kumar, A., and Chaubey, A.R. (2025). Nanotechnology in Enhanced Oil Recovery. In: *Nanotechnology in enhanced oil recovery*, 1e (ed. A.K. Krunal and M. Gangawane), 215–226. Boca Raton: CRC Press. https://doi.org/10.1201/9781032650685-13.

22 Ward-O'Brien, B., McNaughter, P.D., Cai, R. et al. (2022). Quantum confined high-entropy lanthanide oxysulfide colloidal nanocrystals. *Nano Lett* 22: 8045–8051. https://doi.org/10.1021/acs.nanolett.2c01596.

23 Dey, G.R., Soliman, S.S., McCormick, C.R. et al. (2024). Colloidal nanoparticles of high entropy materials: capabilities, challenges, and opportunities in synthesis and characterization. *ACS Nanosci Au* 4: 3–20. https://doi.org/10.1021/acsnanoscienceau.3c00049.

24 Parida, T., Karati, A., Mishra, S. et al. (2022). Low temperature synthesis of multicomponent perovskite by mechanochemical route. *Ceram Int* 48: 6385–6392. https://doi.org/10.1016/j.ceramint.2021.11.181.

25 Wu, H., Lu, Q., Li, Y. et al. (2022). Rapid Joule-heating synthesis for manufacturing high-entropy oxides as efficient electrocatalysts. *Nano Lett* 22: 6492–6500. https://doi.org/10.1021/acs.nanolett.2c01147.

26 Zhao, Z., Chen, H., Xiang, H. et al. (2020). High-entropy $(Y_{0.2}Nd_{0.2}Sm_{0.2}Eu_{0.2}Er_{0.2})AlO_3$: a promising thermal/environmental barrier material for oxide/oxide composites. *J Mater Sci Technol* 47: 45–51. https://doi.org/10.1016/j.jmst.2020.02.011.

27 Liu, J., Ma, C., Wang, L. et al. (2022). Single-phase formation mechanism and dielectric properties of sol-gel-derived $Ba(Ti_{0.2}Zr_{0.2}Sn_{0.2}Hf_{0.2}Ce_{0.2})O_3$ high-entropy ceramics. *J Mater Sci Technol* 130: 103–111. https://doi.org/10.1016/j.jmst.2022.05.012.

28 Kumar, A. and Gangawane, K.M. (2024). Green nanomaterials in energy storage: advancements and challenges. In: *Materials for Boosting Energy Storage. Volume 1: Advances in Sustainable Energy Technologies, Part 12 – Green Nanomaterials in Energy Storage: Advancements and Challenges*, 275–291. American Chemical Society. https://doi.org/10.1021/bk-2024-1477.ch012.

29 Kumar, A. and Gangawane, K.M. (2024). Nanoparticle-modified multifunctional nano carbons—advances in energy storage, materials for boosting energy storage. Advances in sustainable energy technologies, part 7 – nanoparticle-modified multifunctional nano carbons. *Adv Energy Storage* 3: 1–30. https://doi.org/10.1021/bk-2024-1488.ch007.

30 Kumar, A. and Gangawane, K.M. (2024). Recent advances in organic/composite phase change materials (PCM) for energy storage. In: *Materials for Boosting Energy Storage. Volume 3: Advances in Sustainable Energy Technologies*, 309–334. American Chemical Society. https://doi.org/10.1021/bk-2024-1488.ch013.

31 Dąbrowa, J., Olszewska, A., Falkenstein, A. et al. (2020). An innovative approach to design SOFC air electrode materials: high entropy La_{1-x} Sr_x $(Co,Cr,Fe,Mn,Ni)O_{3-\delta}$ ($x = 0$, 0.1, 0.2, 0.3) perovskites synthesized by the sol–gel method. *J Mater Chem A Mater* 8: 24455–24468. https://doi.org/10.1039/D0TA06356H.

32 Kumar, A., Jena, H.M., and Gangawane, K.M. (2025). Iron oxide nanoparticles as efficient adsorbents for malachite green dye removal: isotherms, kinetics, and thermodynamics analysis. *Environ Sci Pollut Res.* https://doi.org/10.1007/s11356-025-36291-w.

33 Nan, H., Lv, S., Xu, Z. et al. (2023). Inducing the cocktail effect in yolk-shell high-entropy perovskite oxides using an electronic structural design for improved electrochemical applications. *Chem Eng J* 452: 139501. https://doi.org/10.1016/j.cej.2022.139501.

34 Mazza, A.R., Skoropata, E., Lapano, J. et al. (2021). Charge doping effects on magnetic properties of single-crystal $La_{1-x}{\times}Sr_x{\times}(Cr_{0.2}{\times}Mn_{0.2}{\times}Fe_{0.2}{\times}Co_{0.2}{\times}Ni_{0.2}){\times}O_3$ $(0{\leq}_x{\leq}0.5)$ high-entropy perovskite oxides. *Phys Rev B* 104: 094204. https://doi.org/10.1103/PhysRevB.104.094204.

35 Farhan, A., Stramaglia, F., Cocconcelli, M. et al. (2022). Weak ferromagnetism in $Tbf_0(Fe_{0.2}{\times}Mn_{0.2}{\times}Co_{0.2}{\times}Cr_{0.2}{\times}Ni_{0.2}){\times}O_3$ high-entropy oxide perovskite thin films. *Phys Rev B* 106: L060404. https://doi.org/10.1103/PhysRevB.106.L060404.

36 Patel, R.K., Ojha, S.K., Kumar, S. et al. (2020). Epitaxial stabilization of ultra thin films of high entropy perovskite. *Appl Phys Lett* 116. https://doi.org/10.1063/1.5133710.

37 Chen, Y.-W., Ruan, J.-J., Ting, J.-M. et al. (2021). Solution-based fabrication of high-entropy $Ba(Ti,Hf,Zr,Fe,Sn)O_3$ films on fluorine-doped tin oxide substrates and their piezoelectric responses. *Ceram Int* 47: 11451–11458. https://doi.org/10.1016/j.ceramint.2020.12.272.

38 Liang, Y., Luo, B., Dong, H., and Wang, D. (2021). Electronic structure and transport properties of sol-gel-derived high-entropy $Ba(Zr_{0.2}Sn_{0.2}Ti_{0.2}Hf_{0.2}Nb_{0.2})O_3$ thin films. *Ceram Int* 47: 20196–20200. https://doi.org/10.1016/j.ceramint.2021.04.026.

39 Kumar, A., Panda, D., Jena, H.M., and Gangawane, K.M. (2025). Recent Advances in Bioelectrochemical systems for higher efficiency in energy production and water purification. In: *Integrated Bioeletrochemical–Constructed Wetland System for Future Sustainable Wastewater Treatment*, 1e (ed. N.K. Kumar Manoj and A.S. Singh Mathuriya), 97–120. Singapore: Springer. https://doi.org/10.1007/978-981-96-2814-8_4.

40 Panda, D., Kumar, A., and K.M. (2024). *Gangawane, Chapter 4 Materials for micro- and nanofabrication, Microfabrication and Nanofabrication*, 77–100. De Gruyter. https://doi.org/10.1515/9783111335537-004.
41 Kumar, A., Panda, D., and Gangawane, K.M. (2024). Chapter 3 Microfabrication: techniques and technology. In: *Microfabrication and Nanofabrication*, 47–76. De Gruyter. https://doi.org/10.1515/9783111335537-003.
42 Bib Khan, J., Kumar Panda, P., Dash, P., and Hsieh, C. (2025). Microwave-assisted synthesis of platinum-free high-entropy alloy catalysts on reduced graphene oxide sheets for enhanced oxygen reduction and evolution reactions, chemistry A. *Eur J* 31. https://doi.org/10.1002/chem.202403863.
43 Panda, P.K., Huang, H.-Y., Dash, P. et al. (2025). Liquid-phase microwave synthesis of platinum-based high-entropy alloy catalysts on carbon supports for electrochemical hydrogen adsorption/desorption and oxygen evolution/reduction reactions. *Int J Hydrog Energy* 111: 536–545. https://doi.org/10.1016/j.ijhydene.2025.02.312.
44 Wang, K., Ma, B., Li, T. et al. (2020). Fabrication of high-entropy perovskite oxide by reactive flash sintering. *Ceram Int* 46: 18358–18361. https://doi.org/10.1016/j.ceramint.2020.04.060.
45 Liu, J., Ren, K., Ma, C. et al. (2020). Dielectric and energy storage properties of flash-sintered high-entropy $(Bi_{0.2}Na_{0.2}K_{0.2}Ba_{0.2}Ca_{0.2})TiO_3$ ceramic. *Ceram Int* 46: 20576–20581. https://doi.org/10.1016/j.ceramint.2020.05.090.
46 Yang, W., Han, Z., and Zheng, G. (2021). The influences of lattice distortion on the antiferroelectric transition and relaxation of oxygen vacancies in high-entropy perovskites $(Bi_{0.2}Na_{0.2}Ba_{0.2}K_{0.2}X_{0.2})TiO_3$ with X = Ca, Sr or La. *Scr Mater* 203: 114096. https://doi.org/10.1016/j.scriptamat.2021.114096.
47 Jin, C., Wang, Y., Dong, H. et al. (2024). A novel spinel high-entropy oxide $(Cr_{0.2}Mn_{0.2}Co_{0.2}Ni_{0.2}Zn_{0.2})3O_4$ as anode material for lithium-ion batteries. *Inorganics (Basel)* 12: 198. https://doi.org/10.3390/inorganics12070198.
48 Bao, J., Jia, W., Zhang, H. et al. (2025). Synergistic enhancement of $Mn2^+$ doping and thermal field treatment on $Cs_2NaBiCl_6$ double perovskite anode performance for lithium-ion batteries. *J Phys Chem C*. https://doi.org/10.1021/acs.jpcc.5c00963.
49 Opra, D.P., Gnedenkov, S.V., Sinebryukhov, S.L. et al. (2020). Doping of titania with manganese for improving cycling and rate performances in lithium-ion batteries. *Chem Phys.* 538. https://doi.org/10.1016/j.chemphys.2020.110864.
50 Ogunniran, K.O., Murugadoss, G., Thangamuthu, R., and Periasamy, P. (2018). All inorganic based $Nd_{0.9}Mn_{0.1}FeO_3$ perovskite for Li-ion battery application: synthesis, structural and morphological investigation. *J Alloys Compd* 766: 1014–1023. https://doi.org/10.1016/j.jallcom.2018.06.340.
51 Saulnier, M., Auclair, A., Liang, G., and Schougaard, S.B. (2016). Manganese dissolution in lithium-ion positive electrode materials. *Solid State Ionics* 294: 1–5. https://doi.org/10.1016/j.ssi.2016.06.007.
52 Banerjee, R., Chatterjee, S., Ranjan, M. et al. (2020). High-entropy perovskites: an emergent class of oxide thermoelectrics with ultralow thermal conductivity.

ACS Sustain Chem Eng 8: 17022–17032. https://doi.org/10.1021/acssuschemeng.0c03849.

53 Zheng, Y., Zou, M., Zhang, W. et al. (2021). Electrical and thermal transport behaviours of high-entropy perovskite thermoelectric oxides. *J Adv Ceram* 10: 377–384. https://doi.org/10.1007/s40145-021-0462-5.

54 Zhang, P., Lou, Z., Qin, M. et al. (2022). High-entropy ($Ca_{0.2}Sr_{0.2}Ba_{0.2}La_{0.2}Pb_{0.2}$)$TiO_3$ perovskite ceramics with A-site short-range disorder for thermoelectric applications. *J Mater Sci Technol* 97: 182–189. https://doi.org/10.1016/j.jmst.2021.05.016.

55 Kumar, A. (2024). Engineered nanomaterials and associated threats in the environment risk assessment strategies, Springer Nature (2024) 515–537. https://doi.org/10.1007/978-981-97-9922-0_24

56 Gu, S., Huang, H.-Y., Lin, Y.-C. et al. (2025). High-entropy alloys with low platinum content: novel catalysts for oxygen reduction and evolution reactions. *Emerg Mater*. https://doi.org/10.1007/s42247-025-01011-0.

57 Xiong, W., Zhang, H., Hu, Z. et al. (2022). Low thermal conductivity in A-site high entropy perovskite relaxor ferroelectric. *Appl Phys Lett* 121. https://doi.org/10.1063/5.0102088.

58 Kumar, A., Dragoe, D., Berardan, D., and Dragoe, N. (2023). Thermoelectric properties of high-entropy rare-earth cobaltates. *J Mater* 9: 191–196. https://doi.org/10.1016/j.jmat.2022.08.001.

59 Lou, Z., Zhang, P., Zhu, J. et al. (2022). A novel high-entropy perovskite ceramics $Sr_{0.9}La_{0.1}(Zr_{0.25}Sn_{0.25}Ti_{0.25}Hf_{0.25})O_3$ with low thermal conductivity and high Seebeck coefficient. *J Eur Ceram Soc* 42: 3480–3488. https://doi.org/10.1016/j.jeurceramsoc.2022.02.053.

60 Chaturvedi, K.R., Kumar, A., Chaubey, A.R. et al. (2025). Fundamentals of nanotechnology for EOR. In: *Nanotechnology in Enhanced Oil Recovery*, 1e (ed. A.K. Krunal and M. Gangawane), 27–38. Boca Raton: CRC Press. https://doi.org/10.1201/9781032650685-2.

61 Chaturvedi, K.R., Kumar, A., and Chaubey, A.R. (2025). Case studies of EOR by using nanoparticles. In: *Nanotechnology in Enhanced Oil Recovery*, 1e (ed. A.K. Krunal and M. Gangawane), 122–126. Boca Raton: CRC Press. https://doi.org/10.1201/9781032650685-8.

62 Chaturvedi, K.R., Kumar, A., and Chaubey, A.R. (2025). CFD of fluid flow for EOR using nanoparticles. In: *Nanotechnology in Enhanced Oil Recovery*, 1e (ed. A.K. Krunal and M. Gangawane), 162–173. Boca Raton: CRC Press. https://doi.org/10.1201/9781032650685-10.

63 Cao, K., Zhu, Y., He, H. et al. (2024). Zero-strain sodium lanthanum titanate perovskite embedded in flexible carbon fibers as a long-span anode for lithium-ion batteries. *ACS Appl Mater Interfaces* 16: 11421–11430. https://doi.org/10.1021/acsami.3c16183.

64 Xu, X., Shao, Z., and Jiang, S.P. (2022). High-entropy materials for water electrolysis. *Energy Technol* 10. https://doi.org/10.1002/ente.202200573.

65 Kumar, A. and Gangawane, K.M. (2023). *Advanced computational approaches for water treatment, in: advanced computational approaches for water treatment*, 1e, 1–12. New York: CRC Press. https://doi.org/10.1201/9781003325147-1.

66 Kumar, A. and Gangawane, K.M. (2023). Molecular dynamics for the membrane process. In: *Advanced Computational Approaches for Water Treatment*, 1e, 13–38. New York: CRC Press. https://doi.org/10.1201/9781003325147-2.

67 Nan, H., Hu, X., and Tian, H. (2019). Recent advances in perovskite oxides for anion-intercalation supercapacitor: a review. *Mater Sci Semicond Process* 94: 35–50. https://doi.org/10.1016/j.mssp.2019.01.033.

68 Zhang, F., He, B., Xin, Y. et al. (2024). Emerging chemistry for wide-temperature sodium-ion batteries. *Chem Rev* 124: 4778–4821. https://doi.org/10.1021/acs.chemrev.3c00728.

69 Eshetu, G.G., Grugeon, S., Kim, H. et al. (2016). Comprehensive insights into the reactivity of electrolytes based on sodium ions. *ChemSusChem* 9: 462–471. https://doi.org/10.1002/cssc.201501605.

70 Wang, L., Song, J., Qiao, R. et al. (2015). Rhombohedral Prussian white as cathode for rechargeable sodium-ion batteries. *J Am Chem Soc* 137: 2548–2554. https://doi.org/10.1021/ja510347s.

71 Wu, X., Deng, W., Qian, J. et al. (2013). Single-crystal FeFe(CN)6 nanoparticles: a high capacity and high rate cathode for Na-ion batteries. *J Mater Chem A Mater* 1: 10130–10134. https://doi.org/10.1039/c3ta12036h.

72 Li, Z., Chen, Z., and Xu, J. (2022). Enhanced energy storage performance of $BaTi_{0.97}Ca_{0.03}O_{2.97}$-based ceramics by doping high-entropy perovskite oxide. *J Alloys Compd* 922: 166179. https://doi.org/10.1016/j.jallcom.2022.166179.

73 Liu, Z., Xu, S., Li, T. et al. (2021). Microstructure and ferroelectric properties of high-entropy perovskite oxides with A-site disorder. *Ceram Int* 47: 33039–33046. https://doi.org/10.1016/j.ceramint.2021.08.204.

74 Yan, B., Chen, K., and An, L. (2023). Design and preparation of lead-free $(Bi_{0.4}Na_{0.2}K_{0.2}Ba_{0.2})TiO_3$-$Sr(Mg_{1/3}Nb_{2/3})O_3$ high-entropy relaxor ceramics for dielectric energy storage. *Chem Eng J* 453: 139921. https://doi.org/10.1016/j.cej.2022.139921.

75 Zhang, Y., Zhang, S., Liu, X. et al. (2024). Synergism effect of Pb doping and microstructure optimization on the superconducting properties of Bi-2212 ceramics. *J Mater Sci Mater Electron* 35: 816. https://doi.org/10.1007/s10854-024-12608-2.

76 Chen, L., Deng, S., Liu, H. et al. (2022). Giant energy-storage density with ultrahigh efficiency in lead-free relaxors via high-entropy design. *Nat Commun* 13: 3089. https://doi.org/10.1038/s41467-022-30821-7.

3

Functional 2D-Based High-Entropy Materials for Energy Storage Applications

Asfaq Ali[1], Sanjeev Verma[2], Pradeep Kumar Panda[3], and Tapas Das[4]

[1] *NIT Rourkela, Department of Chemical Engineering, Rourkela, Odisha, 769008, India*
[2] *NIT Patna, Department of Chemical Engineering & Technology, Patna, Bihar, 800005, India*
[3] *Yuan Ze University, Department of Chemical Engineering and Materials Science, Taoyuan 32003, Taiwan*
[4] *NIT Rourkela, Department of Chemical Engineering, Rourkela, Odisha, 769008, India*

3.1 Introduction

Globally, there has been a rise in the need for energy storage techniques that are both sustainable and efficient, motivated by the growing adoption of renewable energy technologies and the heightened requirement for high-performance energy storage systems [1]. Conventional energy storage materials, like transition metal oxides (TMO_2) and intercalation compounds, often suffer from poor stability, low-rate capability, and structural degradation over time [2, 3]. To overcome these limitations, sophisticated material design methodologies have been explored, resulting in high-entropy materials (HEMs) as prospective functional materials [4]. HEMs, initially made in metallic alloys, include numerous fundamental elements in equimolar proportions, resulting in improved configurational entropy and unique structure–property interactions. The high-entropy method significantly stabilizes unstable solid-solution phases, leading to enhanced durability, thermal endurance, and durability against corrosion [5]. Over the past two decades, this concept has not only extended to metallic systems but also high-entropy ceramics, nitrides, carbides, and chalcogenides, thereby increasing their application across various domains, including catalysis, electronics, and energy storage [6, 7].

The use of a high-entropy method within layered structures has been proven to effectively increase both structural stability and electrochemical properties, resulting in significant improvements in battery performance (Figure 3.1) [8]. These high-entropy layered oxides have a durable host matrix, which is critical for achieving improved long-term cycle and rate capabilities while also increasing the material's air stability to a degree. Despite these encouraging findings, investigation into high-entropy layered oxides is currently situated in its early stages, with more fundamental studies required to properly understand how entropy affects critical aspects such as structural stability, ion diffusion kinetics, and air stability.

High-Entropy Materials for Energy Storage Devices, First Edition.
Edited by Chien-Te Hsieh, Pradeep Kumar Panda, and Arpan Kumar Nayak.

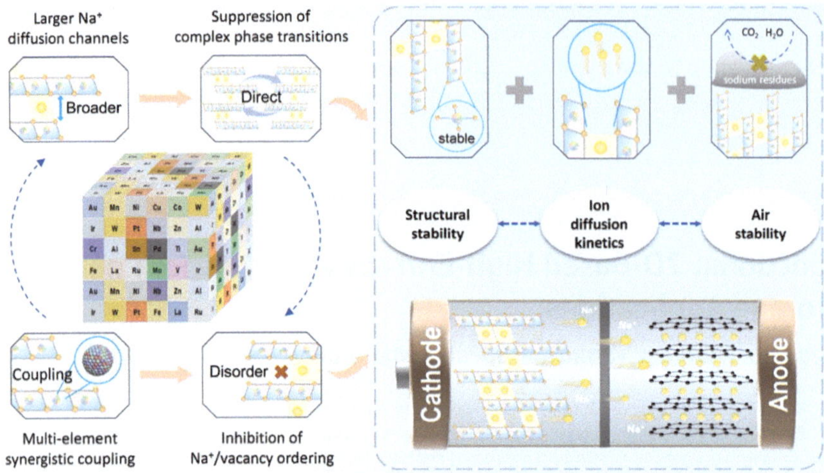

Figure 3.1 Introduction of the high-entropy concept into layered cathodes. *Source:* Reproduced from Ref. [8] / with permission of American Chemical Society.

The current development of HEMs at the nanoscale level has initiated the exploration of entropy-stabilized two-dimensional (2D) materials [9, 10]. These materials, consisting of several primary components in stratified configurations including oxides, chalcogenides, and MXenes, provide a synergy of elevated surface area, adjustable electrical characteristics, and improved electrochemical stability [11, 12]. The interplay between the properties of 2D materials and entropy-driven stabilization facilitates enhanced ion movement, greater availability of active sites for electrochemical processes, and better structural integrity, making them very appealing for energy storage applications [13]. The development of lithium-ion and sodium-ion storage, as well as pseudocapacitive energy storage, has been made possible by their structural adaptability, which permits customized electrochemical characteristics [14–16]. This chapter provides a comprehensive analysis of the evolution of HEMs from bulk to 2D forms, emphasizing their importance in energy storage applications. We analyze the influence of multicomponent disorder on lattice structures and electrical properties, highlighting its effect on electrochemical performance.

This chapter aims to clarify the unique advantages and challenges of 2D HEMs, intending to emphasize their potential to revolutionize energy storage and accelerate the discovery of novel functional materials. Along with the developments and study of individual materials and their electrochemical properties for energy storage applications, this chapter elaborates individual contribution of physicochemical properties in developing high-entropy substances. Recent studies and research for such materials have grabbed the attention of developed nations for fulfilling their energy requirements, reducing dependency on conventional sources. Various parameters, synthesis process, and stabilization of developed materials have been elaborated with details.

3.1.1 Concept of High-Entropy

The concept of high-entropy in HEMs primarily pertains to the configurational entropy (ΔS_{config}) associated with a solid solution that is composed of a single phase, which is essential for maintaining structural stability. The entropy component is generally regarded as the main contributor because of its predominance and the straightforwardness of its calculation [17]. HEMs, especially those utilized as electrode materials, typically consist of both cations and anions. The ΔS_{config} of these materials is defined mathematically as follows:

$$S_{config} = -R\left(\sum_{i=1}^{n} X_i \ln X_i\right)_{cation-site} + \left(\sum_{j=1}^{m} X_j \ln X_j\right)_{anion-site} \tag{3.1}$$

where X_i and X_j signify the molar fractions of the elements residing in the anionic and cationic sites, and R denotes the universal gas constant. An elevation in configurational entropy enhances the stabilization of the solid-solution structure in single phase, a hallmark of HEMs. The classification of a material as a HEM typically adheres to two criteria:

a) The inclusion of a minimum of five primary elemental species
b) A configurational entropy value that surpasses a specified threshold.

Materials are classified according to their entropy values into three categories: HEMs, medium-entropy materials, and low-entropy materials [7, 18].

In addition to entropy factors, the formation of a homogeneous, single-phase solid solution characterized by a homogeneous and random distribution of elements is a critical criterion for classifying a material as a HEM. The examination of the stability and formation of HEMs can be approached from a thermodynamic perspective by employing the Gibbs free energy equation.

$$\Delta G_{mix} = \Delta H_{mix} - T\Delta S_{mix} \tag{3.2}$$

3.1.2 Structural Tunability and Electrochemical Stability of 2D High-Entropy Materials

HEMs have diversified into many categories, including high-entropy alloys (HEAs) and high-entropy ceramics, the latter of which is further categorized by ionic composition into oxides, borides, carbides, nitrides, and silicides. High-entropy oxides (HEOs) have garnered interest due to their electrochemical stability and structural adaptability, positioning them as viable candidates for energy storage applications. The stability of 2D HEMs driven by configurational entropy facilitates distinctive structural tunability, especially in non-oxide systems. In contrast to bulk-phase HEMs, 2D HEMs include various metal and non-metal components, resulting in dynamic charge redistribution, localized lattice strain, and improved electronic characteristics. The parameters influence performance in energy storage applications by improving ion diffusion pathways, raising the availability of active sites, and stabilizing cycle efficiency. By employing entropy-driven phase stabilization, 2D HEMs offer a tailored structure for the development of innovative materials featuring tunable structural and electrochemical properties, bridging traditional high-entropy ceramics with practical energy storage solutions.

3.2 Structural Characteristics and Stabilization of 2D-Based HEMs

The establishment and stabilization of 2D HEMs are under the control of varying degrees of atom combining within the metallic sublattices. A lead-free relaxor ferroelectric ceramic (BNSLBKT-x) with entropy values of 1.64*R*, 1.71*R*, 1.74*R*, and 1.78*R* was synthesized utilizing a typical solid-state reaction technique. High-pure raw materials of Na_2CO_3 (99.9%), Bi_2O_3 (99.9%), $SrCO_3$ (99.0%), K_2CO_3 (99.9%), $BaCO_3$ (99%), La_2O_3 (98%), and TiO_2 (99.9%) (Aladdin Chemical Reagent Co. Ltd., Shanghai) were weighted in the light of a stoichiometric ratio and then ball-mixed with zirconia balls in ethyl alcohol for 12 hours at 450 r min^{-1} [19]. As depicted in Figure 3.2a, entropy-driven stabilization is exhibited by these materials, which results in order–disorder transitions through mechanisms such as substitutional interfaces in non-equimolar systems, short-range ordering via localized clustering, and fully randomized solid solutions that emerge with an increasing number of constituent elements [20]. The configurational entropy term in Gibbs free

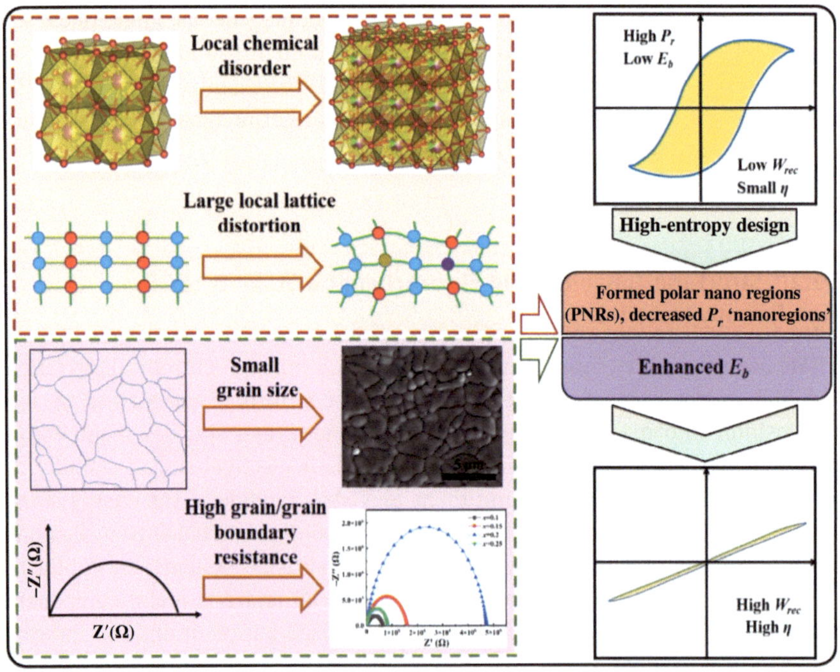

Figure 3.2 Schematic diagram for designing BNSLBKT-*x* ceramics with high-capacitive energy storage via a high-entropy strategy. *Source:* Reproduced from Ref. [19] / with permission of American Chemical Society.

energy equation becomes the primary factor in stabilizing the system as the number of atomic species increases, as represented in Figure 3.2b. This phenomenon is especially evident in high-entropy compounds, where non-metallic elements (denoted as X) occupy interstitial sites within a metallic lattice (denoted as M), resulting in complex M—X coordination bonding states [14].

In contrast to conventional HEAs, the inclusion of non-metallic sublattices in high-entropy compounds results in increased selectivity and structural flexibility. This allows for the integration of elements with a wider range of M—X bond lengths and larger ionic radii, while simultaneously minimizing lattice distortion. Furthermore, the formation of bulk single-phase materials with customized electronic and structural properties can be facilitated by varying atomic occupancy in the X sites, which leads to entropy stabilization [21, 22]. A visual representation of entropy stabilization in 2D HEMs is depicted in Figure 3.2c,d, illustrating the evolution of entropy-stabilized materials from bulk phases to layered 2D systems. This timeline highlights the fundamental mechanisms governing the stabilization of these materials and their significance in tailoring structural and electronic properties.

In order to verify the high configurational entropy of these systems, it is imperative to employ comprehensive characterization techniques. Elemental distribution at the atomic scale is typically analyzed through Scanning Transmission Electron Microscopy (STEM) combined with Energy-Dispersive X-Ray Spectroscopy (EDS) and high-angle annular dark-field (HAADF) imaging, along with X-ray and neutron diffraction methods [23, 24]. Additionally, atomic-scale tomographic methods provide a more comprehensive understanding of the microstructural organization of HEMs. A qualitative method for evaluating the role of entropy in structural stabilization is provided by modifying the stoichiometric ratios beyond equimolar compositions. Furthermore, the elucidation of oxidation states and their impact on the physicochemical properties of high-entropy nanomaterials is facilitated by sophisticated spectroscopic techniques, including X-ray photoelectron spectroscopy (XPS) [25–27].

3.3 Structural Flexibility and Charge Transport in 2D HEMs

2D HEMs differ from conventional HEAs in that they incorporate non-metallic sublattices, which enhance electronic tunability, lattice flexibility, and charge transport kinetics. The assimilation of components with varying bond lengths and ionic radii is facilitated by the diverse M—X coordination, thereby reducing lattice distortion. This structural plasticity serves a critical role in stabilizing disordered frameworks and improving charge transport routes, thereby impacting energy storage capacity.

Figure 3.2 illustrates key structural and charge transport mechanisms in 2D HEMs. In Figure 3.2a, the normalized temperature stability of paired cations in cation-disordered rock salt (DRX) structures highlights entropy-induced phase stabilization. Figure 3.2b–c demonstrate the mechanical straining and distortion in

Table 3.1 Properties of 2D-based high-entropy materials for energy storage.

Material	Crystal structure	Electrical conductivity ($S\,m^{-1}$)	Ionic mobility	Cycle stability (number of cycles)	References
$(CrMnFeCoNi)^3O^4$	Spinel	0.02	High	>1,200 cycles	[28]
(MgNiCoCuZn)O	Rock salt	10^{-3}–10^{3}	Moderate	~500 cycles	[29]
(LiMgCoNiCuZn)O	Rock salt	10^{-1}–10^{1}	High	~600 cycles	[30]

rock salt ceramics and MXene structures, emphasizing the role of lattice flexibility in tuning electrochemical properties. Figure 3.2d–f further explore the electronic structure variations, showing the local changes in bond length, hydrogen evolution reaction (HER) behavior, and binding energy, respectively, revealing how entropy stabilization influences charge mobility and electrochemical activity.

Moreover, entropy stabilization, achieved by regulated atomic occupancy at X sites, enables the development of bulk single-phase materials with customized electrochemical characteristics. Utilizing these distinctive structural and electrical properties, 2D HEMs provide a potential foundation for advanced energy storage applications [24]. To further illustrate the key structural and electrochemical properties of 2D-based HEMs, Table 3.1 summarizes their crystal structure, electrical conductivity, ionic mobility, and cycle stability based on experimental studies.

To further illustrate the key structural and electrochemical characteristics of 2D-based HEMs, Table 3.1 summarizes their crystal structure, electrical conductivity, ionic mobility, and cycle stability based on experimental studies.

3.4 Classification and Synthesis of HEMs

HEMs are categorized according to their configurational entropy and composition. They generally comprise a minimum of five primary components in near-equimolar proportions, resulting in stable, single-phase solid solutions. HEMs are classified into HEAs and high-entropy ceramics, encompassing oxides, borides, carbides, nitrides, and silicides. HEOs have garnered interest for their structural adaptability and electrochemical stability, rendering them appropriate for energy storage applications.

3.4.1 Synthesis Strategies for 2D High-Entropy Materials

All metal compound structures theoretically possess the ability to produce HEMs using configurational entropy at increased temperatures. The stable single-phase HEMs which are synthesized poses a significant challenge due to the marked physicochemical differences among the constituent elements. The synthesis of HEM faces two main challenges: the necessity for stringent synthesis conditions, characterized by high temperatures and significant energy input, and the propensity for materials to experience phase separation. The synthesis of single-phase HEMs is frequently hindered by challenges in achieving integration into a cohesive lattice, attributed to variations in ionic radii among the constituent elements. The selection of elements with minimal variations in ionic radii is crucial for enhancing structural stability. Competing phase transitions at high temperatures impede the development of single-phase solid solutions. When melting at 1,173–1,373 K, the Cr-Fe-Mg-Mn-Ni-O system changes from spinel to rock salt [31]. In high-energy oxides containing elements prone to oxidation, increasing the synthesis temperature may facilitate the change from phase separation to a stable single phase. HEOS with readily reducible elements can be synthesized through precise control of the oxygen partial pressure. An entropy-driven strategy that enhances the concentration of cation species has been demonstrated to effectively stabilize single-phase compounds in systems containing noble metals [32, 33]. Elevating reaction temperature, synthesis duration, and processing conditions is essential for minimizing elemental segregation and phase separation while also facilitating consistent particle growth. The functional properties of HEMs are contingent upon their microscopic structure, necessitating precise synthesis parameters. Nanoscale HEMs frequently exhibit superior physicochemical properties compared to their bulk and thin-film counterparts, establishing them as a significant area of research. Solid-state techniques include mechanochemical synthesis (MS), flash sintering, spark plasma sintering (SPS), and traditional solid-state reactions. Sonochemical synthesis, sol–gel, spray pyrolysis, and co-precipitation combustion are examples of liquid-state techniques. Among the gas-phase techniques is magnetron sputtering deposition [34, 35]. To compare the advantages and limitations of different synthesis strategies, Table 3.2 provides an overview of the commonly used methods for fabricating 2D HEMs, highlighting key reaction conditions, benefits, and challenges.

3.4.2 Synthesis Methods of High-Entropy Oxides (HEOs)

HEOs have been extensively studied, resulting in bulk, nanostructured, and thin-film synthesis methods. The synthesis method greatly impacts HEO structural integrity, phase stability, and function. Solid-state, liquid-state, and gas-phase methods are used to synthesize HEOs. The solid-state reaction is the primary technique for synthesizing bulk HEOs. Metal oxides are pulverized into fine powders, compacted into pellets with a hydraulic press, and calcined at temperatures. Direct calcination of the powder produces bulk and nano-HEOs. By creating localized plasma with a pulsed direct current, SPS accelerates oxide particle densification and activation.

$Sr((Zr_{0.94}Y_{0.06})_{0.2}Sn_{0.2}Ti_{0.2}Hf_{0.2}Mn_{0.2})O_{3-x}$ synthesized by SPS at 1,475 °C for nine minutes had higher structural density than that produced by reaction done at 1,500 °C for two hours [35]. Flash sintering speeds phase transitions in seconds with an electric field. For example, Sr $(Ti_{0.2}Y_{0.2}Zr_{0.2}Sn_{0.2}Hf_{0.2})O_{3-x}$ was synthesized in just one minute at 1,314 °C, while $(Mg_{0.2}Ni_{0.2}Co_{0.2}Cu_{0.2}Zn_{0.2})O$ was generated at ambient temperature in a matter of seconds [41–44].

Table 3.2 Comparison of synthesis methods for 2D-based high-entropy materials.

Synthesis method	Reaction conditions	Advantages	Challenges	Example materials	References
Solid-state reaction	High temperature (>1,000 °C), long reaction time	High purity, good crystallinity	Requires high energy, phase separation risk	(MgNiCo CuZn)O, $(CrMnFe CoNi)_3O_4$	[29, 36, 37]
Sol–gel synthesis	Moderate temperature (600–900 °C), solution-based	Uniform particle distribution, scalable	Longer processing time, high precursor cost	(CoFeNi MnZn)O, $(AlCoFe MnNi)_3O_4$	[29, 38]
Co-precipitation	Room temperature to moderate temperature	Easy to control composition	Post-synthesis calcination required	(MgNiCo CuZn)O	[35, 38]
Spark plasma sintering (SPS)	High pressure, short duration (~10 min)	Densification, reduced grain growth	High equipment cost	$Sr(Zr_{0.94}Y_{0.06})_{0.2}Sn_{0.2}Ti_{0.2}Hf_{0.2}Mn_{0.2}O_{3-x}$	
Mechano-chemical synthesis (ball milling)	Room temperature, mechanical energy	Low cost, no need for high temperature	High defect density, needs post-annealing	(MgNiCo CuZn)O	[39]
Magnetron sputtering	High vacuum, thin-film deposition	High control over structure	Limited to thin films, slow process	(MgZnMn CoNi)O	[40]

A high-energy ball milling process called MS, employs mechanical energy to promote chemical reactions between metal oxides via friction and extrusion, leading to modifications in their physical and chemical characteristics. (MgNiCoCuZn)O was successfully synthesized using MS, demonstrating the development of a rock salt structure (single phase) without the need for supplementary heat treatment [39, 45].

Liquid-phase synthesis techniques, such as spray pyrolysis, co-precipitation, sonochemical-based synthesis (SBM), and sol–gel combustion, facilitate the creation of nanostructured HEOs under comparatively mild conditions. Ultrasonic irradiation of a metal salt solution in SBM facilitates the formation of single-phase HEO without calcination because of acoustic cavitation. The introduction of metal precursor solutions into an alkaline medium with a regulated pH is the first step in reverse co-precipitation, an efficient liquid-phase method. This is followed by filtration, drying, grinding, and calcination [41, 46].

Nano-$(Co_{0.2}Cu_{0.2}Mg_{0.2}Ni_{0.2}Zn_{0.2})O$ was synthesized using this method. Microwave-assisted co-precipitation reduces sintering temperatures through microwave-induced energy absorption, as demonstrated in the synthesis of $(AlCoFeMnNi)_3O_4$ [47, 48]. Spray pyrolysis methods, including nebulized spray pyrolysis (NSP) and flame spray pyrolysis (FSP), entail the aerosolization of metal precursor solutions followed by exposure to elevated temperatures for pyrolysis.

NSP employs a hot-wall reactor for pyrolysis, whereas FSP utilizes a flame environment for decomposition, allowing for meticulous control over particle morphology and composition [49–53].

Sol–gel combustion, known as solution combustion, utilizes metal nitrates as oxidizers and organic compounds (such as glycine) as reducers. A notable decrease in synthesis temperature from over 850 to 185 °C was achieved this method for the formation of $(Co_{0.2}Cu_{0.2}Mg_{0.2}Ni_{0.2}Zn_{0.2})O$ crystals [54, 55].

Gas-phase synthesis techniques, particularly magnetron sputtering deposition, are widely utilized for fabricating thin-film HEOs. This physical vapor deposition (PVD) method involves bombarding a target material with a sputtering gas (e.g. Ar), leading to the deposition of neutral atoms onto a substrate. Thin films of (MgZnMnCoNi)Ox were synthesized via magnetron sputtering at 800 °C, exhibiting enhanced solubility and phase stability because of similar cation radii, stable oxidation states, and well-coordinated structures of the constituent elements [56–58].

Despite progress in HEO synthesis, many challenges remain. Solid-state reactions, especially high-temperature sintering, require high activation energy for elemental mixing, but nanomaterial synthesis often presents challenges. The wide range of oxidation potentials in metallic elements can cause phase separation and impurity formation at high temperatures. Liquid-phase synthesis techniques, while efficient at reduced temperatures, encounter kinetic obstacles that can lead to phase segregation. Consequently, innovative synthesis methodologies are being investigated to overcome these constraints. Innovative methods like the non-equilibrium carbothermal shock technique facilitate the rapid synthesis of micron- or nano-sized particles in milliseconds to seconds [59].

Likewise, Joule-heating synthesis has drastically reduced reaction times to just seconds, so increasing the efficiency of HEO synthesis. Improving material properties, structural homogeneity, and functional efficacy in energy storage applications depends on optimizing synthesis techniques as research in HEOs advances [60].

3.4.3 Synthesis of Other High-Entropy Materials

The synthesis techniques used for HEOs can also be modified for fabrication of other classes of HEMs, comprising nitrides, sulfides, carbides, and phosphates. Stable multi-element compounds with improved structural and functional characteristics have been obtained by effective application of several synthesis approaches. For the synthesis of high-entropy nitrides such as (CoCrCuFeNi)N [61] and (HfNbTiVZr)N [62], for example, deposition done by magnetron sputtering has been used to guarantee homogeneous elemental distribution and great thermal stability. Likewise, high-entropy sulfides such as $Cu_5SnMgGeZnS_9$ and $Cu_3SnMgInZnS_9$ have been synthesized using SPS, thus facilitating fast densification and phase formation under controlled conditions. Leveraging mechanical activation to achieve homogeneous mixing and enhanced reactivity, MS has enabled the preparation of high-entropy carbides including $(Hf_{0.2}Ta_{0.2}Nb_{0.2})C$ [63], and high-entropy sulfides such as $(FeCoNiMnCr)S_1$[56]. Furthermore, solution-based approaches such as co-precipitation have been effectively used for the synthesis of high-entropy phosphates, as demonstrated by $(La_{0.2}Ce_{0.2}Sm_{0.2}Eu_{0.2})PO_4$ [64], thus offering a scalable means for creating new multicomponent materials. These synthesis methods are versatile, making HEMs useful in energy storage, catalysis, and electronics. Advanced synthesis methods like non-equilibrium processing and fast thermal methods will help next-generation functional materials overcome phase stability and compositional homogeneity issues [65, 66].

3.4.4 Parameters Influencing the Electrochemical Performance of High-Entropy Materials

HEMs exhibit substantial potential in energy-related applications, particularly in the areas of energy storage and catalysis. The synthesis of novel HEMs and the assessment of their performance are the primary focus of current research. Nevertheless, the fundamental factors that influence their electrochemical behavior are not given sufficient attention. While configurational entropy plays a crucial role in the stabilization of multi-elemental systems, recent studies indicate that electrochemical performance is not solely dictated by entropy-driven effects [41].

The variations in structural, electronic, and ionic properties indicate the substantial influence of additional parameters on the functionality of HEMs with comparable entropy. In order to facilitate the rational design of HEM-based energy storage systems, researchers have initiated an investigation of critical parameters that influence their electrochemical performance. These encompass the specific functions of individual constituent elements, synthesis methodologies, and elemental composition. It is imperative to comprehend the interplay of these factors in order to enhance reversible capacity, extend cycle life, and optimize electrochemical properties. In this section, we present a comprehensive examination of these parameters, emphasizing their influence on the electrochemical behavior of HEMs and their implications for next-generation energy storage technologies [67].

Figure 3.3 displays the results of several electrochemical tests, which demonstrate that RHEO@TrGO outperforms RHEO in terms of rate performances. The RHEO and RHEO@TrGO initial constant-current charge/discharge curves (versus Li^+/Li) with a current density of 50 mA g^{-1} and within a voltage range of 0.001–3.0 V are shown in Figure 3.3a. RHEO and RHEO@TrGO have initial reversible capacities of 497.30 and 873.55 mA h g^{-1}, respectively. RHEO and RHEO@TrGO have initial coulombic efficiencies of 62.07% and 66.86%, respectively. With little degradation in initial charge/discharge capacity, the RHEO@TrGO electrode exhibits a high capacity. Throughout the charging and discharging operation, the TrGO serves as a transition layer. Compared to RHEO@TrGO, TrGO's initial reversible specific capacity is lower at 252.37 mA h g^{-1} with a coulombic efficiency of 55.78%. This implies that the distinct structure made up of nano-sized (MgCoNiCuZn)O particles and hollow porosity, rather than the inclusion of hollow graphene sphere shells, is the sole cause of the evident performance improvements of RHEO@TrGO at the first cycle. Figure 3.3b displays the RHEO and RHEO@TrGO voltage hysteresis curves. RHEO's platform potential difference during the first cycle is 1.71 V, which is substantially more than RHEO@TrGO's 1.38 V. By comparing with RHEO, this shows that RHEO@TrGO has a clear dynamic improvement. The initial cyclic voltammetry (CV) of RHEO and RHEO@TrGO between 0.01 and 3.0 V is shown in Figure 3.3c

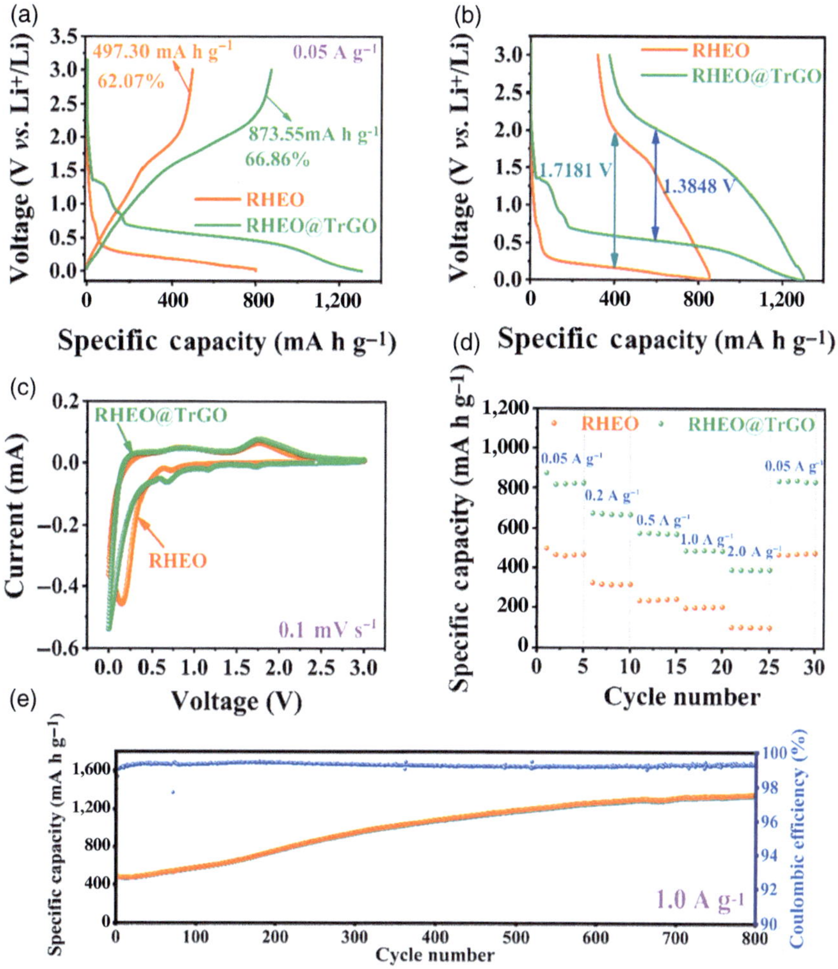

Figure 3.3 (a) First-cycle charge–discharge curves of RHEO and RHEO@TrGO at $0.05\,A\,g^{-1}$, (b) voltage hysteresis curve of RHEO and RHEO@TrGO, (c) CV curves of RHEO and RHEO@TrGO at $0.1\,mV\,s^{-1}$, (d) rate capability of RHEO and RHEO@TrGO, and (e) charge–discharge cycle curves of RHEO@TrGO at $1.0\,A\,g^{-1}$. *Source:* Reproduced from Ref. [68] / with permission of American Chemical Society.

at a scan rate of $0.1\,mV\,s^{-1}$. The creation of the solid electrolyte interphase (SEI) members and the conversion of HEO to Li_2O are responsible for the irreversible peak in RHEO that was seen during the cathodic cycle at 0.25 V. RHEO@TrGO, on the other hand, generated a peak at 0.69 V after going through the identical procedure.

Furthermore, the reduction peaks of TrGO at 1.2 and 1.7 V seen in RHEO@TrGO are consistent with lithiation processes, in which metal and Li_2O are produced as a result of interactions between lithium ions and RHEO@TrGO. An oxidation peak was detected at 1.8 V following the discharge cycle, which is linked to the metal's delithiation reaction, which releases lithium ions and produces metal oxides when it combines with Li_2O. High charging capacities of 822.27, 676.25, 576.04,

488.04, and 402.03 mA h g^{-1} at current densities of 0.05, 0.2, 0.5, 1.0, and 2.0 A g^{-1} are demonstrated by the RHEO@TrGO electrode, as shown in Figures 3.3d. The specific capacity can increase to 830 mA h g^{-1} after the current density is brought back to 0.05 A g^{-1}. At the same current densities, however, the RHEO electrode shows far lower charging capabilities of 453.05, 314.62, 232.94, 197.49, and 97.21 mA h g^{-1}. These numbers are significantly less than those found for RHEO@TrGO. These results imply that RHEO@TrGO has better rate performance, which is explained by the porous structure formed on the TrGO composite's surface. This structure improves conductivity and offers a stable framework, especially at a high current density of 2.0 A g^{-1}. The RHEO@TrGO constant-current charge–discharge curve at 1.0 A g^{-1} is shown in Figure 3.3e. The discharge capacity of RHEO@TrGO continues to rise after 800 cycles and stays over 1,400 mA h g^{-1}.

CV is considered as an efficacious methodology for evaluating the influence of electrode materials on the electrochemical kinetics of lithium-ion batteries (LIBs).

$$i = av^b \tag{3.3}$$

$$\log i = b \log v + \log a \tag{3.4}$$

$$i = (k_1 v) + (k_2 v^{1/2}) \tag{3.5}$$

CV measurements were used to analyze the lithium-ion storage kinetics in order to clarify the capacitance and diffusion-controlled effects of RHEO@TrGO. The CV curves are shown in Figure 3.4a at several scan rates between 0.1 and 1.0 mV s^{-1}. The CV curve shows clear oxidation–reduction peaks as the scan rate rises from 0.1 to 1.0 mV s^{-1}, accompanied by a rise in peak current (i). This can be explained by Eq. (3.1): $i = av^b$, where a and b are empirical parameters. The system goes through a diffusion-controlled process as b gets closer to 0.5. On the other hand, the system switches to a pseudocapacitive control process when b gets closer to 1. The log (i)–log (v) curve can be used to calculate the values of a and b using Eq. (3.3). It has been established that the oxidation and reduction peaks have respective b values of 0.78 and 0.88. These results indicate the presence of mixed capacitance and diffusion-controlled electrode dynamics. The overall current response (i) at a fixed potential (V) can be broken down into two mechanisms: a diffusion-controlled process ($k_2 v^{1/2}$) and a capacitive process ($k_1 v$) in order to further separate their components. The following Eq. (3.3) is used to do this: $I = k_1 v + k_2 v^{1/2}$. It is possible to distinguish between the diffusion-controlled current component, which has a value of $k_2 v^{1/2}$, and the capacitance current, $i_c(V) = k_1 v$, by graphing the curve of $i(V)/v^{1/2}$ against $v^{1/2}$ at a given voltage. As shown in Figure 3.4b, the capacitive contribution accounts for about 87.90% of the overall charge storage at a scan rate of 1.0 mV s^{-1}. The addition of active elements like zinc (Zn) is responsible for the notable pseudocapacitance contribution of rock salt-structured HEOs. The specific capacity of the electrode is greatly increased by the addition of zinc, which can alloy with lithium to generate Li–Zn alloys during the reduction process.

The capacitance contribution rates for RHEO@TrGO are 71%, 72%, 79%, and 88% at scan speeds of 0.1, 0.2, 0.5, and 1.0 mV s^{-1}, as shown in Figure 3.4c. The double-layer capacitance and Faradaic pseudocapacitance brought on by the

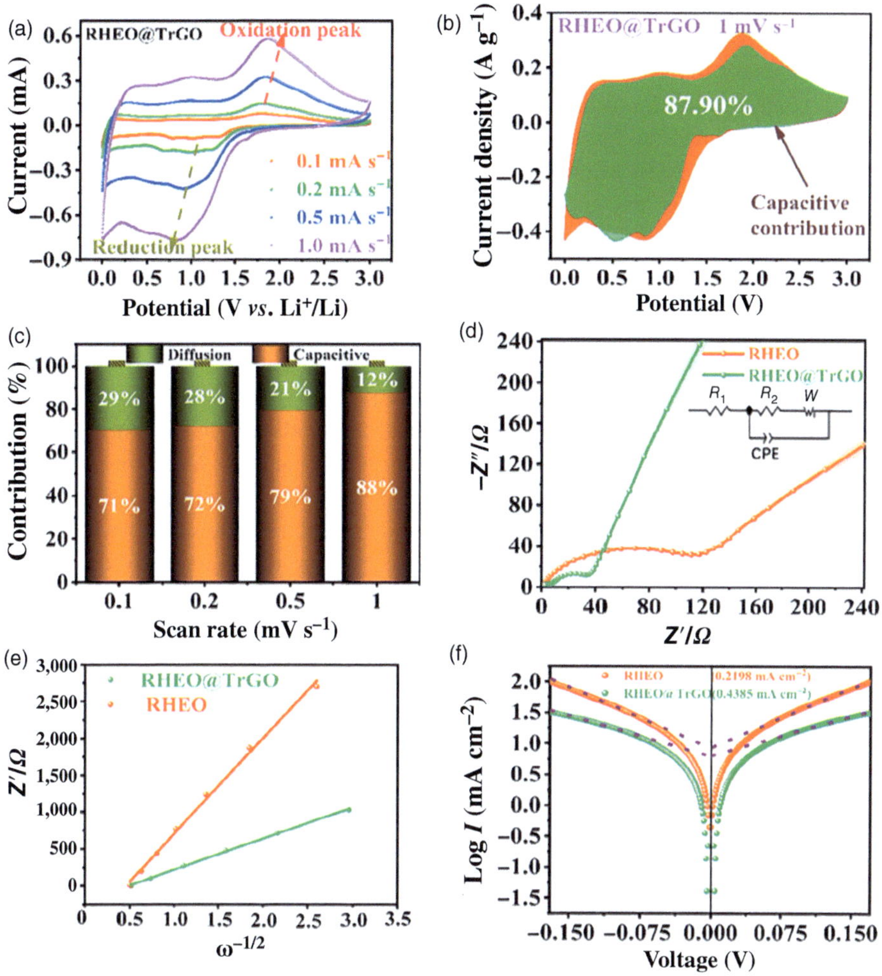

Figure 3.4 (a) Cyclic voltammetry curves of the RHEO@TrGO electrode at different scanning speeds from 0.1 to 1 mA s^{-1}. (b) Capacitive contribution of RHEO@TrGO to the overall charge capacity at 1.0 mV s^{-1}. (c) Contribution ratios of the capacitive process at differing scan rates. (d) Nyquist plots and fitted curves for RHEO and RHEO@TrGO electrodes. (e) Z' plots against $\omega^{-1/2}$ at the low-frequency region of EIS. (f) Tafel curves of RHEO@TrGO and RHEO. *Source:* Reproduced from Ref. [68] / with permission of American Chemical Society.

adsorption and desorption of Li^+ are principally responsible for this significant capacitance. Conversely, the capacitance contributions of RHEO alone were 64, 67, 75, and 80% at the same scan speeds (0.1, 0.2, 0.5, and 1.0 mV s^{-1}). TrGO's porous structure, which improves electron transport during the conversion reaction, is responsible for its larger capacitance contribution as compared to RHEO. Furthermore, RHEO can display pseudocapacitance at high current densities due to TrGO's spherical shell structure, which speeds up charge and discharge rates. TrGO outperforms RHEO with a capacitance contribution rate of 87.90% at a scan rate of 1.0 mv s^{-1}. This illustrates even more how TrGO significantly enhances high-rate performance. Electrochemical impedance spectroscopy (EIS) is a useful technique for determining the electrochemical kinetics of samples. In the Nyquist plot (Figure 3.4d), the semicircular arc in the high-frequency area corresponds to the electrode materials' charge transfer impedance (R_{ct}), and the linear segment at low frequencies represents the diffusion-controlled process. The EIS results were modeled using the analogous circuit shown in Figure 3.4e.

3.4.5 Elemental Composition and Its Influence on Electrochemical Performance

Equimolar elemental ratios are utilized in HEMs to enhance phase stability and configurational entropy. The high-entropy of equimolar HEMs hinders electrochemical cycling redox reactions, even though these materials exhibit structural stability. It is essential to optimize elemental composition to achieve a balance between redox activity and structural integrity, thereby enhancing electrochemical performance.

The entropy of low configuration observed in (NiCoCuZn)O results in a polyphase structure, whereas (Mg_xNiCoCuZn)O ($x = 0, 0.8, 1$) exhibits a stable rock salt phase that exhibits entropy-driven stabilization of multicomponent single-phase systems [69]. The presence of magnesium content significantly influences the functioning of LIBs in the following ways:

a) An increase in magnesium content results in a reduction of both theoretical and reversible capacities.
b) The volume expansion caused by lithiation leads to the cracking of electrodes, yet the presence of electrochemically inert magnesium mitigates particle pulverization. Increased concentrations of magnesium lead to a reduction in particle fragmentation, thereby enhancing structural stability.

To better illustrate the effect of elemental composition on phase stability, electrochemical capacity, and structural integrity, Table 3.3 presents a comparative analysis of different material compositions and their impact on LIB performance.

3.4.6 Influence of Synthesis Processes on Electrochemical Performance

HEMs have different elemental compositions, which require equally different synthesis methods that are quite important for their structural and electrochemical

Table 3.3 Influence of elemental composition on electrochemical performance.

Composition	Phase stability	Reversible capacity (mA h g^{-1})	Structural stability	Redox activity
(NiCoCuZn)O	Multiphase	Low	Poor	High
($Mg_{0.8}$NiCoCuZn)O	Rock salt	Moderate	Good	Moderate
(MgNiCoCuZn)O	Single-phase	High	Excellent	High

Source: References [39, 69].

characteristics. For uses in energy storage, nanoscale structures have many benefits. Their small size and great specific surface area improve the contact between the active electrode material and the electrolyte, promoting fast ion diffusion. Furthermore, guaranteeing excellent structural stability, nanostructuring reduces volumetric expansion and mechanical stress during cycling. On the other hand, high tap-density microparticles allow more important mass loading, enhancing the material's overall reversible capacity [69, 70].

Using solvothermal and electrospinning techniques, the effect of shape on the electrochemical performance of (MgNiCoCuZn)O is studied by synthesis of three different structures: 5MBsg, 5MBa, and 5MBb. For the synthesis of 5MBa and 5MBb, polymeric templates such as polyvinylpyrrolidone (PVP) and polyacrylonitrile (PAN) were used. Whereas 5MBa and 5MBb showed fibrous morphologies, with particle size distributions of 0.2–2 μm and 0.1–1.5 μm, respectively, the resulting 5MBsg material comprised of loosely bound particle aggregates (0.1–9 μm). Using an inverse relationship between particle size and electrochemical performance, electrochemical evaluation of these structures in LIBs revealed that 5MBsg < 5MBb < 5MBa. The larger interfacial area of the fibrous structures with the electrolyte promotes ion exchange. Thus, their enhanced performance is ascribed to both their continuous and aligned paths and their reduction of lithium-ion diffusion distances. Moreover, the broad particle size distribution in 5MBsg produced unequal current distribution and volumetric fluctuations during lithiation/delithiation, thus causing electrode cracking, delamination, and reduced cycling stability. By contrast, the connected fibrous structure reduced mechanical stress and maintained structural integrity over several charge–discharge cycles. Variations in reaction conditions, such as temperature and duration, can significantly affect the crystallinity, particle size, and electrochemical properties of HEMs, even when using the same synthesis method.

3.5 Mechanisms of High-Entropy Structures

Two basic processes help one to appreciate the effect of HES on energy storage materials. First, the high configurational entropy stabilizes the crystal structure, reducing structural fragmentation and lattice distortion in repeated charge–discharge cycles. This structural integrity improves HEM's long-term cycling stability. Second, high-entropy promotes the development of various atomic configurations, improving the density of electrochemically active sites. This then helps ion diffusion and increases the reversible capacity of energy storage devices. The excellent electrochemical performance of HEMs depends on rational design and optimization, as well as a complete awareness of these processes.

3.5.1 Stabilization of Crystal Structure

The stabilization of the crystal structure in HEMs is a result of the interaction between enthalpy and configurational entropy. The configurational entropy that is associated with an increase in the number of elemental species in a system enhances the structural stability by reducing the Gibbs free energy. The system is thermodynamically stabilized when the entropy contribution exceeds the enthalpy. The phase transitions illustrate this phenomenon observed in HEOs, such as $(CrMnFeNiCu)_3O_4$, which undergoes a transformation during cycling from the spinel phase to the rock salt phase [69, 71]. In the same way, $(FeCoNiCrMnZnLi)_3O_4$ undergoes an initial transition from the spinel phase to the rock salt phase, which is followed by the development of an amorphous phase [72].

During Li^+ intercalation and deintercalation, the transition phase of $(CrMnFeCoNi)_3O_4$ was examined at an atomic-scale where the rearrangement of metal–oxygen (M—O) bonds at 0.5 V lithiation leads to the formation of new spinel phases, $Cr_xFe_{3-x}O_4$ and $LiNixCo_{1-x}O_2$, in addition to Mn nanocrystals. Cr, Fe, Co, and Ni precipitate as nanocrystals upon complete lithiation, while the original spinel framework is preserved. The $Cr_xFe_{3-x}O_4$ and $LiNixCo_{1-x}O_2$ phases are restored by the material's reabsorption of metal particles during delithiation. Mn migrates to the $Cr_xFe_{3-x}O_4$ phase, forming $Mn_xCryFe_{3-x-y}O_4$. The reversible transformation between the $Cr_xFe_{3-x}O_4$ and $LiNixCo_{1-x}O_2$ spinel phases is induced by subsequent cycling, demonstrating that high configurational entropy contributes to lattice preservation, thereby ensuring enhanced structural stability. A compositional engineering strategy for regulating the layered structure of TMO_2 with a high-entropy configuration was employed to enhance structural stability and tune interlayer spacing for improved electrochemical performance. They synthesized $NaNi_{0.25}Mg_{0.05}Cu_{0.1}Fe_{0.2}Mn_{0.2}Ti_{0.1}Sn_{0.1}O_2$ (HEO424) by incorporating multiple cations into $NaNi_{0.4}Fe_{0.2}Mn_{0.4}O_2$ (NFM424). Mg^{2+} and Ti^{4+} contribute to structural stabilization, while Cu^{2+} and Sn^{4+} facilitate charge compensation and increase the average operating power. The long-range Na^+/vacancy was disrupted with the introduction of high-entropy ordering and expanded the interlayer spacing within the TMO_2 structure, thereby preventing the formation of the undesirable O3 phase and mitigating Jahn–Teller distortions during the initial charge cycle. Additionally, the reduced cation potential in HEO424 reduced Na^+ interactions, which facilitated the retention of the O3 phase and postponed the O3–P3 transition of phase [28, 73].

The advantages of high-entropy were further demonstrated through a comparative microscopic analysis of NFM424 and HEO424 following cycling. The microstructural degradation of NFM424 was characterized by the development of cracks and the formation of a mixed rock salt phase as a result of the migration of transition metals (TMs) and Na^+. In contrast, HEO424 maintained a well-defined single-crystal structure with no formation of phase of rock salt. The presence of a later phase results in significant lattice parameter variations, which in turn leads to a lattice difference between rock salt and layered phases. Ultimately, this mismatch contributes to structural fatigue and the dissolution of TM species. These results emphasize the vital role of high-entropy engineering in the preservation of crystallographic integrity and the improvement of the electrochemical performance of 2D-based HEMs in energy storage applications [30].

3.5.2 Enhancement of Active Sites in High-Entropy Materials

Lattice disorder and oxygen vacancy generation are the two defect engineering strategies that significantly improve the electrochemical performance of HEMs by rising the reactive sites count. These structural changes enhance ion diffusion kinetics and help to enable better charge storage capacity. The cooperative interaction among several elements helps to explain their remarkable energy storage properties in HEMs. Characterized by the disturbance of the periodic atomic arrangement, lattice disorder produces a disordered lattice structure enhancing electrochemical activity. HEMs are defined by the random but uniform distribution of several elements that generates a unique configuration whereby long-range atomic order coexists with short-range disturbance. By maximizing ion transport routes and increasing the number of accessible active sites, this structural element significantly improves energy density and rate performance.

Modulating the electronic structure of HEMs depends critically on oxygen vacancies, a type of intrinsic flaw brought about by the deficit atoms (oxygen) in the lattice. These vacancies affect essential parameters, including electron cloud distribution, the density of states, and energy band structure, thus affecting the electrochemical behavior of the material. For example, the HEO $(CrMnFeCoNi)_3O_4$ exhibits 0.084 eV band gap at lower level, with −0.072 eV as maximum valence band and minimum value of 0.012 eV at conduction band. 3.407 eV is the value of band gap shown by Co_3O_4. This significant drop in band gap increases the electronic conductivity of HEMs, enabling better charge transfer and storage kinetics [74].

Beyond electronic modulation, oxygen vacancies are also significant for stabilizing intermediate species and improving the active site coordination environment. Mn nanocrystals are seen, and $Cr_xFe_{3-x}O_4$ phases develop during lithiation of (CrMnFeCoNi). Enhanced mobility of the oxygen sublattice made possible by the small particle size and abundant vacancies in $Cr_xFe_{3-x}O_4$ lets Mn atoms be encapsulated during delithiation. This phenomenon questions conventional wisdom regarding transformation anodes, in which only Li^+ diffusion is usually regarded as supporting reversible capacity. Together, the dynamic oxygen mobility, TM migration, and adjustments in elemental valence states improve the reversible capability of HEOs [28]. Additionally, the concentration of oxygen vacancies in $(MgCoNiCuZn)_xLi_{1-x}O$ has been shown to be regulated by the lithium content, which is directly correlated with electrochemical performance [75, 76].

Further underlining the importance of oxygen vacancies in energy storage applications are recent comparative studies between (MgCoNiCuZn)O (MO) and its lithium-substituted counterpart (LiMgCoNiCuZn). Li-MO shows somewhat better performance than MO according to electrochemical analysis, which emphasizes the part oxygen vacancies play in improving lithium storage capacity. Several elements behind this development help to explain the processes:

a) The Fermi energy level is elevated by the high concentration of oxygen vacancies, lowering the barrier of diffusion for Li^+ (0.297 eV in Li-MO vs. 0.381 eV), and increasing the average electron density.

b) Active sites for electrochemical reactions found in oxygen vacancies enhance the surface reactivity of the material and allow embedded pseudocapacitive behavior.
c) Oxygen vacancies improve Li^+ accommodation on Li-MO surface, thus reducing the volume expansion effects during cycling and encouraging the formation of a stable SEI.
d) A negatively charged area develops around the oxygen vacancies, producing a localized electric field that accelerates Li^+ accumulation close to these defects, enhancing charge storage kinetics.

Similarly, in $(Co_{0.2}Cr_{0.2}Fe_{0.2}Mn_{0.2}Ni_{0.2})_3O_4$ (CCFMNO), they found that the several elements occurrence with varying valence states induces a process of charge compensation, resulting in the formation of a significant amount of oxygen vacancies [77]. The deficit in the energy band structure is compensated for by surrounding oxygen atoms, bringing the band for conduction closer to the Fermi level and significantly raising the HEOs' conductivity ($0.02\,S\,m^{-1}$ at ambient temperature). These results highlight, in general, the critical contribution of defect engineering to improve the electrochemical characteristics of 2D-based HEMs. These materials are pretty promising candidates for next-generation energy storage uses since they provide better electrochemical performance than conventional electrode materials by using structural disorder and oxygen vacancies.

In 2016, coin cells with a lithium-metal anode were used to examine the electrochemical performance of pure NM and HE-NM cathodes at room temperature. The charge and discharge patterns were investigated using a voltage range of 2.7–4.3 V vs Li^+/Li. As the temperature rose, the pristine NM cathode's initial specific discharge capacities were 196.68, 207.46, 212.81, and $203.32\,mA\,h\,g^{-1}$, respectively. As a result, the temperature 750 °C was chosen as a comparison. Figure 3.5a reveals that the pristine NM and HE-NM had initial capacities of 212.81 and $206.35\,mA\,h\,g^{-1}$, and initial coulombic efficiencies of 85.76% and 87.96%, respectively. After 100 cycles at 1 C, pristine NM kept 63% of its initial discharge capacity, but HE-NM electrodes retained 76% (Figure 3.5b). Furthermore, the HE-NM had substantially higher rate capability than the pure NM (Figure 3.3c). Differential Scanning Calorimetry (DSC) measurements were used to investigate the thermal stability of virgin NM and HE-NM at 4.3 V. HE-NM had a higher exothermic peak than NM, which was 254.7 °C.

To elucidate the process beyond the increased cycling stability of pristine NM and HE-NM, dQ/dV^{-1} profiles were compared at the 1st, 10th, 25th, and 50th cycles (Figure 3.5d,e). During the charging stage, the pristine NM and HE-NM underwent a succession of phase transformations, shifting from hexagonal to monoclinic to hexagonal and back again (H1 → M → H2 → H3). Microcracks and structural deterioration are mostly caused by the phase change from H2 to H3. As demonstrated in Figure 3.5d,e, from the 1st to the 10th cycle, the pristine NM and HE-NM cathodes behaved similarly, but after the 25th cycle, the pristine NM and HE-NM behaved differently. The tired phase and particle cracks cause the H2 → H3 peak in the NM cathode to be difficult to discern due to its quick fall in intensity. The HE-NM cathode also suffered from declining intensity. However, at the 25th cycle of H2 → H3, the peak was visible, indicating that the HE-NM did not decay as quickly as the pristine NM.

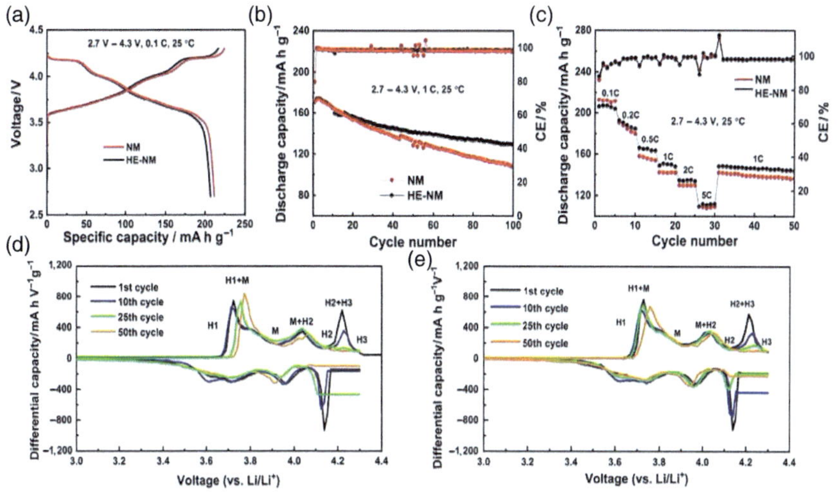

Figure 3.5 (a) Initial charge and discharge curves at 0.1C in the voltage range of 2.7–4.3 V. (b) Cycling performances at 1C and (c) rate performance for the NM and HE-NM in the voltage range of 2.7–4.3 V. The differential capacity (dQ/dV^{-1}) profiles at the 1st, 10th, 25th, and 50th cycles for (d) NM and (e) HE-NM electrodes. *Source:* Reproduced from Ref. [78] / with permission of American Chemical Society.

3.6 Applications of High-Entropy 2D Materials in Energy Storage

Rechargeable batteries are a crucial element of contemporary life since they provide energy to many electronic devices, including portable gadgets and electric vehicles. Due to their unique structural and electrochemical characteristics, HEOs have emerged as promising candidates for energy storage applications. At room temperature, these materials have elevated lithium-ion agility ($>10^{-3}\,S\,cm^{-1}$) and rapid sodium-ion agility ($5\times10^{-6}\,S\,cm^{-1}$), surpassing the ionic conductivity of traditional lithium phosphorus oxynitride (LiPON) by a magnitude of two orders. The monovalent ions' introduction in (MgNiCoCuZn)O triggers a charge compensation process, resulting in the creation of oxygen vacancies that enhances ionic conduction. Due to their high electrical conductivity, HEOs have been extensively studied for application in LIBs/SIBs. Recent research has examined their application as electrode materials in supercapacitors, Solid Oxide Fuel Cells (SOFCs), and lithium-sulfur (Li–S) batteries. Due to their increased ionic transport and structural stability, HEOs might revolutionize energy storage technologies [75].

3.6.1 Lithium-Ion Batteries

LIBs have become the leading energy storage technology owing to their high efficiency, extended cycle life, and lack of a memory effect. The restricted theoretical capacity of traditional electrode materials is a considerable obstacle in satisfying the rising worldwide need for high-energy-density storage solutions. As a result, the current research aims to improve the energy density of LIBs, with HEOs offering a possible path for progress.

3.6.1.1 High-Entropy Oxides as Anode Materials

The intricate structure of HEOs presents a complicated lithium-storing mechanism in LIBs. Numerous research has investigated the electrochemical performance of various HEOs, resulting in differing viewpoints on their lithium storing methods. (MgNiCoCuZn)O as a micron-sized anode material, attaining a notable capacity of 590 mA h g^{-1} with a coulombic efficiency surpassing 99% after 500 cycles at 200 mA g^{-1}. The observed increase in capacity during cycling corresponds with the electrochemical characteristics of conversion-type materials. It is ascribed for large-particle electrodes' activation. A comparative investigation of four oxide-based materials ($S = 1.39R$, medium entropy) showed that the absence of Co led to significant capacity deterioration after just a few cycles, highlighting its essential function in facilitating reversible capacity. The inclusion of five cations in HEOs considerably adds to the configurational entropy, hence enhancing the cycle stability [29, 64].

Recent developments in HEO-based anodes have yielded nanoscale (MgNiCoCuZn)O, demonstrating a remarkable reversible capacity surpassing 920 mA h g^{-1} at 100 mA g^{-1} after 300 cycles. At 3,000 mA g^{-1} current density, the material preserved a capacity of 490 mA h g^{-1}. The enhanced cycling performance and specific capacity of nanoscale (MgNiCoCuZn)O can be ascribed to multiple factors:

i) The inclusion of inert MgO mitigates nanoparticle aggregation and sustains the size of active components, thus ensuring electrode reversibility.
ii) Significant lattice distortion in HEOs prevents particle agglomeration and maintains electrode integrity.
iii) Entropy stabilization improves cycling stability by fostering structural robustness. The nanocrystalline structure enhances atomic rearrangement during charge–discharge cycles, hence boosting the overall cyclability.

Contrary to these results, a study suggested an alternative lithium storage mechanism for rock salt-structured (MgNiCoCuZn)O. Their research indicated that a conversion reaction first converts cations to their metallic forms, resulting in Li_2O, subsequently followed by an alloying/dealloying process involving Zn, Mg, and Li^+. They also noted metal segregation during charging and discharging, resulting in structural deterioration and the ultimate collapse of the rock salt framework, making the process largely irreversible [16, 79].

HEOs with a spinel shape have shown promise as anodes for LIBs owing to their advantageous lithium-ion transport routes. $(CrMnFeCoNi)_3O_4$ was produced by the oxidation of a HEA, attaining 596.5 mA h g^{-1} capacity which is reversible and a coulombic efficiency surpassing 99% at 2,000 mA g^{-1} over 1,200 cycles. Their suggested mechanism of storage for lithium consists of a two-step process:

i) The transformation of MnO_2 and M_2O_3 to MO (where M = Cr, Mn, Fe, Co, Ni).
ii) The subsequent reduction of MO to metallic states. Mn, Ni, and Cr were recognized as essential constituents that improve capacity retention. The existence

of trivalent/quaternary ions, elevated oxygen vacancy concentrations, and a reduced band gap promoted effective electron transport during cycling. Furthermore, the generation of secondary particles from nanoparticle fragmentation induced internal stress, alleviating external mechanical influences and maintaining crystal integrity.

A comprehensive examination of the lithium storage process in $(CrMnFeNiCu)_3O_4$ using fast-scanning X-ray absorption spectroscopy (XAS). Their study indicated that cations in the unaltered material had valence states of $Mn^{2+/3+}$, Cu^{2+}, Cr^{3+}, $Fe^{2+/3+}$, and Ni^{2+}, with chromium and nickel mostly occupying octahedral sites in the spinel structure. The lithiation potential beyond 0.6 V was linked to the reduction of $Mn^{2+/3+}$ to Mn^{2+} and Cu^{2+} to Cu^0. In contrast, successive reductions (Mn^{2+} to Mn^0, Cr^{3+} to Cr^{2+}, $Fe^{2+/3+}$ to Fe^{2+}, and Ni^{2+} to Ni^0) facilitated the emergence of a charging plateau at 0.5 V. The delithiation process was defined by a series of oxidation events, beginning with Mn initiating discharge capacity, succeeded by the concurrent oxidation of Mn, Cr, and Fe, which established 1.5 V of discharge plateau. In comparison, Ni oxidation transpired at 1.67 V, thus aiding in late-stage capacity retention [80].

Copper exhibited inactivity throughout the charge–discharge cycle, adding insignificantly to capacity. The energy storage mechanism of spinel-structured $(Ni_{0.2}Co_{0.2}Mn_{0.2}Fe_{0.2}Ti_{0.2})_3O_4$ (NCMFT) in LIBs. The research demonstrated that Ni and Co favored octahedral coordination, whereas Fe and Mn inhabited tetrahedral sites. NCMFT exhibited a revocable capacity of 560 mA h g^{-1} during 100 cycles at 100 mA g^{-1}, with almost complete capacity retention. Significantly, Ti, existing in a mixed +3/+4 oxidation state, did not engage in the conversion process but was essential in mitigating volume expansion and inhibiting metal nanoparticle aggregation, akin to the inert MgO effect found in (MgNiCoCuZn)O. After 10 cycles, the spinel structure remained preserved, underscoring the stabilizing influence of entropy-driven structural retention.

Besides rock salt and spinel-structured HEOs, perovskite-structured HEOs have been investigated for LIB applications. A high-entropy perovskite oxide (HEPO), $[(BiNa)_{0.2}(LaLi)_{0.2}(CeK)_{0.2}Ca_{0.2}Sr_{0.2}]TiO_3$ was synthesized by a solid-state technique, attaining a capacity of 120.4 mA h g^{-1} after 300 cycles at a current density of 1,000 mA g^{-1}. The incremental enhancement in capacity from 20 to 280 cycles was ascribed to electrolyte infiltration and the development of a membrane (polymer-gel-like). The introduction of La^{3+} and Bi^{3+} ions enhanced Li^+ intercalation at the A-site of HEPO. In comparison to rock salt and spinel-structured HEOs, high-entropy perovskites demonstrated a decreased specific capacity, highlighting the need for more structural optimization in battery development [74, 81, 82].

3.6.2 High-Entropy Oxides Utilized as Cathode Materials

HEOs have attracted considerable interest as cathode materials for LIBs owing to their distinctive single-phase architectures, including numerous cations inside a common anionic framework. The configurational entropy in these systems is mainly determined by cation variety, and subsequent studies have broadened the focus to include several anions, thereby improving entropy stability.

The mechanochemical techniques were used in order to produce a lithium-infused high-entropy fluorine oxide, which was referred to as Li_x $(Co_{0.2}Cu_{0.2}Mg_{0.2}Ni_{0.2}Zn_{0.2})$ OF_x (Li(HEO)F). Their increased electrochemical potentials, which are the consequence of the robust ionic properties of metal–fluoride (M—F) bonding, are usually thought to be feasible candidates for LIB cathodes. Because metal fluorine oxides possess the capacity to establish robust bonds with fluoride ions, this is the case. This enhancement was attributed to the increased configurational entropy. After 150 cycles, Li(HEO)F demonstrated a substantially higher specific capacity than LiNiOF, while the latter experienced a precipitous decline in coulombic efficacy. In contrast, Li(HEO)F demonstrated consistent coulombic efficiency, underscoring the critical role of entropy stabilization in the preservation of reversible capacity and the mitigation of performance decline [83].

A sequence of cation-disordered rock salt materials illustrates that augmenting TM variety mitigated short-range ordering while concurrently improving energy density and rate capability. At a current density of 20 mA g^{-1}, a six-TM DRX cathode, $Li_{1.3}Mn^{2+}{}_{0.1}Co^{2+}{}_{0.1}Mn^{3+}{}_{0.1}Cr^{3+}{}_{0.1}$Ti0.1$Nb_{0.2}O_{1.7}F_{0.3}$ (HEDRX) acquired an extraordinary specific capacity of 307 mA h g^{-1} among these materials. Six-element DRX cathodes routinely exceeded four-element variants, according to comparative studies, which emphasizes the capacity-enhancing effect of compositional complexity. Moreover, the polarization was significantly reduced by the addition of numerous transition elements. The indiscriminate addition of TMs is ineffective due to the potential for the electron transfer efficiency to be reduced, resulting in a decrease in overall capacity if the electrochemically active species are diluted too low. The deliberate integration of redox-active elements into the DRX framework is essential for maintaining a balance between high-energy density and structural stability [84].

Conventional ternary cathodes like $LiNi_{1-x-y}Mn_xCo_yO_2$ (NMC) and $LiNi_{1-x-y}Co_x$ Al_yO_2 (NCA) use cobalt to improve thermal stability and mitigate Li/Ni cation mixing. However, cobalt supply and geopolitical restrictions need cobalt-free alternatives that preserve electrochemical performance. This article presents a Co-free, zero-volume-change layered cathode, zero-strain LiNi0.8$Mn_{0.13}Ti_{0.02}Mg_{0.02}Nb_{0.01}$ $Mo_{0.02}O_2$. Heat at 350 °C caused substantial microcracking in $LiNi_{0.8}Mn_{0.1}Co_{0.1}O_2$ (NMC-811) but relatively minimal void development in HE-LNMO. Studies on structural evolution revealed that HE-LNMO underwent a transition to a spinel phase at 230 °C, eventually evolving into a rock salt structure at 305 °C, resembling the behavior observed in NMC-532. Electrochemical studies showed that HE-LNMO has an initial discharging capacity of 210.1 mA h g^{-1}, equivalent to NMC-811's 208.2 mA h. HE-LNMO outperformed graphite in cycling stability, sustaining over 95% of its capacity after 500 cycles at 2.8–4.2 V. Stability was achieved by suppressing oxygen vacancies around high-valent dopants, reducing redox center oxygen loss. Multicomponent dopant ions preserved electrode structure by preventing cation mixing and rock salt phase transition [85].

The LiFePO4 || SSBTA complete cell displayed a capacity of around 200 mA h g^{-1} over the first three cycles, with a voltage plateau of roughly 3 V (Figure 3.6a,b). This

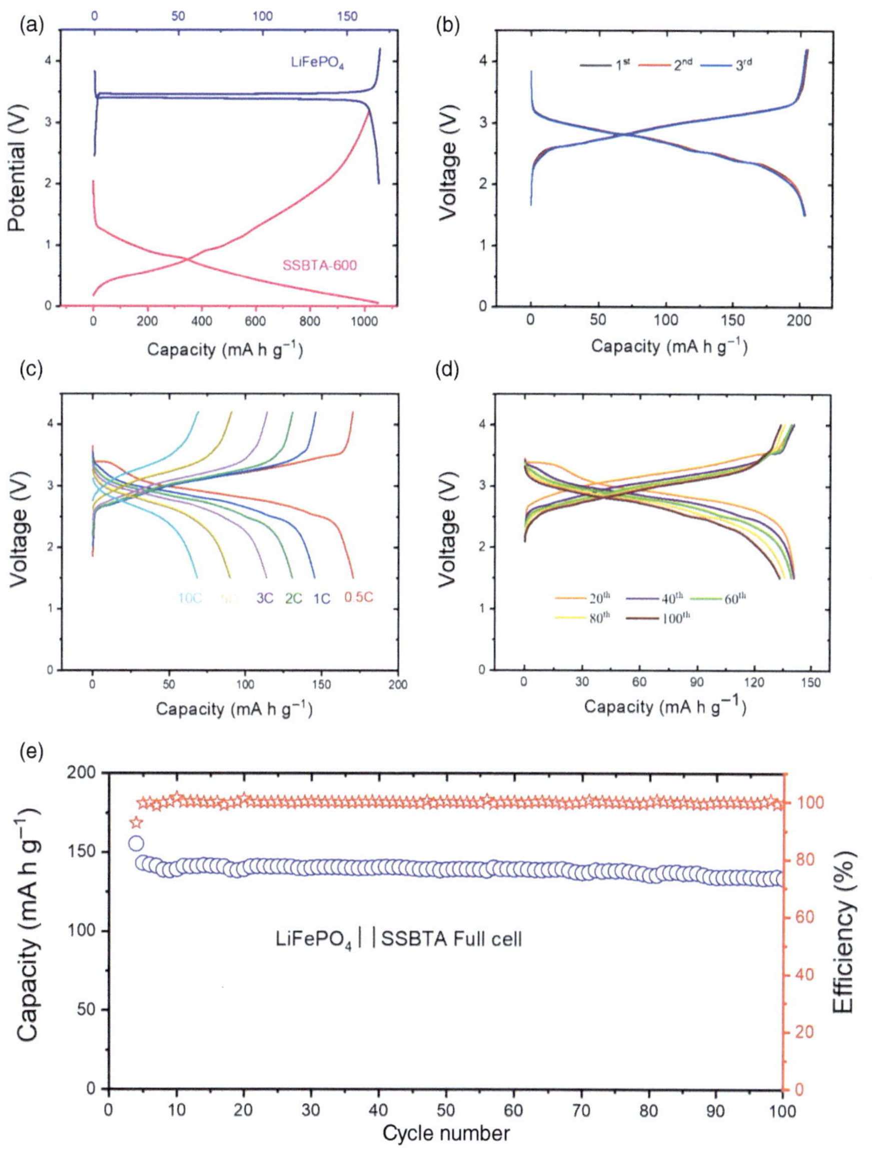

Figure 3.6 The electrochemical performance of $LiFePO_4$||SSBTA full cell: (a) the charge/discharge profiles of cathode $LiFePO_4$ and anode SSBTA countered Li foil; (b) the first three charge/discharge profiles after prelithium in SSBTA process; (c) the rate capability of the full cell at different current density; (d) the charge/discharge profiles during the long-term cyclic performance; (e) the long-term cyclic performance of the full cell. *Source:* Reproduced from Ref. [86] / with permission of American Chemical Society.

performance was attained with a current density of 12 mA g^{-1} (0.1C). Furthermore, the entire cell demonstrated good rate capability, with specific capacities of 145.5, 113.9, 89.9, and 68.4 mA h g^{-1} at C rates of 1C, 3C, 5C, and 10C, respectively (Figure 3.6c). These findings show that the SSBTA-600 anode retains a significant portion of its capacity even at high current densities, which is important for applications that need rapid charge and discharge cycles. Cyclic stability studies at 1C proved the whole cell's robustness, with 89.4% capacity retention after 100 cycles, resulting in a capacity of 134 mA h g^{-1} (Figure 3.6e). The charge/discharge profiles for the 20th, 40th, 60th, 80th, and 100th cycles (Figure 3.6d) show low-voltage fade and capacity loss, indicating that the structural integrity of the SSBTA-600 anode remains intact over extended cycling.

3.6.3 Sodium-Ion Batteries

The sodium-ion batteries (SIBs) exhibit energy storage mechanism parallel to that of LIBs, providing benefits including resource abundance, cost-effectiveness, and improved safety. During intercalation and deintercalation, the increased sodium ionic radius causes electrode material's expansion to a considerable volume, which leads to diminished energy density and shortened cycle life. The notion of HEMs has been investigated to tackle these challenges and enhance electrochemical performance. An O3-type layered oxide $NaNi_{0.12}Cu_{0.12}Mg_{0.12}Fe_{0.15}Co_{0.15}Mn_{0.1}Ti_{0.1}Sn_{0.1}Sb_{0.04}O_2$ demonstrated a retention capacity of 83% over 500 cycles at a rate of 3C, with a coulombic efficiency surpassing 99%. A highly reversible phase transition between the O3 and P3 structures occurred during charge–discharge cycles, with 60% of the capacity retained in the O3 region. In contrast to traditional three-component O3 cathodes, the multicomponent HEO cathode promoted localized interactions between TMs and sodium ions, allowing for charge compensation via changes in oxidation states. This process successfully postponed phase transitions and enhanced cycling stability [87].

The incorporation of five distinct cations within the metal sublattice was responsible for the consistent charge–discharge behavior of a layered $NaMn_{0.2}Fe_{0.2}Co_{0.2}Ni_{0.2}Ti_{0.2}O_2$ composition, which was devoid of potential fluctuations. At a rate of C/10, the material achieved an average discharge capacity of 180 mA h g^{-1}. During the initial charging process, two O3-type crystal structures coexisted; upon the extraction of a minor quantity of sodium, a P3-type phase emerged as a result of sodium vacancies and robust oxygen–oxygen interactions. The O3 phase during discharge, persisted alongside the P3 phase, signifying a partially reversible O3–P3 phase transition. The elevated specific capacity was associated with numerous redox reactions involving Mn, Fe, Co, and Ni [88].

The advancement of P2-structured layered NaxTMO_2 materials is frequently obstructed by the ordered configuration of sodium ions and the interactions between sodium and the TM layer that influence ion mobility. High-entropy structures, distinguished by disordered cation configurations and diminished phase transitions, present a promising solution. A P2-structured $Na_{0.6}(Ti_{0.2}Mn_{0.2}Co_{0.2}Ni_{0.2}Ru_{0.2})O_2$ (P2-TMCNR) electrode demonstrated a structural transformation from the (P2) phase to an intermediate OP4 phase, rather than the typical P2–O2 transition. The enhanced rate capability of P2-TMCNR was ascribed to the absence of sodium ion/vacancy ordering, that improved sodium-ion mobility within the layered structure. Nevertheless, exposure to elevated voltages can cause degradation of the crystal structure, resulting in swift capacity decline [89].

3.6.3.1 Lithium-Sulfur Batteries

Li–S batteries have attracted considerable interest as advanced energy storage systems owing to their remarkably high theoretical capacity (1,672 mA h g^{-1}) and energy density (2,600 Wh kg^{-1}). The practical implementation is impeded by the dissolution of intermediate lithium polysulfides (LiPS) in the organic electrolyte throughout charge–discharge cycles.

The shuttle effect results in the depletion of active material and significant capacity deterioration during extended cycling. Extensive research has concentrated on modifying cathode materials and integrating solid electrolytes to address this issue [90]. Metal oxides exhibit significant polarity, enabling them to anchor LiPS chemically and inhibit their migration [91]. The incorporation of HEOs into Li–S battery cathodes signifies a promising strategy to enhance electrochemical performance. A high-entropy metal oxide, namely (MgCoNiCuZn)O (HEMO-1), was synthesized and integrated into a Ketjen Black (KB) carbon-based sulfur cathode. The initial specific capacities of the KB/S and HEMO-1/KB/S electrodes were comparable; nonetheless, the KB/S electrode exhibited significant capacity degradation, retaining merely 597 mA h g^{-1} following 100 cycles [92].

The HEMO-1/KB/S electrode exhibited a capacity of 768 mA h g^{-1} after 100 cycles and 664 mA h g^{-1} after 200 cycles at a rate of 0.1C. The enhancement was linked to the uniform distribution of various metal active sites in HEMO-1, which promoted the retention of LiPS and improved redox kinetics at the cathode. The Li—O and S—Ni bonding interactions significantly reduced LiPS shuttling, thereby enhancing system stability. A spinel-type HEO $(Mg_{0.2}Mn_{0.2}Ni_{0.2}Co_{0.2}Zn_{0.2})Fe_2O_4$ nanofiber cathode, with an average diameter of around 160 nm, exhibited enhanced performance in Li–S batteries. A comparative study of sulfur-loaded HEO (S/HEO) and sulfur-loaded carbon nanofibers (S/CNFs) at 1.4 g cm^{-2} of sulfur loading demonstrating the S/HEO electrode exhibited superior electrochemical stability. At 1C, S/HEO's initial specific capacity was 879.6 mA h g^{-1}, which notably surpassed that of S/CNF at 544.2 mA h g^{-1}. Following 500 cycles, the S/HEO electrode sustained a capacity of 558.4 mA h g^{-1} and demonstrated an average coulombic efficiency of around 99.4%. It indicates robust chemical binding interactions between metal cations and LiPS, which effectively mitigate the shuttle effect. The measured tap density of S/HEO at 1.92 g cm^{-3} was found to be 2.5 times greater than the conventional sulfur/carbon composites, indicating its potential for high gravimetric and volumetric capacities [93].

3.6.3.2 Supercapacitors

Supercapacitors can be defined as advanced energy storage devices that store charge using electronic double-layer capacitors (EDLCs) or Faradaic charge transfer at the electrode/electrolyte interface. Power density is high in EDLCs, but energy density is low. Despite poor cycle stability, pseudocapacitors store more energy. Thus, supercapacitor technology requires electrode materials that increase energy density and cycle life. By integrating HEOs with nanostructured carbon materials, these constraints may be overcome. Synthesizing Ni^-, Fe^-, Cr^-, Al^-, and Co-based HEOs and chemically depositing them into Carbon nanotube (CNTs) was one method. Using a polyvinyl alcohol/H_2SO_4 hydrogel electrolyte, the HEO-CNT hybrid electrode maintained coulombic efficiency and 100% capacitance retention at $5\,A\,g^{-1}$ over 15,000 charge–discharge cycles as a complete cell. CNT encapsulation modified carbon layer redox characteristics, increasing active site availability and charge transfer kinetics. Structural confinement reduced HEOs' electrolyte exposure, reducing deterioration during continuous cycling [94].

Supercapacitor electrodes benefit from their ability to reduce volumetric expansion during the charge–discharge cycle. We synthesized and tested a range of $La(CoCrFeMnNiAl_x)_{1/(5+x)}O_{3-\delta}$ compositions for supercapacitors ($x = 0.4$, 0.5, 0.6, and 0.7) [58]. Increasing the scan rate increased polarization and ohmic resistance at the electrode/electrolyte interface, moving redox peaks to higher and lower potentials. All compositions maintained capacitance retention of over 85% after 2,000 cycles at $1{,}000\,mA\,g^{-1}$, indicating electrochemical stability. Redox processes need TMO_2. Still, their inclination to agglomerate lowers the number of active ions engaged in charge storage and adds structural flaws such as microcracks, which diminish performance. Al^{3+} inclusion in $La(CoCrFeMnNi)_{0.2}O_{3-\delta}$ reduced TM oxide clumping and increased lattice distortion, improving electrochemical stability. Overdosing Al^{3+} reduced ionic migration routes, reducing capacitance retention over extended cycling [95–97]. 2D high-entropy supercapacitor materials may overcome the constraints of traditional electrode materials. They enable next-generation energy storage devices with improved performance and endurance because of their large surface area, customizable electronic characteristics, and structural stability. More studies should optimize HEM compositions and explore innovative nanostructured designs to increase charge storage processes and electrochemical stability.

During charging, electrolyte ions move toward oppositely charged electrodes, forming an electric double layer; electrodes remain stationary. During discharging, electrons flow from the negative to the positive electrode, and electrolyte ions redistribute to balance charges. The separator prevents short-circuiting while allowing ion passage. The electrolyte facilitates ion transport. HEMs enhance performance by offering multiple redox sites, boosting electrochemical reactions and capacitance. The detailed mechanism has been given in Figure 3.7.

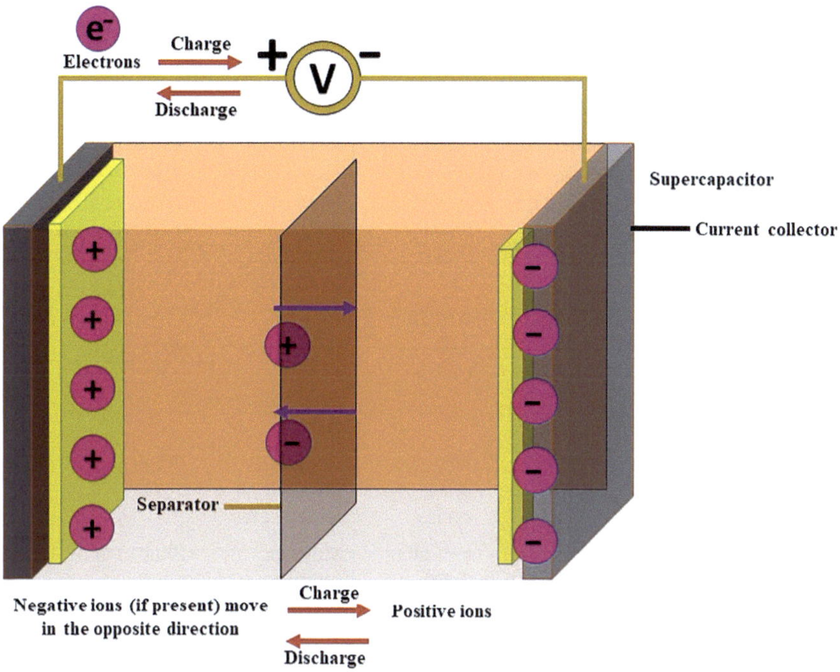

Figure 3.7 Charging–discharging mechanism of supercapacitor.

2D HEMs have emerged as a revolutionary category of materials with considerable promise for energy storage applications. Their distinctive structural attributes, influenced by configurational entropy, provide adjustable electronic properties, optimized charge transport kinetics, and increased stability during electrochemical cycling. Significant progress in the synthesis of 2D HEMs, including solid-state, liquid-state, and gas-phase methodologies, has allowed meticulous regulation of their composition and phase stability. Strategies such as defect engineering, manipulation of oxygen vacancies, and atomic-scale characterization have elucidated the processes that govern their performance. The materials exhibit significant applications in LIBs, supercapacitors, and electrocatalysis due to their high-entropy, which improves ion transport and cycle stability. Despite these advancements, challenges persist regarding synthesis parameters, manufacturing scale, and long-term degradation processes. To enhance the performance of 2D HEM, future investigations should incorporate computational modeling, thorough in situ characterization, and novel processing techniques. The materials discussed have the potential to connect high-entropy ceramics with electrochemical applications, thereby improving energy storage solutions (Table 3.4). This is attributed to their intrinsic entropy-driven stability and structural versatility.

Table 3.4 Recent developments of 2D-based high-entropy materials for supercapacitor applications.

Material composition	Specific capacitance ($F g^{-1}$)	Electrolyte	Cycle stability	Notable features	References	Sources
$(CoCrFeMnNi)_3O_4$	680 at 1 $A g^{-1}$	6 M KOH	92% after 5,000 cycles	High oxygen vacancies, excellent conductivity	Zhang et al., 2021	[98]
$(NiCoMnAlFe)O_x$	623 at 0.5 $A g^{-1}$	1 M Na_2SO_4	90% after 3,000 cycles	Pseudo-capacitive behavior, mixed valence states	Chen et al., 2022	[99]
(MgCoNiCuZn)O	512 at 1 $A g^{-1}$	3 M KOH	94% after 3,000 cycles	Rock salt phase, high surface area	Wang et al., 2021	[100]
(AlCrFeMnTi)O	710 at 2 $A g^{-1}$	2 M H_2SO_4	89% after 5,000 cycles	Superior redox reversibility	Zhao et al., 2023	[101]
(CoNiFeZnMg)O	560 at 0.5 $A g^{-1}$	1 M KOH	91% after 2,000 cycles	Uniform morphology, high conductivity	Singh et al., 2021	[102]
(TiVNbMoW) C MXene	425 at 1 $A g^{-1}$	1 M H_2SO_4	95% after 10,000 cycles	High-entropy MXene, excellent rate capability	Guo et al., 2022	[103]
(CoFeNiCrTi)O	430 at 0.2 $A g^{-1}$	3 M KOH	93% after 2,000 cycles	Homogeneous phase, high charge transfer	Das et al., 2022	[104]

3.7 Conclusions

2D HEMs have the potential to be advanced functional materials for energy storage systems. Due to the presence of numerous fundamental components in nearly equimolar ratios, these materials exhibit substantial configurational entropy, which improves ion transport, electronic characteristics, and structural stability. They are appropriate for lithium-ion, sodium-ion, Li–S, and supercapacitors due to their unique layered architectures, extensive surface area, active sites, and defect interfaces. The chapter delves into the synthesis of 2D HEM, the evolution of its structure, and its electrochemical performance. The synthesis method, elemental composition, oxygen vacancies, and lattice disorder were the primary factors that influenced their behavior. Capacity, rate, and cycle stability are enhanced through defect engineering and entropy-driven phase stabilization. Despite advances, scalable synthesis, long-term stability, and electrochemical material design remain issues. Advanced characterization, computational modeling, and sustainable engineering of 2D HEMS are necessary to optimize their energy storage potential.

3.8 Future Scope

The future advancement of 2D HEMs for energy storage depends on overcoming many scientific and technical obstacles. A crucial aspect is the use of sophisticated in situ and operando characterization methods to elucidate the real-time structural and chemical changes of HEMS during electrochemical cycling. Methods including in situ XAS, transmission electron microscopy (TEM), and electron energy loss spectroscopy (EELS) yield significant insights into phase transitions, redox mechanisms, and degradation behavior in multicomponent systems. Moreover, the extensive compositional design space of HEMs necessitates the growing significance of machine learning and high-throughput screening methodologies. These tools help to find ideal mixtures with particular electrochemical properties and stability [105, 106].

Modern synthesis techniques sometimes use energy-intensive methods, making them inappropriate for mass production. Demand for scalable, inexpensive, and sustainable synthesis methods is rising; among these are non-equilibrium processes such as carbothermal shock synthesis, microwave-assisted synthesis, and solution combustion. A further intriguing approach is the integration of 2D HEMs into solid-state and flexible energy storage systems. In order to facilitate implementations in ubiquitous electronics and next-generation all-solid-state batteries, they must be compatible with solid electrolytes and flexible substrates [22, 107].

References

1 Elalfy, D.A., Gouda, E., Kotb, M.F. et al. (2024). Comprehensive review of energy storage systems technologies, objectives, challenges, and future trends. *Energy Strategy Rev* 54: 101482. https://doi.org/10.1016/j.esr.2024.101482.

2 Reenu, S., Phor, L., Kumar, A., and Chahal, S. (2024). Electrode materials for supercapacitors: a comprehensive review of advancements and performance. *J Energy Storage* 84: 110698. https://doi.org/10.1016/j.est.2024.110698.

3 Liu, Y., Jiang, S.P., and Shao, Z. (2020). Intercalation pseudocapacitance in electrochemical energy storage: recent advances in fundamental understanding and materials development. *Mater Today Adv* 7: 100072. https://doi.org/10.1016/j.mtadv.2020.100072.

4 Li, J., Duan, H., Long, Q. et al. (2024). High-entropy materials: excellent energy-storage and conversion materials in the field of electrochemistry. *Particuology* 92: 42–60. https://doi.org/10.1016/j.partic.2024.04.010.

5 Zhang, Y. (2019). *History of High-Entropy Materials*, 1–33. Singapore: Springer Singapore. https://doi.org/10.1007/978-981-13-8526-1_1.

6 Zhang, J., Liang, J., Li, H. et al. (2024). A review of high-entropy ceramics preparation methods, properties in different application fields and their regulation methods. *J Mater Res Technol* 32: 1083–1117. https://doi.org/10.1016/j.jmrt.2024.07.153.

7 Akrami, S., Edalati, P., Fuji, M., and Edalati, K. (2021). High-entropy ceramics: review of principles, production and applications. *Mater Sci Eng R Rep* 146: 100644. https://doi.org/10.1016/j.mser.2021.100644.

8 Dong, Y., Zhou, Z., Ma, Y. et al. (2024). Layered-structured sodium-ion cathode materials: advancements through high-entropy approaches. *ACS Energy Lett* 5096–5119. https://doi.org/10.1021/acsenergylett.4c02223.

9 Cui, Y., Zhang, Y., Cao, Z. et al. (2022). A perspective on high-entropy two-dimensional materials. *SusMat* 2: 65–75. https://doi.org/10.1002/sus2.47.

10 Panda, D., Kumar, A., and Gangawane, K.M. (2024). Chapter 4 Materials for micro- and nanofabrication. In: *Microfabrication and Nanofabrication*, 77–100. De Gruyter. https://doi.org/10.1515/9783111335537-004.

11 Leani, J.J., Sánchez, H.J., and Pérez, C.A. (2015). Oxide nanolayers in stratified samples studied by X-ray resonant raman scattering at grazing incidence. *J Spectrosc* 2015: 1–7. https://doi.org/10.1155/2015/618279.

12 Pogorielov, M., Smyrnova, K., Kyrylenko, S. et al. (2021). MXenes—a new class of two-dimensional materials: structure, properties and potential applications. *Nanomaterials* 11: 3412. https://doi.org/10.3390/nano11123412.

13 Mei, H., Zhang, Y., Zhang, P. et al. (2024). Entropy engineering of 2D materials. *Adv Sci* 11. https://doi.org/10.1002/advs.202409404.

14 Zhu, Z., Liu, Y., Zhang, H. et al. (2025). Emerging trends in electrochemical energy storage: a focus on low-temperature pseudocapacitors. *Energy Rev* 4: 100118. https://doi.org/10.1016/j.enrev.2024.100118.

15 Kumar, A. and Gangawane, K.M. (2024). Recent advances in organic/composite phase change materials (PCM) for energy storage. In: *Materials for Boosting Energy Storage. Volume 3: Advances in Sustainable Energy Technologies*, vol. 1488, 309–334. American Chemical Society. https://doi.org/10.1021/bk-2024-1488.ch013.

16 Kumar, A. and Gangawane, K.M. (2024). Nanoparticle-modified multifunctional nano carbons—advances in energy storage. In: *Materials for Boosting Energy Storage Volume 3: Advances in Sustainable Energy Technologies, Part 7—Nanoparticle-Modified Multifunctional Nano Carbons—Advances in Energy Storage*, vol. 3, 1–30. https://doi.org/10.1021/bk-2024-1488.ch007.

17 Sarkar, A., Wang, Q., Schiele, A. et al. (2019). High-entropy oxides: fundamental aspects and electrochemical properties. *Adv Mater* 31. https://doi.org/10.1002/adma.201806236.

18 Feng, D., Dong, Y., Nie, P. et al. (2022). CoNiCuMgZn high entropy alloy nanoparticles embedded onto graphene sheets via anchoring and alloying

strategy as efficient electrocatalysts for hydrogen evolution reaction. *Chem Eng J* 430: 132883. https://doi.org/10.1016/j.cej.2021.132883.

19 Mao, P., Guo, Y., Wang, T. et al. (2024). Boosting high electric breakdown strength for excellent energy storage performance in $Bi_{0.5}Na_{0.5}TiO_3$-based lead-free ceramics via a high entropy strategy. *ACS Appl Mater Interfaces*. https://doi.org/10.1021/acsami.4c15339.

20 Sen, S., Palabathuni, M., Ryan, K.M., and Singh, S. (2024). High entropy oxides: mapping the landscape from fundamentals to future vistas. *ACS Energy Lett* 9: 3694–3718. https://doi.org/10.1021/acsenergylett.4c01129.

21 Han, X., Wu, G., Zhao, S. et al. (2023). Nanoscale high-entropy alloy for electrocatalysis. *Matter* 6: 1717–1751. https://doi.org/10.1016/j.matt.2023.03.034.

22 Bolar, S., Ito, Y., and Fujita, T. (2024). Future prospects of high-entropy alloys as next-generation industrial electrode materials. *Chem Sci* 15: 8664–8722. https://doi.org/10.1039/D3SC06784J.

23 Ziehl, T.J., Morris, D., and Zhang, P. (2023). Detection and impact of short-range order in medium/high-entropy alloys. *IScience* 26: 106209. https://doi.org/10.1016/j.isci.2023.106209.

24 Mohammadzadeh, A., Heidarzadeh, A., Becker, H. et al. (2023). Exploring the impact of configurational entropy on the design and development of CoNi-based superalloys for sustainable applications. *J Alloys Compd* 965: 171380. https://doi.org/10.1016/j.jallcom.2023.171380.

25 Hu, R., Jin, S., and Sha, G. (2022). Application of atom probe tomography in understanding high entropy alloys: 3D local chemical compositions in atomic scale analysis. *Prog Mater Sci* 123: 100854. https://doi.org/10.1016/j.pmatsci.2021.100854.

26 He, C., Pan, D., Li, X. et al. (2025). Fresh perspectives and insights into the challenges and opportunities in the emerging high-entropy electrocatalysts. *Coord Chem Rev* 531: 216496. https://doi.org/10.1016/j.ccr.2025.216496.

27 Kumar, A. and Gangawane, K.M. (2024). *Green Nanomaterials in Energy Storage: Advancements and Challenges. Materials for Boosting Energy Storage. Volume 1: Advances in Sustainable Energy Technologies, Part 12—Green Nanomaterials in Energy Storage: Advancements and Challenges*, vol. 1, 275–291. American Chemical Society. https://doi.org/10.1021/bk-2024-1477.ch012.

28 Huang, C.-Y., Huang, C.-W., Wu, M.-C. et al. (2021). Atomic-scale investigation of lithiation/delithiation mechanism in high-entropy spinel oxide with superior electrochemical performance. *Chem Eng J* 420: 129838. https://doi.org/10.1016/j.cej.2021.129838.

29 Sarkar, A., Velasco, L., Wang, D. et al. (2018). High entropy oxides for reversible energy storage. *Nat Commun* 9: 3400. https://doi.org/10.1038/s41467-018-05774-5.

30 Liu, X., Xing, Y., Xu, K. et al. (2022). Kinetically accelerated lithium storage in high-entropy (LiMgCoNiCuZn)O enabled by oxygen vacancies. *Small* 18. https://doi.org/10.1002/smll.202200524.

31 Stygar, M., Dąbrowa, J., Moździerz, M. et al. (2020). Formation and properties of high entropy oxides in Co-Cr-Fe-Mg-Mn-Ni-O system: novel (Cr,Fe,Mg,Mn,Ni)3O_4 and (Co,Cr,Fe,Mg,Mn)3O_4 high entropy spinels. *J Eur Ceram Soc* 40: 1644–1650. https://doi.org/10.1016/j.jeurceramsoc.2019.11.030.

32 Yao, Y., Dong, Q., Brozena, A. et al. (1979). High-entropy nanoparticles: synthesis-structure-property relationships and data-driven discovery. *Science* 2022: 376. https://doi.org/10.1126/science.abn3103.

33 Li, T., Yao, Y., Huang, Z. et al. (2021). Denary oxide nanoparticles as highly stable catalysts for methane combustion. *Nat Catal* 4: 62–70. https://doi.org/10.1038/s41929-020-00554-1.

34 Zhang, J., Yan, J., Calder, S. et al. (2019). Long-range antiferromagnetic order in a rocksalt high entropy oxide. *Chem Mater* 31: 3705–3711. https://doi.org/10.1021/acs.chemmater.9b00624.

35 Biesuz, M., Fu, S., Dong, J. et al. (2019). High entropy $Sr((Zr_{0.94}Y_{0.06})_{0.2}Sn_{0.2}Ti_{0.2}Hf_{0.2}Mn_{0.2})O_{3-x}$ perovskite synthesis by reactive spark plasma sintering. *J Asian Ceram Soc* 7: 127–132. https://doi.org/10.1080/21870764.2019.1595931.

36 Rost, C.M., Sachet, E., Borman, T. et al. (2015). Entropy-stabilized oxides. *Nat Commun* 6: 8485. https://doi.org/10.1038/ncomms9485.

37 Lermontov, S.A., Straumal, E.A., Mazilkin, A.A. et al. (2018). An approach for highly transparent titania aerogels preparation. *Mater Lett* 215: 19–22. https://doi.org/10.1016/j.matlet.2017.12.031.

38 Nguyen, T.X., Patra, J., Chang, J.-K., and Ting, J.-M. (2020). High entropy spinel oxide nanoparticles for superior lithiation–delithiation performance. *J Mater Chem A Mater* 8: 18963–18973. https://doi.org/10.1039/D0TA04844E.

39 Balcerzak, M., Kawamura, K., Bobrowski, R. et al. (2019). Mechanochemical synthesis of (Co,Cu,Mg,Ni,Zn)O high-entropy oxide and its physicochemical properties. *J Electron Mater* 48: 7105–7113. https://doi.org/10.1007/s11664-019-07512-z.

40 Tyagi, P. and Goulet, T. (2018). Nanoscale Tantalum layer impacting magnetic properties of tunnel junction-based molecular devices. *MRS Commun* 8: 1024–1028. https://doi.org/10.1557/mrc.2018.132.

41 Liu, D., Peng, X., Liu, J. et al. (2020). Ultrafast synthesis of entropy-stabilized oxide at room temperature. *J Eur Ceram Soc* 40: 2504–2508. https://doi.org/10.1016/j.jeurceramsoc.2020.01.018.

42 Wang, K., Ma, B., Li, T. et al. (2020). Fabrication of high-entropy perovskite oxide by reactive flash sintering. *Ceram Int* 46: 18358–18361. https://doi.org/10.1016/j.ceramint.2020.04.060.

43 Kumar, A., Gangawane, K.M., and Jena, H.M. (2025). Variations in magnetic behaviour of ferrimagnetic nanoparticles with temperature. *Ind Chem Eng* 1–10. https://doi.org/10.1080/00194506.2025.2471412.

44 Kumar, A., Jena, H.M., and Gangawane, K.M. (2025). Iron oxide nanoparticles as efficient adsorbents for malachite green dye removal: isotherms, kinetics, and thermodynamics analysis. *Environ Sci Pollut Res*. https://doi.org/10.1007/s11356-025-36291-w.

45 Kumar, A., Gangawane, K.M., and Ramanjaneyulu, B. (2022). Ferrofluids for waste-water treatment. In: *International Conference on Chemical, Bio and Environmental Engineering*, 723–744. Springer. https://doi.org/10.1007/978-3-030-96554-9_48.

46 Okejiri, F., Zhang, Z., Liu, J. et al. (2020). Room-temperature synthesis of high-entropy perovskite oxide nanoparticle catalysts through ultrasonication-based method. *ChemSusChem* 13: 111–115. https://doi.org/10.1002/cssc.201902705.

47 Sarkar, A., Djenadic, R., Usharani, N.J. et al. (2017). Nanocrystalline multicomponent entropy stabilised transition metal oxides. *J Eur Ceram Soc* 37: 747–754. https://doi.org/10.1016/j.jeurceramsoc.2016.09.018.

48 Shaw, S.K., Gangwar, A., Sharma, A. et al. (2021). Structural and magnetic properties of nanocrystalline equi-atomic spinel high-entropy oxide (AlCoFeMnNi)3O_4 synthesised by microwave assisted co-precipitation technique. *J Alloys Compd* 878: 160269. https://doi.org/10.1016/j.jallcom.2021.160269.

49 Gautam, A. and Ahmad, M.I. (2021). Low-temperature synthesis of five component single phase high entropy oxide. *Ceram Int* 47: 22225–22228. https://doi.org/10.1016/j.ceramint.2021.04.128.

50 Kumar, A. and Gangawane, K.M. (2022). Effect of precipitating agents on the magnetic and structural properties of the synthesized ferrimagnetic nanoparticles by co-precipitation method. *Powder Technol* 401: 117298. https://doi.org/10.1016/j.powtec.2022.117298.

51 Kumar, A., Panda, D., and Gangawane, K.M.K.M. (2024). Computational modeling on the design of the morphology of aerogels. *Hybrid Aerog Energy Environ Appl* 1: 269. https://doi.org/10.1515/9783111252001-013.

52 Khan, J.B., Panda, P.K., Yang, P.-C. et al. (2024). Microwave synthesis of high-entropy alloy catalysts on graphene oxide sheets for oxygen reduction and evolution reactions. *Int J Hydrog Energy* 53: 999–1008. https://doi.org/10.1016/j.ijhydene.2023.12.024.

53 Panda, P.K., Huang, H.-Y., Dash, P. et al. (2025). Liquid-phase microwave synthesis of platinum-based high-entropy alloy catalysts on carbon supports for electrochemical hydrogen adsorption/desorption and oxygen evolution/reduction reactions. *Int J Hydrog Energy* 111: 536–545. https://doi.org/10.1016/j.ijhydene.2025.02.312.

54 Nguyen, T.X., Su, Y.-H., Hattrick-Simpers, J. et al. (2020). Exploring the first high-entropy thin film libraries: composition spread-controlled crystalline structure. *ACS Comb Sci* 22: 858–866. https://doi.org/10.1021/acscombsci.0c00159.

55 Kumar, A. (2025). Engineered nanomaterials and associated threats in the environment risk assessment. *Strategies* 515–537. https://doi.org/10.1007/978-981-97-9922-0_24.

56 Wu, H., Lu, Q., Li, Y. et al. (2022). Rapid Joule-heating synthesis for manufacturing high-entropy oxides as efficient electrocatalysts. *Nano Lett* 22: 6492–6500. https://doi.org/10.1021/acs.nanolett.2c01147.

57 Gu, S., Huang, H.-Y., Lin, Y.-C. et al. (2025). High-entropy alloys with low platinum content: novel catalysts for oxygen reduction and evolution reactions. *Emerg Mater*. https://doi.org/10.1007/s42247-025-01011-0.

58 Guo, M., Liu, Y., Zhang, F. et al. (2022). Inactive Al_{3+}-doped $La(CoCrFeMnNiAlx)_{1/(5+x)}O_3$ high-entropy perovskite oxides as high performance supercapacitor electrodes. *J Adv Ceram* 11: 742–753. https://doi.org/10.1007/s40145-022-0568-4.

59 Zhang, Z.-H., Liu, Z.-F., Lu, J.-F. et al. (2014). The sintering mechanism in spark plasma sintering—proof of the occurrence of spark discharge. *Scr Mater* 81: 56–59. https://doi.org/10.1016/j.scriptamat.2014.03.011.

60 Bajpai, I., Han, Y.-H., Yun, J. et al. (2016). Preliminary investigation of hydroxyapatite microstructures prepared by flash sintering. *Adv Appl Ceram* 115: 276–281. https://doi.org/10.1080/17436753.2015.1136777.

61 Dedoncker, R., Djemia, P., Radnóczi, G. et al. (2018). Reactive sputter deposition of CoCrCuFeNi in nitrogen/argon mixtures. *J Alloys Compd* 769: 881–888. https://doi.org/10.1016/j.jallcom.2018.08.044.

62 Johansson, K., Riekehr, L., Fritze, S., and Lewin, E. (2018). Multicomponent Hf-Nb-Ti-V-Zr nitride coatings by reactive magnetron sputter deposition. *Surf Coat Technol* 349: 529–539. https://doi.org/10.1016/j.surfcoat.2018.06.030.

63 Sedegov, A., Vorotilo, S., Tsybulin, V. et al. (2019). Synthesis and study of high-entropy ceramics based on the carbides of refractory metals. *IOP Conf Ser Mater Sci Eng* 558: 012043. https://doi.org/10.1088/1757-899X/558/1/012043.

64 Zhao, Z., Chen, H., Xiang, H. et al. (2019). $(La_{0.2}Ce_{0.2}Nd_{0.2}Sm_{0.2}Eu_{0.2})PO_4$: a high-entropy rare-earth phosphate monazite ceramic with low thermal conductivity and good compatibility with Al_2O_3. *J Mater Sci Technol* 35: 2892–2896. https://doi.org/10.1016/j.jmst.2019.08.012.

65 Li, H., Ling, L., Li, S. et al. (2023). High entropy materials—emerging nanomaterials for electrocatalysis. *Energy Adv* 2: 1800–1817. https://doi.org/10.1039/D3YA00305A.

66 Kumar, A., Panda, D., and Gangawane, K.M. (2024). Chapter 3 Microfabrication: techniques and technology. In: *Microfabrication and Nanofabrication*, vol. 11, 47–76. De Gruyter. https://doi.org/10.1515/9783111335537-003.

67 Fukagawa, T., Saito, Y., and Matsuyama, A. (2022). Effect of varying Ni content on hydrogen absorption–desorption and electrochemical properties of Zr-Ti-Ni-Cr-Mn high-entropy alloys. *J Alloys Compd* 896: 163118. https://doi.org/10.1016/j.jallcom.2021.163118.

68 Gao, S., He, H., Zhai, F. et al. (2025). Hollow graphene microsphere in situ deposited with rock-salt-type high-entropy oxide nanoparticles for enhanced rate performance as anodes in lithium-ion batteries. *ACS Appl Energy Mater*. https://doi.org/10.1021/acsaem.4c03317.

69 Wang, S.-Y., Chen, T.-Y., Kuo, C.-H. et al. (2021). Operando synchrotron transmission X-ray microscopy study on (Mg,Co,Ni,Cu,Zn)O high-entropy oxide anodes for lithium-ion batteries. *Mater Chem Phys* 274: 125105. https://doi.org/10.1016/j.matchemphys.2021.125105.

70 Ruan, S., Luo, D., Li, M. et al. (2021). Synthesis and functionalization of 2D nanomaterials for application in lithium-based energy storage systems. *Energy Storage Mater* 38: 200–230. https://doi.org/10.1016/j.ensm.2021.03.001.

71 Patra, J., Nguyen, T.X., Tsai, C. et al. (2022). Effects of elemental modulation on phase purity and electrochemical properties of co-free high-entropy spinel oxide anodes for lithium-ion batteries. *Adv Funct Mater* 32. https://doi.org/10.1002/adfm.202110992.

72 Duan C, Tian K, Li X, Wang D, Sun H, Zheng R, et al. New spinel high-entropy oxides (FeCoNiCrMnXLi)3O4 (X = Cu, Mg, Zn) as the anode material for lithium-ion batteries. *Ceram Int* 2021;47:32025–32. https://doi.org/10.1016/j.ceramint.2021.08.091.

73 Ding, F., Zhao, C., Xiao, D. et al. (2022). Using high-entropy configuration strategy to design Na-ion layered oxide cathodes with superior electrochemical performance and thermal stability. *J Am Chem Soc* 144: 8286–8295. https://doi.org/10.1021/jacs.2c02353.

74 Xiao, B., Wu, G., Wang, T. et al. (2022). High-entropy oxides as advanced anode materials for long-life lithium-ion Batteries. *Nano Energy* 95: 106962. https://doi.org/10.1016/j.nanoen.2022.106962.

75 Bérardan, D., Franger, S., Meena, A.K., and Dragoe, N. (2016). Room temperature lithium superionic conductivity in high entropy oxides. *J Mater Chem A Mater* 4: 9536–9541. https://doi.org/10.1039/C6TA03249D.

76 Kumar, A. and Gangawane, K.M. (2022). Synthesis and effect on the surface morphology & magnetic properties of ferrimagnetic nanoparticles by different wet chemical synthesis methods. *Powder Technol* 410: 117867. https://doi.org/10.1016/j.powtec.2022.117867.

77 Zhao, J., Yang, X., Huang, Y. et al. (2021). Entropy stabilization effect and oxygen vacancies enabling spinel oxide highly reversible lithium-ion storage. *ACS Appl Mater Interfaces* 13: 58674–58681. https://doi.org/10.1021/acsami.1c18362.

78 Choi, S., Feng, W., and Xia, Y. (2024). High entropy and Co-free high nickel based layered $LiNi_{0.9}Mn_{0.1}O_2$ cathode for Li-ion batteries. *ACS Appl Energy Mater* 7: 3339–3346. https://doi.org/10.1021/acsaem.4c00095.

79 Qiu, N., Chen, H., Yang, Z. et al. (2019). A high entropy oxide ($Mg_{0.2}Co_{0.2}Ni_{0.2}Cu_{0.2}Zn_{0.2}O$) with superior lithium storage performance. *J Alloys Compd* 777: 767–774. https://doi.org/10.1016/j.jallcom.2018.11.049.

80 Ghigna, P., Airoldi, L., Fracchia, M. et al. (2020). Lithiation mechanism in high-entropy oxides as anode materials for li-ion batteries: an operando XAS study. *ACS Appl Mater Interfaces* 12: 50344–50354. https://doi.org/10.1021/acsami.0c13161.

81 Kumar, A. and Gangawane, K.M. (2023). Molecular dynamics for the membrane process. In: *Advanced Computational Approaches for Water Treatment*, 1e, 13–38. New York: CRC Press. https://doi.org/10.1201/9781003325147-2.

82 Kumar, A. and Gangawane, K.M. (2023). *Advanced Computational Approaches for Water Treatment*, 1e, 1–12. New York: CRC Press. https://doi.org/10.1201/9781003325147-1.

83 Wang, Q., Sarkar, A., Wang, D. et al. (2019). Multi-anionic and -cationic compounds: new high entropy materials for advanced Li-ion batteries. *Energy Environ Sci* 12: 2433–2442. https://doi.org/10.1039/c9ee00368a.

84 Lun, Z., Ouyang, B., Kwon, D.H. et al. (2021). Cation-disordered rocksalt-type high-entropy cathodes for Li-ion batteries. *Nat Mater* 20: 214–221. https://doi.org/10.1038/s41563-020-00816-0.

85 Zhang, R., Wang, C., Zou, P. et al. (2022). Compositionally complex doping for zero-strain zero-cobalt layered cathodes. *Nature* 610: 67–73. https://doi.org/10.1038/s41586-022-05115-z.

86 Ran, W., Cheng, G., Luo, J. et al. (2025). High-entropy engineering for multivalency-induced stability in snsb-based anodes. *ACS Appl Mater Interfaces*. https://doi.org/10.1021/acsami.4c19698.

87 Zhao, C., Ding, F., Lu, Y. et al. (2020). High-entropy layered oxide cathodes for sodium-ion batteries. *Angew Chem Int Ed* 59: 264–269. https://doi.org/10.1002/anie.201912171.

88 Walczak, K., Plewa, A., Ghica, C. et al. (2022). $NaMn_{0.2}Fe_{0.2}Co_{0.2}Ni_{0.2}Ti_{0.2}O_2$ high-entropy layered oxide – experimental and theoretical evidence of high electrochemical performance in sodium batteries. *Energy Storage Mater* 47: 500–514. https://doi.org/10.1016/j.ensm.2022.02.038.

89 Yang, L., Chen, C., Xiong, S. et al. (2021). Multiprincipal Component P2-$Na_{0.6}(Ti_{0.2}Mn_{0.2}Co_{0.2}Ni_{0.2}Ru_{0.2})O_2$ as a high-rate cathode for sodium-ion batteries. *JACS Au* 1: 98–107. https://doi.org/10.1021/jacsau.0c00002.

90 Abdul Razzaq, A., Yao, Y., Shah, R. et al. (2019). High-performance lithium sulfur batteries enabled by a synergy between sulfur and carbon nanotubes. *Energy Storage Mater* 16: 194–202. https://doi.org/10.1016/j.ensm.2018.05.006.

91 Zhou, L., Danilov, D.L., Eichel, R., and Notten, P.H.L. (2021). Host materials anchoring polysulfides in Li–S batteries reviewed. *Adv Energy Mater* 11. https://doi.org/10.1002/aenm.202001304.

92 Zheng, Y., Yi, Y., Fan, M. et al. (2019). A high-entropy metal oxide as chemical anchor of polysulfide for lithium-sulfur batteries. *Energy Storage Mater* 23: 678–683. https://doi.org/10.1016/j.ensm.2019.02.030.

93 Tian, L., Zhang, Z., Liu, S. et al. (2022). High-entropy spinel oxide nanofibers as catalytic sulfur hosts promise the high gravimetric and volumetric capacities for lithium–sulfur batteries. *Energy Environ Mater* 5: 645–654. https://doi.org/10.1002/eem2.12215.

94 Lal, M.S. and Sundara, R. (2019). High entropy oxides—a cost-effective catalyst for the growth of high yield carbon nanotubes and their energy applications. *ACS Appl Mater Interfaces* 11: 30846–30857. https://doi.org/10.1021/acsami.9b08794.

95 Yin, Y., Zhang, W.-B., Zhang, X.-L. et al. (2023). Low-dimensional high entropy oxide $(FeCoCrMnNi)_3O_4$ for supercapacitor applications. *Dalton Trans* 52: 9005–9016. https://doi.org/10.1039/D3DT00909B.

96 Liang, B., Ai, Y., Wang, Y. et al. (2020). Spinel-type $(FeCoCrMnZn)3O_4$ high-entropy oxide: facile preparation and supercapacitor performance. *Materials* 13: 5798. https://doi.org/10.3390/ma13245798.

97 Nemani, S.K., Torkamanzadeh, M., Wyatt, B.C. et al. (2023). Functional two-dimensional high-entropy materials. *Commun Mater* 4: 16. https://doi.org/10.1038/s43246-023-00341-y.

98 Yin, X., Wan, T., Deng, X. et al. (2021). Corrigendum to “De novo design of polymers embedded with platinum acetylides towards n-type organic thermoelectrics”. *Chem Eng J* 405: 126692. *Chemical Engineering Journal* 2021;421:127825. 10.1016/j.cej.2020.127825.

99 Xie, C., Dan, M., Hu, G. et al. (2022). Piezo-phototronic spin laser based on wurtzite quantum wells. *Nano Energy* 96: 107100. https://doi.org/10.1016/j.nanoen.2022.107100.

100 Maawad, E. and Stark, A. (2021). In situ synchrotron X-ray diffraction study of continuous cooling transformations of TIMETAL 54M. *J Alloys Compd* 881: 160602. https://doi.org/10.1016/j.jallcom.2021.160602.

101 Fan, Y., Chen, Y., Abernathy, H. et al. (2023). Enabling durable hydrogen production and preventing the catastrophic delamination in the solid oxide electrolysis cells by infiltrating $SrFe_2O_{4-\delta}$ solutions into LSM/YSZ-based air electrode. *J Power Sources* 580: 233389. https://doi.org/10.1016/j.jpowsour.2023.233389.

102 Lim, J.-C., Jin, C., Choi, M.S. et al. (2021). Synthesis, morphology, characterisation, and ethanol gas sensing of hierarchical flower-like Co-doped WO_3 nanoplates by solvothermal route. *Ceram Int* 47: 20956–20964. https://doi.org/10.1016/j.ceramint.2021.04.095.

103 Shan, B., Wang, Y., Ji, X., and Huang, Y. (2024). Enhancing low-frequency microwave absorption through structural polarization modulation of MXenes. *Nano Lett* 16: 212. https://doi.org/10.1007/s40820-024-01437-x.

104 Wan, Q., Ding, H., Tang, Y. et al. (2022). The phase regulation of $Bi_4O_5I_2$ nanosheets through the controlled hydrolysis in polyhydric alcohol for the enhanced photocatalytic performances of BPA. *Mater Lett* 326: 132913. https://doi.org/10.1016/j.matlet.2022.132913.

105 Zhang, H. and Yan, J. (2022). Co-benefits of renewable energy development: a brighter sky brings greater renewable power. *Joule* 6: 1142–1144. https://doi.org/10.1016/j.joule.2022.05.017.

106 Wang, C., Ping, W., Bai, Q. et al. (1979). A general method to synthesize and sinter bulk ceramics in seconds. *Science* 2020 (368): 521–526. https://doi.org/10.1126/science.aaz7681.

107 Bib Khan, J., Kumar Panda, P., Dash, P., and Hsieh, C. (2025). Microwave-assisted synthesis of platinum-free high-entropy alloy catalysts on reduced graphene oxide sheets for enhanced oxygen reduction and evolution reactions. *Chem Eur J* 31. https://doi.org/10.1002/chem.202403863.

4

Recent Advancements for High-Entropy Materials for the Dielectric Capacitor

Sushree Sangita Swain[1], Subash Chandra Sahu[2], Arpan Kumar Nayak[3], and Rakesh K. Sahoo[1]

[1] *Department of Physics, Government Women's College, Motijharan, Sambalpur, Odisha 768001, India*
[2] *Department of Chemistry, Government Women's College, Motijharan, Sambalpur, Odisha 768001, India*
[3] *Regional Institute of Education, National Council of Educational Research and Training (NCERT), Mysore, Karnataka 570006, India*

4.1 Introduction

Upon exposure to an external electric field, the molecular or atomic dipoles within a dielectric medium partially align along the field direction, resulting in electric polarization of the material [1, 2]. Substances exhibiting this field-induced dipolar reorientation are categorized as dielectric materials. Since the nomenclature was originally introduced by Michael Faraday, dielectrics have emerged as critical functional materials in contemporary electronic and electromechanical systems, with technological implementations spanning high-permittivity capacitive energy storage devices to field-responsive piezoelectric transducers [3–6].

In recent years, a huge research efforts have been put forward in enhancing the functional performance of dielectric materials. For example, introducing localized polar nano-regions has been shown to improve dielectric energy storage capabilities [7–9]. In piezoelectric systems, the engineering of morphotropic phase boundaries and polymorphic phase boundaries, which enable a reliable and enhanced rate of polarization configurations, has proven effective in increasing piezoelectric coefficients. Nevertheless, the performance enhancements achieved through these strategies remain limited, and the development of innovative design approaches is still highly sought-after [10–12].

A fundamental aspect of optimizing material performance lies in the manipulation of the inherent characteristics of the material itself, which is essential for advancing next-generation applications. Recently, a novel approach based on high-entropy material (HEM) design has garnered arousing attention for its potential to significantly enhance the functional properties of materials [13, 14]. Initially introduced in the context of metallic alloys [15, 16], the high-entropy concept has since been extended to a wide range of material systems, including oxides [17, 18], nitrides [19, 20], chalcogenides [21, 22], and carbides [23, 24]. Unlike conventional

High-Entropy Materials for Energy Storage Devices, First Edition.
Edited by Chien-Te Hsieh, Pradeep Kumar Panda, and Arpan Kumar Nayak.

doping strategies, which are often constrained by solubility limits, high-entropy design permits extensive atomic substitution. This approach enhances the configurational entropy of the system, thereby promoting the stabilization of single-phase crystal structures and enabling the realization of superior dielectric performance. Mathematically expressed as

$$S_{config} = -R\left[\left(\sum_{i=1}^{n} x_i ln x_i\right)_{cation-site} + \left(\sum_{j=1}^{m} x_j ln x_j\right)_{anion-site}\right] \tag{4.1}$$

Here, R denotes the universal gas constant, and $x_{i(j)}$ represents the mole fraction of element i occupying either the cation or the anion sites [25, 26]. According to the configurational entropy S_{config} formula, the entropy increases with the addition of more distinct elements into the system and reaches a maximum when all elements are present in equal mole fractions. Based on this criterion, materials with $S_{config} > 1.5R$ are classified as high-entropy systems, those with $S_{config} < 1.0R$ as low-entropy systems and materials in between as medium-entropy systems [26, 27]. This distinctive structural feature, along with tunable chemical compositions, gives rise to four core effects in high-entropy systems, as outlined by Yeh et al. [28] and illustrated in Figure 4.1.

The unique structural characteristics and chemically tailorable compositions of high-entropy systems give rise to four core effects, as summarized by Yeh et al. [28] and illustrated in Figure 4.1. First, the high-entropy effect contributes to the

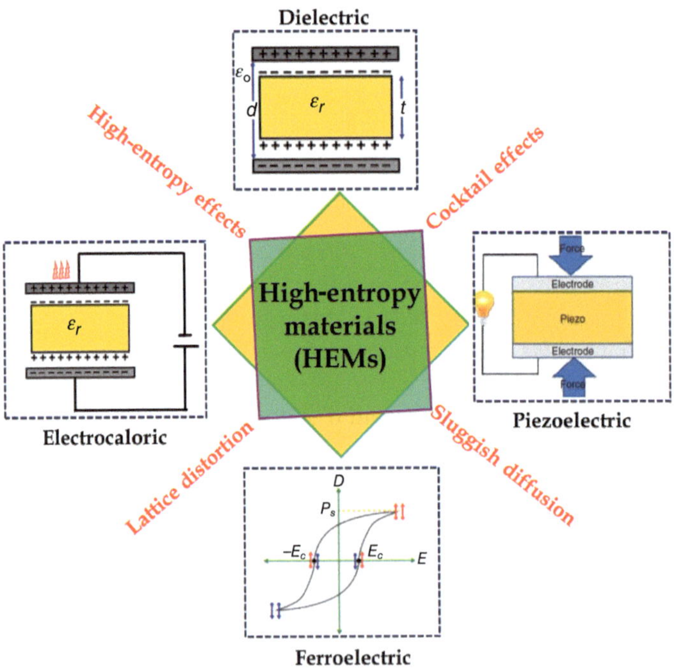

Figure 4.1 Schematic of four core effects in high-entropy systems. *Source:* Adapted from Ref. [28].

stabilization of novel material phases. As multiple elements are introduced, the configurational entropy (ΔS_{mix}) increases, creating a competitive interplay with the enthalpy of mixing (ΔH_{mix}). According to the Gibbs–Helmholtz relation,

$$\Delta G_{mix} = \Delta H_{mix} - T\Delta S_{mix} \tag{4.2}$$

when the entropy term $T\Delta S_{mix}$ exceeds the enthalpic contribution ΔH_{mix}, the Gibbs free energy becomes negative ($\Delta G_{mix} < 0$), indicating thermodynamic favorability and structural stability. This high-entropy effect expands the compositional space of solid solutions, enabling the formation of previously inaccessible single-phase compounds. Second, the introduction of chemically diverse elements induces lattice distortions due to differences in atomic radii, masses, and electronegativities. Such distortions are characteristic of HEMs and can significantly alter their physical properties. Third, sluggish diffusion effects emerge as a consequence of the complex energy landscape and lattice distortions. These effects hinder atomic mobility, suppress grain growth, and promote the stabilization of ultrafine grains or amorphous phases. Fourth, the cocktail effect, arising from the synergistic combination of multiple elements, often leads to emergent and unexpected material properties not attainable in conventional systems. These entropy-driven effects endow HEMs with remarkable property tunability and multifunctionality, thereby enabling a broad range of applications. The above research areas in HEMs are schematically illustrated in Figure 4.1.

Motivated by the breakthroughs observed in high-entropy metallic systems, this strategy has recently been extended to the field of dielectrics, yielding promising results [29–32]. In this chapter, we summarize recent advances in the application of high-entropy design principles to dielectric materials. The focus includes physics, synthesis and developments in HEMs dielectric energy storage, which have been illustrated. Finally, we highlight the current challenges in this field and propose future directions for research to further harness the potential of high-entropy dielectrics.

4.2 Processing Methods and Structural Characteristics of High-Entropy Materials (HEMs)

Figure 4.2 presents various processing methods for HEMs based on their initial physical forms, including powders, fibers, films, and bulk materials. Among these, mechanical alloying of powders is particularly advantageous in zero-dimensional systems due to its superior ability to ensure homogeneous mixing. Fiber fabrication via melt spinning is another viable approach, enabling the formation of continuous HEM fibers. HEM films are primarily utilized in surface-coating applications to enhance surface-dependent properties. Magnetron sputtering, a high-rate vacuum deposition technique, is widely adopted for coating metals and ceramics due to its fast deposition rate, low operational pressure, and excellent coating adhesion. However, the technique is limited by the relatively thin films it produces. Conversely, laser cladding allows the formation of significantly thicker coatings

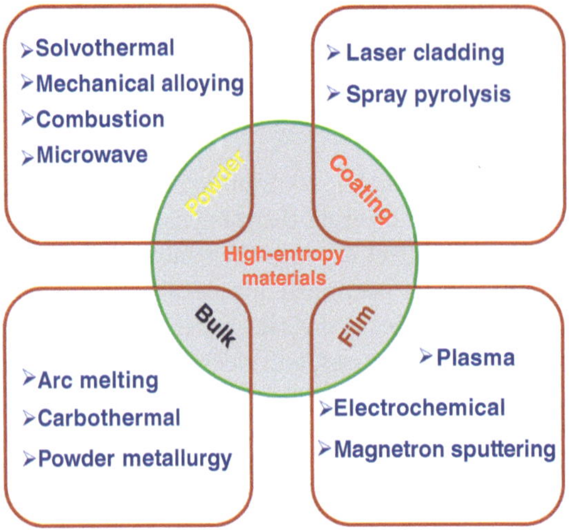

Figure 4.2 Overview of different synthesis methods of HEMs based on their physics form.

(ranging from 50 μm to 2 mm), making it well-suited for robust protective layers. For bulk synthesis, arc melting is frequently employed due to its capacity to produce highly homogeneous microstructures, which contribute to enhanced mechanical performance in structural and functional HEMs.

4.3 Synthesis Techniques for High-Entropy Materials (HEMs)

HEMs necessitate diverse physical and chemical synthesis strategies to effectively modulate thermodynamic parameters, thereby enabling the formation of single-phase solid solutions with uniformly distributed constituent elements. The selection of a synthesis method and the optimization of its specific parameters are crucial in determining the particle size, phase composition, and elemental homogeneity of high-entropy nanomaterials. Figure 4.2 presents an overview of various synthesis techniques employed for fabricating HEMs. High-entropy alloys (HEAs), in particular, have been synthesized using a broad range of methods, including solid-state processing techniques such as mechanical alloying, arc melting, induction melting, casting, and magnetron sputtering. These techniques can be broadly classified into chemical methods such as chemical vapor deposition (CVD) and liquid-phase deposition and physical methods, including physical vapor deposition (PVD), vacuum sputtering, vacuum evaporation, and ion plating [33, 34].

Achieving atomic-level dispersion of multiple elements throughout the crystal lattice is vital for maximizing the configurational entropy of HEMs, a goal often constrained by traditional solid-state techniques. To address this limitation, atom-up synthesis routes—such as sol–gel processing, solvothermal synthesis, molecular precursor strategies, and carbothermal reduction—have been developed. These methods construct the high-entropy lattice directly from atomic or molecular precursors, rather than altering existing crystalline structures.

4.3.1 Arc Melting

Arc melting is an electrothermal technique that generates an electric arc between electrodes or between an electrode and the metal feedstock, typically utilizing direct current for enhanced stability. This technique enables rapid melting and solidification of metal mixtures, producing HEAs with refined microstructures. For instance, Wu et al. synthesized the refractory $Hf_{0.25}Nb_{0.25}Ti_{0.25}Zr_{0.25}$ alloy with a body-centered cubic (BCC) structure via arc melting, followed by cooling in a water-cooled copper mold under argon at 1,300 °C for six hours [35]. Similarly, Liu et al. developed $Fe_{20}Co_{20}Ni_{20}Cr_{20}Mn_{20}$ HEAs with face-centered cubic (FCC) structures by repeated arc melting, demonstrating grain boundary strengthening and Hall–Petch behavior above 850 °C [36].

4.3.2 Powder Metallurgy

Powder metallurgy (PM) involves blending elemental powders, followed by compaction and sintering. PM provides control over microstructure and composition but poses challenges such as high-temperature requirements and potential contamination. Qiu et al. reported a compressive strength of 1,341 MPa and a hardness of 583 HV for AlCrFeNiCoCu HEA [37]. Fu et al. explored Co's influence in $Al_{0.6}NiFeCrCo$ alloys, reporting a compressive strength of 2,150 MPa and hardness of 570 HV [38]. Lui et al. demonstrated PM-derived FeCoCrNi HEAs with 712.5 MPa tensile strength and 56% elongation [39].

4.3.3 Solvothermal Method

Solvothermal synthesis employs high-temperature and high-pressure solvent conditions to produce materials with uniform composition and finely controlled particle size. Jiang et al. employed a one-step solvothermal approach to synthesize $FeCoNiMgCr(OH)(OCH_3)$ for electrochemical water splitting, achieving high catalytic activity due to surface restructuring. Broge et al. [40] developed homogeneous HEA nanoparticles comprising Pt, Ir, Pd, Rh, Ru, Cu, Ni, and Co using a similar approach.

4.3.4 Mechanical Alloying

Mechanical alloying involves the repeated fracturing, cold welding, and rewelding of powder particles within a high-energy ball mill. This leads to atomic-scale mixing and the formation of solid-solution HEAs. Rekha et al. synthesized NiFeCrCoCu-graphene composites using ball milling and ultrasound-assisted exfoliation [40], while Prasad et al. created FeNiCoCrAlMn alloys with FCC and BCC phases, along with minor Mn_3Co_7 inclusions [41].

4.3.5 Carbothermal Shock (CTS) Method

Carbothermal shock (CTS) method is a rapid synthesis method wherein a metal precursor on a carbon substrate is briefly exposed to high temperatures via electric current, producing ultra-small, well-dispersed nanoparticles within milliseconds. Xie et al. synthesized CoMoFeNiCu HEA nanoparticles using CTS, successfully manipulating Co/Mo ratios to overcome mixed-phase restrictions [42]. Yao et al. used CTS to fabricate HEA nanoparticles on activated carbon fibers [43].

4.3.6 Spray Pyrolysis

Spray pyrolysis forms thin films by atomizing a precursor solution and thermally decomposing the droplets onto a substrate. Phakatkar et al. utilized flame spray pyrolysis (FSP) to create oxide-based HEA nanoparticles of Cu, Mn, Fe, Ni, and Zn [44]. Kamecki et al. investigated structural and electrical properties of Mn-Co-Fe-Ni-Cr oxide films fabricated on sapphire and silica substrates [45].

4.3.7 Laser Cladding

Laser cladding leverages focused laser energy to melt and fuse precursor materials, forming HEA coatings with enhanced surface properties. Zhang et al. produced TiZrNbWMo HEA coatings with superior hardness via laser cladding [46]. Ye et al. created AlXFeCoNiCuCr HEAs, targeting enhanced high-temperature performance [47].

4.3.8 Magnetron Sputtering

This PVD method uses ionized gas to sputter atoms from multiple targets onto substrates, forming high-purity, and compositionally controlled HEA thin films. Kim et al. demonstrated TiZrHfNiCuCo-based coatings with tunable hardness and elasticity via reactive DC magnetron sputtering [48]. Another study fabricated $Fe_{33}Co_{30}Ni_{16}Al_7Mn_9W_5$ HEFs at temperatures in the range of 20 °C–600 °C.

4.3.9 Combustion Synthesis

Combustion synthesis utilizes exothermic reactions under high gravity to rapidly form HEAs. Li et al. examined phase evolution in $Al_xCoCrFeNi$ ($x = 0–1$) HEAs, showing transitions from FCC to BCC with increasing Al content [49]. Zhang et al. synthesized a transparent high-entropy ceramic exhibiting a pyrochlore/fluorite structure and conducted a comprehensive analysis of its structural and optical properties [50].

4.3.10 Microwave-Assisted Synthesis

Microwave irradiation enables rapid, uniform heating of precursors, enhancing energy efficiency and structural control. Veronesi et al. synthesized FeCoNiCuAl, FeCrNiTiAl, and $FeCoCrNiAl_{2.5}$ HEAs via microwave-assisted combustion at 2,450–5,800 MHz [51]. Kheradmandfard et al. produced MgCuNiCoZnO nanoparticles with excellent LIB performance [52].

4.3.11 Plasma Sintering

Plasma sintering, particularly spark plasma sintering (SPS), densifies powders rapidly using plasma-generated heat. Toroghinejad et al. synthesized FeCrCoNiCu HEAs, evaluating structural and mechanical properties [53]. Yang et al. enhanced oxidation and corrosion resistance in $FeCoCrNiMoSi_x$ alloys via SPS, achieving superior high-temperature durability [54].

4.3.12 Electrochemical Deposition

Electrodeposition offers precise control over HEA film composition and morphology through electrolytic reduction. Aliyu et al. deposited equiatomic AlCrFeCoNiCu HEA-GO composites onto mild steel, enhancing corrosion resistance in saline environments [55]. Keméηy et al. used non-aqueous electrochemical methods to fabricate AlCoCrFeNi thin films, confirmed by SEM–EDX analysis [56].

4.4 Defining High-Entropy Materials: Composition and Entropy Perspectives

The concept of HEMs is rooted in two primary perspectives: compositional and entropy-based. In 2004, Yeh et al. and Cantor et al. independently introduced the ideas of HEAs and equiatomic multicomponent alloys. Compositionally, HEMs typically consist of five or more principal elements, each present in concentrations ranging from 5% to 35% atomic fraction. From an entropy viewpoint, HEMs are defined based on their configurational entropy, which plays a critical role in stabilizing simple solid-solution phases. With the maturation of HEM theory, entropy-induced atomic disorder has emerged as a key criterion for HEM classification. Interestingly, even ternary and quaternary equimolar or near-equimolar alloys have been found to form stable solid solutions with desirable mechanical and functional properties, prompting their inclusion in the broader HEA category. Unlike traditional alloying approaches that focus on the corners or edges of phase diagrams, the high-entropy strategy encourages exploration of central compositional regions, thereby greatly expanding the design space for novel materials.

4.5 Recently Reported High-Entropy Material

Table 4.1 presents recently reported HEAs from 2021 to 2025, highlighting their synthesis methods and key features [57]. These materials illustrate the growing versatility of HEAs for applications in catalysis, energy storage, and structural design, using varied synthesis techniques and often focusing on phase stability, amorphization, or cost-effective production. Further, the application of the as-synthesized high-entropy sulfide (HES) is highlighted in this table.

Table 4.1 Summary of recent high-entropy oxide (HEO) compositions, synthesis techniques, and their associated properties and applications.

Year	HEO composition	Synthesis method	Properties/ Applications	References
2025	$(La_{0.2}Ba_{0.2}Sr_{0.2}Ca_{0.2}Bi_{0.2})(Ti_{1-x}Sn_x)O_3$	Solid-state reaction	Dielectric behavior	[58]
2025	Pt,Ni,Cu,Al,Sn	Microwave-assisted synthesis	Electrocatalytic performance	[59]
2025	$Pt_{42.9}Al_{6.1}Ni_{5.6}Cu_{22.3}Sn_{23.1}$	Microwave-assisted synthesis	Electrocatalytic activity	[60]
2025	[Pt,Al,Mn,Co,Ni,Cu,Fe]/GO composite	Microwave-assisted synthesis	Electrocatalytic properties	[61]
2024	Graphene quantum dots (C,N,B,P,S,O)	Microwave synthesis	Electrocatalysis	[62]
2024	$Cr_{18.2}Zr_{13.5}Ti_{5.3}Co_{17.8}Ni_{8.4}Cu_{8.5}Fe_{28.3}$ and $Al_{15.3}Mn_{18.2}Ti_{3.7}Co_{23.2}Ni_{5.9}Cu_{5.9}Fe_{27.8}$	Microwave synthesis	Electrocatalysis	[61]
2023	$[Y,Dy,Ce,Nd,La]_2Sn_2O_7$	Co-precipitation	Structural integrity and energy storage	[63]
2023	$[Mn_{0.2}Fe_{0.2}Co_{0.2}Ni_{0.2}Zr_{0.2}]_3O_4$	Microwave synthesis	Fenton-like degradation of tetracycline	[64]
2023	$[1-x][Na_{0.2}Bi_{0.2}Ba_{0.2}Sr_{0.2}Ca_{0.2}]TiO_{3-x}NaNbO_3$ ($x = 0$–0.2)	Solid-state reaction	Structural and dielectric performance	[65]
2023	$(Cr_{0.2}Mn_{0.2}Fe_{0.2}Co_{0.2}Ni_{0.2})_3O_4/C$; $(Mg_{0.2}Co_{0.2}Ni_{0.2}Cu_{0.2}Zn_{0.2})O/C$; $(Cr_{0.2}Mn_{0.2}Fe_{0.2}Co_{0.2}Ni_{0.2})_3O_4$	Sol–gel and electrospinning	Capacitive deionization and structural analysis	[66]
2022	$[La_{0.5}Li_{0.5}]_x[[Bi_{0.5}Na_{0.5}]_{0.25}Ba_{0.25}Sr_{0.25}Ca_{0.25}]_{1-x}TiO_3$ ($x = 0.08$–0.32)	Solid-state method	Dielectric characteristics	[67]
2022	(AlCrFeCoNi)O-carbon composite	Sol–gel auto-combustion	Structural and electrochemical features	[68]
2021	$(AlTiVZrHf)O_x$	Spin coating	Capacitance-voltage response	[69]
2021	$Ba(Ti,Hf,Zr,Fe,Sn)O_3$	Spin coating + hydrothermal treatment	Structural and piezoelectric properties	[70]

4.6 Potential of High-Entropy Materials (HEMs) in Dielectric Energy Storage Devices

Despite the current challenges in processing, integration, and reliability, HEMs exhibit immense potential for next-generation dielectric energy storage technologies. Their unique multi-element architectures and entropy-stabilized phases open new pathways for performance enhancement that are not achievable through conventional materials.

4.6.1 Tunable Dielectric Properties

The vast compositional flexibility of HEMs enables precise tuning of dielectric constants, breakdown strength, and loss tangents. By carefully selecting and combining elements, researchers can create materials with tailored polarization behavior, improved permittivity, and optimized energy density.

4.6.2 Enhanced Phase Stability

HEMs are often thermodynamically stabilized by high configurational entropy, which suppresses phase transitions that can degrade device performance. This intrinsic stability can be particularly advantageous in applications involving extreme thermal or electrical environments, such as pulsed power systems or aerospace capacitors.

4.6.3 Defect Engineering and Doping Tolerance

The inherent disorder and multi-element matrix of HEMs allow for more flexible doping strategies without severely compromising crystal structure or electrical properties. This enables the engineering of point defects and oxygen vacancies to enhance dielectric behavior while maintaining stability.

4.6.4 Potential for Miniaturized and Flexible Devices

With the development of thin-film deposition techniques tailored for HEMs, there is a growing potential for their integration into miniaturized, multilayer capacitors and flexible electronics. Their mechanical robustness and tolerance to lattice distortion also make them candidates for stretchable or wearable energy storage systems.

4.6.5 Compatibility with Emerging Technologies

HEMs could complement other advanced materials in hybrid systems—for example, combining with 2D materials, ferroelectrics, or polymer dielectrics—to achieve multifunctional storage platforms. Their multifunctionality may allow dielectric materials to perform additional roles like sensing, thermal buffering, or structural reinforcement.

4.7 Application of High-Entropy Materials in Dielectric Energy Storage

In response to the applied electric filed, the dielectric capacitors store or release electrical energy based on the dielectric polarization mechanism of the material. Owing to their fast charge–discharge capabilities, they have garnered significant interest for applications in advanced electronics and modern power systems [71, 72]. However, a major limitation remains their relatively low-energy density [20, 73]. To address this, various strategies, such as domain engineering [9], defect regulation [71, 74], interface modification [75], and grain size refinement [76, 77] have been explored (schematically presented in Figure 4.3). Nevertheless, further enhancements are still necessary, particularly through compositional innovations. Recently, the high-entropy design strategy has emerged as a promising approach for improving the energy storage performance of dielectric materials [32, 77]. Since its first application in 2019, numerous studies have followed [30, 78, 79]. However, early designs often exhibited limited energy densities, primarily due to simplistic compositional schemes and an incomplete understanding of the associated mechanisms. In 2022, two independent research groups reported [80, 81] breakthrough results demonstrating that high-entropy engineering could significantly improve energy storage properties in both ceramic and thin-film dielectric systems. These studies now serve as foundational references for the design of high-entropy dielectrics.

A notable advancement was reported by Chen et al. [80], who applied a high-entropy design strategy to (K,Na)NbO_3 (KNN)-based ceramics. By incorporating a range of elements—including Li, Bi, Ba, Sc, Hf, Zr, and Ta—into the K/Na and Nb lattice sites, they introduced substantial compositional complexity. This generated strong random electric fields within the material, effectively disrupting long-range polarization and enhancing dielectric response. Moreover, the substituted ions enabled tuning of phase transition temperatures, facilitating the coexistence of multiple crystal phases at room temperature. As a result, the high-entropy KNN ceramics exhibited a multiphase structure comprising rhombohedral, orthorhombic, tetragonal, and cubic phases, along with pronounced local structural distortions. Experimental findings revealed nanoscale polar-regions (1–3 nm) with

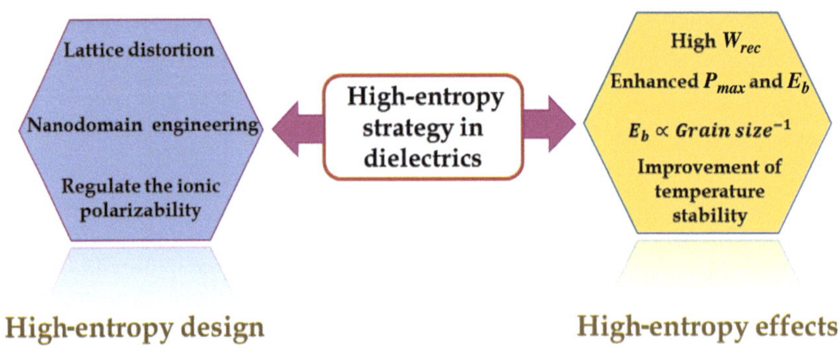

Figure 4.3 Schematic illustration of the high-entropy design strategy for enhancing dielectric energy storage performance.

diverse orientations and slower polarization saturation dynamics under electric fields, contributing to improved energy storage efficiency.

Furthermore, the high-entropy processing led to the formation of uniformly distributed fine grains (~600 nm), which enhanced the breakdown electric field and recoverable energy storage density. Overall, the optimized high-entropy KNN ceramic achieved ultrahigh energy density (≥10 J cm^{-3}) and energy efficiency (≥90%), surpassing the performance of conventional dielectric materials such as $AgNbO_3$, (Bi, Na)TiO_3, $BaTiO_3$, and $NaNbO_3$. A notable example is the work by Yang et al. [81], who investigated a series of entropy-modulated films with the general formula $(Bi_{3.25}La_{0.75})(Ti_{3-x}Zr_x, Hf_x, Sn_x)O_{12}$, which varied from 0.0 to 0.5, corresponding to an increase in configurational entropy from 0.45R to 1.72R. Their findings revealed that increasing atomic-scale disorder through entropy engineering stabilized the pyrochlore phase, which is typically unstable in the Bi_2O_3–TiO_2 system. This stabilization is attributed to a decrease in Gibbs free energy ($\Delta G = \Delta H - T\Delta S$), which renders the phase thermodynamically favorable.

In 2024, Wang et al. [82] reported four distinct ceramic compositions prepared by solid-state reaction method: $Na_{0.5}Bi_{0.5}TiO_3$ (NBT) and with Sr and Sr-La-K-Ba doped derivatives. Figure 4.4a illustrates the schematic of the lattice distortions

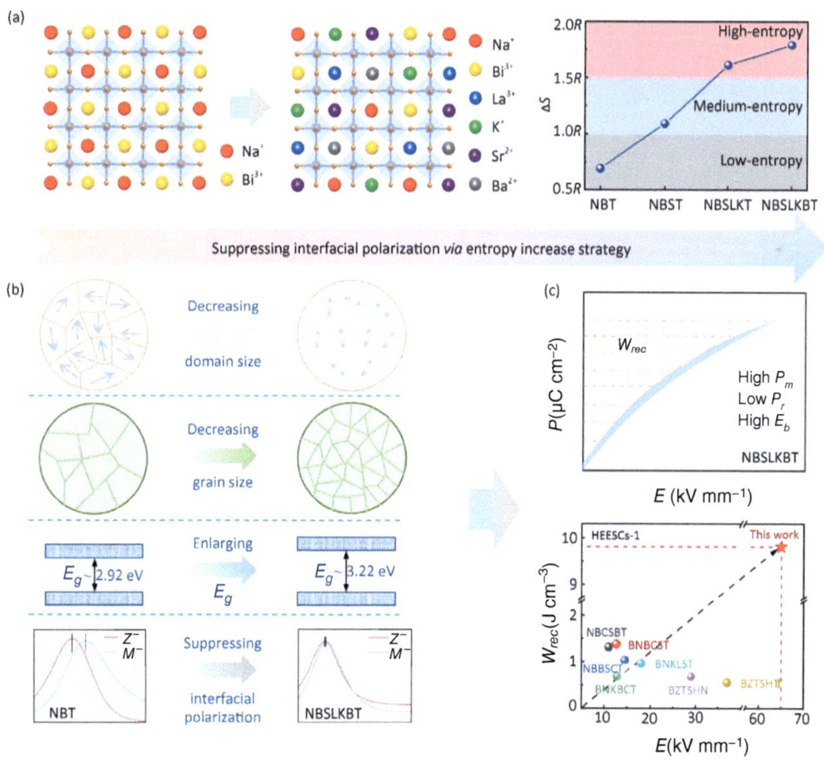

Figure 4.4 The pictorial illustration of the high-entropy strategy in enhancing the charge storage performance of $Na_{0.5}Bi_{0.5}TiO_3$ (NBT) ceramics. *Source:* Reproduced from Ref. [82] / with permission of Elsevier.

Table 4.2 Energy storage characteristics of various NBT-based ceramic systems.

Stoichiometry of the high-entropy ceramic	W_{rec} (J cm^{-3})	η (%)	E (kV cm^{-1})	ΔP (μC cm^{-2})	$\tau_{0.9}$ (ns)	PD (MW cm^{-3})	ε_r (RT)	tan δ (RT)	t (mm)	References
$0.465NBT–0.035CaSnO_3–0.5Sr_{0.7}B_{0.2}TiO_3$	4.0	80	415	38	116	107	2,600	0.013	0.15	[83]
$0.94Bi_{0.4867}Sm_{0.0133}Na_{0.46-1}Sm_{0.0133}TiO_3–0.06BaTiO_3$	4.4	84	230	42	—	—	990	0.035	0.11	[84]
$0.75NBT–0.25[Bi[Mg_{0.5}Hf_{0.5}]O_3]$	4.6	75	310	21	110	39.1	430	0.023	0.12	[85]
$0.62NBT–0.3Sr_{0.7}Bi_{0.2}TiO_3–0.08BiMg_2/_3Nb_1/_3O_3$	7.5	92	470	48	—	—	1,230	0.061	0.15	[86]
$0.85[0.94NBT–0.06BT]–0.15BiMg_{0.5}Hf_{0.5}O_3$	7.8	76	370	35	—	—	1,600	0.230	0.20	[87]
$0.85Bi_{0.47}Na_{0.47}Ba_{0.06}TiO_3–0.15Sr[Al_{0.5}Ta_{0.5}]O_3$	8.3	91	555	39	97	—	1,460	—	0.05	[88]
$0.90[Bi_{0.5}Na_{0.5}]_{0.65}Sr_{0.35}TiO_3–0.10Bi[Mg_{0.5}Zr_{0.5}]O_3$	8.5	86	522	59	—	—	1,320	0.165	0.10	[89]
$0.92[0.6NBT–0.4Sr_{0.7}Bi_{0.2}TiO_3]–0.08BMT$	8.6	93	565	45	—	—	1,350	0.090	0.60	[90]
$NBT–Bi_{0.5}K_{0.5}TiO_3–Sr[Sc_{0.5}Nb_{0.5}]O_3$	9.2	96	385	46	244	—	750	0.014	0.05	[91]
$0.76[0.94NBT–0.06BT]–0.24CaTi_{0.75}Ta_{0.2}O_3$	9.6	88	410	60	—	117	789	0.020	0.11	[92]
0.85BNST–0.15BMT	10.3	97	420	54	1,350	—	615	0.045	0.05	[93]
$0.50[0.75NBT–0.25BT]–0.50BaZrO_3$	13.6	91	660	63	35	313	455	0.070	0.08	[94]
$[Bi_{0.5}Na_{0.5}][Ti_{0.33}Fe_{0.33}Nb_{0.33}]O_3$	13.8	82	640	53	35	136	480	0.019	0.05	[95]
$0.62[0.94NBT–0.06BT]–0.38Ca_{0.7}La_{0.2}TiO_3$	15.1	82	640	61	32	212	710	0.001	0.12	[96]
0.88[0.75NBT–0.25BT]–0.12NN	15.2	91	730	30	38	—	1,896	0.003	0.05	[97]

Abbreviations

- ε_r: Relative dielectric constant (measured at room temperature).
- *tan* δ: Dielectric loss (at room temperature).
- E: Electrical breakdown strength (kV cm^{-1}).
- ΔP: Change in polarization ($\Delta P = P_{max} - P_r$).
- W_{rec}: Recoverable energy density (J cm^{-3}).
- η: Energy storage efficiency (%).
- $\tau_{0.9}$: Charge–discharge time (ns).
- PD: Power density (MW cm^{-3}).
- t: Thickness (mm).

Source: Adapted from Ref. [98].

by mapping the magnitude and direction of cation displacements, respectively. These distortions arise from the random distribution of multiple cation species within the crystal lattice, a hallmark of high-entropy systems. Using Eq. (4.1) the configurational entropies (ΔS_{config}) of these systems are estimated and plotted as shown in Figure 4.4a. The estimated values are 0.69*R*, 1.09*R*, 1.61*R*, and 1.79*R* for $(Na_{1/3}Bi_{1/3}Sr_{1/3})TiO_3$, Sr doped $(Na_{0.2}Bi_{0.2}Sr_{0.2}La_{0.2}K_{0.2})TiO_3$, and Sr-La-K-Ba doped $(Na_{1/6}Bi_{1/6}Sr_{1/6}La_{1/6}K_{1/6}Ba_{1/6})TiO_3$ ceramics spanning over the low–medium–high entropies. Further, applying an entropy-enhancement strategy, the following dielectric and energy storage improvements were observed: (i) enhancement of relaxor behavior due to entropy-induced nanodomain formation, (ii) improvement in dielectric breakdown strength (E_b) attributed to hysteresis diffusion effects and refined grain structure, (iii) increase in band gap (E_g) due to incorporation of highly insulating oxides, and (iv) further enhancement of E_b through suppression of interfacial polarization, as shown in Figure 4.4b. As mentioned above synergistic effects yielded a progressive improvement in energy storage density (W_{rec}), energy efficiency (η), and tolerable electric field (E_b) with increasing configurational entropy. Notably, the NBSLKBT ceramic achieved an ultrahigh W_{rec} of 9.8 J cm^{-3}, an exceptional η of 86.5%, and a remarkable E_b of 65 kV mm^{-1}. As depicted in Figure 4.4c, the $(Na_{1/6}Bi_{1/6}Sr_{1/6}La_{1/6}K_{1/6}Ba_{1/6})TiO_3$ (NBSLKBT) ceramic exhibited the best overall performance among the reported high-entropy energy storage system (Tables 4.2 and 4.3) materials. These values represent enhancements by factors of 16.89, 10.42, and 4.06, respectively, in comparison to the pristine $(Na_{1/3}Bi_{1/3}Sr_{1/3})TiO_3$ (NBT) ceramic. This work validates the efficacy of entropy engineering for optimizing energy storage performance and provides a promising strategy for modulating interfacial polarization in high-entropy oxides (HEOs). In addition, Table 4.3 indicates how the substrate strategically plays a vital role in tuning the entropy and has also shown promise in enhancing the energy storage performance of dielectric thin films.

4.8 Unique Effects in HEAs and Their Influence on Properties

HEAs exhibit four distinctive effects that differentiate them from traditional alloys:

1) *High-entropy effect (thermodynamic)*: The incorporation of five or more elements in near-equiatomic proportions enhances the system's configurational entropy, favoring the formation of single-phase solid solutions over multiphase structures.
2) *Sluggish Diffusion Effect (Kinetic)*: Atomic diffusion rates and phase transformation kinetics are significantly reduced in HEAs. This effect is particularly pronounced in large-grained alloys and contributes to thermal stability and resistance to coarsening.

Table 4.3 Energy storage performance of high-entropy oxide (HEO) thin films on various substrates, adopted from Ref. [98].

Elemental composition	Entropy classification	Substrate	W_{rec} ($J\,cm^{-3}$)	η (%)	E ($kV\,cm^{-1}$)	ΔP ($\mu C\,cm^{-2}$)	ε_r (RT)	Thickness (nm)	References
Ag,Nb,Ta	Low	Si (100)	46.4	80	3,300	34	130	300	[99]
Ba,Zr,Ti	Low	Si	93.5	71	8,500	30	57	415	[100]
Ba,Zr,Ti	Low	ITO-coated layer	40.6	69	4,230	22	86	130	[101]
Ba,Zr,Ti	High due to substrate effect	LSMO + STO + F-Mica	65.1	73	6,150	28	141	130	[102]
Ba,Zr,Ti,Sr	High	Pt-coated Si	89.4	65	6,300	38	151	610	[103]
Ba,Zr,Ti,Sr	High	Pt/Ti/SiO_2/Si	102.0	87	6,200	30	65	100	[104]
Ba,Zr,Ti,Sr	High	Pt (111)/Ti/(100)Si	148.0	91	6,600	49	—	300	[105]
Bi,La,Ti,Zr,Hf,Sn	High	Pt (111)/Ti/SiO_2/Si	182.0	78	6,200	55	80	620	[81]
Bi,Fe,Sr,Ti	High	Nb:STO	51.0	60	3,600	40	225	500	[106]
Ba,Sr,Ti,Bi,Mg,Nb	High	Nb:STO	86.0	73	5,000	51	240	400	[107]
Bi,Fe,Sr,Ti	High	(001)-oriented Nb:STO	112.0	80	4,900	57	250	450	[9]

Abbreviations

- W_{rec}: Recoverable energy density.
- η: Energy storage efficiency.
- E: Breakdown electric field strength.
- ΔP: Polarization difference.
- ε_r: Relative dielectric constant (at room temperature).
- t: Film thickness.
- *LSMO*: $La_{0.7}Sr_{0.3}MnO_3$.
- *STO*: $SrTiO_3$.
- *ITO*: Indium tin oxide.
- *F-Mica*: Fluorophlogopite mica.

Source: Adapted from Ref. [98].

3) *Lattice Distortion Effect (Structural)*: Variations in atomic size, mass, and electronegativity lead to lattice distortions, where bond lengths and angles are irregular. These distortions introduce inhomogeneities across atomic planes, causing peak broadening and intensity reduction in X-ray diffraction (XRD) patterns.
4) *Cocktail Effect (Performance Synergy)*: Unique synergistic properties arise from the complex interactions among multiple alloying elements, leading to enhanced mechanical, thermal, and electrochemical performance.

These combined effects ensure both thermodynamic and kinetic stability of HEA-based catalysts and optimize the electronic structure of active sites, enabling novel functionalities.

4.9 Challenges in HEM Design for Dielectric Energy Storage

4.9.1 Compositional Complexity

The extensive compositional landscape inherent to HEMs, generally comprising five or more principal elements, poses substantial challenges in rational material design. Anticipating dielectric properties across this vast compositional domain is non-trivial due to the nonlinear relationship between composition and property evolution. Experimental investigation of this space is both labor-intensive and cost-prohibitive, while predictive computational methodologies remain in a nascent stage of development.

4.9.2 Phase Stability and Solid-Solution Formation

Formation of a stable single-phase solid solution is critical to ensuring homogeneity in dielectric characteristics. However, the high-entropy strategy does not inherently guarantee thermodynamic phase stability. Phase instability may arise due to the formation of competing neighboring phases and elemental segregation, resulting in dielectric heterogeneity. This adversely affects breakdown strength and energy density, both at the material and device levels.

4.9.3 Structural and Defect Disorder

Alterations in the configuration and phase of HEMs also negatively affect defect states. Microscale defects such as point defects and oxygen vacancies predominantly influence dielectric loss and insulation resistance, more so than macroscale defects. Control over these microscale defects is hindered by the random nature of the local bonding environment in HEMs.

4.9.4 Processing Limitations

Processing significantly impacts material performance. Conventional techniques are often inadequate for optimizing parameters such as temperature, working pressure, and processing time to suppress the emergence of secondary phases and minimize porosity and grain boundary-related defects. Thus, the development of scalable and high-throughput processes is highly desirable at this stage.

4.9.5 Trade-offs in Structure–Property Correlations

Optimizing all three fundamental parameters of a dielectric capacitor simultaneously—(i) dielectric permittivity, (ii) breakdown strength, and (iii) energy storage capability—is both significant and atypical. Generally, materials with high dielectric permittivity exhibit reasonable breakdown strength but suffer from high dielectric losses. In HEMs, this trade-off is exacerbated by unpredictable structural variations arising from the stochastic local bonding environment.

4.9.6 Insufficient Predictive Models

Although a substantial body of theoretical calculations and modeling exists in the literature, there is a paucity of comprehensive theoretical frameworks that can accurately predict the design of high-performance HEMs, especially for dielectric applications. Furthermore, existing models often fall short in calculating configurational entropy and assessing the role of individual components in a multicomponent system.

4.9.7 Thermal and Long-Term Stability

The long-term performance and cycle stability of HEMs in devices remains a major concern. Structural robustness essential for long-term stability is often lacking due to the multicomponent nature of these materials. Additionally, changes in operating temperature exacerbate cycle instability by complicating phase separation and defect control. Moreover, the behavior of HEMs under prolonged high-temperature operation remains inadequately characterized.

4.10 Device-Level Challenges in Incorporating High-Entropy Materials (HEMs)

Despite the promising intrinsic properties of HEMs, their integration into dielectric energy storage systems entails numerous engineering and practical hurdles.

4.10.1 Interface Compatibility

The interface between the current collector and the active material significantly influences performance. Several factors synergistically contribute to this interface effect: (i) thermal expansion coefficient mismatch, (ii) interfacial adhesion, and (iii) chemical reactivity. Thus, a specialized device fabrication technique is highly desirable for effective HEM integration.

4.10.2 Thin-Film Deposition and Scalability

Thin films deposited using PVD, CVD, or liquid precursor methods for dielectric applications require precise control over thickness, film uniformity, and interfacial properties. However, maintaining entropy-stabilized single-phase behavior under these constraints is particularly challenging. Scaling up material synthesis for industrial application remains tedious, complex, and cost-intensive.

4.10.3 Electrical Reliability and Breakdown Performance

While bulk HEMs may exhibit superior breakdown strength, thin-film and device-level performance are highly sensitive to microstructural uniformity, interfacial quality, and defect density. The intrinsic structural disorder of HEMs can intensify localized electric fields, precipitating premature dielectric failure.

4.10.4 Leakage Current and Dielectric Loss

Mitigating leakage current and dielectric losses is crucial for developing energy-efficient devices. In HEMs, compositional disorder, random cation distribution, and defect structures such as oxygen vacancies may create percolation pathways for conduction, reducing storage efficiency.

4.10.5 Thermal Management and Stability

Devices are often subject to dynamic thermal conditions. At elevated temperatures, phase changes and defect migration occur, significantly affecting dielectric integrity. Therefore, ensuring superior thermal stability is essential for long-term device performance.

4.10.6 Integration with Electrodes and Other Functional Layers

In multilayer devices, compatibility between the HEM and electrode materials is crucial. Incompatibility can impair charge injection, storage retention, and overall device performance.

4.10.7 Cost and Process Complexity

Multi-step, high-temperature processing further increases cost and complexity. Additionally, expensive precursors used in HEM synthesis contribute to elevated production costs, limiting large-scale commercialization.

4.11 Temperature-Dependent Conductivity Degradation in HEMs

Reliable, long-term cycle stability of HEM-based devices requires stable electrical conductivity across all thermal regimes. However, it is inherent to HEMs that conductivity tends to degrade with increasing temperature, posing practical challenges for future applications.

4.11.1 Thermal Activation of Charge Carriers

In HEMs, shallow defect levels and diverse cation populations can create unwanted conduction pathways. As operating temperature rises, activated charge carriers including electrons, holes, and oxygen vacancies enhance leakage current and reduce insulation resistance.

4.11.2 Oxygen Vacancy Mobility

Oxygen vacancies at high temperatures play a dual role in dielectric capacitors: they increase ionic conductivity while also contributing to dielectric loss. In particular, oxide-based HEM systems are vulnerable to performance degradation due to the critical role oxygen vacancies play in compromising dielectric integrity and charge storage.

4.11.3 Phase Instability and Diffusion

Multicomponent oxide HEMs comprising alkali ions such as Na^+, Li^+, and K^+ are especially prone to phase separation and elemental interdiffusion under thermal cycling. This results in structural instability, localized conductivity changes, hotspot formation, and ultimately dielectric breakdown.

4.11.4 Degradation of Interfaces and Electrodes

At elevated temperatures, interfacial degradation and increased contact resistance further impair overall energy storage performance.

4.11.5 Trade-off with Thermal Conductivity

HEMs often exhibit low thermal conductivity due to phonon scattering from lattice distortions. While beneficial for thermal insulation, this can lead to localized heating, worsening conductivity degradation and increasing the risk of device failure.

4.12 Dielectric Polarization Response to Temperature Fluctuations in HEMs

Dielectric polarization, a key parameter governing energy storage capacity, is intricately influenced by the complex local structure and multi-element configuration of HEMs, especially under thermal variations. Understanding this response is critical for designing devices that operate under variable thermal environments.

4.12.1 Temperature Dependence of Polarization Mechanisms

Dielectric polarization arises from multiple sources—electronic, ionic, dipolar, and interfacial—each with unique thermal sensitivities. While electronic and ionic contributions are relatively stable with temperature, dipolar and interfacial polarizations are highly temperature-dependent. Thermal agitation may enhance dipole reorientation but can also disrupt alignment, leading to polarization relaxation and reduced net polarization. The high-entropy concept has garnered notable attention in the context of alloys and ceramics. According to thermodynamic principles, temperature correlates directly with entropy. A higher configurational entropy at room temperature can mimic the structural characteristics typically observed at elevated temperatures—namely, enhanced crystallographic symmetry. As schematically depicted in Figure 4.5a, this implies that entropy elevation may induce a high-temperature-like structure even at ambient conditions. Consequently, structural transitions during heating are minimized, enabling superior dielectric stability across wide temperature ranges. For instance, BNT exhibits low symmetry at room temperature, which increases with temperature. This structure–property synergy highlights its potential for dielectric energy storage. Figure 4.5 captures various attributes of such materials [30].

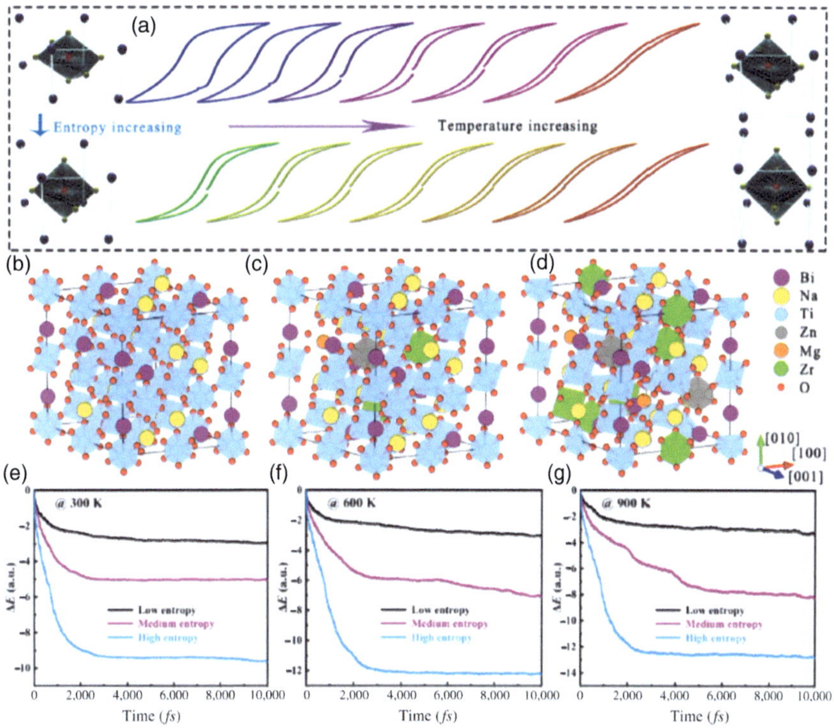

Figure 4.5 (a) Schematic illustration of entropy modulation strategy for enhancing high-temperature dielectric energy storage. The relaxed superlattice configurations of (1−*x*)(0.94$Bi_{0.5}Na_{0.5}TiO_3$-0.06$Ba(Zr_{0.2}Ti_{0.8})O_3$)-x$Bi(Zr_{0.5}Zn_{0.25}Mg_{0.25})O_3$ (BNBZT-xBZZM) with varying *x* (0, 0.08, 0.16, 0.24, 0.32, 0.40) exhibit different configurational entropy (S_{config}): (b) low S_{config}, (c) medium S_{config}, and (d) high S_{config}. Potential energy profiles at (e) 300 K, (f) 600 K, and (g) 900 K are used to evaluate structural dynamics. *Source:* Reproduced from Ref. [30] / with permission of Elsevier.

4.12.2 Relaxor Behavior and Polar Nanoregions (PNRs)

Certain HEMs display relaxor-like characteristics attributed to local chemical and structural heterogeneity, leading to the formation of polar nano-regions (PNRs). These PNRs exhibit strong temperature dependence, manifesting as shifts or broadening of the dielectric peak (Curie point), indicative of diffuse phase transitions. This behavior is advantageous in applications requiring thermally stable dielectric response.

4.12.3 Thermal Fluctuations and Entropy Effects

While configurational entropy enhances phase stability, thermal fluctuations may destabilize long-range dipolar ordering in HEMs. This can induce a ferroelectric–paraelectric transition or reduce remanent polarization in systems subjected to temperature variation.

4.12.4 Effect on Permittivity and Dielectric Loss with Rising Temperature

Initially, dielectric permittivity may increase due to enhanced dipolar alignment but can decline when thermal disorder impairs dipole orientation. Concurrently, dielectric losses tend to rise due to increased dipole relaxation, defect motion, and conductive losses.

4.12.5 Thermal Fatigue and Cycling Effects

Repetitive thermal cycling may induce irreversible changes such as microcracking, phase segregation, and defect redistribution. These phenomena degrade the dielectric performance and energy storage reliability over time.

4.12.6 Implications for Device Design

To ensure consistent dielectric polarization and performance, device design must incorporate the thermal response characteristics of HEMs by:

i) Selecting compositions with thermally stable dielectric responses.
ii) Engineering microstructures to inhibit defect mobility.
iii) Integrating thermal buffer layers to mitigate abrupt temperature-induced stress.

4.13 Conclusion and Future Perspectives

This chapter provides a comprehensive overview of the current status of HEMs for dielectric capacitor applications. It summarizes recent advancements in the design of HEMs, with a particular focus on their use in capacitor-based energy storage.

Additionally, it discusses in detail the various synthesis techniques developed globally for fabricating HEMs intended for dielectric capacitors. Following the literature survey, Section 4.11 elaborates on the device-level challenges associated with integrating HEMs into practical capacitor systems. The key aspects identified in HEM design and device integration are summarized as follows:

i) Conventional synthesis methods often face significant challenges in producing energy-efficient HEMs due to issues such as compositional complexity, phase stability, and solid-solution formation. Moreover, there is a lack of predictive models that can effectively establish structure–property correlations and optimize material performance.
ii) Despite their exceptional intrinsic properties, the integration of HEMs into devices is hindered by several critical issues. These include interface compatibility and mismatch with electrodes, increased leakage currents and dielectric losses, and limited electrical reliability. In addition, scalability, cost, and process complexity remain major obstacles to the widespread adoption of HEMs in commercial applications.
iii) Thermal management within HEM-based materials and devices represents another major challenge. Poor thermal regulation can significantly degrade long-term device performance and reliability.

In conclusion, although HEMs exhibit great promise for dielectric capacitor applications, extensive technological development and exploration are still required. A substantial effort is needed to transition these materials from the laboratory to commercial devices, particularly in terms of design optimization, scalable fabrication, and system integration.

References

1 Yang, B., Liu, Y., Lan, S. et al. (2023). High-entropy design for dielectric materials: status, challenges, and beyond. *J Appl Phys* 133 (11): 1–12.

2 Palani Velayuda Shanmugasundram, H.P., Jayamani, E., and Soon, K.H. (2022). A comprehensive review on dielectric composites: classification of dielectric composites. *Renew Sust Energ Rev* 157: 112075.

3 Pan, H., Li, F., Liu, Y. et al. (2019). Ultrahigh–energy density lead-free dielectric films via polymorphic nanodomain design. *Science* 365 (6453): 578–582.

4 Yang, L., Kong, X., Li, F. et al. (2019). Perovskite lead-free dielectrics for energy storage applications. *Prog Mater Sci* 102: 72–108.

5 Zhao, L., Ke, X., Zhou, Z. et al. (2019). Large electrocaloric effect over a wide temperature range in $BaTiO_3$-modified lead-free ceramics. *J Mater Chem C* 7 (5): 1353–1358.

6 Shi, J., Han, D., Li, Z. et al. (2019). Electrocaloric cooling materials and devices for zero-global-warming-potential, high-efficiency refrigeration. *Joule* 3 (5): 1200–1225.

7 Cheng, H., Ouyang, J., Zhang, Y.-X. et al. (2017). Demonstration of ultra-high recyclable energy densities in domain-engineered ferroelectric films. *Nat Commun* 8 (1): 1999.

8 Liu, J., Li, P., Li, C. et al. (2022). Synergy of a stabilized antiferroelectric phase and domain engineering boosting the energy storage performance of $NaNbO_3$-based relaxor antiferroelectric ceramics. *ACS Appl Mater* 14 (15): 17662–17673.

9 Pan, H., Lan, S., Xu, S. et al. (2021). Ultrahigh energy storage in superparaelectric relaxor ferroelectrics. *Science* 374 (6563): 100–104.

10 Zheng, T., Yu, Y., Lei, H. et al. (2022). Compositionally graded KNN-based multilayer composite with excellent piezoelectric temperature stability. *Adv Mater* 34 (8): 2109175.

11 Huangfu, G., Zeng, K., Wang, B. et al. (2022). Giant electric field–induced strain in lead-free piezoceramics. *Science* 378 (6624): 1125–1130.

12 Li, T., Liu, C., Shi, P. et al. (2022). High-performance strain of lead-free relaxor-ferroelectric piezoceramics by the morphotropic phase boundary modification. *Adv Funct Mater* 32 (32): 2202307.

13 Liu, S., Li, H., Zhong, J. et al. (2022). A crystal glass–nanostructured Al-based electrocatalyst for hydrogen evolution reaction. *Sci Adv* 8 (44): eadd6421.

14 Jiang, B., Yu, Y., Cui, J. et al. (2021). High-entropy-stabilized chalcogenides with high thermoelectric performance. *Science* 371 (6531): 830–834.

15 Zhang, W.-T., Wang, X.-Q., Zhang, F.-Q. et al. (2024). Frontiers in high entropy alloys and high entropy functional materials. *Rare Metals* 43 (10): 4639–4776.

16 Fan, X., Chen, L., Huang, D. et al. (2024). From single metals to high-entropy alloys: how machine learning accelerates the development of metal electrocatalysts. *Adv Funct Mater* 34 (34): 2401887.

17 Kong, X., Yang, L., Meng, F. et al. (2025). High-entropy engineered $BaTiO_3$-based ceramic capacitors with greatly enhanced high-temperature energy storage performance. *Nat Commun* 16 (1): 885.

18 Liu, J., Xiong, X., Li, H. et al. (2024). Application of solution-processed high-entropy metal oxide dielectric layers with high dielectric constant and wide bandgap in thin-film transistors. *Micromachines* 15 (12): 1465.

19 Yang, W., Shen, J., Wang, Z. et al. (2025). Mechanical and electrochemical properties of $(MoNbTaTiZr)_{1-x}N_x$ high-entropy nitride coatings. *J Mater Sci Technol* 208: 78–91.

20 Lu, X., Mu, C., Lu, Y. et al. (2025). Comparative analysis of structural, mechanical and tribological properties of high-entropy nitrides and carbonitrides coatings over a wide temperature range. *J Alloys Compd* 1027: 180591.

21 Zhang, S., Zuo, W., Fu, X. et al. (2025). High-entropy sulfoselenide as negative electrodes with fast kinetics and high stability for sodium-ion batteries. *Nat Commun* 16 (1): 4052.

22 Wu, L. and Hofmann, J.P. (2022). High-entropy transition metal chalcogenides as electrocatalysts for renewable energy conversion. *Curr Opin Electrochem* 34: 101010.

23 Vecchio, K., Curtarolo, S., Kaufmann, K. et al. (2024). Fermi energy engineering of enhanced plasticity in high-entropy carbides. *Acta Mater* 276: 120117.

24 Yang, W., Xu, T., Yang, C. et al. (2024). Selective and bifunctional catalytic electrochemical conversion of organosulfide molecule by high-entropy carbides. *Adv Funct Mater* 34 (49): 2409450.

25 Oses, C., Toher, C., and Curtarolo, S. (2020). High-entropy ceramics. *Nat Rev Mater* 5 (4): 295–309.

26 Sarkar, A., Wang, Q., Schiele, A. et al. (2019). High-entropy oxides: fundamental aspects and electrochemical properties. *Adv Mater* 31 (26): 1806236.

27 Ma, Y., Ma, Y., Wang, Q. et al. (2021). High-entropy energy materials: challenges and new opportunities. *Energy Environ Sci* 14 (5): 2883–2905.

28 Yeh, J.-W. (2013). Alloy design strategies and future trends in high-entropy alloys. *JOM* 65 (12): 1759–1771.

29 Shi, P., Li, R., Li, Y. et al. (2021). Hierarchical crack buffering triples ductility in eutectic herringbone high-entropy alloys. *Science* 373 (6557): 912–918.

30 Qian, J., Ge, G., Yu, Z. et al. (2025). High temperature stable capacitive energy storage up to 320 °C in high-entropy dielectric thin film. *Nano Energy* 135: 110651.

31 Chen, X., Wang, Q., Cheng, Z. et al. (2021). Direct observation of chemical short-range order in a medium-entropy alloy. *Nature* 592 (7856): 712–716.

32 Zhu, X., Xiong, S., Zhu, G. et al. (2024). Dielectric properties and excellent energy storage density under low electric fields for high entropy relaxor ferroelectric $(Li_{0.2}Ca_{0.2}Sr_{0.2}Ba_{0.2}La_{0.2})TiO_3$ ceramic. *J Alloys Compd* 984: 173987.

33 Ren, J., Kumkale, V.Y., Hou, H. et al. (2025). A review of high-entropy materials with their unique applications. *ACHM* 8 (2): 195.

34 Alam, K., Jang, W., Jeong, G. et al. (2024). Synthesis of high-entropy alloys with a tailored composition and phase structure using a single configurable target. *ACS Omega* 9 (1): 1362–1374.

35 Wu, Y.D., Cai, Y.H., Wang, T. et al. (2014). A refractory $Hf_{25}Nb_{25}Ti_{25}Zr_{25}$ high-entropy alloy with excellent structural stability and tensile properties. *Mater Lett* 130: 277–280.

36 Liu, W.H., Wu, Y., He, J.Y. et al. (2013). Grain growth and the Hall–Petch relationship in a high-entropy FeCrNiCoMn alloy. *Scr Mater* 68 (7): 526–529.

37 Qiu, X.-W. (2013). Microstructure and properties of AlCrFeNiCoCu high entropy alloy prepared by powder metallurgy. *J Alloys Compd* 555: 246–249.

38 Lu, Z.P., Wang, H., Chen, M.W. et al. (2015). An assessment on the future development of high-entropy alloys: summary from a recent workshop. *Intermetallics* 66: 67–76.

39 Liu, B., Wang, J., Liu, Y. et al. (2016). Microstructure and mechanical properties of equimolar FeCoCrNi high entropy alloy prepared via powder extrusion. *Intermetallics* 75: 25–30.

40 Broge, N.L.N., Bertelsen, A.D., Søndergaard-Pedersen, F., and Iversen, B.B. (2023). Facile solvothermal synthesis of Pt–Ir–Pd–Rh–Ru–Cu–Ni–Co high-entropy alloy nanoparticles. *Chem Mater* 35 (1): 144–153.

41 Prasad, H., Singh, S., and Panigrahi, B.B. (2017). Mechanical activated synthesis of alumina dispersed FeNiCoCrAlMn high entropy alloy. *J Alloys Compd* 692: 720–726.

42 Xie, P., Yao, Y., Huang, Z. et al. (2019). Highly efficient decomposition of ammonia using high-entropy alloy catalysts. *Nat Commun* 10 (1): 4011.

43 Yao, Y., Huang, Z., Xie, P. et al. (2018). Carbothermal shock synthesis of high-entropy-alloy nanoparticles. *Science* 359 (6383): 1489–1494.

44 Phakatkar, A.H., Saray, M.T., Rasul, M.G. et al. (2021). Ultrafast synthesis of high entropy oxide nanoparticles by flame spray pyrolysis. *Langmuir* 37 (30): 9059–9068.

45 Kamecki, B., Karczewski, J., Cempura, G. et al. (2023). Evaluation of structural and electrical properties of multicomponent spinel oxide thin films deposited via spray pyrolysis technique. *Mater Charact* 203: 113097.

46 Zhang, M., Zhou, X., Yu, X., and Li, J. (2017). Synthesis and characterization of refractory TiZrNbWMo high-entropy alloy coating by laser cladding. *Surf Coat Technol* 311: 321–329.

47 Ye, X., Ma, M., Liu, W. et al. (2011). Synthesis and characterization of high-entropy alloy AlXFeCoNiCuCr by laser cladding. *Adv Mater Sci Eng* 2011 (1): 485942.

48 Kim, Y.S., Park, H.J., Mun, S.C. et al. (2019). Investigation of structure and mechanical properties of TiZrHfNiCuCo high entropy alloy thin films synthesized by magnetron sputtering. *J Alloys Compd* 797: 834–841.

49 Li, R.X., Liaw, P.K., and Zhang, Y. (2017). Synthesis of AlxCoCrFeNi high-entropy alloys by high-gravity combustion from oxides. *Mat Sci Eng A* 707: 668–673.

50 Zhang, K., Li, W., Zeng, J. et al. (2020). Preparation of $(La_{0.2}Nd_{0.2}Sm_{0.2}Gd_{0.2}Yb_{0.2})2Zr_2O_7$ high-entropy transparent ceramic using combustion synthesized nanopowder. *J Alloys Compd* 817: 153328.

51 Veronesi, P., Rosa, R., Colombini, E., and Leonelli, C. (2015). Microwave-assisted preparation of high entropy alloys. *Technologies* 3 (4): 182–197.

52 Kheradmandfard, M., Minouei, H., Tsvetkov, N. et al. (2021). Ultrafast green microwave-assisted synthesis of high-entropy oxide nanoparticles for Li-ion battery applications. *Mater Chem Phys* 262: 124265.

53 Toroghinejad, M.R., Pirmoradian, H., and Shabani, A. (2022). Synthesis of FeCr-CoNiCu high entropy alloy through mechanical alloying and spark plasma sintering processes. *Mater Chem Phys* 289: 126433.

54 Yang, Y., Ren, Y., Tian, Y. et al. (2021). Microstructure and properties of FeCoCrNiMoSix high-entropy alloys fabricated by spark plasma sintering. *J Alloys Compd* 884: 161070.

55 Aliyu, A. and Srivastava, C. (2019). Microstructure-corrosion property correlation in electrodeposited AlCrFeCoNiCu high entropy alloys-graphene oxide composite coatings. *Thin Solid Films* 686: 137434.

56 Keményi, D.M., Miskolcziné Pálfi, N., and Fazakas, É. (2021). Examination of microstructure and corrosion properties of novel AlCoCrFeNi multicomponent alloy. *Mater Today Proc* 45: 4250–4253.

57 Anandkumar, M. and Trofimov, E. (2023). Synthesis, properties, and applications of high-entropy oxide ceramics: current progress and future perspectives. *J Alloys Compd* 960: 170690.

58 Wu, X., Li, W., Huang, Z. et al. (2025). Dielectric and third-order nonlinear optical properties of high-entropy perovskite oxides $(La_{0.2}Ba_{0.2}Sr_{0.2}Ca_{0.2}Bi_{0.2})(Ti_{1-x}Sn_x)O_3$ ($x = 0$~0.5). *Ceram Int* 51: 32090–32099.

59 Gu, S., Huang, H.-Y., Lin, Y.-C. et al. (2025). High-entropy alloys with low platinum content: novel catalysts for oxygen reduction and evolution reactions. *Emerg Mater* 8: 3649–3662.

60 Panda, P.K., Huang, H.-Y., Dash, P. et al. (2025). Liquid-phase microwave synthesis of platinum-based high-entropy alloy catalysts on carbon supports for electrochemical hydrogen adsorption/desorption and oxygen evolution/reduction reactions. *Int J Hydrog Energy* 111: 536–545.

61 Khan, J.B., Panda, P.K., Yang, P.-C. et al. (2024). Microwave synthesis of high-entropy alloy catalysts on graphene oxide sheets for oxygen reduction and evolution reactions. *Int J Hydrog Energy* 53: 999–1008.

62 Liu, K.-C., Kumar Panda, P., Chandra Mallick, B. et al. (2024). Solid-phase microwave synthesis of high-entropy graphene quantum dots as metal-free electrochemical catalysts. *Appl Surf Sci* 648: 159061.

63 Jiang, T., Wu, F., Ren, Y. et al. (2023). Pyrochlore phase $(Y,Dy,Ce,Nd,La)_2Sn_2O_7$ as a superb anode material for lithium-ion batteries. *J Solid State Electrochem* 27 (3): 763–772.

64 Shaw, S.K., Kumari, P., Sharma, A. et al. (2023). Assessment of ionic site distributions in magnetic high entropy oxide of $(Mn_{0.2}Fe_{0.2}Co_{0.2}Ni_{0.2}Zr_{0.2})_3O_4$ and its catalytic behaviour. *Phys B Condens Matter* 652: 414653.

65 Ning, Y., Pu, Y., Wu, C. et al. (2023). Enhanced capacitive energy storage and dielectric temperature stability of A-site disordered high-entropy perovskite oxides. *J Mater Sci Technol* 145: 66–73.

66 Triolo, C., Santangelo, S., Petrovičovà, B. et al. (2023). Evaluation of the specific capacitance of high-entropy oxide-based electrode materials in view of their use for water desalination via capacitive method. *Appl Sci* 13 (2): 721.

67 Zhou, C., Zhang, X., Li, S. et al. (2022). Dielectric and energy storage properties of $(La,Li)_x[(Bi,Na)BaSrCa]_{1-x}TiO_3$ high-entropy perovskite ceramics. *Ceram Int* 48 (17): 24268–24275.

68 Lal, M.S. and Sundara, R. (2022). Multifunctional high entropy oxides incorporated functionalized biowaste derived activated carbon for electrochemical energy storage and desalination. *Electrochim Acta* 405: 139828.

69 Nie, S., Wu, L., Zhao, L. et al. (2021). Entropy-driven chemistry reveals highly stable denary $MgAl_2O_4$-type catalysts. *Chem Catalysis* 1 (3): 648–662.

70 Chen, Y.-W., Ruan, J.-J., Ting, J.-M. et al. (2021). Solution-based fabrication of high-entropy $Ba(Ti,Hf,Zr,Fe,Sn)O_3$ films on fluorine-doped tin oxide substrates and their piezoelectric responses. *Ceram Int* 47 (8): 11451–11458.

71 Kim, J., Saremi, S., Acharya, M. et al. (2020). Ultrahigh capacitive energy density in ion-bombarded relaxor ferroelectric films. *Science* 369 (6499): 81–84.

72 Palneedi, H., Peddigari, M., Hwang, G.-T. et al. (2018). High-performance dielectric ceramic films for energy storage capacitors: progress and outlook. *Adv Funct Mater* 28 (42): 1803665.

73 Li, J., Shen, Z., Chen, X. et al. (2020). Grain-orientation-engineered multilayer ceramic capacitors for energy storage applications. *Nat Mater* 19 (9): 999–1005.

74 Pan, H., Feng, N., Xu, X. et al. (2021). Enhanced electric resistivity and dielectric energy storage by vacancy defect complex. *Energy Storage Mater* 42: 836–844.

75 Sun, Z., Ma, C., Liu, M. et al. (2017). Ultrahigh energy storage performance of lead-free oxide multilayer film capacitors via interface engineering. *Adv Mater* 29 (5): 1604427.

76 Huang, R., Wang, H., Tao, C. et al. (2022). Evolution of polarization crystallites in $0.92BaTiO_3$-$0.08Bi(Ni_{0.5}Zr_{0.5})O_3$ microcrystal-amorphous composite thin film with high energy storage capability and thermal stability. *Chem Eng J* 433: 133579.

77 Ping, X., Pan, X., Meng, B. et al. (2025). Influence of grain refinement on the dielectric properties of B-site doped $BaTiO_3$-based high-entropy ceramics. *Ceram Int* 51: 31378–31386.

78 Zhou, S., Pu, Y., Zhao, X. et al. (2022). Dielectric temperature stability and energy storage performance of NBT-based ceramics by introducing high-entropy oxide. *J Am Ceram Soc* 105 (7): 4796–4804.

79 Yang, W. and Zheng, G. (2022). High energy storage density and efficiency in nanostructured $(Bi_{0.2}Na_{0.2}K_{0.2}La_{0.2}Sr_{0.2})TiO_3$ high-entropy ceramics. *J Am Ceram Soc* 105 (2): 1083–1094.

80 Chen, L., Deng, S., Liu, H. et al. (2022). Giant energy-storage density with ultrahigh efficiency in lead-free relaxors via high-entropy design. *Nat Commun* 13 (1): 3089.

81 Yang, B., Zhang, Y., Pan, H. et al. (2022). High-entropy enhanced capacitive energy storage. *Nat Mater* 21 (9): 1074–1080.

82 Zhao, H., Cao, W., Han, D. et al. (2024). Suppressing interfacial polarization via entropy increase strategy for superior energy-storage performance of $Na_{0.5}Bi_{0.5}TiO_3$-based ceramics. *J Mater* 10 (4): 947–955.

83 Jiang, M., Peng, Z., Zhou, Q. et al. (2024). Superior energy storage performance of BNT-based ferroelectric ceramics based on maintaining high polarization and breakdown strength. *J Adv Dielectr* 14 (04): 2340005.

84 Tang, X., Hu, Z., Koval, V. et al. (2023). Energy storage properties of samarium-doped bismuth sodium titanate-based lead-free ceramics. *Chem Eng J* 473: 145363.

85 Wang, K., Li, W., Tang, X. et al. (2023). High recoverable energy storage density of $Na_{0.5}Bi_{0.5}TiO_3$ lead-free ceramics modified by $Bi(Mg_{0.5}Hf_{0.5})O_3$. *J Adv Dielectr* 13 (3): 2350008.

86 Ji, H., Wang, D., Bao, W. et al. (2021). Ultrahigh energy density in short-range tilted NBT-based lead-free multilayer ceramic capacitors by nanodomain percolation. *Energy Storage Mater* 38: 113–120.

87 Khan, A., Gul, N.S., Luo, M. et al. (2022). Fabrication of a lead-free ternary ceramic system for high energy storage applications in dielectric capacitors. *Front Chem* 10: 01–09.

88 Li, D., Zhou, D., Wang, D. et al. (2023). Lead-free relaxor ferroelectric ceramics with ultrahigh energy storage densities via polymorphic polar nanoregions design. *Small* 19 (8): 2206958.

89 Zhu, X., Gao, Y., Shi, P. et al. (2022). Ultrahigh energy storage density in $(Bi_{0.5}Na_{0.5})_{0.6}5Sr_{0.35}TiO_3$-based lead-free relaxor ceramics with excellent temperature stability. *Nano Energy* 98: 107276.

90 Wang M, Feng Q, Luo C, lan Y, Yuan C, Luo N, et al. Ultrahigh energy storage density and efficiency in $Bi_{0.5}Na_{0.5}TiO_3$-based ceramics via the domain and bandgap engineering. *ACS Appl Mater*. 2021;13(43):51218-29.

91 Zhao, W., Xu, D., Li, D. et al. (2023). Broad-high operating temperature range and enhanced energy storage performances in lead-free ferroelectrics. *Nat Commun* 14 (1): 5725.

92 Li, T., Jiang, X., Li, J. et al. (2022). Ultrahigh energy-storage performances in lead-free $Na_{0.5}Bi_{0.5}TiO_3$-based relaxor antiferroelectric ceramics through a synergistic design strategy. *ACS Appl Mater* 14 (19): 22263–22269.

93 Li, D., Xu, D., Zhao, W. et al. (2023). A high-temperature performing and near-zero energy loss lead-free ceramic capacitor. *Energy Environ Sci* 16 (10): 4511–4521.

94 Liu, H., Sun, Z., Zhang, J. et al. (2023). Chemical design of Pb-free relaxors for giant capacitive energy storage. *J Am Chem Soc* 145 (21): 11764–11772.

95 Chen, L., Yu, H., Wu, J. et al. (2023). Large energy capacitive high-entropy lead-free ferroelectrics. *Nanomicro Lett* 15 (1): 65.

96 Cao, W., Lin, R., Hou, X. et al. (2023). Interfacial polarization restriction for ultrahigh energy-storage density in lead-free ceramics. *Adv Funct Mater* 33 (29): 2301027.

97 Liu, H., Sun, Z., Zhang, J. et al. (2023). Local chemical clustering enabled ultrahigh capacitive energy storage in pb-free relaxors. *J Am Chem Soc* 145 (35): 19396–19404.

98 Zubairi, H., Lu, Z., Zhu, Y. et al. (2024). Current development, optimisation strategies and future perspectives for lead-free dielectric ceramics in high field and high energy density capacitors. *Chem Soc Rev* 53 (21): 10761–10790.

99 Cheng, H., Zhai, X., Ouyang, J. et al. (2023). Achieving a high energy storage density in $Ag(Nb,Ta)O_3$ antiferroelectric films via nanograin engineering. *J Adv Ceram* 12 (1): 196–206.

100 Zhao, F., Wu, Y., Dai, Y. et al. (2021). Silicon-integrated lead-free $BaTiO_3$-based film capacitors with excellent energy storage performance and highly stable irradiation resistance. *J Mater Chem A* 9 (26): 14818–14826.

101 Liang, Z., Liu, M., Shen, L. et al. (2019). All-inorganic flexible embedded thin-film capacitors for dielectric energy storage with high performance. *ACS Appl Mater* 11 (5): 5247–5255.

102 Liang, Z., Ma, C., Shen, L. et al. (2019). Flexible lead-free oxide film capacitors with ultrahigh energy storage performances in extremely wide operating temperature. *Nano Energy* 57: 519–527.

103 Chen, X., Peng, B., Ding, M. et al. (2020). Giant energy storage density in lead-free dielectric thin films deposited on Si wafers with an artificial dead-layer. *Nano Energy* 78: 105390.

104 Wang, K., Ouyang, J., Wuttig, M. et al. (2020). Superparaelectric $(Ba_{0.95},Sr_{0.05})(Zr_{0.2},Ti_{0.8})O_3$ ultracapacitors. *Adv Energy Mater* 10 (37): 2001778.

105 Wang, K., Zhang, Y., Wang, S. et al. (2021). High energy performance ferroelectric $(Ba,Sr)(Zr,Ti)O_3$ film capacitors integrated on Si at 400 °C. *ACS Appl Mater* 13 (19): 22717–22727.

106 Pan, H., Zeng, Y., Shen, Y. et al. (2017). $BiFeO_3$–$SrTiO_3$ thin film as a new lead-free relaxor-ferroelectric capacitor with ultrahigh energy storage performance. *J Mater Chem A* 5 (12): 5920–5926.

107 Fan, Y., Zhou, Z., Chen, Y. et al. (2020). A novel lead-free and high-performance barium strontium titanate-based thin film capacitor with ultrahigh energy storage density and giant power density. *J Mater Chem C* 8 (1): 50–57.

5

Electrokinetics of High-Entropy Materials for Energy Storage Devices

Yukti Setia[1], Nikita Bhatt[1], Sankeerthana Bellamkonda[2], and Malaya K. Sahoo[1]

[1]*Rajiv Gandhi Institute of Petroleum Technology, Department of Sciences & Humanities, Jais, Uttar Pradesh, 229304, India*
[2]*Newcastle University, School of Engineering, Newcastle upon Tyne, NE17RU, UK*

5.1 Introduction

Progress has always accompanied the uncovering and development of new class of materials. Materials, as the center of all scientific and technological innovations, are closely linked to human civilization. In terms of evolution, the discovery of stone materials was one of the most important changes distinguishing early humans from animals. Physicists defined the term "entropy" as a measure of the diversity of the possible configurations of a system which, along with other criteria, would be subjected to thermodynamic restrictions, either at the universal or subatomic levels, [1] "from the universe to the atomic nucleus." Entropy dictates certain molecular behavior—these include reactant adsorption—in addition to translations, rotations, and vibrations of intermediates tethered to active sites, which in turn govern catalytic reaction pathways [2]. It is generally accepted that catalysis is governed by entropy; however, the discussion of catalyst-side entropy, especially configurational entropy due to component mixing, remains deficient. High-entropy materials (HEMs) are a relatively new addition to the numerous materials used to study the effect of entropy as they have a single-phase orientation of crystals made of several elements, which provides broad scope of chemical parameters and limitless combinations of elements. Their adaptability permits the design of compounds to be fine-tuned for specific properties and applications.

In devices like batteries and supercapacitors (SCs), HEMs offer long-lasting performance solutions for electrodes. Notably, they help to diminish the formation of dendrites, a major issue affecting the efficiency and safety of metal batteries [3]. HEMs exhibit unique characteristics, including a strong tendency to construct single-phase solid fluids with simple crystal structures, superior properties compared to their individual components, and the ability to fine-tune functional attributes [4]. Various HEMs have been explored across multiple applications,

High-Entropy Materials for Energy Storage Devices, First Edition.
Edited by Chien-Te Hsieh, Pradeep Kumar Panda, and Arpan Kumar Nayak.

including thermoelectric systems, thermal and sustainable preservations, electrochemical devices, and catalysis [5, 6]. A wide variety of compounds, including alloys, oxides, oxyfluorides, borides, carbides, nitrides, sulfides, and phosphides, are included in these materials [7–16]. HEM's wide compositional flexibility allows for the tuning of physical and chemical characteristics across specific phase spaces, as well as related kinetic effects. One of the most intriguing aspects of HEMs is their role in energy storage systems, where they demonstrate outstanding electrokinetic behavior and durability. High-entropy influences energy storage materials in two keyways. First, it enhances structural stability during charge–discharge cycles, preventing crystal structure collapse and fragmentation for the long-term performance of energy storage devices. Second, it increases active sites available for electrocatalytic reactions, thereby boosting the system's overall reversible capacity. These benefits make HEMs attractive options for energy storage technologies of the future [17, 18]. Understanding the electrokinetic of HEMs is necessary for optimizing their performance in energy storage applications. This chapter explores the fundamental electrochemical mechanisms governing HEMs behavior, concentrating on their role in numerous energy storage devices.

Energy storage systems in electrochemistry operate on distinct principles of charge storage. Energy storage in electrical double-layer capacitors (EDLCs) is accomplished through the separation of charge and electrostatic energy at the electrode–electrolyte interface, taking advantage of large surface area complexes. Fast and reversible redox reactions that increase charge storage capacity are exploited at the electrode surface in pseudocapacitors. HEMs have shown a remarkable potential to be used in energy storage applications, specifically on account of their proficiency to enhance charge retention and cycling stability. This chapter mainly discusses about SCs and lithium-ion batteries (LIBs) in which HEMs improve the structural durability and capacity retention by moderating volume expansion experienced during lithiation/delithiation cycles, and zinc-ion batteries (ZIBs) where it is elucidated that the multicomponent nature of HEMs accelerate diffusion kinetics of zinc ions and mitigate dendrite formation, thereby enhancing the lifespan of the battery, making ZIBs an alternative for future-generation energy storage systems [19, 20].

To fully take advantage of the benefits of HEMs, an advanced comprehension of their electrokinetic mechanisms is imperative. HEMs in electrochemical energy storage is an exceptionally progressing field of research which presents promising scope for the development of batteries delivering high performance and SCs. HEMs assist in the growth of next-lineage energy storage technologies by stabilizing crystal structures, increasing the number of reaction sites, and optimizing electrokinetic properties. This chapter offers a thorough overview of the function of HEMs in energy storage by fusing basic knowledge with real-world applications.

5.2 Fundamentals of Electrokinetics in Energy Storage Devices

SCs are a division of energy storage devices that use Faradaic electrochemical or high-speed electrostatic processes. Active materials like conducting polymers, metal oxides, or high surface porous carbons store the charge mostly at the electrode–electrolyte interface. SCs have a lower specific energy but can deliver substantially rapid charge and discharge rates in minimum time when compared to batteries. In conjunction with their elevated power densities, SCs offer several other benefits over batteries, including great performance stability, extended cycling life, excellent efficiency, and superior operational safety. A key electrochemical distinction between SCs and batteries is their linear voltage increase during sustained current charging, with charge being retained and discharged right away from the electrodes [21].

The charge storage capacity of a capacitive electrode, which remains constant over a specific voltage range, is determined using the following equation:

$$C = \frac{\Delta Q}{\Delta U} \tag{5.1}$$

The voltage received by the electrode is denoted by ΔU, whereas the charge that is stored is denoted by ΔQ. Farads (F) are the units of measurement for capacitance (C), which indicates the capacity to store charge within a specific voltage window where it is measured. On the other hand, a Faradaic reaction is what determines the capacity of an electrode of the battery type, which can be stated in either Coulombs (C) or milliampere-hours (mAh). The energy-repository method of the respective electrodes is divided into two basic groups, which can be used to classify the approach: (i) EDLCS and (ii) pseudocapacitors.

5.2.1 Electric Double-Layer Capacitors

The electric double-layer (EDL) concept was initially introduced and modeled by Von Helmholtz in the 19th century when he examined how opposing charges were distributed at the colloidal particle contact, as shown schematically (Figure 5.1). Due to the vast interfacial surface area and the wide range of charge separation lengths, SCs utilizing the EDL mechanism can reserve significantly high energy. The charge storage mechanism in EDLC can be explained using the Helmholtz double-layer model along with the Gouy–Chapman concept. According to the Helmholtz double-layer concept, an atomic distance separates layers of opposing

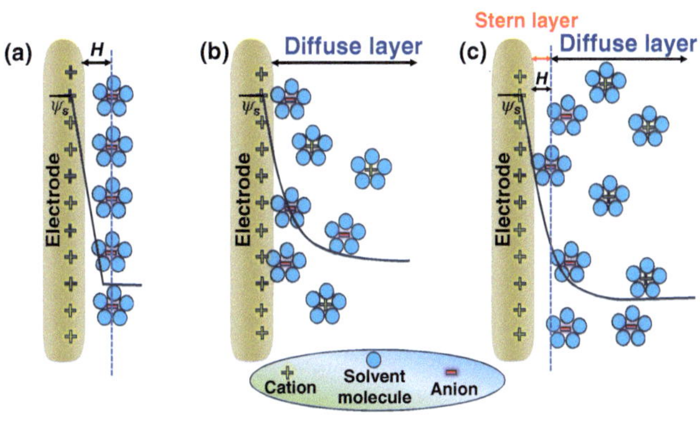

Figure 5.1 The classical model of EDLC.

charges that are formed at electrode–electrolyte interface. The model shows similarity to conventional capacitors with two plates. Gouy and Chapman further altered this straightforward Helmholtz EDL model by taking into account an invariable dispersion of electrolyte ions throughout the electrolyte, propelled by thermal motion, known as the diffuse layer. The Gouy–Chapman model, however, causes the EDL capacitance to be overestimated. When point charge ions are present near the electrode surface, a very high capacitance value would result because two separate arrays of charges have a capacitance that grows in a manner that is inversely variation to the separation between them. Stern subsequently integrated the Gouy–Chapman model along with the Helmholtz model to explicitly identify two ion distribution regions: the diffuse layer and the internal region known as the Stern layer.

EDLCs comprised of two carbon-like electrodes, a separator, and an electrolyte. EDLCs do not transmit charge in the middle of the electrode–electrolyte; instead, they store charge electrostatically, or non-Faradaically, like traditional capacitors. Energy is stored in EDLCs through the utilization of charge formed at the double layer electrochemically. Charge accumulation occurs on the surface as voltage is applied to the electrodes. As a result of the inherent attraction between oppositely charged ions, ion diffusion occurs within the electrolyte, across the separator, and through the pores of the electrode, while the electrodes are designed to prevent these ions from recombining. Because there is negligible charge transfer in the middle of the electrode–electrolyte, the compositional changes that are linked with non-Faradaic reactions are eliminated. This makes EDLC's charge storage extremely reversible, enabling them to attain extremely high cycling stabilities. The specific capacitance of an EDLC is greatly influenced by the outer facet properties of carbon-based substances and the available material surface area used for the electrodes [22]. An EDLC electrode's capacitance is calculated using the following equation (5.2):

$$C = \frac{\varepsilon_r \varepsilon_0}{d} A \tag{5.2}$$

The electrode material's active surface area that the ions from the electrolyte can reach is denoted by A, ε_r represents the relative permittivity linked to the aqueous electrolyte that is being taken, ε_0 represents the vacuum permittivity, and d represents the Debye length (Figure 5.2).

5.2.2 Pseudocapacitors

The charge is transferred through the electrode into the electrolyte in pseudocapacitors, which store charge Faradaically in contrast to electrostatic EDLCs. This occurs via intercalation, redox reactions, and electrosorption. Comparing pseudocapacitors to EDLCs, these Faradaic mechanisms allow them to obtain superior capacitance and energy density. This process is linked to a change in the electrode material's valence state carried on by electron transport. For formal purposes, the term pseudocapacitance describes charge transfer Faradaic operations across the double layer that allow electrode materials with capacitive

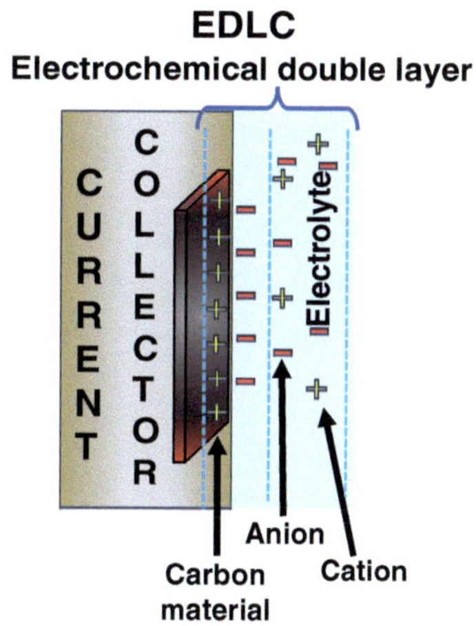

Figure 5.2 The charge storage mechanism for an EDLC.

electrochemical characteristics to retain charge. Within the potential window, their capacitance values are comparatively stable. According to the Eq. (5.3), the charge stored Q is directly variable to the voltage V.

$$Q = CV \tag{5.3}$$

Current (I) is obtained by differentiating charge (Q) with respect to time (t), as shown in Eq. (5.4):

$$I = C\frac{dV}{dt} \tag{5.4}$$

where C, also referred to as specific capacitance and measured in $F\,g^{-1}$ or $F\,cm^{-2}$, is the proportionality constant. The electrochemical behavior of energy storage incorporating pseudocapacitance lies between that of pure electrostatic EDLCs and solid-state diffusion in battery-type materials, where charge storage is dominated by Faradaic processes. Compared to EDLCs, pseudocapacitive materials exhibit higher capacitance but lower power density. Redox charge transfer mechanisms include both the intercalation process and under-potential deposition, such as the deposition of lead on gold [23].

As demonstrated (Figure 5.3), several Faradaic processes can produce electrochemical capacitive characteristics. (i) The occurrence of underpotential deposition is when the flow of ions accumulate at potentials favorable with a reversible redox potential on the 2D metal-electrolyte interface (e.g. H^+ on Pt) [24]; (ii) Redox pseudocapacitance is a Faradaic redox system in which a certain quantity of species that are reduced electrochemically deposited over the outer layer of oxidized species (or vice versa) (e.g. RuO_2 [25] or MnO_2 [26], as well as some conducting polymers); (3) intercalation pseudocapacitance happens when intercalation of ions takes place to a redox-active material without a phase shift as well as on a time period comparable to that of an EDLC (e.g. Nb_2O_5) can be observed. In order to differentiate the behavior of battery-type mechanisms from pseudocapacitive charge storage processes, capacitive features, for instance the dependency of charge accumulation across the potential window, are crucial kinetic features. This is the case despite

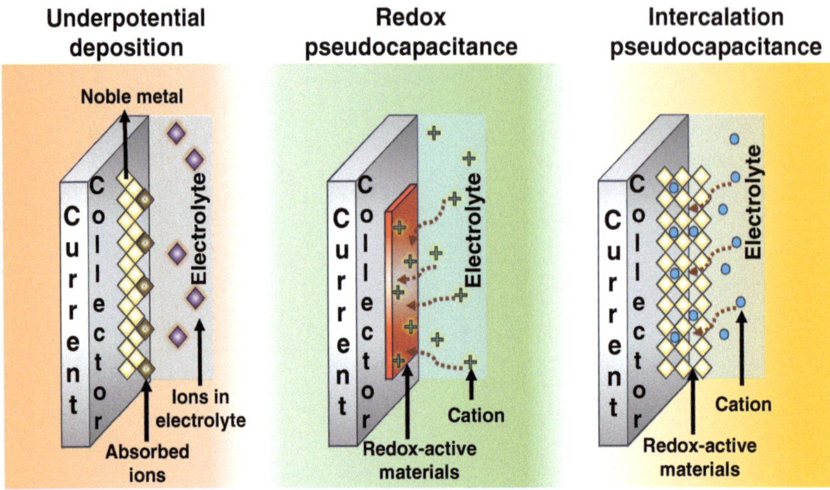

Figure 5.3 The charge storage mechanism of pseudocapacitor.

the fact that the charge storage processes in pseudocapacitive materials are called Faradaic. It is feasible to differentiate the pseudocapacitive charge storage process relative to battery materials by considering the reliance of the current output on the sweep rate. This can be accomplished through the utilization of low scan rate cyclic voltammetry (CV) measurements:

$$I = av^b \tag{5.5}$$

where a and b are configurable parameters, and I denotes the current measured at a sweep rate v. The b is used to evaluate the capacitive and diffusion-controlled contributions. This equation calculates the electrokinetic performance of several nano-electrode materials. A diffusion-controlled Faraday process, is implied by $b = 0.5$, whereas a non-diffusion-controlled capacitive behavior corresponds to $b = 1$. The CV curve for EDLC materials is rectangular, and b is always equal to 1. When the CV curve keeps its optimal rectangular form as the scan rate rises without the anodic and cathodic peaks shifting negatively and positively, respectively, it describes a superior capacitive nature. The CV curve for pseudocapacitive materials is roughly rectangular with a few inflection points, and the b value is near 1. Ion intercalation for Faradaic reactions that are solely battery-type is constrained by solid-state ion diffusion mechanisms. The square root of the scan rate should therefore be proportional to the voltage-meter current, and the specific relationship is explained by Eq. (5.6):

$$i = nFAC \cdot D^{1/2}v^{1/2}\left(\frac{an_aF}{RT}\right)^{1/2} \pi^{1/2}\chi^{(bt)} \tag{5.6}$$

Comparing the two equations above, Cottrell's equation (5.7) is satisfied when the slope $b = ½$, indicating semi-infinite diffusion-controlled charge storage, as observed in bulk battery materials.

$$i = av^{½} \tag{5.7}$$

The response of the capacitive current is linearly dependent on ($b = 1.0$) as per Eq. (5.8):

$$i = vC_dA \tag{5.8}$$

where the capacitance is denoted by C_d. In this instance, charge storage has characteristics like those of a capacitor. Based on previous research and CV findings, a related study shows that the "i" at a specific potential, V, is the sum of two distinct charge storage contributions: a solid-state-dominated Li-ion intercalation and a capacitive contribution. Furthermore, the capacitance contribution ratio between the electrode's two distinct processes at various sweep speeds can be calculated by Eq. (5.9):

$$i = k_1v + k_2v^{1/2} \tag{5.9}$$

where the current contributions from the diffusion control and capacitance areas are denoted by k_1v and $k_2v^{1/2}$.

Electrochemical impedance spectroscopy, denoted as EIS, is a non-destructive method that offers essential understanding into reaction kinetics by distinguishing different processes occurring at an electrode over various timescales. It is among the most illuminating analytical techniques for comprehending the reaction kinetics of the Li^+ through insertion/deinsertion process. For LIBs to be successful, a preliminary solid electrolyte interphase (SEI) must form in the initial cycle, particularly at the time of the fundamental lithiation process. In the following cycles, the SEI morphology can mature. Consequently, it has been believed that the creation of the SEI is a necessary step in the production of Li^+ batteries [27]. Since the SEI is an exterior layer that permits unrestricted ion mobility while separating electron transport, its formation may be examined in situ through the EIS approach. Its primary chemistry consists of the non-soluble breakdown of the electrolyte constituents. In EIS, the semicircle at medium frequencies is represented by the charge transfer resistance (R_{ct}) and its relative double-layer capacitance (C_{dl}). The straight sloping line at the low-frequency end, is the Warburg impedance associated with the combined effect of Li^+ diffusion on the electrode–electrolyte interfaces. Faradaic impedance, which represents cell reaction kinetics and changes significantly with the state of lithiation, is the result of combining charge transfer impedance with Warburg impedance. Additionally, EIS data was used to compute the Li^+ diffusion coefficient (D_{Li^+}) using the given equations:

$$D_{Li^+} = \frac{R^2T^2}{2A^2n^4F^4C^2\sigma_{\omega^2}} \tag{5.10}$$

$$Z' = R_e + R_{ct} + \sigma_\omega \omega^{-0.5} \tag{5.11}$$

where A denotes the electrode's active surface area, T refers to the temperature, F as the Faraday constant, n is the number of electrons, C_0 is the amount of substance concentration of Li^+ in the electrolyte, and R denotes the gas constant [28].

5.3 Mechanistic Insights: Electrokinetics of HEMs in Energy Storage

5.3.1 Supercapacitors

SCs, often referred to as electrochemical capacitors or ultracapacitors, provide capacitances that are numerous orders of magnitude greater than those of traditional capacitors by combining an enormous area of electrode material with slender electrolytic dielectrics. Scientists are now interested in SCs because of their advantage over batteries. SCs are more efficient than batteries, which exhibit a sharp decrease in efficiency as a result of rapid cycles of charging. The electrokinetics in SCs are explained using electrochemical characterization methods as EIS, galvanostatic charge–discharge (GCD), and CV.

Zhu et al. synthesized NiCoMoMnZn-layered double hydroxide (LDH) as an electrocatalytic material for SCs employing an innovative template re-etching approach to enhance energy density [29]. NiCoMoMnZn-LDH serves as a good electrode material for SCs, characterized by a homogeneous elemental distribution, elevated specific surface area, and a permeable, durable structure. Despite the fact that the metal organic framework (MOF) template's partial shape and pore architecture are preserved, the total surface area of the catalyst is significantly greater, which suggests that the electrochemical characteristics of the material can be significantly enhanced by eliminating the organic ligand structure. In NiCoMoMnZn-LDH CV curve, there was easily discernible evidence of a redox peak, indicating that a pseudocapacitive effect is the energy storage mechanism of the materials. The electrochemical reaction during continuous charge/discharge process was given as

$$\mathrm{M(OH)}_x(\mathrm{s}) + y\mathrm{OH}^-(\mathrm{aq}) \leftrightarrow \mathrm{MO}_y(\mathrm{OH})_{x-y}(\mathrm{s}) + y\mathrm{H_2O}(\mathrm{aq}) + y\mathrm{e}^-$$

$$\mathrm{MO}_y(\mathrm{OH})_{x-y}(\mathrm{s}) + (x-y)\mathrm{OH}^-(\mathrm{aq}) \leftrightarrow \mathrm{MO}_x(\mathrm{s}) + (x-y)\mathrm{H_2O}(\mathrm{aq}) + (x-y)\mathrm{e}^-$$

$$(\mathrm{M} = \mathrm{Ni, Co, Zn, Mn, Mo})$$

At 50 $\mathrm{mV\,s^{-1}}$, the CV plot exhibited minimal variation, signifying that the electrochemical durability of the electrode is robust. Furthermore, it is evident from the location of the redox peak that varies with the scan rate, primarily due to the polarization phenomena occurring during the electrode response. The chemical reaction kinetics of NiCoMoMnZn-LDH were studied using CV curves in order to deepen their concept of the energy storage mechanism of the compound. The anodic and cathodic peaks' obtained b values were 0.50 and 0.56, respectively, suggesting that the electrode response combines capacitive along with battery-like charge storage mechanisms in a way that is complementary. They analyzed the rate of contribution of the capacitance of all materials in order to ascertain the amount of the battery effect and the pseudocapacitance influence that occurred throughout the electrochemical reaction process at 5–50 $\mathrm{mV\,s^{-1}}$. Their findings revealed that the pseudocapacitance ratio of all materials escalates with higher scanning speeds. This is because pseudocapacitance was primarily governed by the kinetics of surface reactions, particularly the electron transfer rate at the outermost layer of the electrocatalytic material and also the frequency of the primary acceptor reaction in the diffusion-controlled process, which determines ion transfer within the electrode. As the scanning speed increases, electron transport outpaces ion diffusion, limiting ion penetration into the lattice space and so enhancing the relative contribution of pseudocapacitance. Vadivel et al. synthesized BiOX (X = Br, Cl, I) CO_3, referred to as $BiOXCO_3$ electrode materials, via the solvothermal method [30]. $BiOXCO_3$ (1) represents an equimolar concentration of X and CO_3. $BiOXCO_3$ (2) was synthesized by doubling the amount of bromine, followed by $BiOXCO_3$ (3) with chlorine, $BiOXCO_3$ (4) with iodine, and $BiOXCO_3$ (5) with carbonate, of which $BiOXCO_3$ (5) gives a superior specific capacitance of 645 F g^{-1} at a current density of 1 A g^{-1}. To further analyze the kinetics of $BiOXCO_3$ (5), the correlation between peak current values and scan rate was examined using CV measurements to distinguish the capacitive and diffusion-regulated processes and it exhibited a diffusion contribution of roughly 39% and capacitive contribution of 61% at 25 mV s^{-1}, with the "b" value, i.e. 0.924 (as per Eq. 5.5). The redox process of $BiOXCO_3$ is shown in Figure 5.4.

Yin et al. synthesized porous high-entropy oxides (HEOs) utilizing polyethylene glycol (PEG) as a matrix. They investigated the impact of various types of PEG on the resulting samples [31]. The samples HEO-PEG-400, 1000, and 2000 exhibited simple cubic structures and were susceptible to generating oxygen vacancies compared to the HEO, which had a spinel structure. Among these, HEO-PEG-2000-2 demonstrated superior electrochemical performance, achieving a specific capacitance of 204 F g^{-1} at 0.5 A g^{-1}, with 60% capacitance retention. HEO-PEG-400-2, 1000-2, and 2000-2 exhibited respective b values of 0.52, 0.53, and 0.56. This demonstrated that the battery and capacitor outputs were included in the charge storage techniques of all the electrode materials. At 3 mV s^{-1}, the output ratios of the pseudocapacitance for HEO-PEG-400-2, 1000-2, and 2000-2 were 40%, 47%, and 50%, respectively. The redox reactions that were primarily controlled by the surface and near-surface

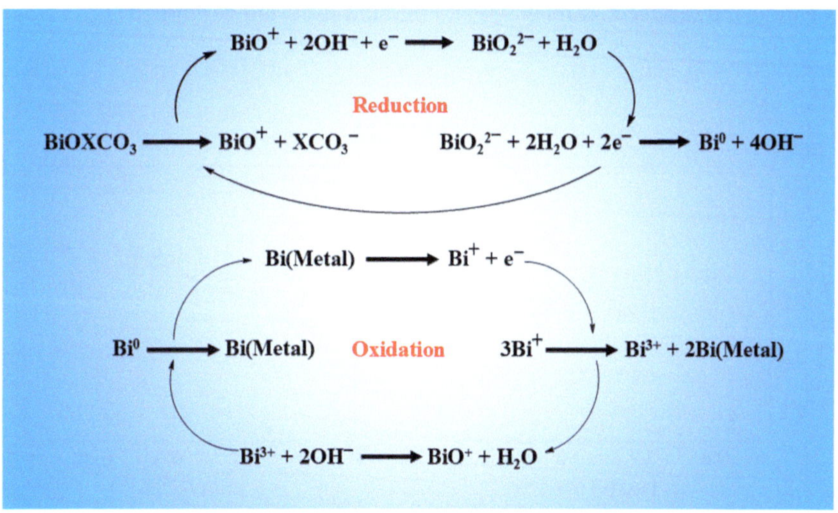

Figure 5.4 Redox mechanism of $BiOXCO_3$. *Source:* Reference [30] / with permission of Elsevier.

determined the charge storage mechanism. HEO-PEG-2000-2 was shown to have the greatest pseudocapacitance contribution rate because of its porous structure with significant active sites for ion redox processes. Mohanty et al. used induction melting to synthesize Fe-Co-Ni-Cu-Zn and ball milled it to minimize its dimensions [32]. The maximum capacitance was found to be 325.17 F g^{-1} using 3 M KOH. The aqueous electrolyte reacts with each surface oxides as follows:

$$NiO + OH^- \leftrightarrow NiOOH + e^-$$

$$CoO_2 + H_2O + e^- \leftrightarrow CoOOH + OH^-$$

$$Fe_3O_4 + 2OH^- \rightarrow 3Fe(OH)_2 + 2e^-$$

$$3Fe(OH)_2 + 2OH^- \rightarrow Fe_3O_4 + 4H_2O + 2e^-$$

$$Cu(II) + OH^- \leftrightarrow CuOH + e^-$$

$$CuOH + OH^- \leftrightarrow CuO + H_2O + e^-$$

$$ZnO + OH^- \leftrightarrow ZnOOH + e^-$$

The slope log(I) vs. log(v) yields the "b" value for this high-entropy alloy (HEA) electrode which was 0.66 (from Eq. 5.5) (Figure 5.5a). This "b" value suggests the presence of a surface-controlled quick Faradaic reaction on the electrode's surface section along with sluggish diffusion-regulated pseudocapacitive reduction/oxidation reaction involving the K^+ ion into Fe-Co-Ni-Cu-Zn HEA. This enhances specific capacitance by enabling ions to be determined by both the inner and outside regions active sites related to multivalent reactions. Therefore, the mesoporous architecture of the material depends on its pseudocapacitive property.

For the purpose of determining the overall charge that was held within the electrode, $Q(t)$, an electrochemical procedure was carried out. This was accomplished by adding the surface-controlled Q_S and the slow diffusion-controlled Q_d.

$$Q(t) = Q_s + Q_d \quad (5.12)$$

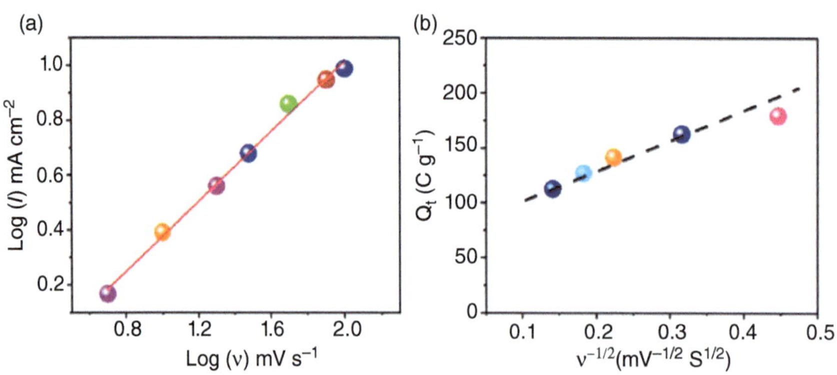

Figure 5.5 (a) Log(I) vs. log(v) for the determination of "b" value, (b) total charge vs. scan rate inverse square root. *Source:* Reference [32] / with permission of Elsevier.

By measuring $Q(t)$ against the sweep rate (inverse square root), the values of Q_S and Q_d were determined.

$$Q(\upsilon) = Q_s + Q_d = Q_{capacitive} + K_1\upsilon^{-1/2} \tag{5.13}$$

At lower scan speeds, the relative contributions of Q_S and Q_d demonstrate that diffusion-controlled processes have larger contributors than Q_S (Figure 5.5b). This indicates that the mesoporous channels of their multivalent quinary ions provide less resistive ionic transport paths, as well as the fact that the material possesses a significant number of Faradaic redox active sites that are accompanied by porous structures that are in the process of slow diffusion control.

Ren et al. studied high-entropy antimonides in microwave oven that are used in homes with atmospheric air for the first time, offering a new low-energy way to create HEM [33]. The battery-supercapacitor hybrid (BSH) device (NiCoFeZnSb//Bi_2O_3@SWCNT) was utilized for this purpose and it achieved 1,850.0 C g^{-1} of capacity. The electrochemical functioning of the high-entropy antimonide electrode was evaluated using the current i with relation to the scan rate v to clarify the charge storage mechanism (from Eq. 5.5). The anode and cathode have respective b values of 0.51 and 0.50 (Figure 5.6a). The findings demonstrate that a diffusion-controlled Faraday mechanism of high-entropy antimonides allows the electrodes to have capacities significantly higher than those of pseudocapacitance and double-layer materials. It was illustrated that the high-entropy antimonide electrode's capacitive contribution is 62.1% at 100 mV s^{-1} (Figure 5.6b). The majority of the overall capacitance was progressively occupied by the surface-controlled capacitance as the scan rate increases. Wei et al. used a combination of solution combustion synthesis and directional freeze-casting to create self-supporting 3D permeable composite electrodes $(CoCrFeMnNi)_3O_4$@Ni referred to as HEO@Ni [34].

They studied the electrochemical reaction kinetics using CV plots at various scan rates to investigate the mechanism of HEO@Ni and its reliance on energy storage. The redox reactions were regulated by semi-infinite diffusion, as evidenced by the b values as 0.469, being comparable to the value of battery-type materials as 0.5,

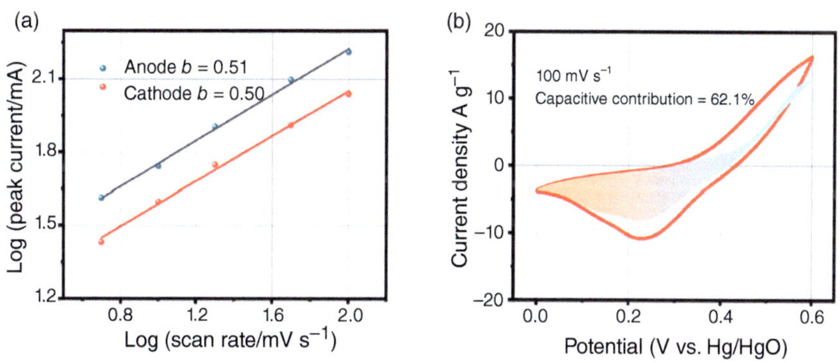

Figure 5.6 (a) Log (current) vs. log (scan rate) for the calculation of "b" value, (b) graph of voltage-ampere response. *Source:* Reference [33] / with permission of Elsevier.

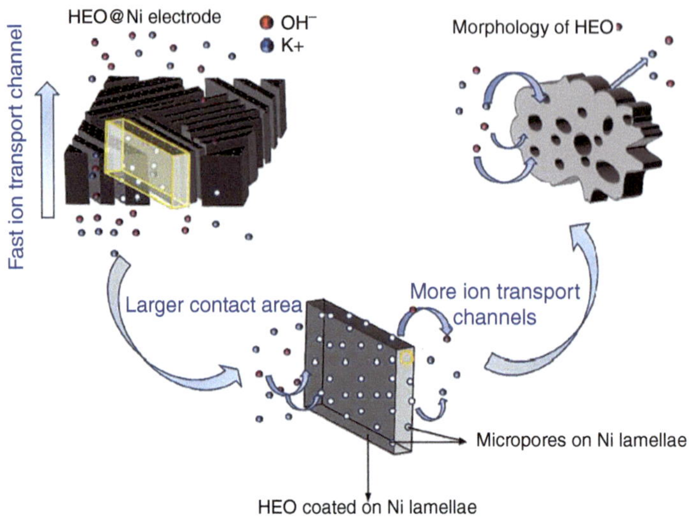

Figure 5.7 Mechanism of charge transfer of HEO@Ni. *Source:* Reference [34] / with permission of Elsevier.

calculated from Eq. (5.5). Their findings indicate that straight-aligned channels significantly minimize electrode contortion while enhancing the electrode–electrolyte interface. This, in turn, promotes efficient interfacial charge transfer and enhances electrokinetics (Figure 5.7). Furthermore, the special lotus flower-like morphology of the HEO coating layer enhanced the micropore quantity, made it easier for the electrolyte to penetrate, and accelerated the active material's rate of activation.

Shen et al. designed HEA nanoparticles with a thin metal oxide layer (MOL) that were attached to carbon that is based on hyper-crosslinked polymer-derived carbon (HCPC) by adsorbing five metal ions in the HCPC, reducing those ions in situ (Fe^{2+}, Co^{2+}, Ni^{2+}, Cu^{2+}, and Sn^{2+}), and then undergoing carbonization [35]. At varying potentials, the b values for HEA-NP@MOL/HCPC-2.0 range from 0.89 to 0.93, indicating the rapid reaction kinetics of the charge storage mechanism. Thorough examination revealed that the sluggish kinetic capacitance falls as the scan rate increases, whereas the capacitance of quick kinetic processes hardly changes. This phenomenon was probably attributable to the observation that the capacitance associated with rapid kinetic processes, which encompassed the capacitive phenomena of the EDL and swift redox reactions occurring at the exterior of the MOL and HCPC substrates, was neither constrained by charge transfer nor governed by diffusion. In contrast, the capacitance contribution was predominantly characterized by redox reactions that were primarily limited by inner diffusion. Table 5.1 summarizes the various kinetic parameters obtained during the electrochemical reactions of HEMs in SCs.

Table 5.1 Summary of electrochemical kinetic parameters for various HEMs in supercapacitors.

HEMs	Synthesis method	Specific capacity ($F\,g^{-1}$)	*b* value	R_s (Ω)	R_{ct} (Ω)	References
NiCoMoMnZn-LDH	Template re-etching	1,810.2	0.50, 0.56	0.62	5.11	[29]
$BiOX_{(X=Br, Cl, I)}\ CO_3$	Solvothermal	645	0.924	—	—	[30]
$(FeCoCrMnNi)_3O_4$	Sol–gel	204	0.52, 0.53, 0.56	1.2–3.2	3.2–27	[31]
FeCoNiCuZn	Induction melting, ball milling	325.17	0.66	—	—	[32]
NiCoFeZnSb	Microwave	3,700	0.51, 0.50	0.67	0.70	[33]
$(CoCrFeMnNi)_3O_4$@Ni	Solution combustion, directional freeze-casting	942	0.469	2.49	2.96	[34]
HEA-NP@MOL/HCPC-2.0	Hyper crosslinking process	495.4	0.89–0.93	8.5	0.4	[35]
Fe-Co-Ni-Cr-Mn	Induction melting, ball milling	450	0.736	7.5	7.2	[47]
$(Ti_{0.2}V_{0.2}Nb_{0.2}Mo_{0.2}W_{0.2})_3AlC_2$	Selective removal of Al	158.43	0.75	0.85	0.02	[48]
$(FeCoNiCuZn)_3O_4$	Induction melting, ball milling	245.7	0.68	—	—	[49]
$(CoCrFeMnNi)_3O_4$	Co-precipitation	239	—	0.96	1.56	[50]
$(Co_{0.2}Cr_{0.2}Fe_{0.2}Mn_{0.2}Ni_{0.2})_3O_4$	Polyacrylamide gel method	384	—	0.54	—	[51]
$(FeCoCrMnZn)_3O_4$	Facile solid-state reaction	340.3	—	1.1	—	[54]
FeNiCoMnMg HEA-NPs/ACNFs	Carbothermal shock	127	—	—	13	[55]

(continued)

Table 5.1 (Continued)

HEMs	Synthesis method	Specific capacity ($F g^{-1}$)	*b* value	R_s (Ω)	R_{ct} (Ω)	References
HEO–CNT	Sol–gel autocombustion	144	—	2.2	1.9	[56]
HEO/f-CSAC	Carbonization, acid functionalization	147.5	—	2.8	3.7	[57]
NiCuFeCoMn	Hydrothermal	1, 241	—	0.8	2	[58]
$Fe_3Cr_3Mn_2NiV$	Mechanical alloying	141.4	0.725	7.8	1,659.2	[59]
$(Fe_3Cr_3Mn_2NiV)O_4$	Oxidation process	294	0.757	11.5	36.4	[59]
HEA-NP@MOL/HCPCs	In situ reduction, carbonization	495.4	0.89–0.93	—	0.4	[60]
$(FeCoNiCrMn)_3O_4$	Dealloying	639	0.649, 0.743	0.65	0.2868	[61]
FeCoNiCrMn	Induction melting, ball milling	386.66	0.65	—	—	[62]
FeCoNiCrZn	Induction melting, ball milling	556	0.71	3.09	—	[63]
$(FeCoNiCuZn)S_2$	Solvothermal, sulfurization	512.67	0.772, 0.814	1.555	0.53	[64]
0.1-NCB@HES	Solvothermal, sulfurization	1, 239.1	0.759, 0.816	0.68	0.108	[64]
MnCuFeNiCo-S	Hydrothermal	—	0.59	0.86	2.37	[65]
$(FeCoNiCuMn)_3O_4$	Liquid-state induction melting	313	0.78	1.0	—	[66]

5.3.2 Li-Ion Batteries

In today's world, LIBs are an essential component of electric vehicles (EVs), microchips, grid storage systems, and mobile electronics. Due to their many benefits, including high-energy density, excellent efficiency, long cycle performance, and minimal self-discharge, LIBs are currently at the forefront of electrochemical energy storage. The high-entropy design idea has significantly broadened the design space for energy storage materials and created scope to overcome the constraint performance of LIBs. For this, various researchers have come up with the different HEMs to explain the electrokinetics in LIBs.

Xiao et al. effectively synthesized $(FeCoNiCrMn)_3O_4$ HEO by oxidizing high-entropy FeCoNiCrMn alloy powders, and they used it as a novel electrode for LIBs [36]. With a capacity of 596.5 mA h g^{-1} and 86.2% of a remarkable capacity retention, $(FeCoNiCrMn)_3O_4$ HEO demonstrated superior cycling activity and durability. The improved electrochemical behavior is assigned to the factors mentioned. First, the amount of ions in the lithium storage method is raised, as well as their theoretical specific capacity, by the comparatively large concentration of trivalent as well as quadrivalent ions in $(FeCoNiCrMn)_3O_4$ HEO. Second, the high concentration of oxygen vacancies creates a large number of pathways for electron transport, which enhances electron mobility and electrochemical performance in general. Third, in the cyclic process, the production of secondary particles made of nanoparticles successfully reduces the volume amplification impact of nanoparticles and enhances structural stability of $(FeCoNiCrMn)_3O_4$ HEO. Fourth, the HEO's electrochemical efficiency is further increased by the small band gap, which makes electron transitions simpler and accelerates electron transport. The following illustrates the charge–discharge process:

$$M_2O_3 + MnO_2 + 4Li^+ \rightarrow 3MO + 2Li_2O$$

$$MO + 2Li^+ \rightarrow M + Li_2O$$

$$(M = Fe, Co, Mn, Cr \text{ and } Ni)$$

The mechanism of improved electrochemical performances was revealed by testing EIS between 0.1 and 100 kHz. The R_{ct} of $(FeCoNiCrMn)_3O_4$ HEO continued to decrease as the cycle number increased, indicating accelerated charge transfer upon electrochemical activation. Upon initial electrochemical activation, its D_{Li^+} dropped abruptly to 7.85×10^{-15} cm^2 s^{-1}. After 25 cycles, D_{Li^+} began to increase gradually, extending to 15.5×10^{-15} cm^2 s^{-1} at 100 cycles. This behavior during the 25 cycles has shown progressive stabilization of the SEI, which reduces the number of lithium-ion diffusion channels, thereby slowing ion diffusion. Furthermore, R_{ct} steadily dropped while D_{Li^+} continued to rise, suggesting an improvement in electron and ion transport kinetics. (Figure 5.8a) illustrates the b values. Since the CV plot exhibits two peaks, two b values were calculated: 0.892 and 0.825. This suggested that the charge storage performance arises from a combination of Faradaic and capacitive processes. Furthermore, as the scan rate increased, the capacitive contribution increased from 40.76% to 82.02% (Figure 5.8b) and this signifies excellent rate performance.

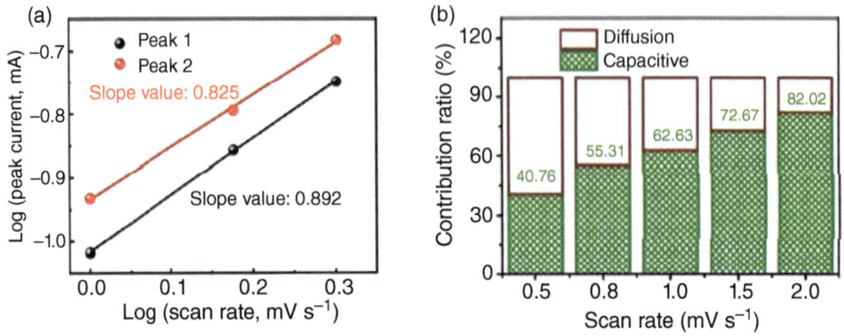

Figure 5.8 (a) Log (current) vs. log (scan rate) for the calculation of "*b*" value, (b) ratios of the diffusion-controlled and capacitive charge contributions. *Source:* Reference [36] / with permission of Elsevier.

Chen et al. used a simple solid-state reaction technique and ball milling to create a novel type of spinel $(Mg_{0.2}Ti_{0.2}Zn_{0.2}Cu_{0.2}Fe_{0.2})_3O_4$ material [37]. The submicrometer-sized particles of the material demonstrated exceptional Li storage capabilities when employed as LIB anodes, providing an impressive reversible capacity of 504 mA g^{-1} at a *j*@100 mA g^{-1} after 300 cycles. In order to examine the electrical conductivity and Li^+ transfer in the material's anode, EIS plots were assessed both before and after 300 cycles and at the open-circuit voltage (OCV), 3.0 V. It was observed that the fresh electrode's semicircle diameter was larger than that recorded after 300th cycles, suggesting a decrease in Li^+ diffusion resistance with cycling. The Li^+ diffusion coefficient (calculated from Eq. 5.10) of the electrode after 300 cycles was 4.716×10^{-15} cm^2 s^{-1}. This improvement explains the electrode's enhanced electrochemical performance. To have a better insight of the kinetics mechanism, the kinetics of lithium uptake within the electrode were investigated and the three-peak slope (*b* values) were determined as 0.97, 0.86, and 0.90, respectively. These values indicate that Li storage was predominantly governed by a surface redox reaction process. From the CV curves, it was observed that the capacitive contribution is 62.9% at 0.1 mV s^{-1} and increases to 93.2% at 1.0 mV s^{-1}. The strong pseudocapacitive contribution leads to the unique framework of the materials and enables Li^+ to interact immediate interaction with the active substance without the need for long-range diffusion pathways, thereby enhancing the electrode's rate capability. Ren et al. utilized Nyquist plots to evaluate the electrical conductivity and Li^+ transfer in (MgCoNiCuZn)O anode in LIBs [38]. Their results showed that the EIS spectra exhibited a smaller semicircle after 390 cycles, indicating improved conductivity of the material. To gain deeper insight into lithium storage behavior, detailed kinetic tests were conducted. The analysis revealed that the *b* values of (MgCoNiCuZn)O were 0.74 and 0.73 at the oxidation and reduction peaks, respectively. The charge storage was generated via an assortment of diffusion control as well as non-diffusion control methods, according to this *b* value. Furthermore, the contribution of diffusion and non-diffusion-controlled capacity was estimated by using Eq. (5.9). From 0.2 to 8 mV s^{-1}, the proportion of capacitive behavior rose significantly from 29.6% to 71.4%. This clearly demonstrated that capacitance effects played an increasingly dominant role in overall capacity at higher scan rates. The redox reaction during charge/discharge process was summarized as follows:

$$\text{HEO} + \frac{33}{20}\text{Li} \leftrightarrow \frac{1}{5}\text{Co}^0 + \frac{1}{5}\text{Cu}^0 + \frac{1}{5}\text{Ni}^0 + \frac{1}{5}\text{MgO} + \frac{1}{20}\text{LiZn}_4 + \frac{8}{5}\text{Li}_2\text{O}$$

Li et al. synthesized a 3D porous $(FeCoNiCuZn)_3O_4$ HEO material for LIBs and studied its electrokinetic using Cottrell's equation [39]. The b values for the anodic as well as cathodic reactions were 0.77 and 0.70, respectively, revealing that both diffusion-controlled and pseudocapacitive contributions affect the electrocatalytic processes of the HEO electrode. To further explore the structural advantages that enhance the electrochemical properties of HEO electrodes for LIBs, D_{Li^+} was calculated using Eq. (5.14):

$$D_{Li^+/Zn^{2+}} = \frac{4}{\pi\tau}\left(\frac{m_B V_m}{M_B S}\right)^2 \left(\frac{\Delta E_s}{\Delta E_t}\right)^2 \tag{5.14}$$

where D_{Li^+} denotes the diffusion coefficient and τ is the length of time that a pulse of applied constant current lasts during electrochemical studies. The molar weight/volume and active mass of the electrode are denoted by M_B, V_m, and m_B, respectively. S represents electrochemical reaction surface area. The open-circuit potential (OCP) difference between two consecutively adjacent processes was denoted by ΔE_s, whereas the voltage difference between the initial and final states inside a single pulse step was represented by ΔE_t. The calculated D^+_{Li}, typically ranging from 10^{-14} to $10^{-9.7}$ $cm^2\ s^{-1}$, reflects the rapid diffusion of Li^+ ions during the lithiation and delithiation. Furthermore, the changes in the resistance of the HEO electrode during prolonged cycling were analyzed. The Nyquist plots displayed the Warburg impedance of the HEO electrode as a linear line toward low-frequency area and the electrode/electrolyte interfacial charge R_{ct} as a dipped semicircle toward high-frequency area. The findings revealed that the HEO electrode exhibited relatively low R_{ct} and consistent linear slopes after taking the 10th and 100th cycles, indicating enhanced Li^+ kinetics. The reaction mechanism of the HEO electrode can be described as a conversion method, represented by the following equations:

Discharge process:

$$n(\text{FeCoNiCuZn})_3\text{O}_4 + x\text{Li}^+ + x\text{e}^- \rightarrow \text{Li}_x^+\text{Fe}_{3n-a}\text{Co}_{3n-b}\text{Ni}_{3n}\text{Cu}_{3n}\text{Zn}_{3n}\text{O}_{4n} + a\text{Fe}^0 + b\text{Co}^0$$

Charge process:

$$\text{Li}_x^+\text{Fe}_{3n-a}\text{Co}_{3n-b}\text{Ni}_{3n}\text{Cu}_{3n}\text{Zn}_{3n}\text{O}_{4n} + a\text{Fe}^0 + b\text{Co}^0 \rightarrow n(\text{FeCoNiCuZn})_3\text{O}_4 + x\text{Li}^+ + x\text{e}^-$$

Yen et al. demonstrated that interfacial engineering in morphology HEO@ polyaniline (HEO-MP) heterojunctions significantly enhances Li-ion storage stability and performance [40]. The material has shown a significant capacity of 325 mA h g^{-1} at 10 A g^{-1}. Furthermore, it exhibited the long-term cycling stability even after 3,200 cycles. To better understand the capacitive and diffusion-controlled contributions in HEO-MP, the kinetics of Li^+ storage were analyzed. The b values for cathodic and anodic peaks were computed to be 0.75 and 0.86, respectively, indicating that charge storage was governed by a combination of capacitive and diffusional processes. Additionally, using Eq. (5.9), the capacitive contribution was found to account for approximately 73.8% of the total charge storage at $0.6\,mV\,s^{-1}$. To study the reaction kinetics electrochemically during the charge as well as the discharge processes, in situ EIS experiments were conducted. These experiments initiated the charge and discharge processes at an OCP of 2.9 V, with a voltage window of 0.01–3.0 V and a current density of $0.1\,A\,g^{-1}$. As the electrode discharged from OCV to 1.5 V, the SEI layer developed, which led to a slight increase in electrode resistance. Li^+ and HEO reacted to form Li_2O and metallic species (Ni, Cu, etc.) when discharged further to 0.5 V, which gradually decreased resistance. During charge–discharge cycle, $LiPF_6$ undergoes the following reaction:

$$LiPF_6 + H_2O \leftrightarrow LiF \downarrow + POF_3 + 2HF$$

The byproduct LiF is non-conductive, and excessive accumulation in the SEI layer hindered Li^+ transport, thereby increasing resistance. Moreover, hydrogen fluoride was capable of dissolving transition metals in HEO, such as Ni^{2+} and Co^{2+}, and reacting with Li_2O:

$$Li_2O + 2HF \rightarrow 2LiF + 2H_2$$

Xiang et al. effectively generated spinel-type $(Al_{0.2}CoCrFeMnNi)_{0.58}O_{4-\delta}$ HEO nanocrystalline powder with a significant amount of oxygen vacancies using the solution combustion synthesis (SCS) process, and it is being investigated as a potential anode active material for LIBs [41]. The electrochemical performance of the material was examined with a Li reference electrode to examine the role of inactive Al^{3+}. CV curves of the Al^{3+}-doped anode revealed cathodic peaks at roughly 1.56 and 0.52 V that corresponded to the depletion of active d block-metal ions along with the making of Li_2O matrix. Following the electrolyte's breakdown, a massive irreversible peak at 0.21 V appears, which was followed by the development of SEI layers on the anode surface. During the first discharge cycle, two minor discharge plateaus at approximately 1.56 and 0.52 V align with the lithiation process of the Al^{3+}-doped HEO as an active material as well as the discharge curve exhibits a slow potential dip to 0.01 V, indicating that the electrolyte breakdown and interfacial lithium storage interaction generate SEI films. More significantly, inactive Al^{3+} doping enhances the nanograin crystallinity, introduces defects, prevents crack propagation, thereby increasing the active surface area for Li intercalation and deintercalation. However, after cycling, the permeable structure no longer exists. EIS was employed to assess Li-ion diffusion kinetics in both spinel-type HEO anodes before cycling and after cycling. The Li-ion diffusion coefficient (D_{Li^+}) for the Al^{3+}-doped anode increased 281.55-fold after 500 cycles, from 2.71×10^{-19} $cm^2\ s^{-1}$ to 7.63×10^{-17} $cm^2\ s^{-1}$ (determined from Eqs. 5.10 and 5.11). This was also 1.85 times more than $(CoCrFeMnNi)_{0.6}O_{4-\delta}$ anode (4.13×10^{-18} $cm^2\ s^{-1}$) after 500 cycles. Consequently, the increased electrical conductivity and Li diffusivity of the Al^{3+}-doped anode during cycling contribute to its enhanced electrochemical performance. Jiang et al. synthesized high-entropy pyrochlore oxide $(Y,Dy,Ce,Nd,La)_2Sn_2O_7$ using the co-precipitation technique [42]. The optimized $M_2Sn_2O_7$/CMC (M = Y,Dy,Ce,Nd,La) system demonstrated a high initial discharge capacity of approximately 766 mA h g^{-1} and outstanding cycle stability. These advantages result from the significant intermolecular interactions offered by the sodium carboxymethyl cellulose (CMC) matrix and the complementary effects of the high-entropy system. The electrochemical reactions involved during lithiation process are as follows:

$$M_2Sn_2O_7 + 8Li^+ + 8e^- \rightarrow 4Li_2O + M_2O_3 + 2Sn$$

$$Sn + 4.4Li^+ + 4.4e^- \rightarrow Li_{4.4}Sn$$

$$Sn + Li_2O \rightarrow SnO_2 + 4Li^+ + 4e^-$$

To further investigate the electrokinetics of the $M_2Sn_2O_7$/CMC system, EIS was performed. The Nyquist plot analysis revealed the Rs value of 1.6 Ω, which was less than the $M_2Sn_2O_7$/PVDF system (1.7 Ω). The R_{ct} values were 21.85 and 63.76 Ω for the $M_2Sn_2O_7$/CMC and $M_2Sn_2O_7$/PVDF systems, respectively, indicating faster electron charge transfer in the $M_2Sn_2O_7$/CMC system. Additionally, the $D_{Li}{}^+$ calculated using Eqs. (5.10) and (5.11) was found to be 3.299×10^{-10} cm^2 s^{-1} for the $M_2Sn_2O_7$/CMC and 2.366×10^{-10} cm^2 s^{-1} for the $M_2Sn_2O_7$/PVDF. This suggests that the enhanced charge transport in the $M_2Sn_2O_7$/CMC system contributes to an improved cycling rate.

5.3.3 Zinc-Air Batteries

Aqueous zinc-air batteries (AZIBS) are perhaps the most advanced of the several kinds of metal-air batteries and have a great capability for energy applications. Theoretically, Zn-air batteries offer a good energy density of 1,086 Wh kg^{-1} (comprising oxygen), approximately quintuple that of Li^+ batteries [43]. Of all primary batteries, Zn-air batteries provide the highest energy density across a wide range of applications. For future electric vehicle applications, they have frequently been promoted as the most practical substitute to LIBs in terms of technology, and they are also cheaper. Current research on AZIBs primarily focuses on identifying suitable cathode materials due to the significant charge density of Zn^{2+}, which leads to excessively strong electrostatic interactions of Zn^{2+} with the cathode crystal structure. Wu et al. synthesized a HEO electrode $(Co_{0.2}Cr_{0.2}Fe_{0.2}Mn_{0.2}Ni_{0.2})_3O_4$ (CCFMnO) to increase the stability of the AZIBs cathode [44]. Electrokinetics of CCFMnO was investigated through charge–discharge curves at 0.1 A g^{-1} (Figure 5.9a). The distinct potential plateau in the charge–discharge curve, attributed to Zn^{2+} intercalation and deintercalation, suggests that CCFMnO has a higher Zn storage capacity and quicker charge transfer kinetics. The remarkable electrochemical activity of CCFMnO is mostly due to the entropy stabilization potency and the difficult interactions of the constituents in the high-entropy cathode, which reinforce its durability and further enhance its electrochemical

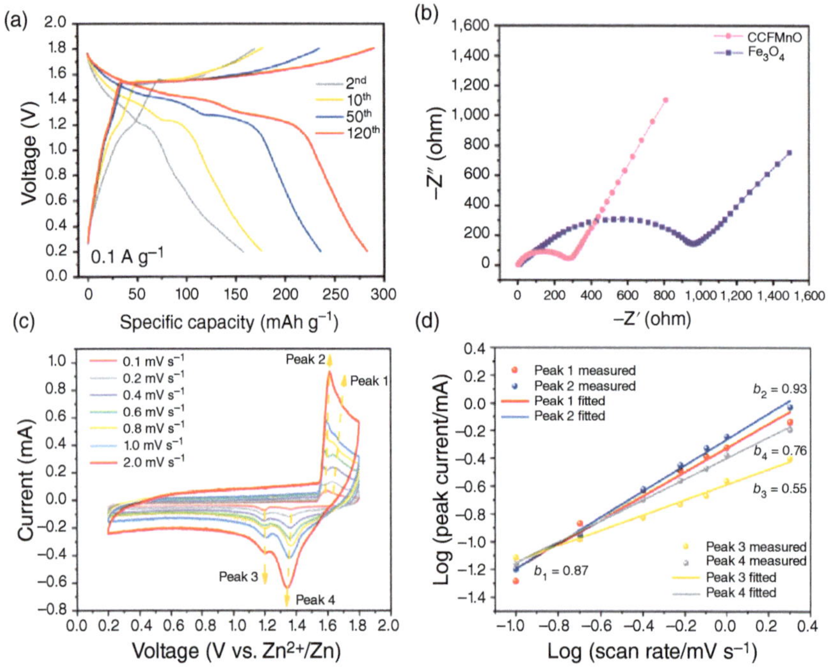

Figure 5.9 (a) Charge–discharge profiles, (b) Nyquist plots, (c) CV curves, (d) log (current) vs. log (scan rate) for the calculation of "*b*" value of CCFMnO. *Source:* Reference [44] / with permission of Elsevier.

behavior. Additionally, the rich pore structure of CCFMnO, in contrast to Fe_3O_4, permits overly reversible Zn^{2+} intercalation and deintercalation, enhancing not only the reaction kinetics of Zn^{2+} but also adding capacity. EIS reaction kinetics showed that the R_{ct} values of Fe_3O_4 and CCFMnO are 960 and 279 Ω, respectively (Figure 5.9b). After 100 cycles at 0.1 A g^{-1}, the R_{ct} values of CCFMnO and Fe_3O_4 are 487 and 1,690 Ω, respectively, confirming that the electrode possesses superior conductivity, facilitating faster electron and ion transfer, thereby enhancing electrochemical behavior. The superior electrocatalytic activity of the CCFMnO cathode was further demonstrated by the reaction kinetics study grounded on CV curves (Figure 5.9c). The CV curve retains its original shape even when the rate increases steadily, suggesting that the electrochemical reaction is reversible due to the CCFMnO cathode's good stability. In contrast, Fe_3O_4 exhibited significant shape distortion. For CCFMnO, the power law formula was used to quantitatively calculate surface capacitance and diffusion control processes. The calculated *b* values for two anodic and two cathodic peaks (by using Eq. 5.5) were 0.87, 0.93, 0.55, and 0.76, respectively (Figure 5.9d), highlighting predominant capacitance-controlled behavior, which supports long-term cycling stability at high current densities and rapid charge storage. At a 2 mV s^{-1}, CCFMnO contributed 77% capacitance, with pseudocapacitance increasing as the scan rate rose, and have notable impact to the total capacitance. The good electrochemical activity of CCFMnO stems from its nano-size, which enhances electrolyte penetration and Zn^{2+} diffusion, its porous structure, which boosts pseudocapacitance, and its high-entropy lattice, which ensures long-term structural stability.

Ding et al. synthesized a sequence of high-entropy-doped $V_2O_5 \cdot 0.48H_2O$ by concurrently adding $Na^+/Al^{3+}/Ni^{2+}/NH^{4+}/F^-$ to the interlayer [45]. This doping not only gives V_2O_5 (VNANONF) structural support to increase its stability, but it also substantially decreases the desolvation loss of the cathode material and improves intercalation kinetics. Additionally, the addition of F^- anion can improve stability of the material and ionic conductivity by strengthening the holding capacity between ions. They synthesized five electrode materials, i.e. VOH, VAO, VANO, VNAOF, and VNANONF. To study the electrocatalytic behavior and charge transfer kinetics of this electrode substance, EIS was conducted. It was observed that with increasing doping ions, the charge transfer resistance (R_{ct}) consistently decreased (VOH → VAO → VANO → VNAOF → VNANONF). These findings suggest that the high-entropy layout does enhance the kinetics of charge transfer.

Further, the distribution of relaxation times (DRTs) through in situ was done (Figure 5.10) to demonstrate the electrokinetics. The DRT graphs identified three key processes: R_s, which is situated at 10^{-5} s; the R_{ct} of electrochemical reactions at $10^{-4}-10^{-1}$ s; and also, Zn^{2+} diffusion of $10^{-2}-10^{-1}$ s. For VOH and VNANONF

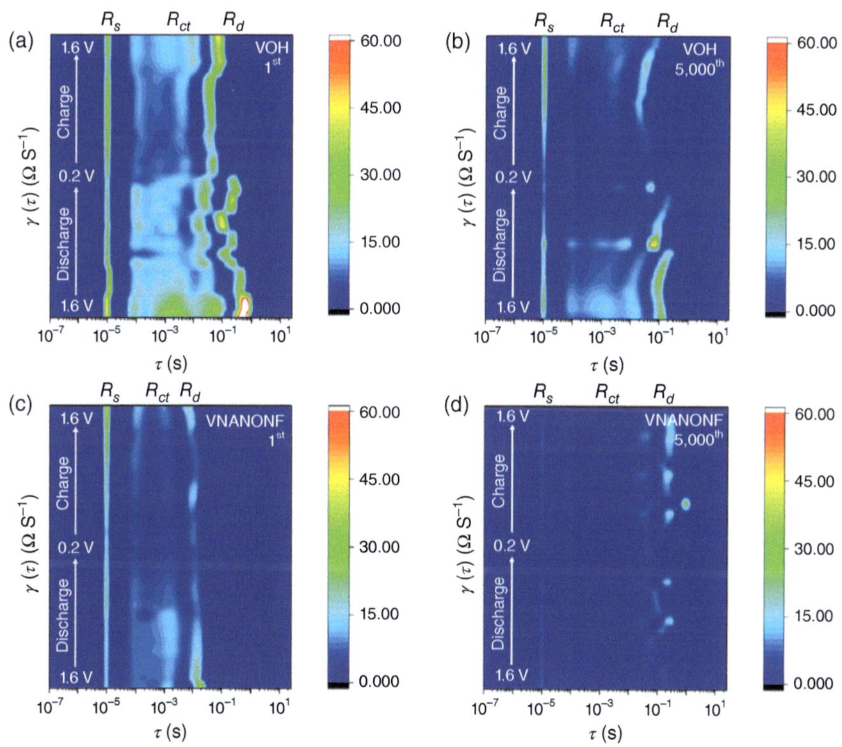

Figure 5.10 (a–d) The in situ relaxation time distribution (DRT) for VOH and VNANONF at the 1st and 5,000th. *Source:* Reference [45] / with permission of Elsevier.

Table 5.2 Summary of electrochemical kinetic parameters for various HEMs in batteries.

Battery	HEM	Synthesis method	Capacity ($mA\,h\,g^{-1}$)	*b* value	R_s (Ω)	R_{ct} (Ω)	$D_{Li^+}/D_{Zn}{}^{2+}/D_K{}^{++}/D_{Na}{}^+$ ($cm^2\,s^{-1}$)	References
Li-ion	$(FeCoNiCrMn)_3O_4$HEO	Ball milling	596.5	0.825, 0.892	—	95.77	46.8×10^{-15}	[36]
	$(Mg_{0.2}Ti_{0.2}Zn_{0.2}Cu_{0.2}Fe_{0.2})_3O_4$	Solid-state reaction, ball milling	504	0.97, 0.86, 0.90	39.68	28.53	1.497×10^{-15}	[37]
	(MgCoNiCuZn)O	Joule heating	150	0.73, 0.74	—	—	$10^{-13} - 10^{-11}$	[38]
	$(FeCoNiCuZn)_3O_4$	Hydrothermal	540	0.70, 0.77	—	—	$10^{-14} - 10^{-9.7}$	[39]
	$(Al_{0.2}CoCrFeMnNi)_{0.58}O_{4-\delta}$	Solution combustion synthesis	634	—	1.87	63.14	2.71×10^{-19}	[41]
	$M_2Sn_2O_7$/CMC	Co-precipitation	766	—	1.6	21.85	3.299×10^{-10}	[42]
	$(NiMnCrCoFe)_3O_4$ ($HEMO_3$)	Hydrothermal	527	0.86	3.6	13.3	3.734×10^{-14}	[52]
	$(Cr_{0.2}Fe_{0.2}Co_{0.2}Ni_{0.2}Zn_{0.2})_3O_4$	Sol–gel	1,022	0.889, 0.873	—	—	—	[53]
	$(Mn_{0.17}Ni_{0.2}Cu_{0.2}Co_{0.23}Zn_{0.2})MoO_4$	Co-precipitation	426	0.60, 0.67	—	31	1.8×10^{-10}	[68]
Li-S	CoNiCuMnZnFe-PBA	Co-precipitation	424.9	0.61, 0.58	3.38	21.85	—	[67]
Zn-ion	$(Co_{0.2}Cr_{0.2}Fe_{0.2}Mn_{0.2}Ni_{0.2})_3O_4$	Solution combustion synthesis	283	0.87, 0.93	—	279	$10^{-8} - 10^{-12}$	[44]
	VNANONF	Doping	546	0.59, 0.84, 0.93, 0.87	—	13.15	$10^{-10.19} - 10^{-8.84}$	[45]
	$Co_{0.6}Ni_{0.6}Fe_{0.6}Mn_{0.6}Cu_{0.6}O_4$	—	115.3	0.81, 0.80	—	—	0.7×10^{-12}	[46]
K-ion	HEA-NPs@NC	Sacrificial template	513	0.726, 0.858	3.91	62.44	—	[69]
	$K_{1.75}Mn_{0.26}Fe_{0.01}Ni_{0.01}Cu_{0.08}Co_{0.09}[Fe(CN)_6]_{0.66} \cdot 0.83H_2O$	Co-precipitation	180	—	—	7360	—	[70]
	$Sb_{1.4}Bi_{0.2}Sn_{0.2}Co_{0.1}Mn_{0.1}Te_3$	Hydrothermal, carbonization	376.5	—	—	2.8	$10^{-12} - 10^{-8}$	[71]
Na-ion	$NaCu_{0.1}Ni_{0.3}Fe_{0.2}Mn_{0.2}Ti_{0.2}O_2$	Pechini	107	—	—	—	$10^{-10} - 10^{-12}$	[72]
	$Cu_{0.88}Sn_{0.02}Sb_{0.02}Bi_{0.02}Mn_{0.02}S_{0.9}Se_{0.1}$	Doping	365.7	0.54, 0.69, 0.73	3.06	5.339	—	[73]
	$Na_4Cr_{0.7}Fe_{0.4}Mn_{0.3}V_{0.3}Al_{0.2}(PO_4)_3$	Sol–gel	131.09	0.95	—	—	$10^{-10} - 10^{-12}$	[74]

cathodes, the first discharging process (Figure 5.10a,b) have shown a gradual decrease in R_{ct} and diffusion values with peaks shifting to lower τ values. This indicates an enhanced charge transfer rate and Zn^{2+} diffusion, driven by progressive Zn^{2+} introduction in VOH and in VNANONF cathodes. Conversely, R_{ct} and diffusion values, as well as their peak positions, go through a reverse procedure during the first charging process. Additionally, when compared to the VOH, it was observed that the R_{ct} and diffusion values along with the peak positions of the electrode showed subtle alteration throughout cycling. By the 5,000th cycle, both cathodes show lower R_{ct} values than their pristine states on account of their structural adaptation for Zn^{2+} insertion/extraction (Figure 5.10c,d). However, VNANONF maintains a consistently lower R_{ct} than VOH, indicating superior charge transfer and Zn^{2+} diffusion. These improvements stem from high-entropy doping, which significantly enhances both electronic and ionic conductivity. As a result, VNANONF exhibits stronger pseudocapacitive behavior, faster charge transfer kinetics, and outstanding rate capability along with the cycling stability, making it a highly advanced cathode for AZIBs. Du et al. synthesized $Co_{0.6}Ni_{0.6}Fe_{0.6}Mn_{0.6}Cu_{0.6}O_4$ (Co-HEO) as a HEO cathode for AZIBs [46]. After an initial activation process, Co-HEO exhibited a large discharge plateau (~1.3 V) and a specific capacity of 302.3 mA h g^{-1}. Its superior performance in terms of rate, compared to monometallic oxides, is because of random distribution of transition metals, which enhances electron compensation efficiency and reduces electrostatic forces that are between Zn^{2+} and its structure. To further explore the "cocktail effect," Cr-HEO ($Cr_{0.6}Co_{0.6}Ni_{0.6}Fe_{0.6}Mn_{0.6}O_4$) was synthesized, showing distinct electrochemical behavior in CV and GCD curves, confirming that element selection influences battery performance. It was observed that Co-HEO exhibited minimal CV peak shifts and reduced overpotential polarization at higher scan rates, with b values of 0.81 and 0.80, indicating efficient electron transfer. Furthermore, the capacitive contribution (59.0–73.8%) and lower R_{ct} from EIS analysis confirmed its superior electronic conductivity and ion transport. Additionally, galvanostatic intermittent titration technique (GITT) analysis showed Co-HEO had a Zn^{2+} ion diffusion coefficient (D_{ion}) of 0.7×10^{-12} cm^2 s^{-1}, consistent with EIS results. These findings highlight Co-HEO's superior diffusion kinetics, making it a promising cathode for AZIBs. Table 5.2 summarizes the various kinetic parameters obtained during the electrochemical reactions of different HEMs in batteries.

5.4 Conclusions and Perspectives

HEMs, characterized by their configurational complexity and multi-elemental compositions in near-equimolar ratios, have appeared as advanced materials for electrochemical energy storage on account of their exceptional physicochemical, structural, and electrokinetic properties. This chapter reviewed an extensive summary of the electrokinetic behavior of HEMs in advanced energy storage systems, including SCs, Li^+ batteries, and AZIBs. The chapter highlights contextualizing entropy in materials science and outlines the structural advantages of HEMs, such as enhanced structural stability, tunable active sites, and resistance to dendrite formation, which are critical for energy storage applications. Electrokinetic phenomena such as ion diffusion, charge transfer resistance, and capacitive contributions are discussed using electrochemical techniques including CV, GCD, EIS, and GITT. In SCs, HEM-based electrodes show hybrid charge storage behavior, combining EDLC and pseudocapacitance. Various studies demonstrate the high surface area, porosity, and multivalent redox activity of HEMs that significantly improve energy density and cycling stability. The "b" value analysis (derived from $I = av^b$) consistently reveals mixed capacitive and diffusion-controlled mechanisms, highlighting the synergistic charge storage nature of these materials. In LIBs, HEMs offer enhanced lithium-ion diffusion coefficients, improved charge transfer kinetics, and superior cycling performance. This is attributed to factors such as oxygen vacancies, multivalent cationic redox reactions, SEI stabilization, and hierarchical porous structures. We review a number of high-entropy spinel and oxide systems, including $(FeCoNiCrMn)_3O_4$, $(MgTiZnCuFe)_3O_4$, and $(AlCoCrFeMnNi)O_{4-\delta})$, with a focus on charge storage kinetics, diffusion coefficients, and lithiation/delithiation processes. HEMs like $(CoCrFeMnNi)_3O_4$ show maintained structural integrity, enhanced Zn^{2+} ion mobility, and decreased charge transfer resistance in ZIBs. High-entropy-doping strategies in $V_2O_5 \cdot 0.48H_2O$ and spinel oxides further modulate interlayer spacing, reduce desolvation loss, and enhance Zn^{2+} intercalation kinetics. DRT analysis further elucidates the dynamic processes underlying ion transport and reaction rates. The enormous potential of HEMs in next-generation energy storage systems is highlighted in the chapter's conclusion. Their entropy-stabilized frameworks offer a special platform for creating electrodes with exceptional performance. Compositional optimization, integration with cutting-edge technologies such as solid-state batteries, and resolving scale-up issues using sophisticated synthesis and characterization methods are some future directions.

References

1 Ying, T., Yu, T., Qi, Y. et al. (2022). High entropy van der waals materials. *Adv Sci* 9 (30): 2203219. https://doi.org/10.1002/advs.202203219.

2 Sun, Y. and Dai, S. (2021). High-entropy materials for catalysis: a new frontier. *Sci Adv* 7 (20): 1600. https://doi.org/10.1126/sciadv.abg1600.

3 Ma, Y., Ma, Y., Wang, Q. et al. (2021). High-entropy energy materials: challenges and new opportunities. *Energy Environ Sci* 14 (5): 2883–2905. https://doi.org/10.1039/d1ee00505g.

4 Oses, C., Toher, C., and Curtarolo, S. (2020). High-entropy ceramics. *Nat Rev Mater* 5 (4): 295–309. https://doi.org/10.1038/s41578-019-0170-8.

5 Liu, K.C., Panda, P.K., Mallick, B.C. et al. (2024). Solid-phase microwave synthesis of high-entropy graphene quantum dots as metal-free electrochemical catalysts. *Appl Surf Sci* 648: 159061. https://doi.org/10.1016/j.apsusc.2023.159061.

6 Khan, J.B., Panda, P.K., Yang, P.C. et al. (2024). Microwave synthesis of high-entropy alloy catalysts on graphene oxide sheets for oxygen reduction and

evolution reactions. *Int J Hydrog Energy* 53: 999–1008. https://doi.org/10.1016/j.ijhydene.2023.12.024.

7 Zhang, Y., Wang, D., and Wang, S. (2022). High-entropy alloys for electrocatalysis: design, characterization, and applications. *Small* 18 (7): 2104339. https://doi.org/10.1002/smll.202104339.

8 Amiri, A. and Shahbazian-Yassar, R. (2021). Recent progress of high-entropy materials for energy storage and conversion. *J Mater Chem A* 9 (2): 782–823. https://doi.org/10.1039/d0ta09578h.

9 Dąbrowa, J., Stygar, M., Mikuła, A. et al. (2018). Synthesis and microstructure of the $(Co,Cr,Fe,Mn,Ni)_3O_4$ high entropy oxide characterized by spinel structure. *Mater Lett* 216: 32–36. https://doi.org/10.1016/j.matlet.2017.12.148.

10 Pankratova, D., Giacomelli, S.M., Yusupov, K. et al. (2023). Co-Cr-Fe-Mn-Ni oxide as a highly efficient thermoelectric high-entropy alloy. *ACS Omega* 8 (16): 14484–14489. https://doi.org/10.1021/acsomega.2c08278.

11 Joshi, A., Chakrabarty, S., Akella, S.H. et al. (2023). High-entropy Co-free O3-type layered oxyfluoride: a promising air-stable cathode for sodium-ion batteries. *Adv Mater* 35 (51): 2304440. https://doi.org/10.1002/adma.202304440.

12 Qureshi, T., Khan, M.M., and Pali, H.S. (2024). high-entropy borides—challenges and opportunities. *J Mater Sci* 59 (34): 15921–15991. https://doi.org/10.1007/s10853-024-10064-1.

13 Hossain, M.D., Borman, T., Oses, C. et al. (2021). Entropy landscaping of high-entropy carbides. *Adv Mater* 33 (42): 2102904. https://doi.org/10.1002/adma.202102904.

14 Wen, T., Ye, B., Nguyen, M.C. et al. (2020). Thermophysical and mechanical properties of novel high-entropy metal nitride-carbides. *J Am Ceram Soc* 103 (11): 6475–6489. https://doi.org/10.1111/jace.17333.

15 Nguyen, T.X., Su, Y.H., Lin, C.C., and Ting, J.M. (2021). Self-reconstruction of sulfate-containing high entropy sulfide for exceptionally high-performance oxygen evolution reaction electrocatalyst. *Adv Funct Mater* 31 (48): 2106229. https://doi.org/10.1002/adfm.202106229.

16 Lu, Y., Kang, Q., Dong, F. et al. (2025). Metalloid phosphorus induces tunable defect engineering in high entropy oxide toward advanced lithium-ion batteries. *Adv Funct Mater* 35 (3): 2413782. https://doi.org/10.1002/adfm.202413782.

17 Liu, D., Guo, P., Pan, H., and Wu, R. (2024). Emerging high-entropy compounds for electrochemical energy storage and conversion. *Prog Mater Sci* 18: 101300. https://doi.org/10.1016/j.pmatsci.2024.101300.

18 Ma, J. and Huang, C. (2023). High entropy energy storage materials: synthesis and application. *J Energy Storage* 66: 107419. https://doi.org/10.1016/j.est.2023.107419.

19 Ren, J., Kumkale, V.Y., Hou, H. et al. (2025). A review of high-entropy materials with their unique applications. *Adv Compos Hybrid Mater* 8 (2): 195. https://doi.org/10.1007/s42114-025-01275-4.

20 Sarkar, A., Velasco, L., Wang, D.I. et al. (2018). High entropy oxides for reversible energy storage. *Nat Commun* 9 (1): 3400. https://doi.org/10.1038/s41467-018-05774-5.

21 Dissanayake, K. and Kularatna-Abeywardana, D. (2024). A review of supercapacitors: materials, technology, challenges, and renewable energy applications. *J Energy Storage* 96: 112563. https://doi.org/10.1016/j.est.2024.112563.

22 Shao, Y., El-Kady, M.F., Sun, J. et al. (2018). Design and mechanisms of asymmetric supercapacitors. *Chem Rev* 118 (18): 9233–9280. https://doi.org/10.1021/acs.chemrev.8b00252.

23 Wang, J., Dong, S., Ding, B. et al. (2017). Pseudocapacitive materials for electrochemical capacitors: from rational synthesis to capacitance optimization. *Natl Sci Rev* 4 (1): 71–90. https://doi.org/10.1093/nsr/nww072.

24 Sudha, V. and Sangaranarayanan, M.V. (2002). Underpotential deposition of metals: structural and thermodynamic considerations. *J Phys Chem B* 106 (10): 2699–2707. https://doi.org/10.1021/jp013544b.

25 Hu, C.C., Chang, K.H., Lin, M.C., and Wu, Y.T. (2006). Design and tailoring of the nanotubular arrayed architecture of hydrous RuO_2 for next generation supercapacitors. *Nano Lett* 6 (12): 2690–2695. https://doi.org/10.1021/nl061576a.

26 Toupin, M., Brousse, T., and Bélanger, D. (2004). Charge storage mechanism of MnO_2 electrode used in aqueous electrochemical capacitor. *Chem Mater* 16 (16): 3184–3190. https://doi.org/10.1021/cm049649j.

27 Zhang, S.S., Xu, K., and Jow, T.R. (2006). EIS study on the formation of solid electrolyte interface in Li-ion battery. *Electrochim Acta* 51 (8-9): 1636–1640. https://doi.org/10.1016/j.electacta.2005.02.137.

28 Cao, Q., Zhang, H.P., Wang, G.J. et al. (2007). A novel carbon-coated $LiCoO_2$ as cathode material for lithium ion battery. *Electrochem Commun* 9 (5): 1228–1232. https://doi.org/10.1016/j.elecom.2007.01.017.

29 Zhu, Z., Zhang, Y., Kong, D. et al. (2024). A novel high entropy hydroxide electrode material for promoting energy density of supercapacitors and its efficient synthesis strategy. *Small* 20 (20): 2307754. https://doi.org/10.1002/smll.202307754.

30 Vadivel, S., Dharmavarapu, V.S., Sujita, P. et al. (2025). A robust approach for designing a bismuth high entropy material ($BiOXCO_3$) as a novel electrode material for supercapacitor applications. *Mater Res Bull* 184: 113225. https://doi.org/10.1016/j.materresbull.2024.113225.

31 Yin, Y., Zhang, W.B., Chen, B. et al. (2024). Porous powder synthesis of $(FeCoCrMnNi)_3O_4$ high entropy ceramic and its supercapacitor applications. *Ceram Int* 50 (7): 10292–10304. https://doi.org/10.1016/j.ceramint.2023.12.341.

32 Mohanty, G.C., Gowda, C.C., Gakhad, P. et al. (2023). Iron-cobalt-nickel-copper-zinc (FeCoNiCuZn) high entropy alloy as positive electrode for high specific capacitance supercapacitor. *Electrochim Acta* 470: 143272. https://doi.org/10.1016/j.electacta.2023.143272.

33 Ren, L., Liu, J., Liu, X. et al. (2023). Rapid synthesis of high-entropy antimonides under air atmosphere using microwave method to ultra-high energy density supercapacitors. *J Alloys Compd* 967: 171816. https://doi.org/10.1016/j.jallcom.2023.171816.

34 Wei, L.Z., Shi, S.J., Guo, M.Y. et al. (2025). Self-supporting high-entropy-oxide@ layered-Ni electrode for high-performance supercapacitors. *J Energy Storage* 110: 115351. https://doi.org/10.1016/j.est.2025.115351.

35 Shen, E., Song, X., Chen, Q. et al. (2021). Spontaneously forming oxide layer of high entropy alloy nanoparticles deposited on porous carbons for supercapacitors. *ChemElectroChem* 8 (1): 260–269. https://doi.org/10.1002/celc.202001289.

36 Xiao, B., Wu, G., Wang, T. et al. (2022). High-entropy oxides as advanced anode materials for long-life lithium-ion Batteries. *Nano Energy* 95: 106962. https://doi.org/10.1016/j.nanoen.2022.106962.

37 Chen, H., Qiu, N., Wu, B. et al. (2020). A new spinel high-entropy oxide $(Mg_{0.2}Ti_{0.2}Zn_{0.2}Cu_{0.2}Fe_{0.2})_3O_4$ with fast reaction kinetics and excellent stability as an anode material for lithium ion batteries. *RSC Adv* 10 (16): 9736–9744. https://doi.org/10.1039/d0ra00255k.

38 Ren, R., Xiong, Y., Xu, Z. et al. (2024). Fast synthesis of high-entropy oxides for lithium-ion storage. *Chem Eng J* 479: 147896. https://doi.org/10.1016/j.cej.2023.147896.

39 Li, Y., Chen, Z., Liu, J. et al. (2023). Novel high entropy oxide as anode for high performance lithium-ion capacitors. *Ceram Int* 49 (23): 38439–38447. https://doi.org/10.1016/j.ceramint.2023.09.173.

40 Yen, J.Z., Yang, Y.C., and Tuan, H.Y. (2022). Interface engineering of high entropy Oxide@ Polyaniline heterojunction enables highly stable and excellent lithium ion storage performance. *Chem Eng J* 450: 137924. https://doi.org/10.1016/j.cej.2022.137924.

41 Xiang, H.Z., Xie, H.X., Chen, Y.X. et al. (2021). Porous spinel-type $(Al_{0.2}CoCrFeMnNi)_{0.58}O_{4-\delta}$ high-entropy oxide as a novel high-performance anode material for lithium-ion batteries. *J Mater Sci* 56: 8127–8142. https://doi.org/10.1007/s10853-021-05805-5.

42 Jiang, T., Wu, F., Ren, Y. et al. (2023). Pyrochlore phase $(Y, Dy, Ce, Nd, La)_2Sn_2O_7$ as a superb anode material for lithium-ion batteries. *J Solid State Electrochem* 27 (3): 763–772. https://doi.org/10.1007/s10008-022-05369-7.

43 Li, Y. and Dai, H. (2014). Recent advances in zinc–air batteries. *Chem Soc Rev* 43 (15): 5257–5275. https://doi.org/10.1039/c4cs00015c.

44 Wu, Z., Lv, Q., Sun, W. et al. (2025). Novel high-entropy oxides with ultra-long cycle life as cathode materials for aqueous zinc-ion batteries. *J Power Sources* 634: 236488. https://doi.org/10.1016/j.jpowsour.2025.236488.

45 Ding, X., Zhu, Q., Fan, Y. et al. (2025). High-entropy V-based cathode for high-capacity and long-life aqueous zinc-ion battery. *Nano Energy* 136: 110701. https://doi.org/10.1016/j.nanoen.2025.110701.

46 Du, K., Liu, Y., Yang, Y. et al. (2023). High entropy oxides modulate atomic-level interactions for high-performance aqueous zinc-ion batteries. *Adv Mater* 35 (51): 2301538. https://doi.org/10.1002/adma.202301538.

47 Mohanty, G.C. and Verma, A. (2025). Supercapacitive study of a high entropy alloy and green carbon composite electrode with a wide potential window (1.3 V) and its application towards an aqueous symmetric device. *New J Chem* 49 (2): 521–529. https://doi.org/10.1039/d4nj04306e.

48 Yan, Y., Huang, X., Yan, X. et al. (2025). New ultra-high cycle stable accordion-like high entropy MXene with improving energy storage performance of supercapacitors. *Adv Funct Mater* 2420578. https://doi.org/10.1002/adfm.202420578.

49 Mohanty, G.C., Das, S., and Verma, A. (2024). Fabrication of aqueous asymmetric supercapacitor device by using spinel type $(FeCoNiCuZn)_3O_4$ high entropy oxide and green carbon derived from plastic wastes. *Ceram Int* 50 (23): 48938–48947. https://doi.org/10.1016/j.ceramint.2024.09.152.

50 Talluri, B., Aparna, M.L., Sreenivasulu, N. et al. (2021). High entropy spinel metal oxide $(CoCrFeMnNi)_3O_4$ nanoparticles as a high-performance supercapacitor electrode material. *J Energy Storage* 42: 103004. https://doi.org/10.1016/j.est.2021.103004.

51 Li, L., Ji, P., Geng, C. et al. (2023). Facile synthesis of high-entropy $(Co_{0.2}Cr_{0.2}Fe_{0.2}Mn_{0.2}Ni_{0.2})_3O_4$ nanopowders and their electrochemical properties as supercapacitor electrode. *J Energy Storage* 73: 109182. https://doi.org/10.1016/j.est.2023.109182.

52 Wang, X.L., Jin, E.M., Sahoo, G., and Jeong, S.M. (2023). High-entropy metal oxide $(NiMnCrCoFe)_3O_4$ anode materials with controlled morphology for high-performance lithium-ion batteries. *Batteries* 9 (3): 147. https://doi.org/10.3390/batteries9030147.

53 Yang, X., Wang, H., Song, Y. et al. (2022). Low-temperature synthesis of a porous high-entropy transition-metal oxide as an anode for high-performance lithium-ion batteries. *ACS Appl Mater Interfaces* 14 (23): 26873–26881. https://doi.org/10.1021/acsami.2c07576.

54 Liang, B., Ai, Y., Wang, Y. et al. (2020). Spinel-type $(FeCoCrMnZn)_3O_4$ high-entropy oxide: facile preparation and supercapacitor performance. *Materials* 13 (24): 5798. https://doi.org/10.3390/ma13245798.

55 Xu, X., Du, Y., Wang, C. et al. (2020). High-entropy alloy nanoparticles on aligned electronspun carbon nanofibers for supercapacitors. *J Alloys Compd* 822: 153642. https://doi.org/10.1016/j.jallcom.2020.153642.

56 Lal, M.S. and Sundara, R. (2019). High entropy oxides—a cost-effective catalyst for the growth of high yield carbon nanotubes and their energy applications. *ACS Appl Mater Interfaces* 11 (34): 30846–30857. https://doi.org/10.1021/acsami.9b08794.

57 Lal, M.S. and Sundara, R. (2022). Multifunctional high entropy oxides incorporated functionalized biowaste derived activated carbon for electrochemical energy storage and desalination. *Electrochim Acta* 405: 139828. https://doi.org/10.1016/j.electacta.2021.139828.

58 Verma, A., Kim, K.H., Mathur, S., and Lee, D. (2022). Interdependence of the electrical performance of NiCuFeCoMn multi-structure carbonates as electrode material for supercapacitors. *J Alloys Compd* 922: 166222. https://doi.org/10.1016/j.jallcom.2022.166222.

59 Başgöz, Ö., Güngör, A., Güler, Ö., and Erdem, E. (2025). High-Entropy Alloys and Oxides as Supercapacitor Electrodes: A Structural and Electrochemical Perspective for Energy Storage. *Adv Sustain Syst* 2500201. https://doi.org/10.1002/adsu.202500201.

60 Zhang, D., Xu, S., Li, T. et al. (2023). High-entropy oxides prepared by dealloying method for supercapacitors. *ACS Appl Eng Mater* 1 (2): 780–789. https://doi.org/10.1021/acsaenm.2c00198.

61 Mohanty, G.C., Gowda, C.C., Gakhad, P. et al. (2023). High energy density liquid state asymmetric supercapacitor devices using Co–Cr–Ni–Fe–Mn high entropy alloy. *Mater Adv* 4 (17): 3839–3852. https://doi.org/10.1039/D3MA00327B.

62 Mohanty, G.C., Gowda, C.C., Gakhad, P. et al. (2024). Effect of Zn/Mn on the supercapacitor behavior of high-entropy FeCoNiCrZn/Mn alloy. *Energy Adv* 3 (12): 2972–2985. https://doi.org/10.1039/D4YA00376D.

63 Chen, K., Xie, H., Yang, H. et al. (2025). Construction of crystalline/amorphous heterostructures to enhance the supercapacitor performance of a high-entropy sulfide. *J Alloys Compd* 1015: 178921. https://doi.org/10.1016/j.jallcom.2025.178921.

64 Wang, S., Ji, Y., Zhang, B. et al. (2025). Walnut-like high-entropy sulfides via facile route for enhanced supercapacitor performance. *J Energy Storage* 110: 115336. https://doi.org/10.1016/j.est.2025.115336.

65 Mohanty, G.C., Das, S., and Verma, A. (2024). Synthesis of $(FeCoNiCuMn)_3O_4$ spinel-high entropy oxide and green carbon from agricultural waste for supercapacitor application. *RSC Adv* 14 (46): 33830–33842. https://doi.org/10.1039/D4RA05204H.

66 Du, M., Geng, P., Pei, C. et al. (2022). High-entropy Prussian blue analogues and their oxide family as sulfur hosts for lithium-sulfur batteries. *Angew Chem Int Ed* 61 (41): e202209350. https://doi.org/10.1002/anie.202209350.

67 Beere, H.K., Upadhya, H., Maiti, U.N., and Ghosh, D. (2025). Exploring high-entropy molybdate as a highly stable anode for Li-ion batteries. *ACS Appl Energy Mater* 8: 4711–4721. https://doi.org/10.1021/acsaem.5c00389.

68 Zhang, R., Xue, H., Du, D. et al. (2025). Regulation of configurational entropy to realize long cycle lifespan of high entropy alloy anodes for potassium batteries. *Adv Funct Mater* 35 (26): 2422218. https://doi.org/10.1002/adfm.202422218.

69 Yan, W., Feng, X., Min, X. et al. (2024). High-entropy Prussian Blue analogues as high-capacity cathode material for potassium ion batteries. *J Electrochem Soc* 171 (8): 080529. https://doi.org/10.1149/1945-7111/ad6d93.

70 Wang, Z., Qiao, S., Ma, M. et al. (2025). High-entropy conversion-alloying anode material for advanced potassium-ion batteries. *ACS Nano* 19: 15148–15160. https://doi.org/10.1021/acsnano.5c03792.

71 Lin, C.C., Liu, H.Y., Kang, J.W. et al. (2022). In-situ X-ray studies of high-entropy layered oxide cathode for sodium-ion batteries. *Energy Storage Mater* 51: 159–171. https://doi.org/10.1016/j.ensm.2022.06.035.

72 Zhang, S., Zuo, W., Fu, X. et al. (2025). High-entropy sulfoselenide as negative electrodes with fast kinetics and high stability for sodium-ion batteries. *Nat Commun* 16 (1): 4052. https://doi.org/10.1038/s41467-025-59078-6.

73 Zhu, P., Wang, Y., Li, J., and Jin, Y. (2024). A strategy of high entropy enabling advanced NASICON sodium-ion cathodes with high stabilization and multicationic redox. *Energy Storage Mater* 71: 103650. https://doi.org/10.1016/j.ensm.2024.103650.

6

Importance of High-Entropy Materials for Energy Storage Applications

Jala Bib Khan[1,2], Pradeep Kumar Panda[1], Pranjyan Dash[3], and Chien-Te Hsieh[1]

[1] *Yuan Ze University, Department of Chemical Engineering and Materials Science, Taoyuan 32003, Taiwan*
[2] *National Taiwan Ocean University, Department of Optoelectronics and Materials Technology, Keelung 20224, Taiwan*
[3] *National Taipei University of Technology (Taipei Tech), Department of Chemical Engineering and Biotechnology, Taipei 10608, Taiwan*

6.1 Introduction

Historically, the alloying approach has mostly included incorporating minimal amounts of various substances as dopants within the fundamental material to create a new combination with enhanced capabilities. This method allows for the development of limited kinds of alloys, and substantial study has already investigated these alloys [1]. The distinctive and beneficial catalytic capabilities of the alloy are a result of the clear interaction between the several elements. As research into the alloying impact continues, however, it becomes clear that different species are incompatible with one another, and phase separation always restricts the degree to which the structure and composition could be tuned. To build a functioning interface domain with exact command, it is necessary to avoid impurity formation, which might reduce the active location's surface accessibility. Therefore, preserving the single-phase configuration is crucial for facilitating atomic mixing of diverse components and acquiring synergistic qualities unattainable from a singular component [2–4]. Yeh et al. proposed the notion of high-entropy alloys (HEAs), asserting that the configurational entropy of combining escalates with the incorporation of higher equimolar primary components into the alloy environment [5]. Typically, high-entropy materials (HEMs) are defined as materials that include a minimum of five metal cations within equimolar or nearly equimolar ratios, each having a unique radius and electronegativity. Numerous environmental along with energy systems have shown that HEMs are effective catalysts, ever since Dai et al. conducted the first investigation on their catalytic applications. This was followed by an increase in the number of HEMs reported in various fields. Carbides, silicides, borides, nitrides, sulfides, as well as non-oxide HEMs became known as

High-Entropy Materials for Energy Storage Devices, First Edition.
Edited by Chien-Te Hsieh, Pradeep Kumar Panda, and Arpan Kumar Nayak.

areas of interest in recent years [6–9]. Due to their structural benefits as well as the resultant high-entropy impacts (cocktail effect, high-entropy effect, severe lattice distortion effect, and sluggish diffusion effect), HEMs demonstrate outstanding properties compared to the majority of conventional metallic compounds in various fields, such as supercapacitors, thermoelectric conduction, high-temperature corrosion and oxidation resistance, catalytic oxidation, and thermoelectric conversion [10–14]. With the advancement of human civilization and science, along with technology, conventional fossil fuels are currently extensively extracted and consumed. The outcome will not only precipitate a shortage of energy but will also significantly impact individuals' lives as well as their health resulting from global warming plus pollution emissions. To address these issues, researchers have been pursuing novel, eco-friendly, and dependable energy storage along with conversion technologies [15–17]. Due to the intermittent nature of sources of clean energy, it is obligatory to explore higher-efficiency energy storage technologies with extended cycle life, specifically rechargeable batteries, fuel cells, and supercapacitors [18, 19]. Rechargeable batteries work by facilitating the redox process, which involves the reversible insertion and removal of metal ions from the electrode substance. Improving specific surface area and conductivity, inhibiting dendrite formation, and promoting its interfacial transfer of charge kinetics are all possible outcomes of size layout, functional substance compounding, along with heteroatom addition as applied to rechargeable batteries [20, 21]. Energy storage units known as supercapacitors use a polarized electrolyte to store energy. A supercapacitor's rapid charge rate and extended cycle life are benefits of its outermost redox reactions and rapid ion adsorption/desorption. The integration of pseudocapacitive materials, like metal oxides or conducting polymers, into carbon substrates, the construction of multilayer porous architectures, and heteroatom doping could enhance the electrochemical efficiency of supercapacitors [22, 23]. All energy storage applications have obstacles, which HEMs address well.

6.2 Fundamentals of High-Entropy Materials

6.2.1 Definition

HEMs are sometimes referred to as HEAs [24]. Controversy exists about the meaning of HEAs, with two definitions currently commonly acknowledged. The first concept was introduced in 2004, based upon compositional criteria. A HEA is characterized by the presence of at least five central metal components, each constituting an atomic proportion ranging from 5% to 35%. The preceding description inherently encompasses alloys with a minimal atomic proportion of elemental additions, even those with less than 5%. The above description could possibly be articulated as follows: $N_{major} \geq 5$, $5\ \text{at.\%} \leq x_i \leq 35\ \text{at.\%}$, $N_{minor} \geq 0$, $x_j \leq 5\ \text{at.\%}$.

N_{major} and N_{minor} denote the quantities of both the main and the supporting components, respectively. "x_i" and "x_j" represent the atomic proportions of the main and secondary components, respectively. According to the description, the HEA system might consist of a certain equimolar ratio, several non-equimolar components, or many minor components. This description emphasizes the utilization of elevated mixing entropy for promoting the development of solid-solution phases. The additional interpretation is predicated on the notion of entropy. HEAs are defined as an alloy with a mixing entropy above 1.61R within their randomized solution form, whereby "R" represents the universal gas constant. This description could potentially be articulated as $S_{mix} \geq 1.61R$.

Entropy is an essential property of state in thermodynamics that describes disorder. The higher the level of mixedness, the larger the system's entropy. Boltzmann's thermodynamics statistical concept states that the entropy of an arrangement can potentially be articulated as $S = k \ln W$, where "k" denotes the Boltzmann constant, which is associated with the molar gas constant R ($R = 1.38 \times 10^{13}\,\mathrm{J\,K^{-1}}$), and "$W$" signifies the thermodynamic probability, representing the entire number of microstates included inside the macrostate. The entropy measurement of the entity increases as the augmentation of the quantity of microstates, and the entropy serves as a quantification of the quantities of microstates inside the entity [25–27].

6.2.2 Classification

The relentless investigation by individuals has enhanced the variety of HEMs via the occurrence of high-entropy compounds. Alongside HEMs of alloys (HEAs) and oxides (HEOs), sulfides (HESs), hydroxides (HEHOs), phosphides (HEPs), carbides (HECs), nitrides (HENs), borides (HEBs), fluorides (HEFs), and selenides (HESEs) enhance the utility of HEMs in electrocatalysis, rendering the high-entropy approach an efficacious technique for modulating the adsorption power of an electrocatalytic active location [2, 28].

6.2.3 Key Impacts

In 2006, a foundational comprehension of HEMs was introduced through the "four key effects" such as sluggish diffusion, high-entropy, cocktail, and severe lattice distortion effects, and these are frequently employed to elucidate the mechanisms underlying various distinctive phenomena related to HEMs. Many studies have confirmed the effects' veracity after extensive testing throughout the course of time [29].

6.2.4 High-Entropy Effect

High entropy denotes atomic-scale chemical chaos, indicating that the atoms within an alloy exist in a disordered configuration. The high-entropy impact might clarify the mutual breakdown of various components in HEA to generate a singular phase. The elevated mixed entropy reduces the Gibbs free energy (ΔG_{mix}) ($\Delta G = \Delta H - T\Delta S$) for the overall system, hence stabilizing the phase. The elevated mixed entropy could mitigate electronegativity differences, enhancing element compatibility and facilitating the production of simple body-centered cubic/face-centered cubic (BCC/FCC), or hexagonal closed pack (HCP) arrangements while preventing the emergence of specific metal combinations [6, 30].

6.2.5 Lattice Distortion Effect

When describing atomic locations in a HEA, the term "lattice distortion" describes how much they deviate from the optimum lattice arrangement. When various types of elements combine to create a single-phase solid-state arrangement, the differing atomic radii cause interactions that compress and elongate the atoms, leading to a deviation in atomic organization from the repeated symmetry of the structure of a crystal, thereby producing lattice distortion. Excessive lattice distortions within the solid-state solution may induce phase shifts or amorphization. Lattice distortions inside HEAs readily generate disorders and enhance the energized electronic enrichment around the Fermi level, resulting in larger active locations, accelerated electron transport, and improved electrochemical movement [31, 32].

6.2.6 Sluggish Diffusion Effect

Sluggish diffusion denotes an occurrence in sluggish interior diffusion plus phase transition rates, which hinder the processes of grain coarsening, recrystallization, and other structural transformations at elevated temperatures. The equimolar or roughly equimolar proportions of substances in HEAs allow each primary atom to function as both solute and solvent, leading to enhanced atomic diffusion obstacle, a slower pace of atomic diffusion, along with elevated activation energy enabling atomic diffusion [33].

6.2.7 Cocktail Effect

The cocktail phenomenon is often viewed to be a synergistic interaction, whereby elemental components enhance one another's performance. Specific components serve distinct tasks; hence, HEAs exhibit exceptional qualities by manipulating the kinds and percentages of the constituent elements. Low-density HEAs are often manufactured using lightweight elements, such as aluminum, lithium, and magnesium as primary constituents, while resistant HEAs utilize refractory metal substances including vanadium, niobium, chromium, tungsten, and zirconium as its major components [34].

6.3 Synthesis

HEMs provide synthesis versatility via two predominant strategies: "top-down" and "bottom-up." The "top-down" strategy, similar to solid-phase manufacturing, demands high temperatures plus prolonged reaction durations, which necessitate supplementary operations such as grinding bulk HEMs into nanoparticles. In contrast, the "bottom-up" approach, which entails atomic-level amalgamation of precursors, is better appropriate for the synthesis of HEMs. This approach facilitates more temperate circumstances [35, 36]. This section examines the various synthesis processes used for the manufacture of HEMs.

6.3.1 Physical Vapor Deposition Technique

Physical vapor deposition (PVD) facilitates the fabrication of HEA films using magnetron co-sputtering using elemental feedstock onto a substrate. Utilizing ionic liquids (ILs) instead of conventional solid substrates facilitates the acquisition of HEA-NPs having adjustable size and morphology. ILs, characterized by reduced pressures of vapor and melting temperatures, function as stabilizers and suspension mediums for nanoparticle development [37].

6.3.2 Arc Discharge Technique

Utilizing the arc discharge technique, homogeneously blended multinary metal powders serve as targets, and the very high temperatures generated by the arc discharge may vaporize the bulk targets, yielding HEA-NPs during the rapid cooling phase [38].

6.3.3 Mechanical Alloying Technique

HEMs could be produced using physical combining as well as mechanical alloying techniques via high-energy milling of metal particulates in conjunction with milling additives. Rekha et al. originally described an alloy containing HEA-NPs plus multilayer graphene produced using a ball milling procedure subsequent to graphene exfoliation [39].

6.3.4 Wet Chemical Technique

In the wet chemical process, HEMs are often made in a solution by combining capping agents, which control crystal development and aggregation, with co-reducing metal salts. Singh and Srivastava initially documented such effective synthesis of non-noble NiFeCrCuCo-based HEA-NPs utilizing the wet chemical approach [40].

6.3.5 Alloying and Dealloying Technique

A co-alloying and dealloying process was developed for creating HEMs containing a nanochannel-rich morphology by integrating melt processing, rapid quenching, and selective chemical etching. Five- to eight-porous HEA frameworks with interconnected nano-ligament dimensions of 2–3 nm, including Al, Au, Cu, Ni, Pd, Pt, Mo, Fe, and Co, have been successfully synthesized using this technique by Sun et al. [41].

6.3.6 Thermal Shock Technique

The thermal shock approach entails the fast heating as well as cooling of metallic precursors in order to synthesize HEMs of diminutive dimensions. In 2018, Hu et al. effectively created HEA-NPs for the first time employing the carbon thermal shock technique [42].

6.3.7 Laser Scanning Ablation Technique

The laser scanning ablation (LSA) technique uses a high-energy pulsed laser beam to vaporize source materials, which then recondense to produce nanoparticles that are deposited onto a substrate. Zou et al. devised an LSA technique by employing an ultrafast pulsed laser for fabricating a database of HEAs plus HEMs nanoparticles, including phosphides, oxides, nitrides, sulfides, and borides, according to ambient circumstances [43].

6.3.8 Fast-Moving Bed Pyrolysis Technique

In the fast-moving bed pyrolysis (FMBP) process, feedstocks such as metal precursors or metal salts are continually supplied to a reactor operating at elevated temperatures. When metals are subjected to high temperatures, they undergo decomposition, resulting in the creation of HEMs in powdered form. Gao et al. illustrated the adaptable and straightforward FMBP method for synthesizing MnCoCuRhPdSnIrPtAu HEA-NPs, which are uniformly distributed over graphene oxide and possess an exceptionally tiny size of 2 nm [44].

6.3.9 Template-Assisted Technique

Distinguishing the steps of morphology as well as contents command has been recognized as a crucial approach to get a universal manufacturing process for nano-HEMs. The approach provides an efficient strategy in which the template governs the morphology, while the chemical composition is established by precursor reduction, deposition, and the process of diffusion. Huang et al. introduced a global template-directed synthetic technique using a combination of ethylene glycol with *N*,*N*-dimethylacetamide [45].

6.3.10 Microwave-Assisted Shock Technique

Microwave irradiation facilitates rapid heating plus cooling, creating optimal circumstances for the production of nanoscale HEMs. Yang et al. produced IrFeCoNiCu-based HEA-NPs deposited onto the carbon paper with a microwave-assisted shock technique [46].

6.3.11 Pulsed Microwave-Assisted Technique

The pulsed microwave-assisted methodology (PM) is an efficient, time-saving, and straightforward method for synthesizing HEMs. Khan, Jala Bib, et al. employed the PM technique for the first time to fabricate HEMs with minimal Pt content and noble metal-free HEMs [14, 47].

6.3.12 Hydro/Solvothermal Technique

The hydro/solvothermal approach is extensively used in the synthesis of nanomaterials, providing advantages such as high purity, superior dispersion, and dimension regulation. Chen et al. prepared HE-MOF by combining Ni, Co, Zn, and V acetylacetonate salts with an organic ligand in an equimolar ratio inside a water/ethanol mixed solvent mixture at 120 °C for 20 hours [48].

6.3.13 Sol–Gel Technique

Hydro/solvothermal methods are often intricate to execute and may result in materials exhibiting elemental discrimination. Ma et al. presented a straightforward sol–gel technique to fabricate the homogenous NiCoFePB nano-electrocatalysts. Through the use of acid-catalyzed hydrolysis along with condensation processes at ambient temperature, they successfully inhibited heterogeneous precipitation and achieved a uniform elemental arrangement across the nanoscale, yielding high-value-added compounds. The synthesized nano-HEM had a network architecture comprised of nanoparticle aggregates and revealed amorphous characteristics; the incorporation of B and P facilitated electron redistribution among the energetic metals [49]. Table 6.1 summarizes the advantages of different synthesis procedures.

Table 6.1 Summary of HEMs Different Synthesis Techniques and Their Advantages.

Synthesis	Advantages	References
Physical vapor deposition	Adjustable size and morphology	[37]
Arc discharge technique	Vaporize the bulk targets, yielding HEA-NPs	[38]
Mechanical alloying technique	Atomic scale uniformity	[39]
Wet chemical technique	Better compositional homogeneity at the atomic level	[40]
Alloying and dealloying technique	Creation of highly porous nanostructure	[41]
Thermal shock technique	Homogeneous single-phase morphology	[42]
Laser scanning ablation technique	Precise control over microstate and composition	[43]
Fast-moving bed pyrolysis technique	Homogeneous phase formation	[44]
Template-assisted technique	Control over particle size and uniform morphology	[45]
Microwave-assisted shock technique	Creation of optimal conditions for nanoscale HEMs	[46]
Pulsed microwave-assisted technique	Fast, easy, and cost-effective	[47]
Hydro/solvothermal technique	High purity, superior dispersion, and dimension regulation	[48]
Sol–gel technique	Uniform elemental dispersion	[49]

6.4 Applications

The expanding domain of HEMs is becoming more significant in energy storage owing to its distinctive features and prospective uses. HEMs differ from traditional materials owing to their compositional complexities as well as high configurational entropy, resulting in notable physical and chemical characteristics [50]. This section presents current advancements in the energy storage applications of HEMs.

6.4.1 Lithium-Ion Battery

Within lithium-ion batteries (LIBs), a predominant category of high-energy metal anodes is HEOs, succeeded by HECs, HEAs, HESs, HEFs, as well as HEPs. Regarding various high-energy materials, their distinct characteristics in energy storage and voltage output, along with cycle performance within LIBs, are attributable to variations concerning oxidation–reduction potential together with material structure. Comparative analysis revealed that the majority of HEMs include Mn, Ni, Cr, Co, Zn, Fe, Mg, and other elements, which are advantageous for multi-electron processes. Sun et al. found that the $(Cr0.2Mn_{0.2}Fe_{0.2}Co_{0.2}Ni_{0.2})_3O_4$ material used for electrodes had a capacity stability rate of 100% despite 5,000 cycles of operation at 5 A g^{-1} [51]. The comprehension of the inherent structure of crystals and lithium storing function of HEO is limited, impeding their future advancement and use. The $(FeCoNiCrMn)_3O_4$ HEO was synthesized to tackle these issues, demonstrating remarkable cycle durability, an impressive reversible capacity reaching 596.5 mA h g^{-1}, plus an 86.2% capacity retention through 1,200 rounds at 2.0 A g^{-1}. The prolonged operational durability might be traced to its peculiar lattice arrangement along with a minimal energy gap [52]. Using the co-precipitation process, Li et al. created $Zn_{0.5}Co_{0.5}Mn_{0.5}Al_{0.5}Mg_{0.5}O_4$ featuring a six-metal spinel configuration. The material retained 81% of its original performance through 5,000 repetitions under 2 A g^{-1} [53]. Owing to the elevated redox potential present in certain HEMs, their applications as anodes in lithium-ion complete batteries, this is often essential to provide a suitable voltage window. Yan et al. constructed a complete battery based on HEO-MP//NCM, employing NCM as a cathode. Operating within a voltage window of around 0.8–4 V resulted in a capacity of 329 mA h g^{-1} throughout 120 repetitions [54]. Multi-anionic plus multi-cationic substances are synthesized using efficient mechanochemistry utilizing a newly developed HEO containing multiple transition metal cations to serve as the starting material, together with lithium fluoride or sodium chloride as the other reacting substances, resulting in the production of materials where lithium or sodium has been incorporated. The entropy-stabilized oxyfluoride $Li^x(Co^{0.2}Cu^{0.2}Mg^{0.2}Ni^{0.2}Zn^{0.2})OF^x$, which contains lithium, shows an operating voltage of 3.4 V relative to Li^+/Li, rendering it suitable as an active material for the cathode. This novel substance has superior lithium storage capabilities compared to traditional oxyfluorides owing to entropy stability, which generally enables the customization of cycle performance via unprecedented variations in component elements [55]. To mitigate the challenges of chemomechanical alongside thermal fluctuations, as well as the restricted cycle life in commonly used Ni-rich cathodes, an ultrahigh nickel, Co-free layered cathode, $LiNi_{0.88}Mn_{0.03}Mg_{0.02}Fe_{0.02}Ti_{0.02}Mo_{0.02}Nb_{0.01}O_2$ (designated as HE-SC-N88), was synthesized. HE-SC-N88, resulting from the advantages of SC- and HE-doping, possesses an absence of grain boundaries and a stable structure featuring reduced lattice strain, which prevents physical breakdown, reduces secondary reactions, and mitigates oxygen depletion. Thus, this specific HE-SC-N88 cathode reveals outstanding electrochemical behavior, particularly its sustained operational integrity under rigorous testing in both half-cell and fully assembled battery systems, and the postponed alterations in its structure due to oxygen departure during the process of being heated [56].

6.4.2 Potassium-Ion Battery

The scarcity of resources plus the inequitable allocation of lithium provides a compelling need to innovate rechargeable batteries with other carriers for charging. Potassium-ion batteries (KIBs) rank highest among alternatives due to their plentiful raw ingredients, relatively large density of energy, rapid ion transportation kinetics inside the electrolyte, along with inexpensive cost [57]. Vacuum solid-state synthesis of a 2D $CoVMnFeZnPS_3$ ($HEPS_3$) with thickness varying between 6 and 10 nm was performed to explore the influence of the randomness of the configuration regarding the cycling efficiency of 2D metal phosphorous trichalcogenides (MPS_3) while used as anodes within potassium-ion gadgets. $HEPS_3$ facilitates effective potassium-ion movement and intercalation around the electrode's interface, attributed to the impacts by high-entropy via the combination of diverse metallic species around K^+ energy binding, demonstrating a remarkable reversible capacity of 524 $mA\,h\,g^{-1}$, a tremendous high-rate performance of about 10 $A\,g^{-1}$, along with outstanding durability during cycling for more than 1,000 repetitions [58]. The implantation of five distinct transition metals (Fe^{2+}, Cu^{2+}, Co2+, Mn^{2+}, and Ni^{2+}) that occupy identical nitrogen sites for coordination yields a configurational entropy above 1.5R, aimed at improving the electrochemical efficiency of high-entropy Prussian blue cathode components. The HEPB-1 cathode material exhibits superior electrochemical properties, showing first discharge values of 86.69 and 74.51 $mA\,h\,g^{-1}$ (including a charge recovery of 75.2% through 100 cycles under current densities of 20 and 100 $mA\,g^{-1}$, in that order) [59]. Layered oxides with a manganese base constitute an attractive cathode selection in economical PIBs. Ti, Mg, Ni, and Fe in equal atomic proportions were successfully incorporated into the manganese-based layered oxide's transition metal layers aiming to produce an entropy-stabilized material, specifically $K_{0.6}Ni_{0.05}Fe_{0.05}Mg_{0.05}Ti_{0.05}Mn_{0.725}O_2$ (HE-KMO, with an entropy of 1.17R). So, the results show that HE-KMO's multilayered makeup is more stable than the conventional low-entropy $K_{0.6}MnO_2$ (LE-KMO, with an entropy of 0.66R) over repeated cycles as well as in the presence of humidity. The entropy-stabilized HE-KMO demonstrates excellent cycle stability and exceptional rate capability due to its stable high-entropy stacking and the solid-solution K+ storage mechanism. It's important to point out that even following 3,000 cycles, under a current load of 5,000 $mA\,g^{-1}$ can retain its capacity stability percentage of 86% [60]. In contrast to bielemental metal compounds, the synergistic interactions among different components in HEAs could significantly boost potassium storing performance of these materials. The HEA-CNFs deployed as an electrode in KIBs demonstrated an exceptional charge capability of 1,400 mA h g^{-1} during 800 charge–discharge repetitions at a rate of 0.5 A g^{-1}. Furthermore, being a self-sustaining KIBs anode, HEA-CNFs demonstrated a stable battery charge storage of 280 mA h g^{-1} despite 200 repetitions during the rate of 0.2 A g^{-1}, indicating significant promise for potassium-ion storage [61].

6.4.3 Sodium-Ion Battery

Sodium-ion batteries (SIBs) have been engineered as viable substitutes for LIBs in response to the increasing appetite for energy and worries about the availability of lithium minerals [62]. Na layered oxides consistently have slow kinetics and detrimental phase changes during deep desodiation in the O_3 structure, resulting in poor-rate capability as well as substantial capacity loss. To address such constraints, a $Na_{0.83}Li_{0.1}Ni_{0.25}Co_{0.2}Mn_{0.15}Ti_{0.15}Sn_{0.15}O_{2-\delta}$ (named as MTS15) was developed, including increased O–Na–O slab distance, which improves Na^+ diffusion rates and structural durability. The synthesized MTS15 cathode demonstrates remarkable rate capability (76.7% retained capacity at 10 C), durable cycling (87.2% capacity persistence despite 200 cycles), a reversible charge capacity of about 109.4 mA h g^{-1}, commendable full-cell effectiveness (84.3% capacity persistence despite 100 cycles), and outstanding air stability [63]. The minimal poisoning and substantial specific capacity associated with layered oxides of manganese have piqued their focus as potential cathode components for SIBs. Poor structural stability owing to Jahn–Teller distortion plus difficult structural change, as well as slow intrinsic Na^+ migration, hinder implementation. The synthesis of P2-$Na_{0.67}Mn_{0.6}Cu_{0.08}Ni_{0.09}Fe_{0.18}Ti_{0.05}O_2$ improves the cycle durability along with high-voltage capability. It accomplishes a deep desodiation and provides a storage of about 158.1 mA h g^{-1}, which is equivalent to 0.61 Na, having 98.2% of early-stage coulombic efficiency. Together, the cationic as well as anionic redox processes are responsible for the charge adjustment [64]. High stability in structure and efficient Na+ ion transport has captivated vanadium-based sodium superionic conductors (NASICON) to an unprecedented degree. Unfortunately, at high voltages (about 4.0 V), the limited electronic conductivity along with irreversible phase shift of V^{4+}/V^{5+} cause $Na_4VM(PO_4)_3$ to experience low-rate capability and capacity degradation. NVFP-HE NASICON, a material with a high-entropy of $Na_4VFe_{0.6}Mn_{0.2}Cr_{0.1}(CoMgAl)_{0.1}(PO_4)_3$, is intended to achieve strong ionic and electronic conductivity, along with great rate capability even at elevated voltage, by the inherent effect of multi-metals. In addition to a large discharge capability of about 141.98 mA h g^{-1} at 1C, this NVFP-HE cathode showcases an outstanding capacity of about 85.77 mA h g^{-1} even operating at a 50C rate. Even after 10,000 repetitions at 50C rate, 66.7% of the original capacity can be maintained with NVFP-HE. It should be emphasized that the drop in performance due to the fixed phase transition at 4 V is attenuated by the high-entropy. One should observe that at 0.2C, the NVFP-HE//HC complete cell produces 81 mA h g^{-1} [65].

6.4.4 Lithium-Air Battery

Rechargeable metal-air batteries have been considered the utmost plausible choices for storing energy and have attracted considerable scientific interest and examination. The advanced lithium-air (Li-O_2) battery possesses an impressive theoretical specific energy of approximately 3,500 Wh kg^{-1}, derived from the reaction between O_2 plus Li to produce Li_2O_2. Nevertheless, contemporary improved metal-air batteries have several obstacles, including limited practical capacity and density, minimal energy conversion yield, and inadequate cycle life [66]. A comprehensive d-band method was used by a high-entropy cathode catalyst (HESe) to enhance the cycle life of LiO_2 batteries. The extensive distribution of d-band results in the orbital fusion between the metal 3d plus O 2p in O_2, LiO_2 and Li_2O_2, which could enhance the entry of electrons during discharge as well as lower the energy hurdle for transmission of charge and gas component desorption during charging. The HESe cathode possesses a very prolonged cycle life exceeding 480 cycles during ambient air despite any protective measures in LiO_2, showcasing it's extremely effective multiphase catalytic capacity for Li_2O_2 [67]. The ruthenium-on-lanthanum $La(Mn_{0.2}Co_{0.2}Fe_{0.2}Ni_{0.2}Cr_{0.2})O_3$ perovskite oxide, which contains one atom of ruthenium (Ru), operates very well and remains stable in LiO_2 batteries. Establishing Ru-O-M (where M is an element such as Mn, Co, Fe, or Ni) bridges allows for rapid electron mobility from Ru to HEPO through heteronuclear coordination. This effectively redirects electrons inside the Ru@HEPO, leading to improved electrocatalytic function and the rate of transferring charge across the interface. The LiO_2 batteries made of Ru@HEPO exhibit outstanding electrochemical properties, including an elevated energy efficiency (Figure 6.1a), a minimal charge movement resistance (Figure 6.1b), a high-rate capability, a minimal overpotential of about 0.52 V at the rate of 100 mA g^{-1} (Figure 6.1c), plus 345 cycles of long-lasting durability were achieved at 300 mA g^{-1} (Figure 6.1d) [68].

6.4.5 Zinc-Air Battery

The creation of economical, high-performance, and reliable dual-functional catalysts toward the oxygen reduction as well as evolution processes (ORR/OER) is crucial for the commercialization of zinc-air (Zn-O_2) batteries. An HEA (CrMnFeCoNi) based bifunctional ORR/OER (Figure 6.2a) catalyst exhibits exceptional OER effectiveness by having a minimal overpotential of about 265 mV to achieve a current density of 10 mA cm^{-2}, plus demonstrates brilliant ORR efficiency with a half-potential value of about 0.78 V as well as an onset potential value of about 0.88 V. Zn-O_2 batteries are fabricated using CrMnFeCoNi as the air electrode and a zinc sheet functioning as the negative electrode. The resulting devices exhibit an open-circuit potential of 1.489 V (Figure 6.2b), reaching 90% of the ideal value (1.66 V), along with a maximum power output of 116.5 mW cm^{-2} (Figure 6.2c) and a gravimetric capacity of 836 mA h g^{-1}. Performance remains consistent over 10 days of uninterrupted operation (Figure 6.2d), corresponding to 720 charge–discharge cycles at 8 mA cm^{-2}, and extends up to 16.6 days—equivalent to 1,200 cycles, when operated at 5 mA cm^{-2} [69]. FeCoNiMoW nanoparticles manufactured for Zn-O_2

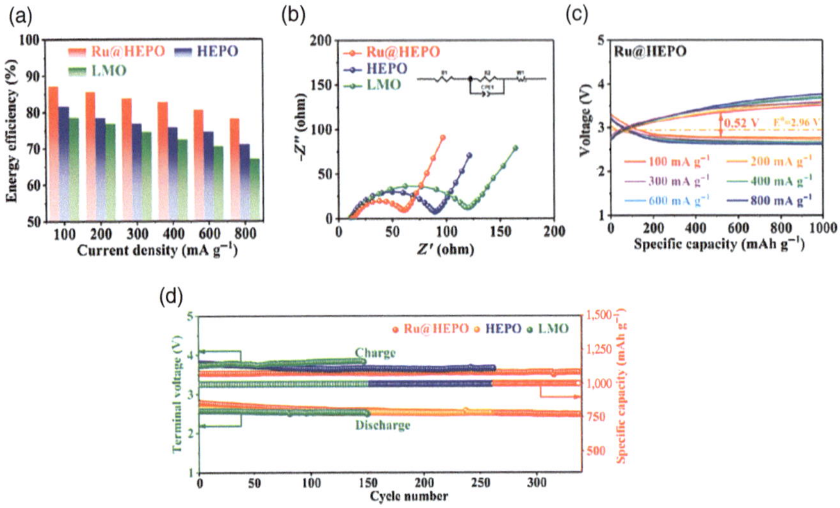

Figure 6.1 (a) Energy efficiency bars for various catalysts, (b) Nyquist plots of various catalysts, (c) Charge/discharge curves of Ru@HEPO, and (d) Stability curves of various catalysts. *Source:* Reference [68] / with permission of John Wiley & Sons.

batteries have high-entropy characteristics, minor lattice distortions, and altered electronic structures, resulting in enhanced OER efficacy. Aqueous Zn-O_2 batteries using this HEA have an open-circuit voltage value of about 1.59 V, a maximum output density value of about 116.9 mW cm^{-2}, a value of 857 mA h $g_{Zn}{}^{-1}$ in terms of specific charge capacity (Figure 6.2e), and an outstanding endurance for more than 660 hours of uninterrupted practical operations (Figure 6.2f) [70].

6.4.6 Aluminum-Air Battery

The high-entropy spinel oxide (CMNFCO), fabricated utilizing a straightforward hydrothermal process, exhibits superior ORR functionality by having a good onset potential value of 0.88 V, as well as a half-wave potential value of about 0.76 V (Figure 6.3a), plus a remarkable ORR endurance, retaining 91.5% of current after a 33-hour chronoamperometry test (Figure 6.3a), operating at the value of 0.287 V. Conversely, CMNFCO exhibited subpar OER efficacy, featured by a minimal overpotential value of 411 mV to achieve 10 mA cm^{-2}. An Al-air (Al-O_2) battery utilizing CMNFCO as a cathode electrocatalyst, demonstrated a maximum output density value of about 148.8 mW cm^{-2} (Figure 6.3c) [71]. High-alloy aluminum containing Al-Li-Mg-Zn-Cu, produced by deformation-driven alterations, is recommended for use as anodes in main aluminum-air batteries, aiding in the suppression of hydrogen evolution self-corrosion, minimizing the polarization, plus improving the discharge potential. The energy plus power density attained were 2.73 Wh g^{-1} and 88.3 mW cm^{-2}, respectively, surpassing those of pure aluminum (Figure 6.3d) [72].

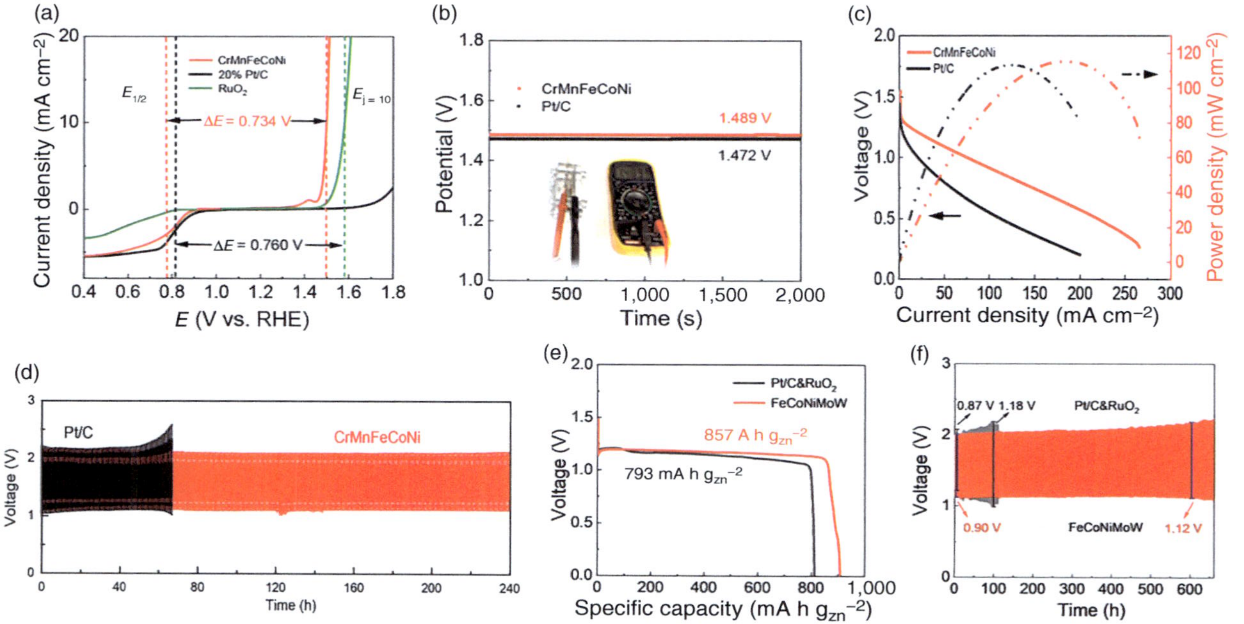

Figure 6.2 (a) OER/ORR curves, (b) open-circuit voltages, (c) power density curves, and (d) stability curves of various catalysts. *Source:* Reproduced with permission from Ref. [69] / Elsevier, (e) specific capacity curves, and (f) stability curves of various catalysts. *Source:* Reference [70] / with permission of John Wiley & Sons.

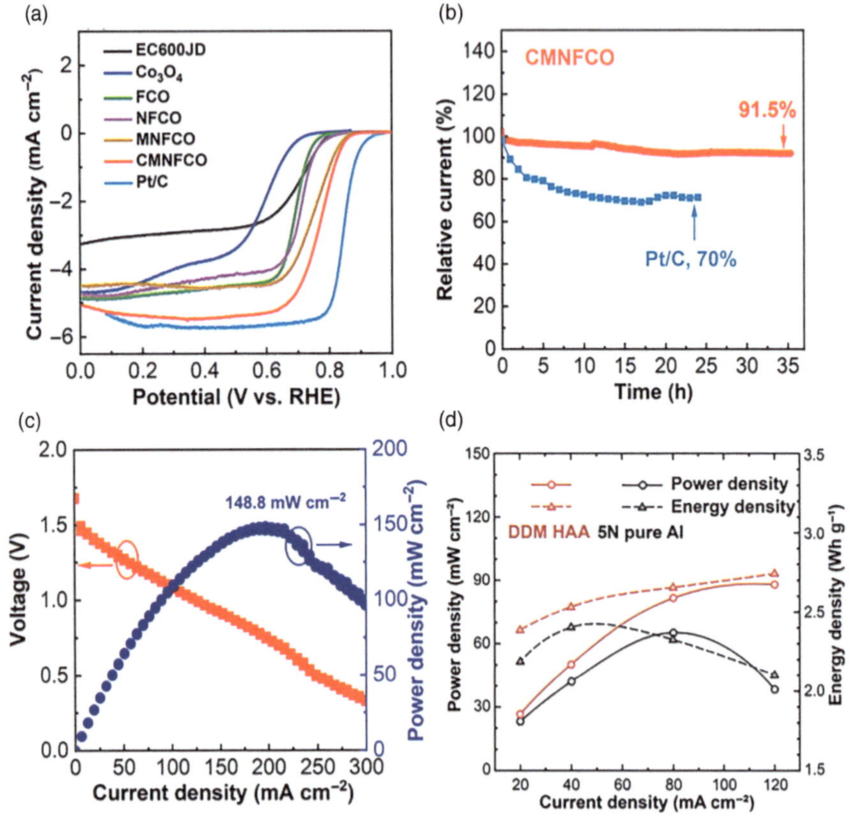

Figure 6.3 (a) ORR curves, (b) stability curves, (c) power density curves of various catalysts. *Source:* Reference [71] / with permission of Elsevier, (d) energy and power density curves of various catalysts. *Source:* Reference [72] / with permission of Elsevier.

6.4.7 Supercapacitors

HEMs' use in supercapacitors had shown a good capacitive performance. A hybrid unsymmetrical liquid-state gadget was constructed employing FeCoNiCuMn plus green graphene to be electrodes. The gadget achieved a large capacitance of $83.22\,F\,g^{-1}$ at a current density of $2\,A\,g^{-1}$ using the green source (Figure 6.4a). The gadget exhibited a specific energy of $33.4\,Wh\,kg^{-1}$ and a specific power of $1.7\,kW\,kg^{-1}$ (Figure 6.4b) [73]. The Fe-Co-Ni-Cu-Zn HEA possesses a maximum gravimetric capacitance of $325.17\,F\,g^{-1}$ at the rate of $1\,A\,g^{-1}$ in sole electrode configuration with a good electrochemical stability (Figure 6.4c), and it achieves a maximum energy density of $23.82\,Wh\,kg^{-1}$ while using a power output of $325\,W\,kg^{-1}$ (Figure 6.4d), attributable to the presence of many active sites that facilitate improved electrochemical redox processes (Figure 6.4e) [74]. The CoCrNiFeMn bulk HEA, synthesized by induction melting process and then ball milled to produce nanoparticles, achieves a maximum specific capacitance value of $386.66\,F\,g^{-1}$ using a value of $5\,mV\,s^{-1}$ inside a 3 M KOH electrolyte inside a three-electrode configuration. An asymmetric supercapacitor system using liquid-state CoCrNiFeMn and

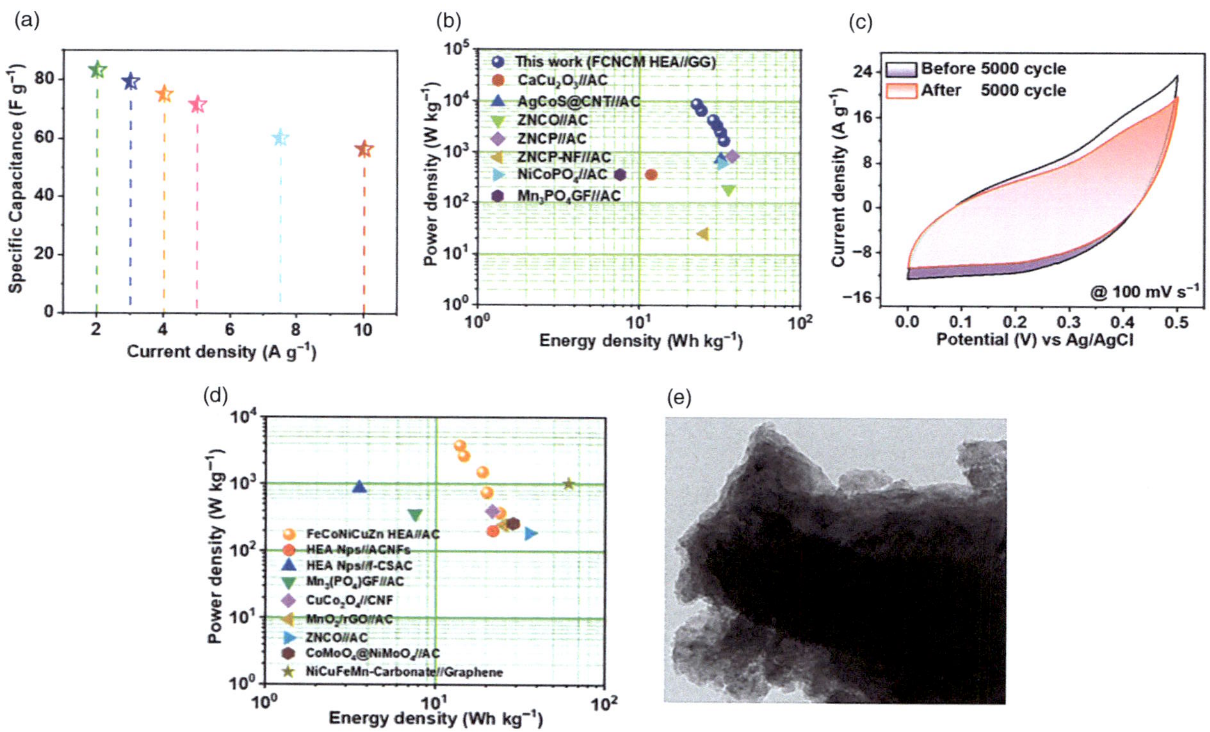

Figure 6.4 (a) Specific capacity vs current density curves, and (b) Ragone plot comparison, (c) cyclic voltammetry stability curves, *Source:* Reference [73] / John Wiley & Sons / CC BY 4.0, (d) Ragone plot comparison, and (e) HRTEM image for the Fe-Co-Ni-Cu-Zn HEA. *Source:* Reference [74] / with permission of Elsevier.

activated carbon was constructed, and the stored energy powered an LED operating at 1.5 V. The energy density value of the asymmetric supercapacitor gadget was 21 Wh kg^{-1} with a maximum power output of 307 W kg^{-1} [75].

6.4.8 Vanadium Redox Flow Battery

The vanadium redox flow battery (VRFB) serves as the premier option for storing massive amounts of renewable energy. The electrode, a crucial component of the VRFB, significantly impacts both the energy efficacy plus the overall price of the entire system. A homogeneous coating of VNbMoTaW HEA oxide (HEAO) was applied to the GF surface by modulating O_2 gas supply, significantly enhancing the reaction rates across the VRFB electrode. The VRFB utilizing the HEAO-altered graphite electrode, namely the $HEAO_{50}$ electrode, exhibited an enhanced energy efficiency of 80.50% at 100 mA cm^{-2}, surpassing the VRFB with the unchanged graphite electrode by 9.49% [76]. A new $(BiZrMoWCeLa)O_2$ HEO calcined at 750 °C, designated HEO-750, exhibits enhanced electrocatalytic function for the V^{3+}/V^{2+} and VO^{2+}/VO_2^+ redox pairs in comparison to cells constructed using alternative materials. The charge–discharge tests further validate that VRFBs employing the HEO-750 catalyst exhibit exceptional coulombic efficiency, voltage efficiency, and with a current density of 80 mA cm^{-2}, the energy efficiencies were 97.22%, 87.47%, and 85.04%, whereas at 160 mA cm^{-2}, these values changed to 98.10%, 74.76%, and 73.34%, respectively. Furthermore, after 500 operational periods, no noticeable deterioration occurs. The findings are ascribed to the calcination treatment with heat, promoting the development of a novel single-phase fluorite arrangement, hence enhancing the redox responses associated with the V redox couples. Moreover, an elevated surface area, enhanced wettability, plus many O_2 vacancies may provide additional electroactive sites, hence augmenting the electrochemical ability, charge movement of redox reactions, and durability of the VRFB electrodes [77]. A deformity-rich $(CoCrFeMnNi)_3O_4$ HEO electrocatalyst featuring a single-phase spinel architecture demonstrates superior catalytic effectiveness plus reversibility regarding the VO^{2+}/VO_2^+ redox pair. Due to its distinctive single-phase spinel architecture plus plentiful vacancies of O_2, the VRFB flow cell employing the HEO-modified heat-treated graphite felt (HEO-HGF) electrode attains an exceptional energy efficiency of 79.23% and 70.58% at 160 and 240 mA cm^{-2}, respectively, as well as demonstrates commendable endurance over 500 cycles [78].

6.4.9 Dielectric Capacitors

The dielectric capacitor acts as a fundamental electrical component in contemporary electronics plus pulse power technologies. Concurrently, with the evolving trends of integration, downsizing, plus diversification, the need for enhanced energy storage density as well as durability in dielectric capacitors is escalating. A unique high-entropy perovskite oxide, designated as (1–x)NBBSCT-xNN, was engineered to enhance temperature dielectric durability along with energy storage efficiency by integrating relaxor plus antiferroelectric properties. The ideal makeup

Table 6.2 Summary of HEMs Different Applications and Their Advantages.

Applications	Advantages	References
Lithium-ion battery	High-energy density	[51]
Potassium-ion battery	Inexpensive, rapid ion transportation, plentiful raw ingredients	[57]
Sodium-ion battery	High availability	[62]
Lithium-air battery	High theoretical efficiency	[66]
Zinc-air battery	High-energy density	[69]
Aluminum-air battery	High-energy density with low cost	[72]
Supercapacitors	Good capacitive performance	[73]
Vanadium redox flow battery	Decoupling energy and power capacity	[76]
Dielectric capacitors	Rapid charging with a small size	[79]

of $x = 0.2$ demonstrates a substantial energy storage density of 3.51 J cm^{-3}, with broad thermal stability, exceptional frequency stability, plus a rapid discharge rate ($t_{0.9}$ = 55.2 ns). This mostly results from the improvement of relaxing behavior along with the elevation of energy density (E_b) owing to the reduction of particle size [79]. The $Ba_{0.2}(KBi)_x(SrCa)_{(0.4-x)}TiO_3$ HECs, engineered to enhance energy storage, exhibited exceptional performance, attaining an energy storage density of 4.35 J cm^{-3} and an effectiveness of 89.5% against a 320 kV cm^{-1} field is applied. The specimen demonstrated outstanding endurance regarding temperature from 20 to 160 °C as well as 1–200 Hz of frequency. The exceptional output was ascribed to the enhanced polarization plus potential withstand capability with the introduction of the entropy idea [80]. A high-entropy relax or ferroelectric ceramic $(Li_{0.2}Ca_{0.2}Sr_{0.2}Ba_{0.2}La_{0.2})TiO_3$, manufactured by the conventional solid-state reaction process, demonstrated an extensive microstructure characterized by a single-phase perovskite framework plus reduced grain size. The distinctive microstructure plus superior protection not simply expanded the energy gap but also reduced residual polarization, leading to a greater E_b of 1.51 J cm^{-3} along with an efficiency of 89.6% despite a minimal field power value of about 260 kV cm^{-1}. The material demonstrated outstanding temperature plus frequency durability, along with fast charging and discharging capabilities (about 35 ns) [81]. Table 6.2 summarizes applications with these advantages.

6.5 Challenges and Limitations

Recent years have seen a surge in interest in high-entropy materials (HEMs), especially HEOs and HEAs, because of their distinctive features, which include improved longevity, adjustable electronic frameworks, and configurational entropy stabilization. Although they show great potential for energy storage applications (e.g. catalysis, batteries, and supercapacitors), however, further development is required before they can be widely adopted. The following is an exhaustive list of these difficulties.

6.5.1 Synthesis Challenges

Complexity: Usually, HEMs have five or more components in almost equal amounts. Extensive experimental and computational modeling is required to identify the ideal composition for individual energy storage applications. It is challenging to forecast phase stability, the structure of the crystal, and operational features because of the wide compositional space.

Temperature: For the uniformity and crystallinity of numerous HEMs, high-temperature methods of synthesis are necessary. These include solid-state processes, sintering, and arc melting. Grain expansion, reduced area of surface, as well as active sites might result from these energy-intensive activities.

Phase: It is difficult to synthesize and operate with a material that has a single phase and a high-entropy. The effectiveness might be negatively affected by phase segregation, which might happen due to small changes in stoichiometry or manufacturing conditions.

Scalability: Numerous contemporary methods for synthesizing HEMs are tailored for laboratory use and are not optimally suited for industrial mass production. Developing cost-effective scale synthesis technologies remains very challenging.

6.5.2 Structural and Chemical Instability

Limited Long-Term Stability: While HEMs generally exhibit more stability than single-component materials, some compositions may degrade under severe electrochemical conditions, including high voltages or environments with raised or diminished pH values. Capacity fading may occur, for example, owing to changes in oxidation state or the dissolution of transition metals during cycling.

Kinetic Limitations: If the material lacks sufficient ionic or electronic conductivity, the sluggish ion or electron transport kinetics may be attributed to the complex multi-element composition of HEMs.

Surface Reconstruction: Surface reorganization during electrochemical cycling may lead to the formation of new phases or passivation layers on HEMs, altering their behavior. This incidence does not consistently provide a favorable influence.

6.5.3 Performance Limitations

Moderate Activity: Despite the fact that HEMs are designed to take the role of noble metals for storing energy functions, they often have poorer catalytic activity compared to benchmark elements such as Pt or IrO_2, for example, in oxygen evolution/reduction processes. Tuning the mixture, doping, and applying nanostructuring precisely is necessary to achieve similar effectiveness, which increases the complexity.

Trade-Off Between Activity and Stability: Enhancing the catalytic effectiveness of HEMs often requires the incorporation of doping, defects, or vacancies, which may compromise their stability over time.

Low Specific Capacity: Due to their intricate crystal structures and the absence of lithium insertion/extraction sites, HEMs may exhibit lower specific capacities for battery applications compared to conventional materials like $LiCoO_2$ or graphite.

6.5.4 Computational and Experimental Challenges

Predictive Modeling: Accurately predicting the properties of HEMs by computational methods, such as density functional theory (DFT), is challenging because to the multitude of elements and potential combinations. In the realm of HEMs, improved screening and AI methodologies are still in their nascent phases.

6.5.4.1 Characterization Difficulties

Due to the multitude of components in HEMs, characterization techniques such as X-ray diffraction (XRD), X-ray photoelectron spectroscopy (XPS), and transmission electron microscopy (TEM) become more intricate. When signals from many elements converge, it may be challenging to elucidate the local chemistry and structure.

6.5.5 Economic and Environmental Concerns

Cost of Raw Materials: Although HEMs strive to lessen dependence on noble metals, the supply chain restrictions and high cost of some rare-earth elements (such as La and Ce) utilized in HEMs pose challenges.

Environmental Impact: The mining and processing of transition and rare-earth metals may result in significant ecological repercussions. The development of ecologically sustainable synthesis routes and recycling methods for HEMs is essential but remains unexamined.

6.5.6 Application Specific Challenges

Battery Application: In LIBs the reversible capacity is limited, and the rates of lithium diffusion are inadequate. SIBs have similar issues with ion mobility and structural integrity during cycling.

Supercapacitor: Due to their reduced surface area and permeability, HEMs may exhibit a lower specific capacitance compared to traditional materials like activated carbon or metal oxides.

6.5.7 Lack of Standardization

A clear methodology for evaluating HEMs is not yet established for several energy storage applications. Comparing studies is challenging because of the significant variability in variables, such as electrochemical efficacy, configurational entropy, and phase purity.

6.6 Future Prospective

HEMs are a promising class of materials with unique features that have garnered significant interest in energy storage applications; yet, several obstacles and unexplored potential remain in future research directions to fully comprehend and capitalize on their capabilities.

6.6.1 Advancements in Material Design and Synthesis

The identification of novel HEM compositions using computational modeling and high-throughput screening will significantly contribute to the progression of energy storage applications. AI and machine learning may facilitate the prediction of more stable and electrochemically active HEM compositions, while innovative synthesis methods can enhance the attributes of these materials.

6.6.2 Optimizing Electrochemical Efficiency

Diverse strategies, including doping, heterostructure engineering, defect creation, and the incorporation of carbon supports, can enhance redox reactions, electronic and ionic conductivity, and charge storage capacities. Thus, employing various strategies with different HEM compositions could boost the overall effectiveness.

6.6.3 Scalability and Cost-Effectiveness

The commercial utilization of HEMs in energy storage applications remains challenging due to intricate synthesis methods and elevated material costs; therefore, future research on HEMs should prioritize simplified synthesis techniques and the use of cost-effective materials suitable for large-scale production.

6.6.4 Prolonged Stability and Resilience

In practical applications, the stability of HEMs during repeated cycles is essential for their mass production. Comprehensive investigations are necessary to elucidate the degradation mechanisms, including structural collapse, phase segregation, and surface passivation associated with HEMs, in order to achieve long-term cycling stability. Various strategies, such as surface modification, coating, and protective layers, can be employed to attain enduring stability.

6.6.5 Green and Sustainable Energy Solutions

The environmental effect must be considered while aiding with HEMs, including the use of eco-friendly synthesis techniques, biodegradable materials, and recycling procedures to achieve sustainable objectives.

6.6.6 Multifunctional and Hybrid Energy Storage Systems

The distinctive properties of HEMs differentiate them from other materials, and future research should focus on using HEMs in hybrid energy storage devices that function as both batteries and supercapacitors.

6.7 Conclusions

HEMs are a novel and transformational category of developing materials with significant promise to address the global energy dilemma. Their distinctive characteristics, including adjustable qualities, high stability, and varied elemental compositions, render them exceptionally ideal for energy storage applications, such as supercapacitors, dielectric capacitors, and batteries. In the transition to green energy and the substitution of fossil fuels with renewable sources, energy storage devices are essential, and HEMs facilitate the attainment of this objective. HEMs exhibited exceptional electrochemical and mechanical capabilities as electrode materials, offering substantial energy storage capacity, extended life cycles, and robust stability under adverse environments. Their multitasking capabilities in various energy storage devices, including the provision of multiple redox-active sites, extensive active surface area, and structural stability in batteries, as well as the enhancement of energy storage efficiency in dielectric capacitors and the provision of tunable conductivity, high surface area, and pseudocapacitive behavior for high power density, render them an optimal choice for next-generation energy storage devices.

However, the proliferation of HEMs is complicated. A comprehensive knowledge of structural features, degradation mechanisms, environmental issues, and financial considerations remains unresolved. The utilization of computational modeling and AI, coupled with novel synthesis techniques and material compositions, may enhance the performance of HEMs in large-scale manufacturing.

Concerning energy storage, HEMs possess a promising future. Through innovation and ongoing study, these materials possess the capacity to address the challenges of renewable energy storage.

References

1 Amiri, A. and Shahbazian-Yassar, R. (2021). Recent progress of high-entropy materials for energy storage and conversion. *J Mater Chem A* 9 (2): 782–823.

2 Wu, J., Wang, H., Liu, N. et al. (2024). High-entropy materials in electrocatalysis: understanding, design, and development. *Small* 20 (43): 2403162.

3 Ma, Y., Ma, Y., Wang, Q. et al. (2021). High-entropy energy materials: challenges and new opportunities. *Energy Environ Sci* 14 (5): 2883–2905.

4 George, E.P. and Ritchie, R.O. (2022). High-entropy materials. *MRS Bull* 47 (2): 145–150.

5 Yeh, J.W., Chen, S.K., Lin, S.J. et al. (2004). Nanostructured high-entropy alloys with multiple principal elements: novel alloy design concepts and outcomes. *Adv Eng Mater* 6 (5): 299–303.
6 Deng, C., Wang, T., Wu, P. et al. (2024). High entropy materials for catalysis: a critical review of fundamental concepts and applications. *Nano Energy* 120: 109153.
7 Wang, Z., Li, Z.T., Zhao, S.J., and Wu, Z.G. (2021). High-entropy carbide ceramics: a perspective review. *Tungsten* 3 (2): 131–142.
8 Qureshi, T., Khan, M.M., and Pali, H.S. (2024 Sep). high-entropy borides—challenges and opportunities. *J Mater Sci* 59 (34): 15921–15991.
9 Novikov, V., Stepanov, N., Zherebtsov, S., and Salishchev, G. (2022). Structure and properties of high-entropy nitride coatings. *Metals* 12 (5): 847.
10 Gu, X., Guo, X.B., Li, W.H. et al. (2024). High-entropy materials for application: electricity, magnetism, and optics. *ACS Appl Mater Interfaces* 16 (40): 53372–53392.
11 Veselkov, S., Samoilova, O., Shaburova, N., and Trofimov, E. (2021). High-temperature oxidation of high-entropic alloys: a review. *Materials* 14 (10): 2595.
12 Oueldna, N., Sabi, N., Aziam, H. et al. (2024). High-entropy materials for thermoelectric applications: towards performance and reliability. *Mater Horiz* 11 (10): 2323–2354.
13 Hussain, I., Lamiel, C., Ahmad, M. et al. (2021). High entropy alloys as electrode material for supercapacitors: a review. *J Energy Storage* 44: 103405.
14 Khan, J.B., Panda, P.K., Yang, P.C. et al. (2024). Microwave synthesis of high-entropy alloy catalysts on graphene oxide sheets for oxygen reduction and evolution reactions. *Int J Hydrog Energy* 53: 999–1008.
15 Sebestyén, V. (2021 Nov). Renewable and sustainable energy reviews: environmental impact networks of renewable energy power plants. *Renew Sust Energ Rev* 151: 111626.
16 Yoro, K.O., Daramola, M.O., Sekoai, P.T. et al. (2021). Update on current approaches, challenges, and prospects of modeling and simulation in renewable and sustainable energy systems. *Renew Sust Energ Rev* 150: 111506.
17 Zhang, Y., Wang, D., and Wang, S. (2022). High-entropy alloys for electrocatalysis: design, characterization, and applications. *Small* 18 (7): 2104339.
18 Ma, J. and Huang, C. (2023). High entropy energy storage materials: synthesis and application. *J Energy Storage* 66: 107419.
19 Xu, X., Shao, Z., and Jiang, S.P. (2022). High-entropy materials for water electrolysis. *Energy Technol* 10 (11): 2200573.
20 Olabi, A.G., Abbas, Q., Shinde, P.A., and Abdelkareem, M.A. (2023). Rechargeable batteries: technological advancement, challenges, current and emerging applications. *Energy* 266: 126408.
21 de Vasconcelos, L.S., Xu, R., Xu, Z. et al. (2022). Chemomechanics of rechargeable batteries: status, theories, and perspectives. *Chem Rev* 122 (15): 13043–13107.

22 Sharma, S. and Chand, P. (2023). Supercapacitor and electrochemical techniques: a brief review. *Resul Chem* 5: 100885.

23 Şahin, M.E., Blaabjerg, F., and Sangwongwanich, A. (2022). A comprehensive review on supercapacitor applications and developments. *Energies* 15 (3): 674.

24 Brechtl, J., Lee, C., and Liaw, P.K. (2023). High-entropy materials: fundamentals and applications. *J Mater Res Technol* 23 (1).

25 Zhang, Y. (2019). High-entropy materials. *Springer Nat Singap Pte Ltd* 2: 215–232.

26 Cantor, B. (2024). *Fundamentals of Multicomponent High-Entropy Materials*. Oxford University Press.

27 Nene, S.S. (2024). *Fundamentals of High Entropy Alloys. High Entropy Alloys: A Machine-Generated Literature Overview*, 5–30. Singapore: Springer Nature Singapore.

28 Xiang, H., Dai, F.Z., and Zhou, Y. (2023). *High-Entropy Materials: From Basics to Applications*. Wiley.

29 Hsu, W.L., Tsai, C.W., Yeh, A.C., and Yeh, J.W. (2024). Clarifying the four core effects of high-entropy materials. *Nat Rev Chem* 8 (6): 471–485.

30 Tomboc, G.M., Kwon, T., Joo, J., and Lee, K. (2020). High entropy alloy electrocatalysts: a critical assessment of fabrication and performance. *J Mater Chem A* 8 (30): 14844–14862.

31 Wang, H., He, Q., Gao, X. et al. (2024). Multifunctional high entropy alloys enabled by severe lattice distortion. *Adv Mater* 36 (17): 2305453.

32 Aidhy, D.S. (2024). Chemical randomness, lattice distortion and the wide distributions in the atomic level properties in high entropy alloys. *Comput Mater Sci* 237: 112912.

33 Mehta, A. and Sohn, Y.H. (2021). Fundamental core effects in transition metal high-entropy alloys: "high-entropy" and "sluggish diffusion" effects. *Diffus Found* 29: 75–93.

34 Qi, L. and Guan, J. (2024). Electronic structure modulation of high entropy materials for advanced electrocatalysis. *Green Energy Environ* 10 (5): 917–936.

35 Zhang, L., Jia, J., and Yan, J. (2024). Challenges and strategies for synthesizing high performance micro and nanoscale high entropy oxide materials. *Small* 20 (28): 2309586.

36 Jiang, S., Yu, Y., He, H. et al. (2024). General synthesis of composition-tunable high-entropy amorphous oxides toward high efficiency oxygen evolution reaction. *Small* 20 (28): 2310786.

37 Lotfi-Khojasteh, E., Elmkhah, H., Nouri, M., and Mayrhofer, P.H. (2024). Atomic radius mismatch: a key parameter for design and synthesis of high-entropy physical vapor deposition coatings. *Adv Eng Mater* 26 (6): 2301934.

38 Ritter, T.G., Pappu, S., and Shahbazian-Yassar, R. (2024). Scalable synthesis methods for high-entropy nanoparticles. *Adv Energy Sustainab Res* 5 (8): 2300297.

39 Li, H., Zhu, H., Zhang, S. et al. (2020). Nano high-entropy materials: synthesis strategies and catalytic applications. *Small Struct* 1 (2): 2000033.

40 Sun, X. and Sun, Y. (2024). Synthesis of metallic high-entropy alloy nanoparticles. *Chem Soc Rev*.

41 Chen, P.C., Liu, X., Hedrick, J.L. et al. (2016). Polyelemental nanoparticle libraries. *Science* 352 (6293): 1565–1569.

42 Liu, Y., Tian, X., Han, Y.C. et al. (2023). High-temperature shock synthesis of high-entropy-alloy nanoparticles for catalysis. *Chin J Catal* 48: 66–89.

43 Wang, B., Wang, C., Yu, X. et al. (2022). General synthesis of high-entropy alloy and ceramic nanoparticles in nanoseconds. *Nat Synth* 1 (2): 138–146.

44 Gao, S., Hao, S., Huang, Z. et al. (2020). Synthesis of high-entropy alloy nanoparticles on supports by the fast moving bed pyrolysis. *Nat Commun* 11 (1): 2016.

45 Li, M., Huang, C., Yang, H. et al. (2023). Programmable synthesis of high-entropy nanoalloys for efficient ethanol oxidation reaction. *ACS Nano* 17 (14): 13659–13671.

46 Maulana, A.L., Chen, P.C., Shi, Z. et al. (2023). Understanding the structural evolution of IrFeCoNiCu high-entropy alloy nanoparticles under the acidic oxygen evolution reaction. *Nano Lett* 23 (14): 6637–6644.

47 Bib Khan, J., Kumar Panda, P., Dash, P., and Hsieh, C.T. (2025). Microwave-assisted synthesis of platinum-free high-entropy alloy catalysts on reduced graphene oxide sheets for enhanced oxygen reduction and evolution reactions. *Chem Eur J* 31 (11): e202403863.

48 Sun, Y., Wu, W., Yu, L. et al. (2023). Asymmetric acidic/alkaline N_2 electrofixation accelerated by high-entropy metal–organic framework derivatives. *Carbon Energy* 5 (3): e263.

49 Yan, X., Zhou, Y., and Wang, S. (2025). Nano-high entropy materials in electrocatalysis. *Adv Funct Mater* 35 (2): 2413115.

50 Lokhande, V., Malavekar, D., Kim, C. et al. (2024). Order within disorder: unveiling the potential of high entropy materials in energy storage and electrocatalysis. *Energy Storage Mater* 15: 103718.

51 Li, C., Chen, X., Li, J. et al. (2024). High entropy anodes in batteries: from fundamentals to applications. *Energy Storage Mater* 9: 103468.

52 Xiao, B., Wu, G., Wang, T. et al. (2022). High-entropy oxides as advanced anode materials for long-life lithium-ion Batteries. *Nano Energy* 95: 106962.

53 Li, S., Peng, Z., and Fu, X. (2023). $Zn_{0.5}Co_{0.5}Mn_{0.5}Fe_{0.5}Al_{0.5}Mg_{0.5}O_4$ high-entropy oxide with high capacity and ultra-long life for Li-ion battery anodes. *J Adv Ceram* 12 (1): 59–71.

54 Yen, J.Z., Yang, Y.C., and Tuan, H.Y. (2022). Interface engineering of high entropy oxide@ polyaniline heterojunction enables highly stable and excellent lithium ion storage performance. *Chem Eng J* 450: 137924.

55 Wang, Q., Sarkar, A., Wang, D. et al. (2019). Multi-anionic and-cationic compounds: new high entropy materials for advanced Li-ion batteries. *Energy Environ Sci* 12 (8): 2433–2442.

56 Liang, L., Su, M., Sun, Z. et al. (2024). High-entropy doping promising ultrahigh-Ni Co-free single-crystalline cathode toward commercializable high-energy lithium-ion batteries. *Sci Adv* 10 (25): eado4472.

57 Min, X., Xiao, J., Fang, M. et al. (2021). Potassium-ion batteries: outlook on present and future technologies. *Energy Environ Sci* 14 (4): 2186–2243.

58 Chien, P.W., Chang, C.B., and Tuan, H.Y. (2023). High-entropy two-dimensional metal phosphorus trichalcogenides boost high-performance potassium ion storage devices via electrochemical reconstruction. *Energy Storage Mater* 61: 102853.

59 Yan, W., Feng, X., Min, X. et al. (2024). High-entropy prussian blue analogues as high-capacity cathode material for potassium ion batteries. *J Electrochem Soc* 171 (8): 080529.

60 Cai, Y., Liu, W., Chang, F. et al. (2023). Entropy-stabilized layered $K_{0.6}Ni_{0.05}Fe_{0.05}Mg_{0.05}Ti_{0.05}Mn_{0.725}O_2$ as a high-rate and stable cathode for potassium-ion batteries. *ACS Appl Mater Interfaces* 15 (41): 48277–48286.

61 Zheng, D., Zha, J., Wang, Y. et al. (2025). $Co_{0.2}Sb_{0.2}Fe_{0.2}Mn_{0.2}Ni_{0.2}$ high-entropy alloy carbon nanofiber as anode for lithium/potassium ion batteries. *J Mater Sci* 12: 1–4.

62 Cai, X., Yue, Y., Yi, Z. et al. (2024). Challenges and industrial perspectives on the development of sodium ion batteries. *Nano Energy* 26: 110052.

63 Wang, H., Gao, X., Zhang, S. et al. (2023). High-entropy Na-deficient layered oxides for sodium-ion batteries. *ACS Nano* 17 (13): 12530–12543.

64 Liu, Z., Liu, R., Xu, S. et al. (2024). Achieving a deeply desodiated stabilized cathode material by the high entropy strategy for sodium-ion batteries. *Angew Chem Int Ed* 63 (29): e202405620.

65 Ahsan, M.T., Qiu, D., Ali, Z. et al. (2024). Unraveling the fast Na diffusion kinetics of NASICON at high voltage via high entropy for sodium-ion battery. *Adv Energy Mater* 14 (4): 2302733.

66 Wen, B., Huang, Y., Jiang, Z. et al. (2024). Exciton dissociation into charge carriers in porphyrinic metal-organic frameworks for light-assisted Li-O_2 batteries. *Adv Mater* 36 (32): 2405440.

67 Li, X., Zhang, G., Zhang, D. et al. (2024). A high-entropy cathode catalyst with multiphase catalytic capability of Li_2O_2 and Li_2CO_3 enabling ultralong cycle life in Li–air batteries. *Energy Environ Sci* 17 (21): 8198–8208.

68 Du, D., He, H., Zheng, R. et al. (2024). Single-atom immobilization boosting oxygen redox kinetics of high-entropy perovskite oxide toward high-performance lithium-oxygen batteries. *Adv Energy Mater* 14 (17): 2304238.

69 He, R., Yang, L., Zhang, Y. et al. (2023). A CrMnFeCoNi high entropy alloy boosting oxygen evolution/reduction reactions and zinc-air battery performance. *Energy Storage Mater* 58: 287–298.

70 He, R., Yang, L., Zhang, Y. et al. (2023). A 3d-4d-5d high entropy alloy as a bifunctional oxygen catalyst for robust aqueous zinc–air batteries. *Adv Mater* 35 (46): 2303719.

71 Du, Y., Yang, F., Li, J. et al. (2024). Electrocatalytic performances of high-entropy spinel oxide $(Cr_{1/5}Mn_{1/5}Ni_{1/5}Fe_{1/5}Co_{1/5})3O_4$ for oxygen reduction/evolution reactions in alkaline electrolyte. *J Alloys Compd* 1004: 175923.

72 Xie, Y., Meng, X., Mao, D. et al. (2022). Deformation-driven modification of Al-Li-Mg-Zn-Cu high-alloy aluminum as anodes for primary aluminum-air batteries. *Scr Mater* 212: 114551.

73 Mohanty, G.C., Chowde Gowda, C., Gakhad, P. et al. (2024). Enhanced energy density of high entropy alloy (Fe-Co-Ni-Cu-Mn) and green graphene hybrid supercapacitor. *Energy Storage* 6 (4): e668.

74 Mohanty, G.C., Gowda, C.C., Gakhad, P. et al. (2023). Iron-cobalt-nickel-copper-zinc (FeCoNiCuZn) high entropy alloy as positive electrode for high specific capacitance supercapacitor. *Electrochim Acta* 470: 143272.

75 Mohanty, G.C., Gowda, C.C., Gakhad, P. et al. (2023). High energy density liquid state asymmetric supercapacitor devices using Co–Cr–Ni–Fe–Mn high entropy alloy. *Mater Adv* 4 (17): 3839–3852.

76 Tiwari, K., Wang, C.H., Lou, B.S. et al. (2024). High entropy alloy oxide coating of VNbMoTaWOx as a novel electrode modification of vanadium redox flow batteries. *J Energy Storage* 94: 112344.

77 Demeku, A.M., Kabtamu, D.M., Chen, G.C. et al. (2024). High-entropy oxide of $(BiZrMoWCeLa)O_2$ as a novel catalyst for vanadium redox flow batteries. *ACS Appl Mater Interfaces* 16 (8): 10019–10032.

78 Xiao, X.H., Kabtamu, D.M., Demeku, A.M. et al. (2024). Defect-rich high-entropy spinel oxide catalyst for efficient vanadium redox flow battery. *J Power Sources* 597: 234178.

79 Ning, Y., Pu, Y., Wu, C. et al. (2023). Enhanced capacitive energy storage and dielectric temperature stability of A-site disordered high-entropy perovskite oxides. *J Mater Sci Technol* 145: 66–73.

80 Bi, Z., Zhou, S., Ye, J. et al. (2025). $BaTiO_3$-based high-entropy ceramics for enhanced capacitive energy storage performance. *Ceram Int* 51 (12): 16052–16060.

81 Zhu, X., Xiong, S., Zhu, G. et al. (2024). Dielectric properties and excellent energy storage density under low electric fields for high entropy relaxor ferroelectric $(Li_{0.2}Ca_{0.2}Sr_{0.2}Ba_{0.2}La_{0.2})TiO_3$ ceramic. *J Alloys Compd* 984: 173987.

7

Noble-Metal-Based High-Entropy Oxides for Energy Storage Applications

Dibyananda Majhi, Shreeganesh Subraya Hegde, and Subrahmanyam Challapalli

Indian Institute of Technology Hyderabad, Kandi, Energy and Environmental Laboratory, Department of Chemistry, IITH Rd, near NH-65, Sangareddy, Telangana, 502285, India

7.1 Introduction

Energy and environmental issues are global problems over the last few decades. Energy generation from non-renewable resources not only causes depletion of their abundance but also results in serious environmental impacts. Development of effective, greener energy generation technologies is highly essential to feed the global energy demand [1, 2]. Recent research efforts in nanoscience and nanotechnology have been instrumental in advancing the development of high-performance green energy technologies, reflecting a growing focus on harnessing innovative materials and methods to enhance sustainability and efficiency in energy solutions for storage systems [3]. Despite extensive research on materials for advanced supercapacitors and battery electrodes in recent decades, replacing traditional materials such as graphene-based materials, activated carbon, and metal oxide/sulfide materials remains challenging. This is because commercialization demands a wide range of criteria, including cost-effectiveness, improved electrical conductivity, active surface area, porosity, long-term cyclability, a stable range of potential, and thermal performance. Among the diverse array of materials, high-entropy oxides (HEOs) and noble-metal-based high-entropy oxides (NM-HEOs) have garnered broad interest across various applications owing to their remarkable properties [4]. Due to their excellent electrochemical properties and stability, these particular materials have been widely used in energy storage and generation systems. These materials exhibit different crystal structures depending on the composition of constituent elements, such as rock salt, spinel, and perovskite and possess exceptional structural stability. Due to their multifunctional activity, e.g. tunable electrochemical properties and multi-electron redox properties, these materials are highly suitable for energy storage devices. These special classes of materials have been widely investigated for Li-ion, Na-ion, Li-S, and other batteries

High-Entropy Materials for Energy Storage Devices, First Edition.
Edited by Chien-Te Hsieh, Pradeep Kumar Panda, and Arpan Kumar Nayak.

like metal-air batteries. In addition, HEOs and NM-HEOs are very suitable for high-capacitance electrodes for supercapacitor applications. Incorporating noble metals like Pt, Pd, etc., can further enhance the intrinsic electronic conductivity and active sites. This further resulted in the enhancement of charge transfer kinetics and cycling stability. In noble-metal-based HEOs, the high-entropy tenability and catalysis/conductivity give superior storage performance and stability. In this chapter, the application of HEOs and NM-HEOs for various battery and supercapacitor applications will be discussed in particular. Sections below provide descriptions for synthesis, structural aspects, applications for battery and supercapacitor, current challenges, and future perspectives of this special class of materials.

7.2 High-Entropy Oxides and Noble-Metal-Based High-Entropy Oxides

Study on high-entropy materials (HEMs) constituted by at least five or more elements is an old concept. HEMs are generally a homogeneous mixture with multi-elemental compositions (nearly equimolar ratio of constituent elements) stabilized by high configurational entropy in a single crystal lattice. High-entropy alloys (HEAs), which are a class under HEMs, were reported by Yeh and Cantor in 2004 [5, 6]. This was the first report on a crystalline and multicomponent HEM. However, in these reports, they could not provide the detailed crystal structure and a general synthesis protocol using various compositional aspects. In 2016, a significant advancement was achieved by Mirkin and coworkers, where an array of different elements was combined using confined nano-reactors to get nanoparticles [7]. After this discovery, HEMs are explored in many directions and have been demonstrated for applications in various fields like battery, hydrogen storage, thermoelectric, etc., [8]. The development in HEMs led to many inventions, and these materials are later on termed as high-entropy (HE), which is being used for HE oxides (HEOs) and sulfides (HESs). Among these HEOs is a special class of materials with especially widespread applications in catalysis and energy storage devices. These materials exist in different crystal structures depending on the multi-elemental compositions [8]. These materials have plenty of special properties to be useful for the above-mentioned applications with excellent efficacy. These materials have many compositional permutations/combinations and still have a huge scope to explore for a variety of applications. Particularly, HEOs are created when there is a solid solution of ≥5 elements that are present within one sublattice. The configurational entropy (S_{config}) of HEOs surges with an increase in the number of constituent elements. When the S_{config} is more than $1.5R$, stable single-phase crystals are possible, which usually requires a high-temperature synthesis condition. For a system with cations, the S_{config} is maximum when all

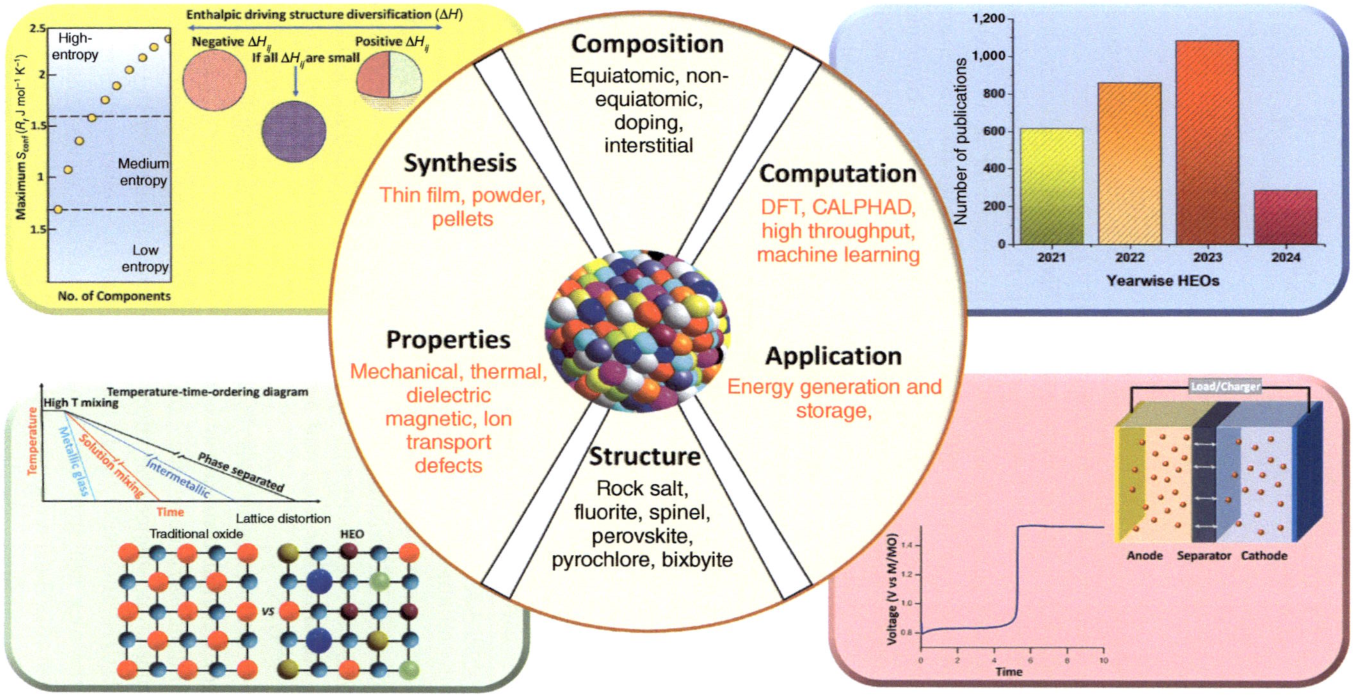

Figure 7.1 Pictorial illustration for fabrication of HEOs; their theoretical study assisted material fabrication, properties study, and applications for energy generation and storage. *Source:* Reference [8] / American Chemical Society / CC BY 4.0.

are taken in equimolar quantity and reaches a value of 1.61R. The S_{config} can be obtained by the following formulas (7.1) and (7.2):

$$S_{conf} = -R\left\{\left(\sum_{a=1}^{M} x_a \ln x_a\right)_{cation-site} + \left(\sum_{b=1}^{N} y_b \ln y_b\right)_{anion-site}\right\} \tag{7.1}$$

$$S_{HEO-Conf} = -R + \left(\sum_{a=1}^{M}(x_a \ln x_a)\right)_{cation-site} \tag{7.2}$$

where R = ideal gas constant, and x_a and y_b are mole fractions of constituent elements in the cationic and anionic sites, respectively [9].

These oxides are also known as entropy-stabilized oxides. Unlike traditional doped oxides, these materials have four important effects, including sluggish diffusion, large configurational entropy, cocktail effect, and lattice distortion [10–13]. Lattice distortion with high severity resulted from the size mismatch in the presence of elements exhibiting different sizes. This effect consequently pushes the energy barrier to a higher value for long-range diffusion. The decreased diffusion effect of these unique materials makes it kinetically favorable. As a result, HEOs and NM-HEOs are defect-tolerant. For example, after oxide vacancies (or any defects), these materials retain their crystal structures, which helps in high cycling stability during electrochemical performance in batteries/electrochemical cells [10, 14]. In addition, these materials have high ion transport properties and electronic conductivity. Relative to traditional oxides, these materials have several cations in a multivalent state, which enhances the electronic conductivity [15]. A schematic illustration has been given in Figure 7.1, which demonstrates a structural, synthetic methodology, properties, applications, and theoretical aspects for HEOs materials. Various HEOs and NM-HEOs have been explored for energy storage devices, particularly for battery and supercapacitor applications, owing to above-mentioned special properties of these materials. Soon after the discovery of lithium/sodium-substituted superionic conductivity was explored, e.g. (MgCoNiCuZn)O, a growing research focus/innovation has been put forward for the construction of HEOs/NM-HEOs and composites for battery applications. These materials have an increasing demand for various rechargeable batteries (lithium-ion, lithium-sulfur, etc.) and are utilized as cathode, anode, as well as solid-state electrolytes.

Various theoretical studies have also been carried out for HEO-based materials for probing the new material formation and characterizing different properties of these materials for initial assessment, where these materials can be used with improved properties. Ting et al. studied the effect of Mn concentration on the structural aspect of the $(CrFeMnCoNi)O_x$ and $(MgZnMnCoNi)O_x$. Lower concentration of Mn favored the rock salt structure, whereas at increased concentration, spinel phase stabilized [8]. Thus, a combined theoretical and experimental analysis can effectively be used for these materials for thorough investigations.

7.3 Synthesis Methods for High-Entropy Oxides and Noble-Metal-Based High-Entropy Oxides

HEOs and NM-HEOs are fascinating materials for advanced energy storage devices and have a growing demand for current as well as future research goals. The incorporation of noble metal atoms like Pt, Pd, Ru, and sometimes copper into the HEO matrices enhances the properties to a greater extent than the base materials. Taking into consideration the increasing attention for these particular materials, many challenges and opportunities are encountered. Achieving a homogeneously distributed multi-metal centers with intense care about the structural stability and phase purity and optimizing their properties for various energy generation/storage applications is a tedious task for the scientific community. To fulfil all these demands, various synthesis procedures have been adopted, which are pictorially presented in Figure 7.2.

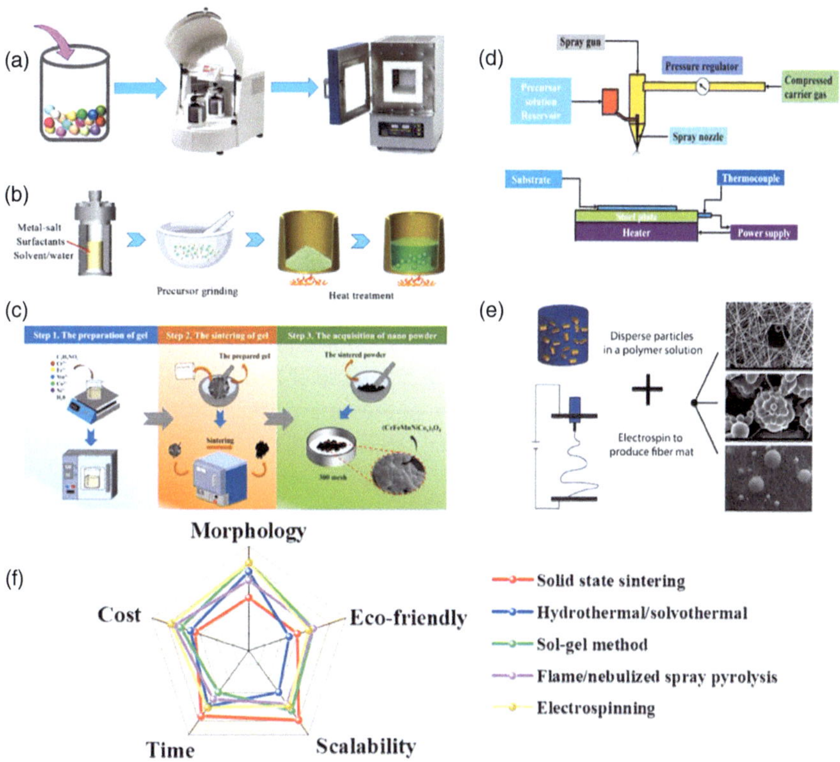

Figure 7.2 Different synthesis protocols for preparation of HEOs/NM-HEOs and their comparison: (a) solid-state sintering, (b) solvothermal, (c) sol-gel, (d) flame spray and nebulized spray pyrolysis, (e) electrospinning method, and (f) comparison of different synthetic methods. *Source:* Reproduced from Ref. [16] / John Wiley & Sons / CC BY 4.0.

The inclusion of noble metals into the HEO matrices further needed more percussion to avoid segregation and volatility of the precursors to maintain a single-phase crystal structure. Many optimizations, such as temperature variation, heating rate, and post-synthetic treatment, are employed to address these challenges. Overall, the cautious assortment and optimization of the preparation method are pivotal for achieving the structural, functional, and chemical attributes of both high-entropy oxides and noble-metal-based high-entropy oxides, and thus making these materials highly efficient for various applications, including energy storage devices. A few important synthesis procedures, which are mostly adopted for the synthesis of this particular class of materials, are stated below in detail.

7.3.1 Solid-State Sintering

This method follows conventional ceramic routes for the synthesis of HEOs and NM-HEOs (Figure 7.2a). This is quite a simple experimental procedure and can be used for industrial production [17, 18]. Stoichiometric amounts of various metallic oxide precursors are thoroughly mixed by ball milling, and the resultant powder is calcined at high temperature for long hours to get a single-phase crystal structure of HEOs/NM-HEOs. Various parameters like reaction temperature, heating and cooling rate, mixing time, etc., play a significant role in getting single-phase HEOs/NM-HEOs. Otherwise, it may lead to the formation of undesired agglomeration of a particular phase, formation of mono-metal oxides of a particular element and nanocrystalline overgrowth [19]. It is a relatively simple procedure for materials synthesis, but it has the shortcoming of high temperature annealing (often 900–1,200 °C) and long reaction time (often 12–24 hours). For e.g. $(FeCoNiCrMn)_3O_4$ was synthesized by taking an equimolar quantity of metal oxide precursors and processed for planetary ball milling and then calcined at 900 °C for 12 hours of time duration. For preparing noble-metal-based systems, this protocol is not suitable because of the segregation of these particular elements and also may lead to the spate phase formation.

7.3.2 Solvothermal Method

Solvothermal is employed through the use of any solvent where precursors are dissolved and then reacted at different temperatures and pressures in a closed vessel (Figure 7.2b). When water is used as a solvent, then it is called a hydrothermal method. It is an efficient, simple, and precise method for the preparation of HEOs/NM-HEOs. Morphology-controlled reactions can also be performed using this method easily by just changing temperature, pressure, and reaction time [20, 21]. In this method, stoichiometric quantities of metal salt precursors are taken and dissolved in solvents to prepare a homogeneous solution. Then, this mixture is heated at high temperature and pressure. The as-obtained materials can be obtained by vacuum drying. Unlike previously mentioned solid-state method, this method led to the formation of uniform particle size materials and lower sintering temperatures. For example, $(CoCuFeMnNi)_3O_4$ material is prepared by the solvothermal method (Temperature-400 °C) for two hours [22]. Although these methods have many advantages, the use of environmentally hazardous solvents and need for further treatment to get rid of impurities are quite tedious for large-scale preparation. Also, scale-up production may need many optimization steps as the precursor salts may agglomerate and quite difficult to get a homogeneous mixture on a large scale.

7.3.3 Sol-Gel Synthesis

This is the most frequently used method for the synthesis of HEOs/NM-HEOs (Figure 7.2c). It involves multiple processing steps such as (i) formation of a sol of multiple components, (ii) colloidal suspension of NPs in liquid media, (iii) formation of a gel structure, followed by (iv) heat treatment to get the resultant HEOs/NM-HEOs which are then ground well for further characterization/application. This method is useful in getting a uniform particle size and requires a lower processing temperature compared to the conventional solid-state method. Various structures of HEOs/NM-HEOs have been obtained using this particular method. An HEO $(Cr_{0.2}Mn_{0.2}Fe_{0.2}Co_{0.2}Zn_{0.2})_3O_4$ is prepared using the sol-gel method, which has a spinel oxide structure [23]. A perovskite-type oxide, $(La_{0.6}Sr_{0.4})(CoFeMnNiMg)O_3$, is prepared using this reaction protocol [24]. In general, this method has prodigious potential for the synthesis of this type of material with controlled morphology and particle size. This method can also be useful for the scale-up synthesis and has a great future perspective in optimizing the synthesis protocol to get HEOs/NM-HEOs with enhanced performance.

7.3.4 Spray Pyrolysis

Nebulized spray pyrolysis (NSP), along with flame spray pyrolysis (FSP), are commonly used techniques in industries for the preparation of HEOs/NM-HEOs by atomizing the precursor solutions (Figure 7.2d). This is a single-step process, where the reactant solution is atomized that leads to fine droplets by using a carrier gas which are collected and subsequently heated on a steel plate. Such a continuous process can generate nearly hundreds of milligrams of nanoparticles of HEOs in a minute. The size of the nanoparticle can be manipulated by optimizing the temperature, flow rate of carrier gas, and precursor concentration. A (CoCuMgNiZn)O material is prepared using these spray pyrolysis techniques [25]. Noting the fast-processing time and uniform particle size production, this method has high industrial significance.

7.3.5 Electrospinning Method

This particular method is highly versatile and can have potential in producing HEOs/NM-HEOs nanofibers (Figure 7.2e). The electrospinning method is useful in producing nanofibers with a specific diameter, surface area, and alignment [26]. A sub-1-nanometre nanowire of $CoZnCuNiFeZrCeO_x$ is prepared using this method.

This is an efficient method in the preparation of a controlled structure, which is very useful for tuning the electrochemical properties. However, this method has experimental limitations in large-scale production due to process parameter complexity and also in complex composition control.

Above all, the selection of different synthesis methods can be employed depending on the specific need and purpose. Efficiency/characteristics of the different synthesis methods are presented in Figure 7.2f. For example, when large-scale production is needed, the solid-state method can be employed. In addition, if we are considering copper as one of the noble metal atoms, many strategies have been employed for the preparation of copper-based NM-HEOs and have been used for a variety of energy applications. It is practically very tough for the synthesis of other noble-metal-based HEOs (Pt, Au, Pd, etc.) due to the segregation behavior of these noble ions used. However, alternative routes like (i) preparation of HEOs and (ii) loading different noble metals can be done for fabricating stable and highly efficient NM-HEOs materials. An ultra-small high-entropy alloy cluster of Ru, Ir, Cu, Ag, Au, Pd, and Pt on $(AlNiCoFeCrMoTi)_3O_4$ support is fabricated [27]. This material is then used for bifunctional OER/ORR, which has future scope for Zn-air batteries. Hence, plenty of scope is available in this direction in preparing these noble metal-based HEOs and testing for various energy generation/storage applications.

7.4 Noble-Metal-Based High-Entropy Oxides for Energy Applications

HEOs and NM-HEOs have been used as a promising material for energy generation and storage. The extensive applications of these materials in above-mentioned field pertain to their tunable electronic properties due to unique combinations of various metal centers. Further, these materials with single/multiple noble metal centers like Pt, Pd, Ru, etc., impose exceptional thermal and chemical stability due to entropy-stabilized structures. This diversity in the composition with synergistic effects gives an excellent redox behavior and oxygen mobility for these materials, which can be used extensively in the field of batteries, showing improved (oxygen reduction reaction [ORR] and oxygen evolution reaction [OER] kinetics [28–30]. Furthermore, these materials with improved electrochemical properties have been extensively used in metal-ion batteries and supercapacitor applications. In the following two subsections, the application of these materials, particularly for battery and supercapacitor applications, is elaborately discussed.

7.4.1 Noble-Metal-Based High-Entropy Oxides for Battery Application

Previous sections discussed the important synthesis protocol and structural and electrochemical properties of HEOs and NM-HEOs. These materials have notable emerging applications in rechargeable batteries with the excellent properties of vast elemental space, multi-electron transfer mechanism, retention of unique crystal structure, and adjustable electrochemical properties [16]. These materials are used as cathode, anode, as well as act as a solid-state electrolyte in various metal-ion and metal-air batteries.

7.4.1.1 Anode Material

The energy storage mechanism in HEOs can be understood by taking into consideration the structural aspects and electronic as well as electrochemical properties. HEOs' constructed anode materials are sprightly, depending on the number of electrons transported by their molecular formula unit. Diverse elemental composition in it creates a synergistic effect, which impacts the mechanism of energy storage in lithium-ion batteries (LIBs). Close interaction of lithium ion and the metal cations in the HEOs, the electronic modification during the lithiation/delithiation process, leads to unique charge storage properties as compared to single-component oxides. For example, the charge storage mechanism in $Sn_{0.8}(Co_{0.2}Mg_{0.2}Mn_{0.2}Ni_{0.2}Zn_{0.2})_{2.2}O_4$ is illustrated in Figure 7.3 [16].

In these aspects, it is very crucial to choose different elements and also figure out the important role of the particular constituent species for the HEOs anode. Various elements, including Mn, Ni, Cu, etc., are redox-active components for lithium storage [31, 32]. Zinc can help in LiZn nanophase alloy formation to enhance the storage capacity where some inactive components like Al and Mg can be helpful in retaining the small size of active components, increasing configurational entropy, preserving the structural integrity, and blocking the agglomeration of secondary phase [33, 34]. Sarkar et al systematically investigated the structural properties and

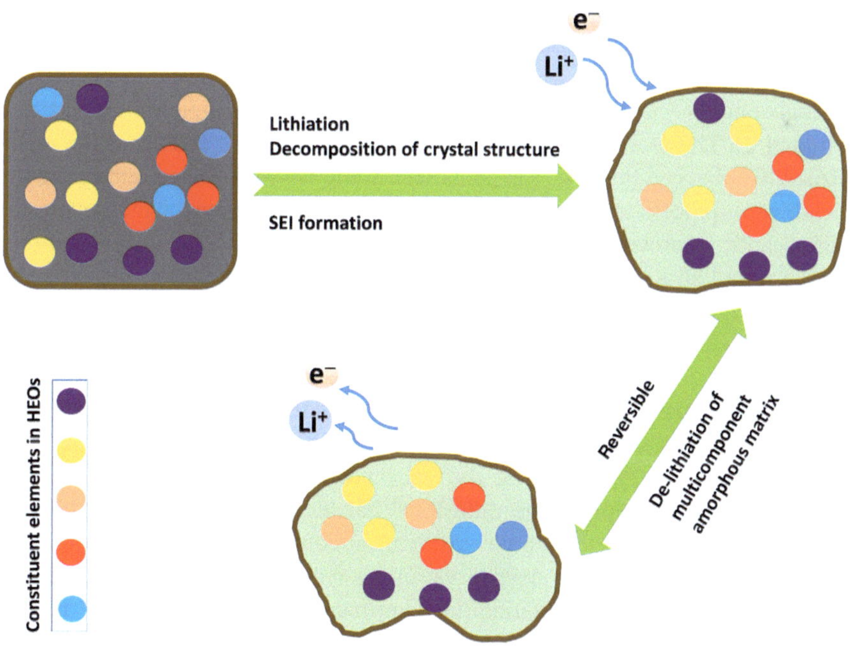

Figure 7.3 A schematic illustration of the lithium storage mechanism in $Sn_{0.8}(Co_{0.2}Mg_{0.2}Mn_{0.2}Ni_{0.2}Zn_{0.2})_{2.2}O_4$ HEO.

performance of $(Mg_{0.2}Co_{0.2}Ni_{0.2}Cu_{0.2}Zn_{0.2})O$ transition metal (TM) HEOs, which are used as anode material in LIBs [35]. A study revealed that removing single elements from the lattice structure significantly decreases the configurational entropy to 1.39*R*, indicating the absence of single-phase existence. Furthermore, the electrochemical performance is found to be highly dependent on the composition of the TM-HEO. Specifically, the absence of the Co drastically reduced the cyclic stability to 10 cycles, highlighting its critical role in the electrochemical stability of the HEO. Similarly, the absence of Zn in the lattice resulted in the formation of separate phases of LiO, CoO, and CuO upon Li extraction. In contrast, the presence of magnesium contributed substantially to the stability of the rock salt structure, enhancing the cycling performance of the electrode. The formation of the inactive MgO preserves the original lattice parameters of the active metal oxides during the discharge and charging process by preventing agglomeration. Furthermore, Chen and coworkers investigated the impact of surface area and particle size of TM-HEOs on electrochemical Li-storage application by varying the ball milling time during the synthesis procedure [36]. Decreasing the average particle size with increased surface area of the TM-HEOs rock salt structure, its structural stability was significantly enhanced, leading to improved Li-ion diffusion and storage performance. Chen and coworkers synthesized $[Ni_{0.2}Co_{0.2}Mn_{0.2}Fe_{0.2}Ti_{0.2}]_3O_4$ TM-HEO in an equimolar mixture ratio that consists of heterovalent cations of Fe^{3+} and Ti^{4+}, specifically designed to eliminate the structural stabilizing effect of inactive MgO. Interestingly, during the charging process, the Ti ions did not undergo any conversion reactions, maintaining their oxidation states between Ti^{3+} and Ti^{4+}. This led to the formation of a stable $LiTi_2O_4$ spinel phase, which significantly enhances the structural stability of the HEO [37]. Nowadays, non-equimolar ratios of Co, Ni, Fe, Cr, and Mn are used to develop TM-HEOs. Nguyen and coworkers synthesized a single-phase $(FeCoNiCrMn)_3O_4$ spinel structure with multiple valence states of Mn and Cr. The presence of variable oxidation states of TM cations in the crystal structure enhanced the electrical conductivity and was distributed during the first delithiation process at a low potential. Moreover, the entropy of the system was significantly improved due to the uniform distribution of the multivalent cations in the spinel structure [38], resulting in a stable anode platform. In contrast, a structural transformation from crystalline to amorphous nature was observed by Wang et al. during the charging process of $(FeCoNiCrMn)_3O_4$ HEO anode [22, 39]. The formation of an amorphous structure during the charging process effectively suppressed elemental segregation and enhanced the stability of the electrode. The oxygen vacancies in the HEOs enhance the electrochemical kinetics of the electrode material by increasing the imbalanced charge separation, increasing the Li storage by reducing the diffusional barrier, and providing an additional catalytically active site for the electrochemical conversion reactions [40]. Lokcu et al demonstrate that the oxygen vacancies enhance Li incorporation in the $(MgCoNiZn)_{1-x}Li_xO$ by facilitating the conversion of Co ions from Co^{2+} to Co^{3+} as the charge compensation mechanism [41]. Similarly, Mozdzierz and coworkers synthesized $(MgCoCuNiZn)_{1-x}Li_xO$ with various contents of Li loading (ranging from 0, 0.05, 0.10, 0.15, 0.20, 0.25, and 0.30) [42]. Increasing the Li loading leads to the shrinkage of the unit cell, accompanied by charge compensation of TM cations from an oxidation state of (II) to (III) state, resulting in the shrinkage of ionic radii and the entire volume of the crystal, consequently lead to the creation of increased oxygen vacancies in the unit cells. This local disturbance in the crystal lattice was observed in the HEO materials, resulting in improved performance of the rechargeable batteries.

7.4.1.2 Cathode Materials

HEO-derived advanced cathodes are highly prospective candidates for future generation rechargeable batteries with improved structural stability and higher energy density compared to conventional cathodes. A specific combination of more than one transition element in the HEOs helps them to serve as layered hydroxide materials where specific ions like Fe^{3+}, Ni^{2+} play a major role in the electrochemical cycling process and greatly impact the storage properties. In addition, Al^{3+}, Mg^{2+}, Ti^{4+}, etc., are electrochemically inert but impose structural stability [16]. Incorporation of particular elements like Cu^{2+}, Sb^{5+} helps in increasing the energy output of the battery. Metal ions like Li^+, Zn^{2+} have a tendency to facilitate the anion (oxygen) redox reaction within the material structure. This anionic reaction offers an additional capacity boost along with the conventional cationic redox reaction, which results in increased overall storage capacity of the HEO-based cathode materials. For instance, $Na_{2/3}Li_{1/6}Fe_{1/6}Co_{1/6}Ni_{1/6}Mn_{1/3}O_2$ cathode material with a high configurational entropy of 1.56*R* was developed by Yao and coworkers with improved performance and cyclic life [43]. Similarly, a sodium-deficient $Na_{0.83}Li_{0.1}Ni_{0.25}Co_{0.2}Mn_{0.15}Ti_{0.15}Sn_{0.15}O_{2-\delta}$ cathode with an entropy factor of 1.75*R* was synthesized by Wang and coworker for enhanced sodium ion transport [44]. Ding et al. synthesized $NaNi_{0.25}Mg_{0.05}Cu_{0.1}Fe_{0.2}Mn_{0.2}Ti_{0.1}Sn_{0.1}O_2$ with an entropy of 1.83 R, effectively hindering hexagonal-to-monoclinic transitions and forming a stable O3' phase [45]. Additionally, Joshi and coworkers developed a cobalt-free HEO cathode, $Na_{0.9}Li_{0.1}Ni_{0.4}Fe_{0.2}Mn_{0.44}Ti_{0.04}Mg_{0.02}O_{1.9}F_{0.1}$, maintaining its structure by inhibiting the O3-O3′ phase transformation during cycling [46]. In summary, these approaches demonstrate how the configuration entropy and composition of TMO slabs can be manipulated to enhance the electrochemical performance and structural stability of the HEO cathode.

7.4.1.3 Electrolyte Material

Electrolyte materials are the major and important component of the battery systems, which allow the movement of ions between the cathode and anode. So, the electrolyte materials used should have specific properties like high ionic conductivity, a wide potential window, as well as good chemical and thermal stability. Taking into consideration these factors, high-entropy electrolytes (HEEs) exhibit better activity than in comparison to traditional electrolytes. In particular, (i) the increased configurational entropy of HEEs improves thermal stability through a reduction in the system's Gibbs free energy; (ii) HEEs allow controlling the solvation structure to optimize ion transport dynamics; and (iii) it decreases the activation energy barrier

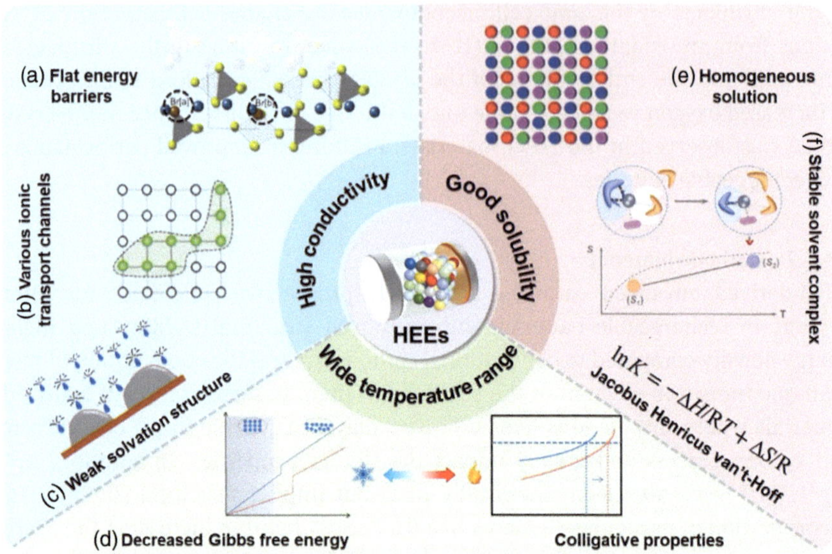

Figure 7.4 The three main roles of HEEs and the underlying mechanisms of high-entropy in boosting the performance of electrolytes. (a) Schematic of flat energy batteries, (b) figure of various transport channels, (c) schematic of weak solvation structure, (d) schematic representation of decreased Gibbs free energy, (e) figure of homogeneous solution, and (f) schematic representation of stable solvent complex. *Source:* Reproduced from Ref. [47] / with permission of American Chemical Society.

for ion migration by establishing an extended network of intercalation sites, thus significantly improving ionic conductivity [47]. As such, HEEs are very promising for creating electrolytes with reduced volatility, increased ionic conductivity, better cyclic stability, and stable performance at high temperatures. These properties make HEO-based electrolytes highly attractive candidates for next-generation battery systems (Figure 7.4).

The introduction of Li or Na into HEO structures causes a high density of oxygen vacancies by a charge compensation mechanism, significantly improving ionic conductivity and thereby opening the way for the creation of all-solid-state electrolytes. Pioneering this strategy, Berardan and coworkers prepared $(Mg,Co,Ni,Cu,Zn)_{1-x}Li_xO$ by inserting monovalent Li ions into the HEO structure, achieving 10^{-3} S cm^{-1} ionic conductivity at room temperature [48]. Currently, $Li_7La_3Zr_2O_{12}$ (LLZO) ionic HEO is diversely used as a solid electrolyte [49]. However, the conductance of such electrolyte was enhanced by substituting higher valence noble ion cations like Ta^{5+}, Sb^{5+}, W^{6+}, and Nb^{5+} [50–53]. These substitutions generate additional lithium vacancies and elevate the Li-ion concentration, significantly enhancing ionic conductivity and structural stability.

Although the HEOs give excellent efficacy for energy storage applications, the inclusion of noble metals like Pd, Au, Ru, and Pt can further elevate their properties as well as stability. But incorporation of these elements in HEOs is highly challenging and still a prime research focus. Sometimes, copper and osmium are also included in the noble metal list [54]. Recently, many copper-based HEOs have been explored for LIBs. Performance analysis of different HEOs used as anodes in LIBs is given in Table 7.1.

7.4.2 Noble-Metal-Based High-Entropy Oxides for Supercapacitor Application

Batteries and supercapacitors are promising energy storage devices used in diverse fields [62]. Both are electrochemical storage devices, but the storage mechanism is quite different, and hence their performance. Both of these devices have their merits for particular applications, but supercapacitors offer various benefits over batteries. They particularly have high power density, charging-discharging capabilities with good cyclic stability, useful for good long-term application [63]. Several parameters and components of supercapacitors impact the performance of the energy storage devices, e.g. electrodes used, type of electrolytes, and voltage window used [64]. Hence, significant research efforts have been put forward to develop novel and efficient electrode materials, which are the major components of a supercapacitor. This resulted in the design of a variety of supercapacitor electrodes taking into consideration their structural aspects, electron conductivity, and ion transport properties [65].

Based on the electrochemical energy storage mechanism, or the charge–discharge mechanism, these supercapacitors can be classified into three distinct types: electric double-layer capacitors (EDLCs), pseudocapacitors, and hybrid capacitors. In the first category of supercapacitor, the storage mechanism depends on the accumulation of ions onto the electrode materials (non-Faradaic), which are generally electrically conducting. Hence, the surface area of the electrodes majorly impacts the storage capacity of EDLC-based supercapacitor [66, 67]. Unlike the above-mentioned category, pseudocapacitors are based on the electrostatic charge separation at the electrolyte–electrode interfaces (Faradaic), which involves a reversible redox process [68]. This redox reaction, which involves the transfer of electrons between the active electrode–electrolyte interfaces, significantly improves the capacitance and energy density of pseudocapacitors compared to EDLCs [68]. Electrode materials used in pseudocapacitors are diverse and include a range of substances such as TM-based materials, conducting polymers, and redox-active heteroatom-doped carbon materials, each contributing unique properties to enhance performance [68]. The final category is hybrid capacitors, which follow both EDLC and pseudocapacitive mechanisms in their energy storage processes, as this type consists of both materials [69]. Recently, among the diverse array of materials developed, HEOs have gained a lot of interest in supercapacitor applications owing to their distinctive structure and remarkable properties [70]. So far, V_2O_5, MnO_2, Fe_2O_3, Co_3O_4, NiO, and RuO_2 have been the most commonly utilized metal-oxide-based pseudocapacitive electrode materials. However, they also suffer from several limitations, including low conductivity, high resistance, and

Table 7.1 Performance analysis of HEOs for anode in lithium-ion batteries.

Crystal structure	HEOs	Capacity (initial) @0.1 A g^{-1}	Cycle retention (number of cycles)	Current density performance	Fabrication procedure	References
Spinel	$(Mg_{0.2}Ti_{0.2}Zn_{0.2}Cu_{0.2}Fe_{0.2})_3O_4$	1,261 mA h g^{-1}	96.2%@2 A g^{-1} (800)	272 mA h g^{-1} @2 A g^{-1}	Solid-state method	[18]
Spinel	$(CrFeMnCoMg)_3O_4$	395 mA h g^{-1}	95%@2 A g^{-1} (800)	183 mA h g^{-1} @2 A g^{-1}	Solution combustion	[55]
Spinel	$(FeCuNiCrMn)_3O_4$	—	99%@0.5 A g^{-1} (250)	451 mA h g^{-1} @2 A g^{-1}	Hydrothermal	[56]
Rock salt	$(Mg_{0.8}CoNiCuZn)O$	850 mA h g^{-1}	~100%@0.1 A g^{-1} (100)	206 mA h g^{-1} @3 A g^{-1}	Solid-state sintering	[57]
Spinel	$(FeCoNiCrMn)_3O_4$	1,645 mA h g^{-1}	86%@2 A g^{-1} (1,200)	482 mA h g^{-1} @3 A g^{-1}	Solid-state method	[58]
Spinel	$(FeCoNiZnMnLi)_3O_4$	665 mA h g^{-1}	84%@0.1 A g^{-1} (100)	225 mA h g^{-1} @2 A g^{-1}	Solid state sintering	[59]
Spinel	$(Al_{0.2}CoCrFeMnNi)_{0.58}O_{4-\delta}$	893 mA h g^{-1}	—	634 mA h g^{-1} @3 A g^{-1}	Solution combustion	[60]
Perovskite	$[(Bi,Na)_{1/5}(La,Li)_{1/5}(Ce,K)_{1/5}Ca_{1/5}Sr_{1/5}]TiO_3$	125.9 mA h g^{-1}	~100%@ 1 A g^{-1} (300)	70 mA h g^{-1}@ 3 A g^{-1}	Solid-state sintering	[61]

Table 7.2 Recent developments in HEOs for supercapacitance and their performance.

Material	Synthesis method	Specific capacitance	Current density	Electrolyte	Capacitance retention	No. of cycles	References
$(FeCoCrMnNi)_3O_4$	Sol–gel	$204\ F\ g^{-1}$	$0.5\ A\ g^{-1}$	1 M KOH	60.4%	10,000 cycles	[70]
$(FeCoCrMnNi)_3O_4$	Sol–gel	$332.2\ F\ g^{-1}$	$0.3\ A\ g^{-1}$	1 M KOH	67%	5,000 cycles	[71]
$(FeCoCrMnZn)_3O_4$	Facile solid-state reaction	$340.3\ F\ g^{-1}$	$1\ A\ g^{-1}$	1 M KOH	69%	1,000 cycles at $5\ A\ g^{-1}$	[72]
$(FeCoCrMnMg)_3O_4$	Facile solid-state reaction	$193.7\ F\ g^{-1}$	$1\ A\ g^{-1}$	1 M KOH	—	—	[73]
$(CrMnFeCoNi)_3O_4$	Reverse co-precipitation	$239\ F\ g^{-1}$	$0.5\ A\ g^{-1}$	2 M KOH	76%	1,000 cycles at $1\ A\ g^{-1}$	[74]
$(Co_{0.2}Cr_{0.2}Fe_{0.2}Mn_{0.2}Ni_{0.2})_3O_4$	Facile polyacrylamide gel	$384\ F\ g^{-1}$	$1\ A\ g^{-1}$	6 M KOH	60%	2,000 cycles at $5\ A\ g^{-1}$	[75]
$(FeCoNiCrMn)_3O_4$	Facile oxidation treatment	$639\ F\ g^{-1}$	$1\ A\ g^{-1}$	2 M KOH	80.77%	1,000 cycles at $10\ A\ g^{-1}$	[76]
$La(CoCrFeMnNiAl_{0.5})_{1/(5.5)}O_{3-\delta}$	Coprecipitation - calcination	$353.65\ F\ g^{-1}$	$1\ A\ g^{-1}$	—	88.61%	2,000 cycles	[77]
$(Al_{0.2}Co_{0.2}Cr_{0.2}Mn_{0.2}Ni_{0.2})_3O_4$	Reverse Co-precipitation	$318\ mA\ h\ g^{-1}$	$1\ A\ g^{-1}$	KOH	—	—	[78]
$(Zr_{0.5}Ti_{0.5}Ce_{0.5}Hf_{0.5})O_7$	Sol–gel	$703.3\ F\ g^{-1}$	—	1 M Na_2SO_4 + 66.7 wt% Sucrose solution	97.9%	10,000 cycles at $15\ A\ g^{-1}$	[79]
$(Fe_3Cr_3Mn_2NiV)O_4$	Mechanical alloying	$294\ F\ g^{-1}$	—	6 M KOH	—	5,000 cycles	[80]

degradation of the electrodes during long cycles [71]. As a result, a novel material design approach known as HE has emerged. This concept involves incorporating multiple components, typically five or more, to enhance the configurational entropy and improve performance characteristics. Liang and coworkers synthesized a single-spinel-structure powder of $(FeCoCrMnZn)_3O_4$ using a simple solid-state reaction route at 900 °C. Furthermore, the supercapacitor performance of a $(FeCoCrMnZn)_3O_4$ was evaluated after coating on a Ni foam electrode. At a current density of 1 $A\,g^{-1}$, the composite electrode exhibited a specific capacitance of 340.3 $F\,g^{-1}$ in a 1 M KOH electrolyte [72, 73].

Bhusankar Talluri and coworkers synthesized a supercapacitor electrode material composed of spinel $(CrMnFeCoNi)_3O_4$ nanoparticles through the reverse co-precipitation method. The electrode achieved an impressive capacitance value of 239 $F\,g^{-1}$ with a specific energy of 24.1 $Wh\,kg^{-1}$ at 0.5 $A\,g^{-1}$. Notably, it retained 38% of its performance even as the current density surged from 0.5 to 25 $A\,g^{-1}$. Furthermore, after completing a thousand charge–discharge cycles, the electrode preserved 76% of its initial capacitance, while maintaining a commendable coulombic efficiency of 86%, underscoring its long-term stability and efficiency [74]. Researchers used the polyacrylamide gel method to prepare $(CoCrFeMnNi)_3O_4$ for supercapacitor applications. These electrode materials in 6 M KOH showed a specific capacitance of 384 $F\,g^{-1}$ at 1 $A\,g^{-1}$ current density. At 10 $A\,g^{-1}$ current density, the electrode retained 72% of its initial capacitance, ensuring excellent rate capability [75]. Thus, HEOs regarded as an excellent class of materials, garnering significant attention in the realm of energy storage applications, particularly supercapacitor applications. However, further advancements and optimization are still necessary to fully unlock their potential. Noble metal atoms can also be included in the HEOs, where they can perform with extraordinary performance with improved stability. Recent developments in HEOs for supercapacitance and their performance metrics (Table 7.2).

7.5 Current Challenges and Future Perspectives

There is growing attention on the development of HEOs and NM-HEOs-based materials for energy storage devices, e.g. batteries and supercapacitors, due to their unique electrochemical properties and structural stability. Despite its many advantages also, these materials also face a lot of challenges in the synthesis of pure phases and application with enhanced efficiency. The traditional wet chemistry route suffers from incomplete homogeneous mixing of constituent elements or impurity phases. Thus, it is very difficult to maintain the proper stoichiometry of the materials. The solid state method also suffers from high temperature as well as long-term annealing of the samples to get a phase-pure HEO material. For battery application, it suffers from slow ionic mobility, blocking of the migration pathway due to high cationic disorder, and low capacity. In addition, these materials in supercapacitors also fall short of meeting the growing demands of modern devices. Although there are existing challenges and a need for further research, addressing the drawbacks by combining pseudo-capacitive high-entropy materials with carbon materials offers notable benefits, making them a promising solution for advancing supercapacitor technology.

Although the materials have many hurdles, they hold great promise for their exceptional stability and structural/property tunability. By solving the issue related to noble-metal-based HEOs synthesis, these materials will have great future potential for metal-ion and metal-air batteries. Along with battery applications, forming a composite with these materials can have a better future for advanced supercapacitor applications, as well as other energy applications.

7.6 Conclusions

HEOs and NM-HEOs are excellent materials for next-generation energy storage devices with optimal activity and stability present scenario. HEOs materials possess a unique multi-elemental composition, resulting in tunable properties and structural stability. These materials have diverse applications in batteries, including cathode, anode, and solid-state electrolyte. These features make it special than the traditional battery materials. Incorporating the noble metal into the HEOs matrices to form NM-HEOs further induces improved catalytic activity, electronic conductivity, and surface-active sites, which results in an excellent battery material. These materials have been explored and have great future potential for lithium-ion, sodium-ion, metal-air, and multivalent-ion batteries. Simultaneously, a large surface area, rapid ion/electron transport, and defect engineering possibilities make it a good candidate for supercapacitor application. Among the diverse array of materials, HEOs and noble metal HEOs have garnered significant interest across various applications owing to their remarkable properties. But integrating these HEO materials with carbon components can significantly improve the potential for high-performance supercapacitors, contributing to the advancement of green energy storage technologies. Despite many challenges, working on designing an effective synthesis method, scalability, processing, playing with the compositional aspect, defect engineering, and studying the mechanistic aspects can lead to the development of multifunctional and high-performance energy storage devices.

Funding Statement

This book chapter was supported by a National Postdoctoral Fellowship program from the Science and Engineering Research Board (SERB), India (File No. PDF/2023/002768).

Author Contributions

DM and CS developed the study concept, designed the study, wrote the manuscript, and completed the revision. SH is involved in writing the supercapacitor portion.

Conflict of Interest

The authors declare no competing interests.

References

1 Khan, S.A.R., Zia-Ul-Haq, H.M., Ponce, P., and Janjua, L. (2023). Re-investigating the impact of non-renewable and renewable energy on environmental quality: A roadmap towards sustainable development. *Res Policy* 81: 103411.
2 Suait, M.S., Ludin, N.A., and Sopian, K. (2025). Introduction to renewable energy technologies and energy transition strategies. In: *Renewable Energy Technologies and Strategies in the Global Energy Transition*, ACS Symposium Series, vol. 1499 (ed. M.S. Suait, S. Jarimi, and N. SAM), 1–14.
3 Krishnamoorthy, K., Pazhamalai, P., and Kim, S.J. (2018). Two-dimensional siloxene nanosheets: novel high-performance supercapacitor electrode materials. *Energy Environ Sci* 11 (6): 1595–1602.
4 Mohanty, G.C., Chowde Gowda, C., Gakhad, P. et al. (2024). Enhanced energy density of high entropy alloy (Fe-Co-Ni-Cu-Mn) and green graphene hybrid supercapacitor. *Energy Storage* 6 (4): e668.
5 Yeh, J.W., Chen, S.K., Lin, S.J. et al. (2004). Nanostructured high-entropy alloys with multiple principal elements: novel alloy design concepts and outcomes. *Adv Eng Mater* 6 (5): 299–303.
6 Cantor, B., Chang, I.T.H., Knight, P., and Vincent, A.J.B. (2004). Microstructural development in equiatomic multicomponent alloys. *Mater Sci Eng A* 375–377: 213–218.
7 Chen, P.C., Liu, X., Hedrick, J.L. et al. (2016). Polyelemental nanoparticle libraries. *Science* 352 (6293): 1565–1569.
8 Sen, S., Palabathuni, M., Ryan, K.M., and Singh, S. (2024). High entropy oxides: mapping the landscape from fundamentals to future vistas. *ACS Energy Lett* 9 (8): 3694–3718.
9 Amiri, A. and Shahbazian-Yassar, R. (2021). Recent progress of high-entropy materials for energy storage and conversion. *J Mater Chem A* 9 (2): 782–823.
10 Fu, M., Ma, X., Zhao, K. et al. (2021). High-entropy materials for energy-related applications. *iScience* 24 (3): 102177.
11 Ma, Y., Ma, Y., Wang, Q. et al. (2021). High-entropy energy materials: challenges and new opportunities. *Energy Environ Sci* 14 (5): 2883–2905.
12 Sarkar, A., Wang, Q., Schiele, A. et al. (2019). High-entropy oxides: fundamental aspects and electrochemical properties. *Adv Mater* 31 (26): 1806236.
13 Sarkar, A., Breitung, B., and Hahn, H. (2020). High entropy oxides: the role of entropy, enthalpy and synergy. *Scr Mater* 187: 43–48.
14 Shin, D., Chae, S., Park, S. et al. (2023). Rational engineering of high-entropy oxides for Li-ion battery anodes with finely tuned combustion syntheses. *NPG Asia Mater* 15 (1): 1–14.

15 Wang, K., Hua, W., Huang, X. et al. (2023). Synergy of cations in high entropy oxide lithium ion battery anode. *Nat Commun* 14 (1): 1487.

16 Ran, B., Li, H., Cheng, R. et al. (2024). High-entropy oxides for rechargeable batteries. *Adv Sci* 11 (25): 2401034.

17 Xiao, B., Wu, G., Wang, T. et al. (2023). Enhanced Li-ion diffusion and cycling stability of ni-free high-entropy spinel oxide anodes with high-concentration oxygen vacancies. *ACS Appl Mater Interfaces* 15 (2): 2792–2803.

18 Chen, H., Qiu, N., Wu, B. et al. (2020). A new spinel high-entropy oxide $(Mg_{0.2}Ti_{0.2}Zn_{0.2}Cu_{0.2}Fe_{0.2})_3O_4$ with fast reaction kinetics and excellent stability as an anode material for lithium ion batteries. *RSC Adv* 10 (16): 9736–9744.

19 Dong, Q., Hong, M., Gao, J. et al. (2022). Rapid synthesis of high-entropy oxide microparticles. *Small* 18 (11): 2104761.

20 Mao, A., Xiang, H.Z., Zhang, Z.G. et al. (2019). Solution combustion synthesis and magnetic property of rock-salt $(Co_{0.2}Cu_{0.2}Mg_{0.2}Ni_{0.2}Zn_{0.2})O$ high-entropy oxide nanocrystalline powder. *J Magn Magn Mater* 484: 245–252.

21 Liu, X., Xing, Y., Xu, K. et al. (2022). Kinetically accelerated lithium storage in high-entropy (LiMgCoNiCuZn)O enabled by oxygen vacancies. *Small* 18 (18): 2200524.

22 Wang, D., Liu, Z., Du, S. et al. (2019). Low-temperature synthesis of small-sized high-entropy oxides for water oxidation. *J Mater Chem A* 7 (42): 24211–24216.

23 Petrovičovà, B., Xu, W., Musolino, M.G. et al. (2022). High-entropy spinel oxides produced via sol-gel and electrospinning and their evaluation as anodes in Li-ion batteries. *Appl Sci* 12 (12): 5965.

24 Tang, L., Yang, Y., Guo, H. et al. (2022). High configuration entropy activated lattice oxygen for O_2 formation on perovskite electrocatalyst. *Adv Funct Mater* 32 (28): 2112157.

25 Sarkar, A., Djenadic, R., Usharani, N.J. et al. (2017). Nanocrystalline multi-component entropy stabilised transition metal oxides. *J Eur Ceram Soc* 37 (2): 747–754.

26 Chen, S., Wang, Y., Pu, G. et al. (2023). High-entropy materials: controllable synthesis, deep characterization, electrochemical energy application, and outlook. *Energy Fuel* 37 (1): 36–57.

27 Jin, Z., Zhou, X., Hu, Y. et al. (2022). A fourteen-component high-entropy alloy@oxide bifunctional electrocatalyst with a record-low ΔE of 0.61 V for highly reversible Zn–air batteries. *Chem Sci* 13 (41): 12056–12064.

28 Khan, J.B., Panda, P.K., Yang, P.C. et al. (2024). Microwave synthesis of high-entropy alloy catalysts on graphene oxide sheets for oxygen reduction and evolution reactions. *Int J Hydrog Energy* 53: 999–1008.

29 Panda, P.K., Huang, H.Y., Dash, P. et al. (2025). Liquid-phase microwave synthesis of platinum-based high-entropy alloy catalysts on carbon supports for electrochemical hydrogen adsorption/desorption and oxygen evolution/reduction reactions. *Int J Hydrog Energy* 111: 536–545.

30 Khan, J.B., Panda, P.K., Dash, P., and Hsieh, C.T. (2025). Microwave-assisted synthesis of platinum-free high-entropy alloy catalysts on reduced graphene

oxide sheets for enhanced oxygen reduction and evolution reactions. *Chem Eur J* 31: e202403863.

31 Li, H., Duan, Y., Zhao, Z. et al. (2023). Atomic-scale storage mechanism in ultra-small size $(FeCuCrMnNi)_3O_4$/rGO with super-stable sodium storage and accelerated kinetics. *Chem Eng J* 469: 143951.

32 Qiu, N., Chen, H., Yang, Z. et al. (2019). A high entropy oxide ($Mg_{0.2}Co_{0.2}Ni_{0.2}Cu_{0.2}Zn_{0.2}O$) with superior lithium storage performance. *J Alloys Compd* 777: 767–774.

33 He, Q., Li, J., Liu, W. et al. (2024). High entropy oxides for electrochemical energy storage and conversion: a critical review. *J Power Sources* 619: 235207.

34 Li, X., Zhang, W., Lv, K. et al. (2024). Research progress on high-entropy oxides as advanced anode, cathode, and solid-electrolyte materials for lithium-ion batteries. *J Power Sources* 620: 235259.

35 Sarkar, A., Velasco, L., Wang, D. et al. (2018). High entropy oxides for reversible energy storage. *Nat Commun* 9 (1): 3400.

36 Chen, H., Qiu, N., Wu, B. et al. (2019). Tunable pseudocapacitive contribution by dimension control in nanocrystalline-constructed $(Mg_{0.2}Co_{0.2}Ni_{0.2}Cu_{0.2}Zn_{0.2})O$ solid solutions to achieve superior lithium-storage properties. *RSC Adv* 9 (50): 28908–28915.

37 Chen, T.Y., Wang, S.Y., Kuo, C.H. et al. (2020). In operando synchrotron X-ray studies of a novel spinel $(Ni_{0.2}Co_{0.2}Mn_{0.2}Fe_{0.2}Ti_{0.2})_3O_4$ high-entropy oxide for energy storage applications. *J Mater Chem A* 8 (41): 21756–21770.

38 Nguyen, T.X., Patra, J., Chang, J.K., and Ting, J.M. (2020). High entropy spinel oxide nanoparticles for superior lithiation–delithiation performance. *J Mater Chem A* 8 (36): 18963–18973.

39 Wang, D., Jiang, S., Duan, C. et al. (2020). Spinel-structured high entropy oxide $(FeCoNiCrMn)_3O_4$ as anode towards superior lithium storage performance. *J Alloys Compd* 844: 156158.

40 Tomboc, G.M., Zhang, X., Choi, S. et al. (2022). Stabilization, characterization, and electrochemical applications of high-entropy oxides: critical assessment of crystal phase–properties relationship. *Adv Funct Mater* 32 (43): 2205142.

41 Lokcu, E., Toparli, Ç., and Anik, M. (2020). Electrochemical performance of $(MgCoNiZn)_{1-x}Li_xO$ high-entropy oxides in lithium-ion batteries. *ACS Appl Mater Interfaces* 12 (21): 23860–23866.

42 Mozdzierz, M., Dąbrowa, J., Stępień, A. et al. (2021). Mixed ionic-electronic transport in the high-entropy $(Co,Cu,Mg,Ni,Zn)_{1-x}Li_xO$ oxides. *Acta Mater* 208: 116735.

43 Yao, L., Zou, P., Wang, C. et al. (2022). High-entropy and superstructure-stabilized layered oxide cathodes for sodium-ion batteries. *Adv Energy Mater* 12 (41): 2201989.

44 Wang, H., Gao, X., Zhang, S. et al. (2023). High-entropy Na-deficient layered oxides for sodium-ion batteries. *ACS Nano* 17 (13): 12530–12543.

45 Ding, F., Zhao, C., Xiao, D. et al. (2022). Using high-entropy configuration strategy to design na-ion layered oxide cathodes with superior electrochemical performance and thermal stability. *J Am Chem Soc* 144 (18): 8286–8295.

46 Joshi, A., Chakrabarty, S., Akella, S.H. et al. (2023). High-entropy co-free O3-type layered oxyfluoride: a promising air-stable cathode for sodium-ion batteries. *Adv Mater* 35 (51): 2304440.

47 Ren, K.F., Liu, H., Guo, J.X. et al. (2024). Working principles of high-entropy electrolytes in rechargeable batteries. *ACS Energy Lett* 9 (6): 2960–2980.

48 Berardan, D., Franger, S., Meena, A.K., and Dragoe, N. (2016). Room temperature lithium superionic conductivity in high entropy oxides. *J Mater Chem A* 4 (24): 9536–9541.

49 Kuo, C.H., Wang, A.Y., Liu, H.Y. et al. (2022). A novel garnet-type high-entropy oxide as air-stable solid electrolyte for Li-ion batteries. *APL Mater* 10 (12): 121104.

50 Zhu, Y., Connell, J.G., Tepavcevic, S. et al. (2019). Dopant-dependent stability of garnet solid electrolyte interfaces with lithium metal. *Adv Energy Mater* 9 (12): 1803440.

51 Inada, R., Takeda, A., Yamazaki, Y. et al. (2020). Effect of postannealing on the properties of a Ta-doped $Li_7La_3Zr_2O_{12}$ solid electrolyte degraded by Li dendrite penetration. *ACS Appl Energy Mater* 3 (12): 12517–12524.

52 Ramakumar, S., Satyanarayana, L., Manorama, S.V., and Murugan, R. (2013). Structure and Li^+ dynamics of Sb-doped $Li_7La_3Zr_2O_{12}$ fast lithium ion conductors. *Phys Chem Chem Phys* 15 (27): 11327–11338.

53 Liu, X., Gao, M., Liu, Y. et al. (2019). Improving the room temperature ionic conductivity of Al-$Li_7La_3Zr_2O_{12}$ ceramics by Ba and Y or Ba and W co-doping. *Ceram Int* 45 (10): 13488–13495.

54 Medici, S., Peana, M., Nurchi, V.M. et al. (2015). Noble metals in medicine: latest advances. *Coord Chem Rev* 284: 329–350.

55 Luo, C., Bi, J., Xie, L. et al. (2023). Electrochemical properties of four novel high-entropy spinel oxides used as lithium-ion battery anodes synthesized by the glycine-nitrate SCS method. *J Mater Sci* 58: 8005–8021.

56 Nguyen, T.X., Tsai, C.C., Patra, J. et al. (2022). Co-free high entropy spinel oxide anode with controlled morphology and crystallinity for outstanding charge/discharge performance in lithium-ion batteries. *Chem Eng J* 430: 132658.

57 Wang, S.Y., Chen, T.Y., Kuo, C.H. et al. (2021). Operando synchrotron transmission X-ray microscopy study on (Mg,Co,Ni,Cu,Zn)O high-entropy oxide anodes for lithium-ion batteries. *Mater Chem Phys* 274: 125105.

58 Xiao, B., Wu, G., Wang, T. et al. (2022). High-entropy oxides as advanced anode materials for long-life lithium-ion Batteries. *Nano Energy* 95: 106962.

59 Tian, K., Duan, C., and Ma, Q. (2022). High-entropy chemistry stabilizing spinel oxide $(CoNiZnXMnLi)_3O_4$ (X = Fe,Cr) for high-performance anode of Li-ion batteries. *Rare Metals* 41 (4): 1265–1275.

60 Xiang, H.-Z., Xie, H.-X., Chen, Y.-X. et al. (2021). Porous spinel-type $(Al_{0.2}CoCrFeMnNi)_{0.58}O_{4-\delta}$ highentropy oxide as a novel high-performance anode material for lithium-ion batteries. *J Mater Sci* 56: 8127–8142.

61 Yan, J., Wnag, D., Zhang, X. et al. (2020). A high-entropy perovskite titanate lithium-ion battery anode. *J Mater Sci* 55: 6942–6951.

62 Husain, Z., Shakeelur Raheman, A.R., Ansari, K.B. et al. (2022). Nano-sized mesoporous biochar derived from biomass pyrolysis as electrochemical energy storage supercapacitor. *Mater Sci Energy Technol* 5: 99–109.
63 Gopalakrishnan, A. and Badhulika, S. (2020). Effect of self-doped heteroatoms on the performance of biomass-derived carbon for supercapacitor applications. *J Power Sources* 480: 228830.
64 Zhang, S. and Pan, N. (2015). Supercapacitors performance evaluation. *Adv Energy Mater* 5 (6): 1401401.
65 Wang, F., Wu, X., Yuan, X. et al. (2017). Latest advances in supercapacitors: from new electrode materials to novel device designs. *Chem Soc Rev* 46 (22): 6816–6854.
66 Ni, M., Zhou, L., Liu, Y., and Ni, R. (2023). Advances in the synthesis and applications of porous carbon materials. *Front Chem* 11: 1205280.
67 You, X., Misra, M., Gregori, S., and Mohanty, A.K. (2018). Preparation of an electric double layer capacitor (EDLC) using miscanthus-derived biocarbon. *ACS Sustain Chem Eng* 6: 318–324.
68 Fleischmann, S., Mitchell, J.B., Wang, R. et al. (2020). Pseudocapacitance: From fundamental understanding to high power energy storage materials. *Chem Rev* 120: 6738–6782.
69 Chatterjee, D.P. and Nandi, A.K. (2021). A review on the recent advances in hybrid supercapacitors. *J Mater Chem A* 9: 15880–15918.
70 Yin, Y., Zhang, W.B., Chen, B. et al. (2024). Porous powder synthesis of $(FeCoCrMnNi)_3O_4$ high entropy ceramic and its supercapacitor applications. *Ceram Int* 50: 10292–10304.
71 Yin, Y., Zhang, W.B., Zhang, X.L. et al. (2023). Low-dimensional high entropy oxide $(FeCoCrMnNi)_3O_4$ for supercapacitor applications. *Dalton Trans* 52 (26): 9005–9016.
72 Liang, B., Ai, Y., Wang, Y. et al. (2020). Spinel-type $(FeCoCrMnZn)_3O_4$ high-entropy oxide: facile preparation and supercapacitor performance. *Materials* 13 (24): 5798.
73 Yiliang, W., Yunlong, A.I., Shuwei, Y. et al. (2021). Facile synthesis and supercapacitor performance of M_3O_4(M=FeCoCrMnMg) high entropy oxide powders. *J Inorg Mater* 36 (4): 425.
74 Talluri, B., Aparna, M.L., Sreenivasulu, N. et al. (2021). High entropy spinel metal oxide $(CoCrFeMnNi)_3O_4$ nanoparticles as a high-performance supercapacitor electrode material. *J Energy Storage* 42: 103004.
75 Li, L., Ji, P., Geng, C. et al. (2023). Facile synthesis of high-entropy $(Co_{0.2}Cr_{0.2}Fe_{0.2}Mn_{0.2}Ni_{0.2})_3O_4$ nanopowders and their electrochemical properties as supercapacitor electrode. *J Energy Storage* 73: 109182.
76 Zhang, D., Xu, S., Li, T. et al. (2023). High-entropy oxides prepared by dealloying method for supercapacitors. *ACS Appl Eng Mater* 1: 780–789.
77 Guo, M., Liu, Y., Zhang, F. et al. (2022). Inactive Al^{3+}-doped $La(CoCrFeMnNiAl_x)_{1/(5+x)}O_3$ high-entropy perovskite oxides as high performance supercapacitor electrodes. *J Adv Ceram* 11: 742–753.

78 Sreenivasulu, N., Kumar, U.N., Madhav, K.M.V.V. et al. (2022). Structural and electrochemical investigations on nanocrystalline high entropy spinel oxides for battery-like supercapacitor applications. *ChemistrySelect* 5: e202104015.

79 Tang, P., Cao, Y., Li, H. et al. (2022). The preparation of high-performance aqueous supercapacitor with high-entropy pyrochlore-type electrode and super-concentrated electrolyte. *Ceram Int* 28: 2660–2669.

80 Basgoz, O., Gungor, A., Guler, O., and Erdem, E. (2025). High-entropy alloys and oxides as supercapacitor electrodes: a structural and electrochemical perspective for energy storage. *Adv Sustain Syst* 2500201.

8

Noble-Metal-Free High-Entropy Oxides for Energy Storage Applications

Biraj K. Satpathy

Indian Institute of Technology, Palaj, Gandhinagar, Materials Engineering Department, Gandhinagar, Gujarat 382355, India

8.1 Introduction

Energy sources globally are typically categorized into two primary groups: the extensively used fossil fuels, which consist solely of coal, oil, and natural gas, and the increasingly adopted clean energy sources, which primarily consists of solar, wind, and hydropower. The energy consumption worldwide is predicted to have increased two-fold to 28 TW or 20 billion tons of oil produced yearly as observed over the last few years [1]. But the overuse of these conventional fossil fuels causes resource depletion as well as the release of greenhouse gases during burning, which exacerbates environmental problems like global warming and altering the climate. This helped foster the advancing sustainable energy technologies, along with the quest for renewable sources of energy as research topics in recent years. In this regard, to harness and store energy, numerous technologies, including batteries, dielectric capacitors (DC), and supercapacitors (SCs), to mention a few, have been employed that are increasingly gaining more and more recognition. Hence, developing functional materials is utmost crucial in order to achieve these advanced applications having enhanced capacity and longevity.

Lately, there has been a notable emphasis on a rising trend to an exciting category of materials named high-entropy materials (HEMs). These materials, due to their distinct structure, controlled basic configuration, and tunable qualities, have garnered significant interest in recent times.

8.1.1 Mechanism of HEMs

The concept of high-entropy materials was first introduced in two significant research studies in 2004. The idea of high-entropy alloys (HEAs) was initially proposed and described by Yeh et al. [2]. Not long after, Cantor et al. released a study that was comparable to Yeh et al.'s [3]. This basically comprised five-element single-phase solid solution Cr-Mn-Fe-Co-Ni HEAs. Since then, scholars have

High-Entropy Materials for Energy Storage Devices, First Edition.
Edited by Chien-Te Hsieh, Pradeep Kumar Panda, and Arpan Kumar Nayak.

begun investigating HEMs. This was on the observation that when extra equimolar primary components are added to the alloy system, the configurational entropy of mixing (ΔS_{mix} from Eq. (8.1)) rises [3].

$$\Delta S_{mix} = -R\sum_{i=1}^{n} c_i \ln c_i = -R\sum_{n} \frac{1}{n} \ln \frac{1}{n} = R \ln n \tag{8.1}$$

where R indicates the gas constant, n represents the number (elements), and c_i denotes the component concentration i.

According to this concept, a system comprising a minimal combination of equimolar elements (≥5) has a thermodynamic contribution of high ΔS_{mix} to the Gibbs free energy (ΔG_{mix}) at high temperature, making it adequate to offset the enthalpy generation (ΔH_{mix}) of intermetallic compounds while producing a single-phase solution consisting of several kinds of elements ($\Delta G_{mix} \leq 0$ as per Eq. (8.2)). Subsequently, the notion of HEA was broadened to include multi-elemental alloys with $\Delta S_{mix} > 1.5R$ (entropy-based by definition) and alloys made up of no fewer than five elements, ranging from 5% to 35% compositions (composition-based by definition) [4]. Similarly, the oxide counterparts of HEAs are known as high-entropy oxides (HEOs). Alike HEAs, HEOs based on their composition are defined as materials with oxygen and no less than five major cations in equimolar or nearly equimolar amounts, having the same atomic content of around 5% and 35% [5].

$$\Delta G_{mix} = \Delta H_{mix} - T\Delta S_{mix} \tag{8.2}$$

Multiple components with varying electronegativity and radii are generally randomly placed within the lattice to form the typical, i.e., HEA and HEO systems [6, 7]. This results in a short-range order (SRO) structure along with a lowering in configurational entropy (ΔS_{config}) which is affected by the positioning of elements within the system. But in HEOs, the oxygen atom filling the anionic sites prevents the formation of SRO structure, which raises the ΔS. Accordingly, the configurational entropy equation defines HEOs as follows:

$$\Delta S_{config} = -R\left[\left(x\sum_{a=1}^{M} x_a \ln x_a\right)_{A-site} + \left(y\sum_{b=1}^{N} y_b \ln y_b\right)_{B-site} + \left(z\sum_{o=1}^{P} z_o \ln z_o\right)_{oxygen}\right]$$

where x_a, y_b, and z_a denote the mole fractions (elements a, b, and c) that, in an $A_xB_yZ_o$ oxide system, occupy the A, B, and O^{2-} sites. R stands for the universal gas constant, and M, N, as well as P account for the number of elements that occupy the A, B, and O^{2-} sites [8].

In accordance with HEA principles, these exceptional qualities, which are commonly employed in energy–energy applications, are the consequence of four main factors, i.e., increased entropy, inefficient diffusion, lattice distortion, and synergistic effect as shown in Figure 8.1.

HEOs create appealing characteristics that surpass their basic elements, such as the ability to modify their functional aspects and their affinity toward single-phase solid solutions, along with basic crystal structures. At higher temperatures, both configurational entropy (Eq. (8.1)) and the rigidity of the single-phase system are

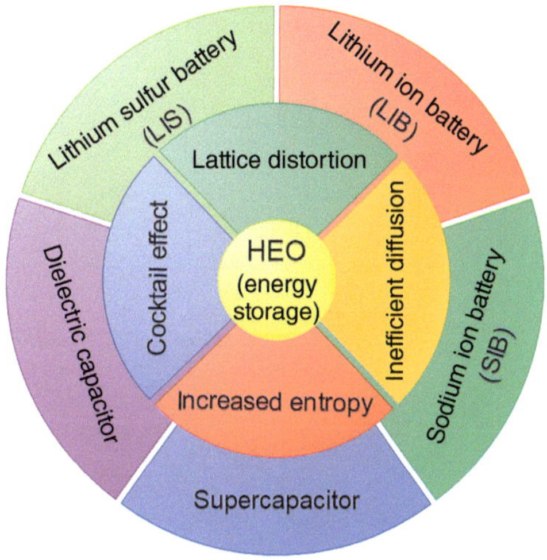

Figure 8.1 Schematic illustration of core properties and energy storage applications of HEOs.

increased due to several near-equimolar elements. Atoms of various sizes occupying lattice positions at random make up the crystal lattice of an HEO. Substantial lattice distortion results from these slightly off-balance lattice sites and atoms' haphazard locations. This leads to increased hardness, reduced thermal and electrical conductivity, and sluggish diffusion, which is another consequence that delays atomic mobility and replacements. Last, when a mixture contains several constituents with different characteristics, the cocktail effect of HEOs takes place. Such an alloy's final characteristic is unpredictable and not equal to the combined value of each of its constituent parts [9]. Because of the elements' synergistic behavior, HEOs with distinctive characteristics are developed, which are very advantageous for a variety of electrochemical applications.

There are now several different HEO crystal structures known to exist, such as (i) rock salt, (ii) fluorite (F-HEO), (iii) perovskite, (iv) spinel, (v) bixbyite, (vi) pyrochlore, and (vii) layered O_3-type structures. Because of its ease of creation, rock salt HEO (R-HEO) was discovered initially and has been the subject of the most research. In the case of F-HEO and R-HEO, there is just one Wyckoff location in there to contain cations and can produce a structure that is evenly distributed and has outstanding chemical stability, giving them a lot of promise for use in electrochemical processes [10]. Additionally, the many cation sites seen in perovskite (PE-HEO) and spinel-type HEO (SP-HEO) structures result in a more complicated atom arrangement, providing a great deal of room for composition and property tailoring. Although they have also been shown to exhibit exceptional thermostability, other crystalline structures, including fluorite [11], bixbyite [12], pyrochlore [13], as well as O_3-layered structure [14] are less common in the field of electrochemistry. There is no doubt that the specific phase-formation processes behind these HEOs vary greatly depending on their unique compositions, element mixtures, and crystal structures, as shown in Figure 8.2. However, because of the metals' uneven distribution at the cation sublattice, resulting in even distribution of

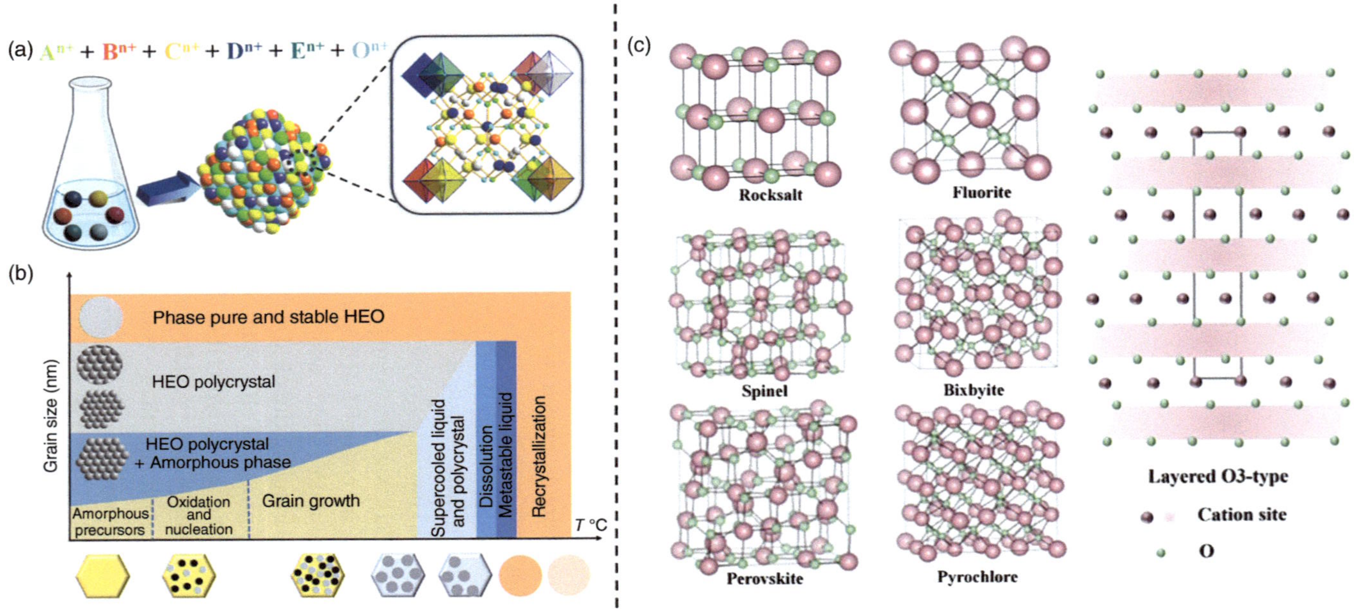

Figure 8.2 Crystal structural formation (a, b) and classification of common HEOs (c). *Source:* Reference [16] / American Chemical Society / CC BY 4.0.

metal oxide solid solutions, they have all demonstrated improved high-temperature stability [15].

HEOs' exceptional stability has fueled the rapid expansion of this field of study, especially in electrochemical applications. The rapid rise of the HEO as a subject of interest is being aided by the widespread adoption of several syntheses as well as processing techniques, proving to offer highly reliable material systems.

8.1.2 Advanced Synthesis Techniques

In general, HEO synthesis involves two steps: (i) blending metals (or their precursors) uniformly and (ii) high-temperature interdiffusion of elements. Numerous synthetic techniques, such as Co-precipitation, solvo/hydrothermal approach, solution combustion, pyrolysis, mechanochemistry, sonochemistry, and sintering, are examples of lesser-energy-demanding wet-chemistry techniques that have been successfully used to prepare HEO NPs with multiple elements and crystal structures as shown in Table 8.1 [5]. However, at higher synthetic temperatures, entropy stabilization is likely to dominate the free energy, which typically results in bigger particle sizes. As a result, developing high-efficiency synthesis techniques is still challenging [36].

8.1.2.1 Solid-State Methods

The most widely used techniques for creating HEOs are solid-state procedures. Typically, mechanical grinding is used to first activate the metal granule. Metal oxide precursors can be efficiently diminished in size by high-energy ball milling,

Table 8.1 Synthesis routes and types of products of HEOs for energy storage.

End product	Routes	Parameters (temperature, time, speed, flow rate)	Applications	References
Powder	Solid-state/ball milling	3–6 hours; 400 rpm	LIBs; DC	[17, 18]
	Co-precipitation	Agent: NH_3; $C_6H_5Na_3O_7 \cdot 2H_2O$	SC	[19, 20]
	Spray pyrolysis	Gas used: CH_4/O_2,	LIBs	[21, 22]
	Hydro/solvothermal	Temp/time: 140 °C/ 5 hours	LIB, SC	[23, 24]
	Solution combustion/ sol gel	Glycine, NH_3	SC	[25, 26]
	Thermal plasma	Gas used: Ar and CO_2	SC	[27]
	Dealloying	NaOH; $(NH_4)_2SO_4$	SC, LIB	[28, 29]
Pellet	Pellet conventional sintering	Temp: >1,000 °C	DC	[30, 31]
	SPS	Temp: >1,200–1,400 °C	LIB, SIB, DC	[32, 33]
Thin films	MSD	Gas: Ar	SC, DC	[34, 35]

which yields a fine powder. The impact- and friction-induced heating promotes atomic diffusion rapidly to produce entropy-stabilized transformed metal oxides. The nucleation formation process of HEO crystals is initiated by calcining (high-temperature annealing) the recovered reactants (powdered form) after the metal precursors have been mechanically activated. There are basically three types of sintering techniques, namely—conventional sintering (CS), spark plasma sintering (SPS), and reactive flash sintering (RFS)—that are typically employed to create HEOs. Because CS is characterized by high temperatures and lengthy reaction periods—typically exceeding eight hours—it may result in particle aggregation, the development of additional phases, contaminants, or otherwise serious surface flaws, particularly while in the cooling phase. To overcome these problems, alternative sintering techniques, such as RFS and SPS, have been developed. These techniques, on the other hand, are powered by electrical current and pressure, provide more efficient cooling, reduced holding times and less sintering temperatures. SPS, also known as field-assisted sintering, is distinguished by the direct use of pulsed current to powder samples for rapid heating and its quicker reaction time compared to CS (usually a few minutes). This technique mixes the metal powder using uniaxial pressure and high-pulsed DC. Fast heating rates during the hot-pressing process in SPS may cause metals to densify more quickly and produce fewer surface flaws. For instance, Ji et al. developed the newest form of high-entropy (pyrochlore phase) ceramic $(Gd_{0.2}Sm_{0.2}Dy_{0.2}Er_{0.2}Yb_{0.2})_2Ti_2O_7$ using SPS technique with improved hardness and density at 1,300 °C, respectively [37]. Additionally, RFS generates intense heating, which facilitates the densification of metal oxides in a time- and energy-efficient manner. To instantly form and densify the HEOs, the powder obtained mechanically is first subjected to moderate-temperature sintering initially prior to being exposed to a strong electrical field. Compared to the CS and SPS methods, the RFS not only employs shorter reaction durations (i.e., 10 °C min^{-1}) but it is also more efficient in obtaining HEO with adequate densification since it offers preferable heating and cooling rates than SPS [15]. A pure SP-HEO phase was prepared and sintered (in a single step) by Tyagi et al. $(Mg_{0.2}Ti_{0.2}Zn_{0.2}Cu_{0.2}Fe_{0.2})_3O_4)$ via RFS for battery applications [38]. Gordón et al. preferred the sintering technique to investigate the viability of creating quinary equimolar HEOs. Upon evaluating different compositions under atmospheric pressure, they resulted in the creation of two main phases: spinel as well as rock salt. This led to producing a new HEO $(Co_{0.2}, Cu_{0.2}, Mg_{0.2}, Mn_{0.2}, Ni_{0.2})O$ prepared under N_2 atmosphere [39]. Another work by Xu et al. implemented a two-step RFS (TSRFS) process skillfully varying their power and time to obtain an optimized densification, grain growth, and homogeneous microstructures. Unlike traditional one-step RFS, TSRFS produced a pyrochlore-type HEO $(La_{0.2}Nd_{0.2}Sm_{0.2}Eu_{0.2}Gd_{0.2})_2Zr_2O_7$ of theoretical density of around 98.15% with a grain size nearly halved to about 684 nm [40]. Experimental studies by Ning et al. show that applying an external electric field during flash sintering which could help enhance atomic diffusion and densification, while simultaneously inhibiting excessive grain growth and preventing elemental volatilization. Consequently, a dense high-entropy ceramics with composition $(Na_{0.2}Bi_{0.2}Ba_{0.2}Sr_{0.2}Ca_{0.2})TiO_3$ could show an exceptional energy storage performance [30]. Similarly, another group formed and sintered magnetic spinel-type $(Mn_{0.2}Co_{0.2}Ni_{0.2}Cu_{0.2}X_{0.2})Fe_2O_4$ (X = Fe,Mg) HEOs in a single step via RFS. This enabled them to understand the crystal structure and correlate it with their magnetic characteristics in accordance with the RFS technique [41].

8.1.2.2 Wet-Chemical Methods

These techniques use less energy than solid-state techniques and are frequently employed in the production of HEOs because they make it simple for regulating the morphology, size of particles, as well as distribution of elements. In general, these involve metal salts that are often made as a uniform solution integrating elements at the atomic level. Thereafter, to create single-phase HEOs, the resulting mixture is subsequently calcined at a temperature that is comparatively lower to remove the solvents along with other reagents. This section showcases a brief overview of many popular HEO synthesis techniques.

Spray Pyrolysis Spray pyrolysis is a cost-efficient technique used for chemically making powders or coating thin films. Usually, an ultrasonic atomizer is used to spray a solution consisting of salts of multi-metal cations. Gas–liquid interfaces separate the resulting aerosols, which are then quickly pyrolyzed by flame or high-temperature ovens before forming big drops. That type of a dispersion condition corresponds to the atomic dispersion of glycine combustion or sol-gel. In order to facilitate the sintering stage, this method is focused on atomizing a combination of mixed-metal precursors and minimize the atoms' migration distance and obstacles [36]. There are two types of spray pyrolysis: flame spray pyrolysis or FSP and nebulized spray pyrolysis or NSP. Several factors such as the size and distribution of the droplets, as well as the atomization rate vary depending on the atomizer being used. An ultrasonic nebulizer is used to turn metal precursor solution into aerosol mists during the NSP process. To achieve quick heat treatment, the generated mists are transported to a tube furnace by oxygen flow. However, FSP synthesizes HEO using a high-temperature flame. The FSP method produces a self-sustaining spray flame by dissolving the metal constituents in a flammable solution. In order to form the nanosized powder, the atomized droplets that include the metal precursor solution undergo evaporation and condensation. FSP provides an optimal combustion enthalpy density associated with the flame producing the HEO having greater purity of phase and surface area in contrast to the NSP process [21]. Brandt et al. reported based on an application as advanced anode active materials were prepared through liquid-feed flame spray pyrolysis or LF-FSP. The group demonstrated two new HEO compositions consisting of Zn and Cu in place of "Co" from $Co[(CoNiMnFeTi)_3O_4]$, with Zn and Cu considering the detrimental effects on the economy and supply chain of the whole material while keeping the electrochemical performance intact [42]. Another interesting report by Luo et al. showed an improved continuous-flow synthesis method using FSP, in which varied air flow rates are used for quick quenching in order to change the pace of cooling and regulate the lattice strain of HEO-NPs [43]. This spray pyrolysis method appears to be a good fit for not only developing HEO powder materials but also for creating and investigating thin-film HEOs based on spinel as well as for researching novel compositions that incorporate transition metal cations. A work reported by Kamecki et al. prepared a thin film of thickness around 500 nm as (Mn, Co, Fe, Ni, $Cr)_3O_4$ having homogeneous distribution of elements [44]. Luo et al. suggested a micron-droplet-confined approach using FSP in order to accomplish the continuous-flow formation of self-supporting HEA-NPs. Here, they showed by using kinetic and thermodynamic regulations (high temperature and rapid cooling); binary to septenary alloys may develop from droplets to <10 nm nanoparticles (NPs) in less than 5 ms. A single-phase solid solution using the immiscible elements AuPtPdRuIr HEA-NPs was developed in this way [45].

Reverse Coprecipitation Another popular wet-chemical technique for creating HEOs is reverse coprecipitation (RCP). This is simpler than spray pyrolysis; it just entails dissolving a minimum of five metal salts to form a homogeneous solution (water or other appropriate solvents), then adding a precipitating agent, followed by drying the mixture to get a solid combination. Following drying, the combination powder is typically heated to between 500 and 900 °C. Talluri et al. reported the first time that pure SP-HEO $(CrMnFeCoNi)_3O_4$ NPs prepared via RCP have been used in SCs [19]. Adding to that, another report on nanocrystalline SP-HEO $(Al_{0.2}Co_{0.2}Cr_{0.2}Mn_{0.2}Ni_{0.2})_3O_4$ SCs was made by Sreenivasulu et al. [25].

Solution Combustion Synthesis Solution combustion synthesis (SCS) or the *sol-gel* method is an exothermic reaction performed among an oxidizer (such as metal nitrates) and reducing agents or fuel (basically organic compounds such as glycine, ethylene glycol, urea, and citric acid) to prepare the sol. The precursor, which undergoes subjected to calcination to produce the oxide powder, is obtained by ageing this intermediate product at low temperatures. This has many advantages, e.g., the formation of nanopowder with proper crystal structure, no post-calcination, rapid rate of heating, less energy consumption and processing interval. Although formation of HEOs requires high-temperature calcination, optimizing the temperature and duration of combustion or the variables involving strongly the rate of chemical reaction rates provides immense control over determining the HEO phase in the synthesis process. An interesting work shown by Wang et al. demonstrated a straightforward synthesis process to overcome these obstacles, which not only explored efficiently but could also build gel precursors with superior porosity. Here, $La(Mn_{0.2}Fe_{0.2}Co_{0.2}Ni_{0.2}Cu_{0.2})O_3$ PE-HEO with a significant specific surface area along with lots of mesopores was produced using a metal–organic framework (MOF) gel technique presenting a straightforward and effective method for SCs [46]. In addition, there are also reports on low-temperature $(FeCoCrMnNi)_3O_4$ HEO with significantly increased energy density (103.8 Wh kg^{-1}) shown as a promising electrode material for improved SCs properties [47]. The presence of magnesium stabilizes the electrochemical processes in the HEO, whereas the other transition metals (Cr, Mn, Fe, Cu) contribute to the multi-electron redox system. This novel $(Mg_{0.21}Cr_{0.21}Mn_{0.21}Fe_{0.21}Cu_{0.16})_3O_4$ HEO single phase was attained by Chillal et al. because the molar quantity of precursor components was detailed, highlighting the value of precise stoichiometry [48], Gupta et al. examined how two distinct synthesis techniques—sol-gel and RCP affect the new $(CuNiFeMnCo)_3O_4$ SP-HEO phase change, microstructure, ionic state shift, and relevance to electrochemical charge storage. While the sample produced with the RCP approach contained spherical particles, the sample created using the sol-gel method exhibited a porous shape [49].

Hydrothermal and Solvothermal Methods These are simple, energy-efficient, and controllable processes to prepare various materials with varying morphologies. Proper optimization of reaction conditions like pressure, temperature, time, pH, and the precursor/additive concentration value helps in controlling materials' surface morphology, size, and crystallinity [15]. Krawczyk et al. presented a novel and effective process for creating the multicomponent PE_HEO with the composition $(Gd_{0.2}Nd_{0.2}La_{0.2}Sm_{0.2}Y_{0.2})CoO_3$. This process involved liquid nitrogen quenching followed by an altered coprecipitation hydrothermal process [35]. Similarly, using the rapid microwave solvothermal approach, nano HEOs $(FeCoNi_2CrMn)_3O_4$) were developed by Wang et al. The microwave operation ensured the uniform distribution of NPs quickly [50]. Another study by Li et al. showed an efficient synthesis $(FeCoNiCuZn)_3O_4$ HEO material by simple hydrothermal method, demonstrating notable performance in a lithium-ion hybrid capacitor (LIHC). This was the first time where HEO has been implemented in LIHC [51].

All these solid-state and solution-based methods result in HEOs mostly in their powder form. However, other techniques involve the formation of HEOs as thin films comprise epitaxial growth methods using substrates like magnetron sputtering deposition (MSD) and pulsed-laser deposition.

8.1.2.3 Magnetron Sputtering Deposition

This subtype of physical vapor deposition (PVD) is regarded as an economical method that offers fine control in tuning the size, shape, and/or distribution of constituent elements. The system comprises a vacuum chamber, a target source (the required elemental alloy), a substrate (like an Si wafer), and a sputtering gas (like Ar/O_2). In this process, sputtered atoms are deposited constantly upon the substrate due to the continuous bombardment of gaseous ions present in plasma (Ar) onto the target with the help of the electric field, then annealing, that eventually makes the final film. It is also well recognized that HEOs can help the miniaturization of future gadgets and information devices. Wang et al. developed a negative temperature coefficient thermistor thin films preferring co-sputtering combinatorial synthesis and oxidation by annealing to create high-entropy $Co_{21}Mn_{17}Fe_{19}Ni_{21}Zn_{22}O_{4\pm\delta}$ films. The system was loaded with three targets, each measuring 80 mm in diameter and 3 mm in thickness: MnCoFe (1 : 1 : 1), Ni, and Zn (purity: 99.99 % for each) [52]. Similarly, Khan et al., in their study, used three distinct oxygen flow fractions to develop HEO thin films with a composition of $AlCoCrCu_{0.5}FeNi$ via radio frequency (RF)-reactive magnetron sputtering [53]. Zenkin et al. in their research showed that not only addition of Mg influenced the entropy of the materials, but they also observed that inclusion of aluminum could

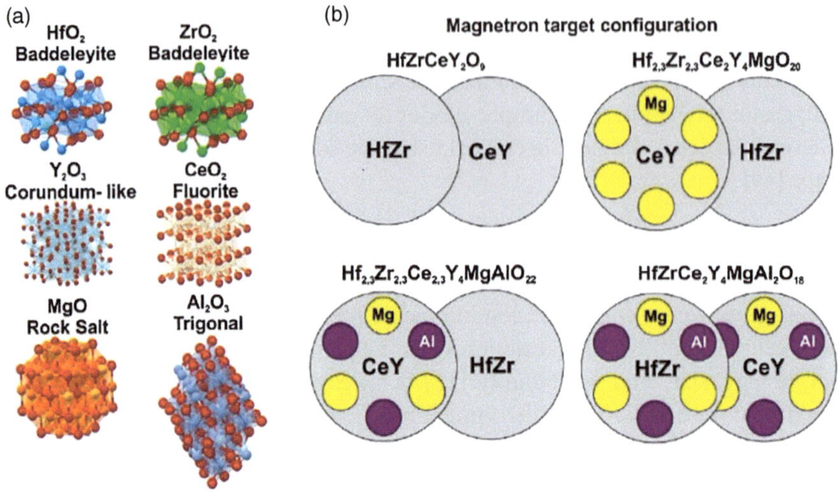

Figure 8.3 The crystal structures depiction of the chosen target (a); setup for depositing the HfZrCeYMgAlO system (b). *Source:* Reprinted (adapted) from the Ref. [54]. Copyright (2024), with permission from Elsevier.

greatly improve the coating's mechanical properties and crystallinity. Two circular unequal magnetron targets fitted with Hf-Zr (50/50 at.%) and Ce-Y (33/66 at.%) were used to sputter HfZrCeYMgAlO HEO films, as shown in Figure 8.3 [54].

8.1.2.4 Pulsed Laser Deposition PLD

PLD is another PVD method mainly done in a vacuum environment in which the desired material is vaporized or ablated in tiny parts when a pulsed laser is emitted at it at a high energy (either for ns or fs <15 seconds). This makes up the plasma plume (resembling atomic vapor) which eventually deposits onto a heated base as a thin film. A work by Guo et al. used PLD with a ceramic target having a rock salt structure [$(Co_{0.2}Ni_{0.2}Cu_{0.2}Mg_{0.2}Zn_{0.2})O$] in their work to create HEO thin films. Here, they used fluxes at an increasing rate to study the grain growth [55]. In another work, like Wang et al. demonstrated, a maximum number, i.e., eight, have ever been included in an HEO thin film comprised $LiCr_{1/6}Mn_{1/6}Fe_{1/6}Co_{1/6}Ni_{1/6}Cu_{1/6}O_2$ as a layered rock salt structure. Here, they used PLD to create (001)-oriented epitaxial thin films as shown in Figure 8.4 [56].

Another group by Niculescu et al. studied the local structure of $(MgNiCuCoZn)_{0.167}(MnCr)_{0.083}O$ in light of its source HEO, i.e., $(MgNiCuCoZn)_{0.2}O$, using PLD. In addition, they created a number of thin films at progressively higher partial pressures of oxygen, resulting in different forms [27].

8.1.2.5 Some Other Processes

Thermal Plasma Route The thermal plasma technique uses a high-temperature plasma arc to generate extremely reactive and energetic species that are able to react rapidly, exhibit rapid quenching rates, and can be processed quickly to produce highly pure NP. Here, plasma becomes the primary source of certain chemical

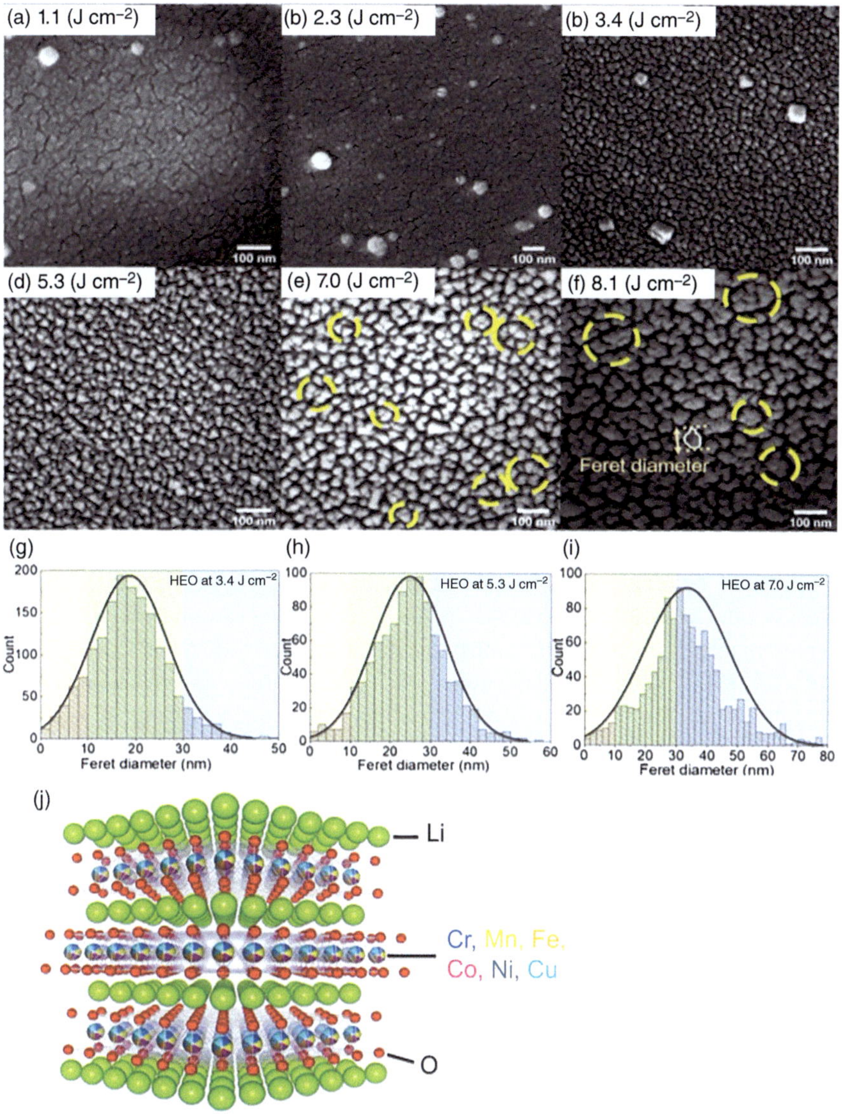

Figure 8.4 Images obtained from SEM of the HEO thin films' top view obtained via PLD at various fluxes having energy (J cm^{-2}) as (a) 1.1 – (f) 8.1; (g)–(i) indicate the size variation in selected grains and islands at 3.4, 5.3, and 7.0. Representation of $LiMO_2$ structure (j). To draw the figure, Visualization for Electronic and Structural Analysis (VESTA1) is used. *Source:* Reprinted (adapted) from the Refs. [56, 57] Copyright (2022) American Chemical Society.

radicals that propel the creation of the intended product. This method represents a single-step synthesis producing NPs that involves several steps like vaporization, nucleation, and then condensation. Optimal vaporization of the precursor components is induced by the high temperature as well as high-enthalpy plasma jet, whereas homogenous nucleation is facilitated by the high quenching rate. Rapid

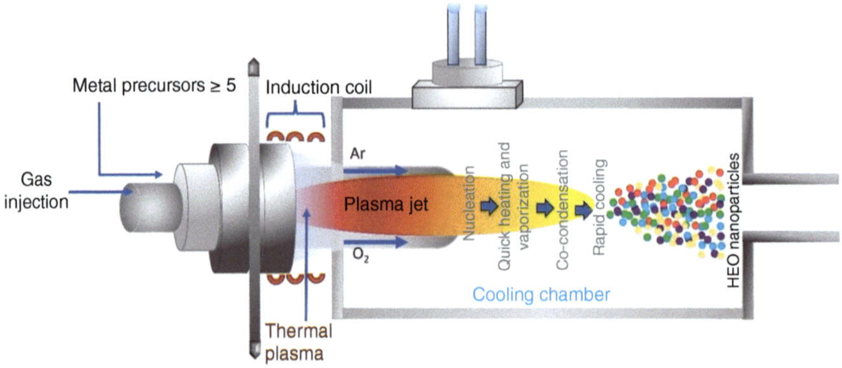

Figure 8.5 Schematic representation of the thermal plasma reactor for the synthesis of HEO nanoparticles.

and large-scale manufacturing, high purity and quench rates, and the absence of post-treatment are the benefits of employing thermal plasma for NP synthesis. The formation of HEO NPs using thermal plasma technology has demonstrated a number of benefits over other traditional methods. These include very high ranges or rates of temperature (2,000–20,000 K), enthalpies (around 10^6–10^8 J kg^{-1}), energy densities (~10^6–10^7 J m^{-3}), heat flux densities (~10^7–10^9 W m^{-2}), quenching (~10^6–10^8 K s^{-1}), and processing rates [58]. For example, as per the report by Pasupathi et al., the thermal plasma jet, which was produced using Ar and CO_2 gases, had a high temperature (~16,800 K), a lengthy plasma plume (0.2 m), and a high enthalpy (1,450 kJ kg^{-1}). The fragmented solution of precursor was injected inside the plasma region after producing a steady plasma jet. The presence of an ample supply of oxygen in the surroundings allows the generated plasma energy to disintegrate and evaporate the material precursor that serves as the feedstock as metal oxide vapor. A considerable quantity of metal oxides that had been vaporized condensed on the chamber wall that was cooled by water, as shown in Figure 8.5. Plasma processing variables, including feeder location, atomizing pressure, gas combination and flow rates, precursor feeding rate, and plasma power, were adjusted to maximize output yield while consuming the least amount of discharge power [59].

An HEO material $(NiFeMnCuZn)_3O_4$ as a single phase has been created for the first time by the same group under a discharging power of 12 kW using a carbonaceous thermal plasma (Ar/CO_2) gas [60]. Recently, a DC thermal plasma torch with a low current of discharge employed at atmospheric pressure, (thermal plasma method), to make phase-pure (Mn, Fe, Co, Ni, and Cu)$_3$O$_4$ HEO NPs [57]. In another study, they also used a carbonaceous (Ar-$CO_2CH_4^-$) thermal plasma form to produce HEOs NPs $(Ni_{0.2}Co_{0.2}Cr_{0.2}Mn_{0.2}V_{0.2})_3O_4$ infused with carbon in a single step [61].

Dealloying The method of dealloying is a popular procedure done among elements in alloys that involves the removal of the active formulations or elements in a selective manner using chemical or electrochemical techniques in accordance with variations in chemical behavior of elements or compositions in it. Additionally, the remaining components naturally undergo migration and diffusion to generate a distinct microstructure. By controlling the dealloying and subsequent treatment procedures, this method can accomplish proper regulation of morphology, structure, and space order. There are several different types involved in dealloying, for example, vapor, chemical, and electrochemical dealloying [62]. Zhang et al., after successfully etching the $FeCoNiCrMnAl_{95}$ alloy, a simple oxidation process, yielded a homogeneously distributed SP-HEO, namely, $(FeCoNiCrMn)_3O_4$ [28]. A typical dealloying process was employed by Jin et al. for making $(AlCoFeMoCr)_3O_4$ nanoporous SP-HEO [63]. Pei et al. prepared an HEO based on the rock salt (NiCoCuFeMn)O by in situ derivatization on HEA (hp-HEA/HEO) through the dealloying process and high-temperature solution treatment, employing a novel free-standing hierarchical porous structure technique [64]. Some studies on Nb showed that the presence of Nb will help interact with other elements, i.e., atoms make the Nb-rich intermetallic Laves phase, helping to refine grains, enhance corrosion resistance, and boost yield as well as hardness [29]. Lin et al. prepared (CoCrFeNiNb) HEO using chemical dealloying and melt-spinning as a self-supporting nanoporous material [65].

8.1.3 Characterization

The enhanced properties of HEMs arise mainly due to their unique compositions and fascinating effects such as lattice deformation, imperfections or defects, SROs, separation of phase, and local assembly. Despite the fact that the increased compositional and structural complexity of HEMs makes it difficult to accurately characterize surrounding structures, significant advancements have been made through sophisticated methods and tools. The developed characterization procedures involve˙ microscopy, X-ray spectroscopy (XAS), and absorption, along with neutron diffraction, have been primarily utilized for evaluating HEMs and are described here.

8.1.3.1 X-ray and Neutron Diffraction

These tools are most preferably used or proficient ways to study unknown materials for their phase formations as well as determining their crystal structures. Out of all the diffraction-based techniques, powder X-ray diffraction (XRD) is basically employed to acquire both quantitative and qualitative information on the basis of the crystal structures obtained for a material. Detailed crystal structural information that includes lattice parameters, crystallite size, preferred orientation or position, crystallinity, lattice strain, along with packing proportion of all atoms (element-wise) in a lattice is possible to obtain through both the Rietveld refinement approach and obtaining detailed data from XRD. However, weak and broadening diffraction peaks arise due to the presence of light-weight elements (H, Li, Be, and B) and intense lattice distortion along with nano-crystalline domains in HEMs; hence, analyzing such sample structures becomes arduous using in-laboratory diffraction instruments. Therefore, this disadvantage can be effectively eliminated by using advanced diffraction techniques that comprise sources that are high intensity and monochromatic, as in synchrotron or neutron diffraction instrumentation, which provide atomic-level information to examine defects, lattice strain, site order and disorder, atomic displacement, and structure distortion. Furthermore, the combination of that scattering information with pair-distribution function (PDF) allows for a finer assessment that deals with the atomic arrangements present locally of High-Entropy Compounds (HES) having intricate crystal lattices as well as samples with amorphous structures that are difficult for standard XRD to identify. Kirsch et al. developed five prototype compositions and determined the synthetic accessibility. These consist of four $A_2Mn_4O_{10}$ compositions where A represents Nd, Er, Y, Eu, Sm, Ce, and Bi, and one $Bi_2M_4O_9(Bi_2(Al_{0.25}Ga_{0.25}Fe_{0.25}Mn_{0.25})_4O_9)$. Thorough analysis involved neutron powder diffraction (NPD) and XRD accompanied by Rietveld, X-ray total scattering PDF analysis, scanning transmission electron microscopy (STEM)–energy-dispersive X-ray spectroscopy (EDX), and spectroscopy (both infrared, as well as Raman); it was validated that the materials were, in fact, mixed solid solutions. The HEOs are formed by heating a nanostructured precursor that we create using a sol-gel technique. Later on, they also used XAS and *in situ* XRD to monitor this formation [66].

8.1.3.2 Electron Microscope

Techniques of electron microscopy typically involve two types, i.e., one is scanning electron microscopy (SEM), while the other one is transmission electron microscopy (TEM) which are basically implemented to explore the features of surface, size, and morphological characteristics of the materials. Furthermore, STEM, electron energy loss spectroscopy (EELS), high-angle annular dark field (HAADF), low-angle annular dark field (LAADF), annular bright field (ABF) [67], energy-filtered transmission electron microscopy (EFTEM), along with related EDX, are some other sophisticated instruments combined with SEM and TEM. These are considered to be the easily accessed methods for analyzing particular structures as well as the elemental composition and distribution. While TEM is normally focused upon a more constrained area with greater resolutions, SEM is usually used to investigate the three-dimensional properties of a wider part of samples. More significantly, TEM is an effective method for analyzing the crystal structures in a smaller sample region. Defects like vacancies and dislocations, as well as lattice distortion linked to HECs, can be readily identified and measured in detail using TEM and STEM images with high resolution, which are especially useful for studying the sites of atoms and the lattice plane fringes. Similarly, the TEM technique known as selected area electronic diffraction (SAED) determines the diffraction pattern as well as the crystal structure of HECs [67]. Su et al. use an *in-situ* TEM approach that includes image analysis and electron diffraction (ED) with spectroscopy to examine the lithiation and delithiation patterns of an interesting R-HEO material $(Co_{0.2}Ni_{0.2}Mn_{0.2}Zn_{0.2}Fe_{0.2})_3O_{3.2}$ at an atomic level. This approach enables us to monitor the morphological change of particular HEO NPs alongside the continual transformations of phase caused due to electrochemical activity, as shown in Figure 8.6 [68].

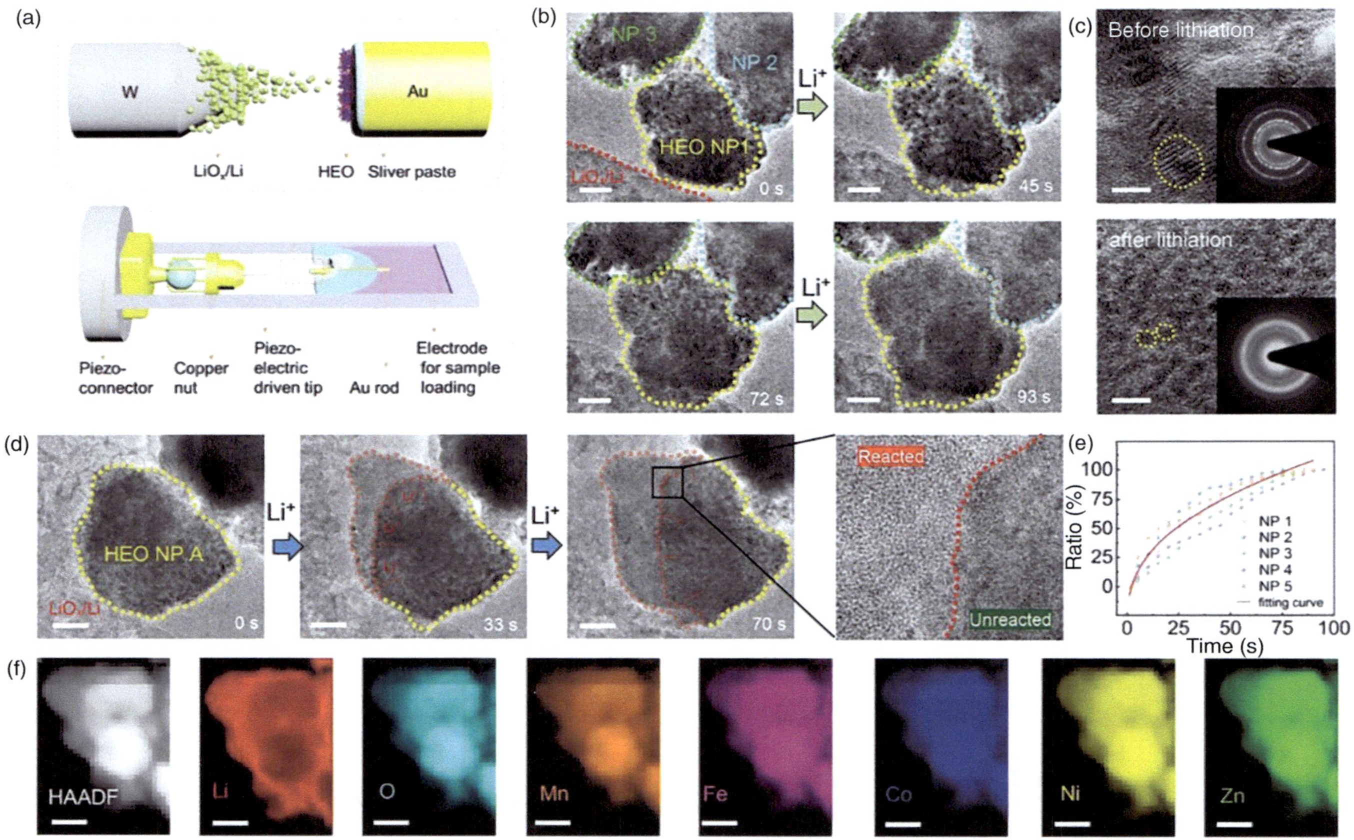

Figure 8.6 Electrochemical lithiation of rocksalt $M_3O_{3.2}$ HEO via in situ TEM at single-NP level. (a) Representation of the experimental design in a schematic manner. (b) Time-series TEM images (Scale bar, 50 nm) of $M_3O_{3.2}$ NPs during lithiation. (c) High Resolution TEM and SAED images (Scale bar, 5 nm) of $M_3O_{3.2}$ NPs pre- and post-lithiation. (d) Set of TEM images (Scale bar, 50 nm) showing the advancement of lithiation reaction over a certain time frame for a single $M_3O_{3.2}$ NP. (e) Plotting of the ratio (R) of the reacted parts' projected areas over the whole NP against lithiation time (t). (f) Elemental mapping images of a lithiated $M_3O_{3.2}$ NP obtained in HAADF-STEM as well as STEM-EELS. Scale bar, 50 nm. *Source:* Reproduced from Ref. [68] / with permission of John Wiley & Sons.

Also, in addition to examining regional structures, STEM, in conjunction with EELS or Energy Dispersive Spectroscopy (EDS), and/or 3D-APT can effectively evaluate the compositional uniformity of HEMs by obtaining elemental distribution and valences at atomic resolution [69].

8.1.3.3 X-ray Spectroscopy and Absorption

Besides the discussed characterizations (diffraction/microscopy), XAS and absorption are primary methods used to investigate the chemical nature of each element in HECs. Effective techniques for examining the coordination environments associated with local atoms and the chemical states of the specimens are X-ray absorption spectroscopy (XAS), i.e., X-ray absorption near-edge structure (XANES), and extended x-ray absorption fine structure (EXAFS) spectra. While the latter offers comprehensive details on chemical conditions of various atoms in HEMs, the former aids in determining the oxidation state as well as composition of the elements, comprising cations and anions. Besides cations' synergistic influence in the HEO $Mg_{0.2}Co_{0.2}Ni_{0.2}Cu_{0.2}Zn_{0.2}O$ through the electrochemical reaction with, lithium based on a comprehensive valence states analysis, Wang et al. also carried out a thorough characterization of the material's microstructure and valence state of the elements at various cycling states of the HEO-based anode. By employing XAS that includes EXAFS Fourier transforms or EXAFS-FT with XANES, the elements' anticipated valence states were ascertained at various electrochemical reaction states as shown in Figure 8.7 [70].

Generally, X-ray photoelectron spectroscopy (XPS) is an extensively operated technique to assess the chemical environment along with oxidation states of samples; like XANES, however, XPS can only limit its signal detection to the specimen's surface and subsurface regions that are a few nanometers in size. Because surface atoms are mostly identified as the active sites for the majority of electrochemical activity, the precision of XPS results is still an indispensable technique, even though it is comparatively less accurate than that of XANES in determining the surface information of HECs.

8.1.4 HEOs for Electrochemical Energy Storage

Electrochemical energy storage (EES) is transforming our daily lives. There are multiple advanced EES devices, namely rechargeable lithium-ion batteries (LIBs), lithium–sulfur batteries (LISs), sodium–ion batteries (SIBs), DCs, as well as SCs that have gained significant interest as they deliver high-energy/power densities (Table 8.2). In fundamental terms, cathodes or otherwise anodes with high voltage or capacity might be used for building high-energy batteries. Nonetheless, safety and cycling concerns with these kinds of batteries continue to be significant obstacles to overcome. HEOs own special characteristics because of the numerous metals in crystallographic locations and entropy stabilization. HEOs have a large solid solution limit owing to the high-entropy effect, which also makes them more stable and helps them get over cycle limitations. Furthermore, multi-metal doping produces lattice distortion, which helps to strengthen and harden materials.

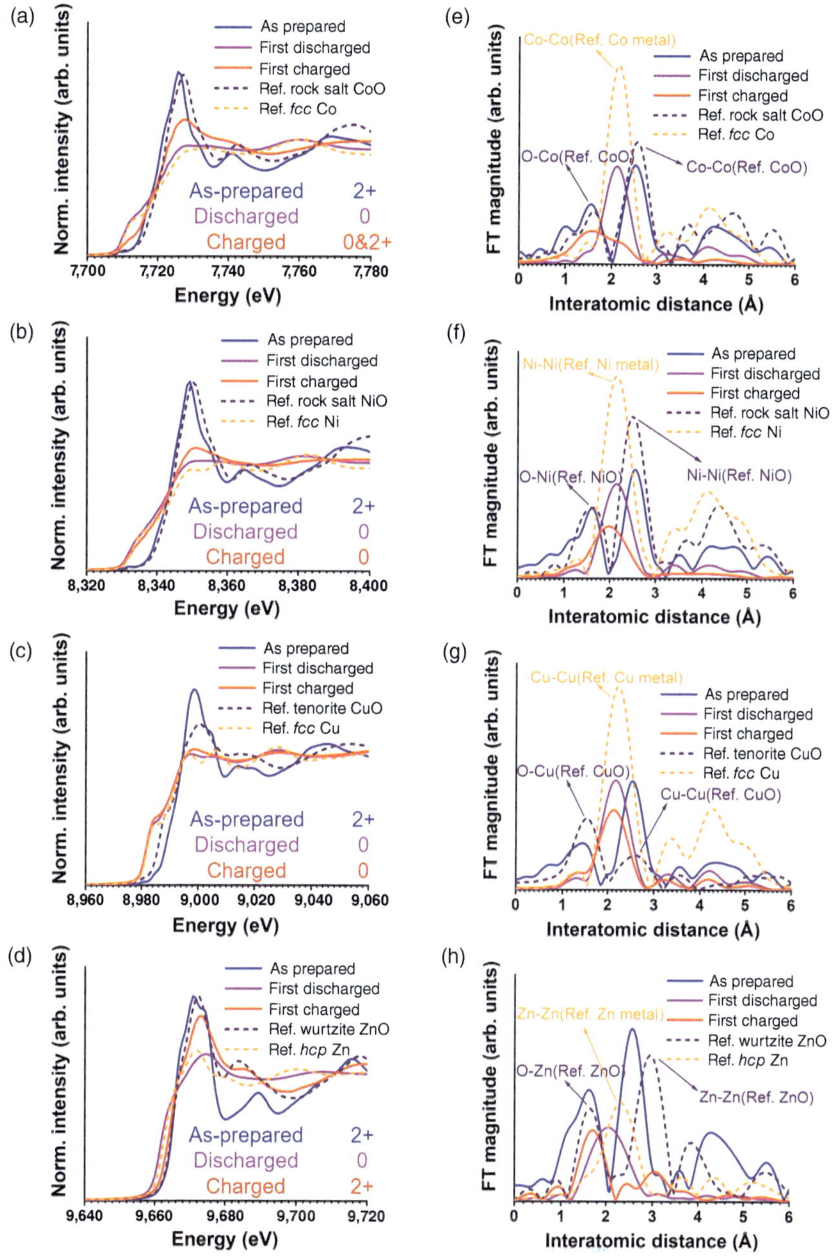

Figure 8.7 XANES analysis of Ni, Co, Cu, and Zn (a–d); relevant EXAFS–FT (e–h); dash lines are references computed using standard samples supplied by the beamline, whereas solid lines show the experimental data. The terms "first discharged" and "first charged" describe samples that have been charged and discharged to 3.0 and 0.01 V (against Li/Li+). *Source:* Reference [70] / Springer Nature / CC BY 4.0.

Table 8.2 Electrochemical properties of HEOs applied to different energy storage devices.

HEO	Structure	Performance	Application	References
$(FeCoNiCrMn)_3O_4$	Spinel	~597 mA h g^{-1}/86% after 1,200 cycles at 2.0 A g^{-1}.	LIB/anode	[71]
(MgCoNiCuZn)O	Rock salt	~525 mA h g^{-1}/150 mA h g^{-1} after 2,600 cycles at 10 A g^{-1}.	LIB/anode	[72]
$Mg_{0.2}Co_{0.2}Ni_{0.2}Cu_{0.2}Zn_{0.2}O$	—	~1,585 mA h g^{-1}/920 mA h g^{-1} after 300 cycles at 0.1 A g^{-1}.	LIB/anode	[73]
$(P_xA_{1-x})B_2O_4$ (A/B = Cr,Mn,Fe,Co,Ni)	Spinel	~2,536 mA h g^{-1} after 400 cycles at 1 A g^{-1}.	LIB/anode	[74]
$(CrMnFeNiMg)_3O_4$	Spinel	~857 mA h g^{-1}/80% after 300 cycles at C/10.	LIS/cathode	[75]
$FeCoNiO_x$–$MnCrO_x$	—	~623 mA h g^{-1}/after 500 cycles at 2C.	LIS/cathode	[76]
$P'2Na_{0.59}Mn_{0.90}Ti_{0.02}Cu_{0.02}Ni_{0.02}Co_{0.02}Fe_{0.02}O_{1.95}F_{0.0}$		~224 mA h g^{-1}/134 mA h g^{-1} after 300 cycles at 1 A g^{-1}.	SIB/cathode	[77]
$Na(MgCu)_{1/12}(NiCoFeMnTi)_{1/6}O_2$	O_3-type	~146 mA h g^{-1}/83% after 700 cycles at 0.1 A g^{-1}.	SIB/cathode	[78]
$Na_{0.83}Li_{0.1}Ni_{0.25}Co_{0.2}Mn_{0.15}Ti_{0.15}Sn_{0.15}O_2^-$	O_3-type	~109 mA h g^{-1}/87% after 200 cycles at 10C.	SIB/cathode	[79]
$(Ba_{0.8}Bi_{0.2})(Ti_{0.58}Zr_{0.14}Sc_{0.18}Mg_{0.04}Nb_{0.05}Ta_{0.01})O_3$	—	W_{rec} of ~5 J cm^{-3} and η of ~94%.	DC	[80]
$(Ba_{0.8}Bi_{0.2})(Ti_{0.58}Zr_{0.14}Sc_{0.18}Zn_{0.04}Nb_{0.05}Ta_{0.01})O_3$	—	W_{rec} of ~3.9 J cm^{-3} and η of ~96%.	DC	[80]
$(1-x)(0.9BaTi_{0.97}Ca_{0.03}O_3$–$0.1Bi_{0.55}Na_{0.45}TiO_3)$–$xBi(Li_{0.2}Y_{0.2}Mg_{0.2}Ti_{0.2}Ta_{0.2})O_3$	—	W_{rec} of ~4.9 J cm^{-3} and η of ~91%.	DC	[81]
$(Li_{0.2}Ca_{0.2}Sr_{0.2}Ba_{0.2}La_{0.2})TiO_3$	Perovskite	W_{rec} of ~1.5 J cm^{-3} and η of ~90%.	DC	[82]
$(Mn,Fe,Co,Ni,Cu)_3O_4$	Spinel	431 F g^{-1}/~92% after 10,000 cycles at 3 A g^{-1}.	SC	[58]
$(Co_{0.2}Cr_{0.2}Fe_{0.2}Mn_{0.2}Ni_{0.2})_3O_4$	Spinel	384 F g^{-1}/~60% after 2,000 cycles at 5 A g^{-1}.	SC	[83]
$La(Mn_{0.2}Fe_{0.2}Co_{0.2}Ni_{0.2}Cu_{0.2})O_3$	Perovskite	~609 F g^{-1}/~84% after 10,000 cycles at 5 A g^{-1}.	SC	[46]

HEOs are therefore highly regarded for these kinds of applications because of their slow diffusion and synergistic impact, which significantly affects their relaxation, thermal conductivity, and structural and chemical stability.

8.1.4.1 Anode

Since transition metal oxides (TMOs) possess higher theoretical capacities than traditional graphite anodes, they have been studied extensively as active anode supports for LIBs. Nevertheless, TMOs' uses are severely limited by the quick capacity loss driven by their significant phase shift alteration over cycling. Thus, a lot of effort has been put into the engineering of metal oxides having structural reliability and thus can be employed as conversion anodes during LIB cycle operations. One promising method for resolving this problem is to engineer the configurational entropy to produce high-entropy phases [69]. A HEO named as $(Co_{0.2}Cu_{0.2}Mg_{0.2}Ni_{0.2}Zn_{0.2})O$, synthesized by Wang et al., demonstrated excellent cycle stability over 50 cycles for LIB as anode. The cell could retain ~300 mA h g^{-1} discharging capacity from 446 mA h g^{-1} (initial capacity) and obtained 240 Wh kg^{-1} as energy density [84]. Another intriguing conversion anode for LIBs: the results of the design and comprehension of a hitherto unknown entropy-balanced SP-HEO, $(Co_{0.2}Mn_{0.2}V_{0.2}Fe_{0.2}Zn_{0.2})_3O_4$ are presented by Hou et al. This oxide exhibits a slight volume change character in addition to highly reversible capacities of ~900 and 500 mA h g^{-1} at 0.2 and 3 A g^{-1} for approximately 500 and 2,000 cycles, respectively, [85]. An interesting conclusion drawn by Ghigna et al. showed that $Mg_{0.2}Co_{0.2}Ni_{0.2}Cu_{0.2}Zn_{0.2}O$ is no longer appropriate as LIB anode despite favorable findings in several literature. As per their study, they reported that it is very much vital to look into other phase stabilization techniques, such as appropriately adjusting crucial elements like metal type, cation quantity, stoichiometry, and morphology, as well as investigating other structures, like spinel, where another lithiation/delithiation mechanism takes place [86]. Although cobalt has been an essential element used for batteries for years, its precarious worldwide supply issue is typically hampered by geopolitical meddling. There have been a lot of other issues regarding the use of cobalt such as it is expensive, hazardous, and rare. Therefore, Co-based HEOs are not suitable for widespread use [87]. Hence, Patra et al. decided to create a range of Co-free spinel type HEOs (HESO) anodes since it significantly affected the anodes' electrochemical performance. The implications of HEOs' chemical structures being phase pure and resulting lithiation/delithiation characteristics were studied by using $(CrNiMnFeCu)_3O_4$ as the HEO anode. It exhibited nearly no capacity decline after 400 cycles, along with a remarkably high-rate capacity of 480 mA h g^{-1} delivered at 2,000 mA g^{-1} [75]. A high-capacity porous nanoarchitecture $(Cr_{0.2}Fe_{0.2}Co_{0.2}Ni_{0.2}Zn_{0.2})_3O_4$ HEO anode was successfully made using such a facile sol gel technique, yielding a capacity of 1,022 mA h g^{-1} (1 A g^{-1}) even after 1,000 cycles [88]. Hollow graphene spheres serving as an effective framework for electrolyte conduction, comprising in situ infused (MgCoNiCuZn)O NPs having rock-salt-type structure, present insightful information for improving the characteristics of HEO anode material as an anode as observed by Gao et al. Additionally, the reversible capacities obtained at 1 A g^{-1} continue to increase until they reach around 1,400 mA h g^{-1} over 800 cycles [89]. There are also reports on a nanosized $(FeCoNiCrMn)_3O_4$ HEO-cyclized polyacrylonitrile (cPAN) electrode that shows noticeably improved rate and cycle performance because the in situ generated cPAN coating strengthens the structural and interfacial stability in addition to improving electrical conductivity [90].

8.1.4.2 Cathodes

LIBs are essential for enabling environmentally friendly energy storage applications and operating e-vehicles (EVs) in the perpetual global transition to a carbon-free society with the goal to mitigate climate change. The demand necessitates to develop more energy-dense and less expensive materials. Other problems including irreversible phase transitions, capacity loss, and voltage decay that could reduce LIBs' energy density and raise safety concerns, as well as poor rate performance brought on by slow reaction kinetics, are major challenges that must be tackled for increasing their commercial viability. Thus, the idea of high-entropy can be used to overcome these issues. The ease of integrating multiple components provides an obvious advantage by enabling a modified material form to alter the cathode properties. To prove this, recently, a work by Kim et al. showed that incorporating non-equimolar HEOs like $(Li_{1.15}Na_{0.05}Ni_{0.19}Mn_{0.56}Fe_{0.02}Mg_{0.02}Al_{0.02}O_{1.97}F_{0.03})$ could not only stabilize structurally and improve phase stability but also improve the Li^+ ion diffusion kinetics and rate capability [91]. Huang et al. chose Mn as an essential component for making the cathodes for LIBs, for Mn is inexpensive, abundant, and capable of producing large amounts of energy. They prepared two types of lithium-rich structures of manganese-based HEO materials, in which $Li_xMn_{0.6}Ni_{0.1}Co_{0.1}Fe_{0.1}Al_{0.1}O_y$, was layered ($x$ = 1.5, named as HEO-L), whereas the inadequate lithium variant resembled a spinel-layered hybrid (x = 0, named as HEO-H). The HEO-H had superior capacity retention along with nearly 100% average voltage retention after 100 cycles, whereas the Mn-rich HEO-L demonstrated greater capacity and an average voltage retaining ability of 93% [92]. A high-entropy layered oxide (HEO5's, $LiNi_{0.2}Co_{0.2}Mn_{0.2}Fe_{0.2}Al_{0.2}O_2$) rapid capacity fading mechanism in the first cycle has been studied by Zheng et al. The primary cause of this degradation is the formation of an M_3O_4 surface phase, which also prevents further lithiation. This is a crucial finding for the future creation of high-entropy multilayer oxides [93]. A novel approach toward resolving the structural stability of cations for high-Ni-layered TM oxide cathodes and not compromising the high Ni content's sole capacity input is high-entropy doping. In this regard, Liang et al. reported a low-electronegativity cationic high-entropy doping as an effective approach for making the high-Ni, Co-less layered cathode $(LiNi_{0.8}Mn_{0.12}Al_{0.02}Ti_{0.02}Cr_{0.02}Fe_{0.02}O_2)$. It could not only enhance the capacity substantially but also show an improved cycling stability by assessing multiple mixtures of high-entropy dopant substitutions for Co [94]. Another report by Yuan et al. also showed an enhancement in capacity and cyclic stability with a coating layer of HEO $((La_{0.2}Nd_{0.2}Sm_{0.2}Eu_{0.2}Gd_{0.2})_2Zr_2O_7)$ on $LiNi_{0.8}Co_{0.1}Mn_{0.1}O_2$ (NCM811) [95].

Zhao et al. showed that layered O_3-type $NaNi_{0.12}Cu_{0.12}Mg_{0.12}Fe_{0.15}Co_{0.15}Mn_{0.1}Ti_{0.1}Sn_{0.1}Sb_{0.04}O_2$ demonstrates exceptional rate capability (ca. 80 % hold at 5.0C rate) and prolonged stable cycling (ca. 83 % holding for 500 repetitions). This Na-ion cathode's high-entropy allows it to adapt to changes in immediate interactions during Na^+ (de)intercalation [96]. Similarly, a report by Wang et al. demonstrated an important study in which an Na-deficient O_3-type cathode $Na_{0.83}Li_{0.1}Ni_{0.25}Co_{0.2}Mn_{0.15}Ti_{0.15}Sn_{0.15}O_{2-\delta}$, having a larger O–Na–O slab separation, leads to improved diffusion kinetics and structural rigidity. Moreover, the entropy-tuned cathode notably demonstrates remarkable rate capability (~77% capacity hold at 10C), stability (~87% hold for 200 repetitions), optimum full-cell results (~84% capacity hold for 100 repeating cycles), and outstanding air stability [79]. Another interesting study by Yao et al. showed a design ($Na_{2/3}Li_{1/6}Fe_{1/6}Co_{1/6}Ni_{1/6}Mn_{1/3}O_2$) featuring superlattice stabilization that can improve the performing rate of layered cathodes and raise their cycle life. It has excellent properties such as energy density (531 Wh kg^{-1}), reversible capacity (~171 mA h g^{-1} at 0.1C), quick charging (78 mA h g^{-1} at 10C), and stability (~89% retention in capacity at 1C at 90 repetitive cycles) [97].

For practical Li-S batteries, Tian et al. proposed a method of producing extremely potent hosts while considering cathode density, Lithium Poly Sulphides entrapment, as well as electrocatalytic performance into account. Incorporating HEO nanofibers ($(Mg_{0.2}Mn_{0.2}Co_{0.2}Ni_{0.2}Zn_{0.2})Fe_2O_4$) as a carbon-free sulfur catalytic host may improve the slow electrochemical kinetics of sulfur redox and help Li-S batteries achieve better energy densities. Thus, the S/HEO composite could obtain a rate efficiency high up to ~632 mA h g^{-1} obtained at 5C rate, strong stability for 500 continuous cycles at 1C rate, along with a large specific capacity, i.e., 1368.7 mA h g^{-1} at 0.1C rate [98]. A single-phase HEO $(Ni_{0.2}Co_{0.2}Cu_{0.2}Mg_{0.2}Zn_{0.2})O$ or HEO850 is synthesized by Raza et al. using a self-sacrificial template of MOFs) and a unique low-temperature annealing technique. This was subsequently incorporated into the sulfur host. At 0.5C, it produces an excessive initial specific discharge that is around 1,244 mA h g^{-1}, and after about 800 redox cycles, it remains about 784 mA h g^{-1} with little deterioration [99]. $La_{0.8}Sr_{0.2}(Cr_{0.2}Mn_{0.2}Fe_{0.2}Co_{0.2}Ni_{0.2})O_3$ (HE-LSMO), a high-entropy perovskite oxide, demonstrates quick electrochemical kinetics of liquid–solid transformation activities. It achieves a large areal capacity (~7 mA h cm^{-2}) in addition to superior cycling capability at 0.1C rate. A good performance was obtained for C/10 to C/2 rates, while an extra 300 cycling led to a retaining capacity of up to 80%, also demonstrated by the ideal sulfur cathode by the HEO $(CrMnFeNiMg)_3O_4$. Moreover, despite the long cycles, the cathode exhibits a strong peak and reversible discharging rates of 857 and 552 mA h g^{-1} cycled at a rate of C/10 [100].

8.1.5 Dielectric Capacitor

DCs find extensive application in a variety of categories, including distributed power networks, microwave communication, and hybrid electric cars as a key unit of high-power electronic amenities. This is due to DCs' excellent power density and very fast charge–discharge rates; they retain energy by displacing bonded charged components as opposed to using chemical reactions like batteries and solid oxide fuel cells do. The DCs' recoverable energy storage density (W_{rec}) can be estimated using the formula, $W_{rec} = \int_{P_r}^{P_m} EdP$, where E represents the applied electric field, P_m equals the maximum polarization, while P_r represents the remnant polarization. However, ceramic capacitors' poor W_{rec} typically $<4\,J\,cm^{-3}$ largely hinders their range of applications and their advancement toward smart devices and miniaturization. During the charge-discharge operation, a significant amount of the energy that has been stored can be released that could lead to a high energy storage efficiency (η). The design of efficient lead-free energy-storing capacitors accompanied with very high W_{rec} and η offers tremendous market potential in addition to environmental sustainability [18]. A set of $(1-x)Bi_{0.5}(Na_{0.9}Li_{0.1})_{0.5}TiO_{3-x}Sr(Sc_{0.5}Nb_{0.25}Ta_{0.25})O_3$ ceramics-based high-entropy DCs was made by Wang et al., demonstrating a high η (90%) and an outstanding W_{rec} of $7.6\,J\,cm^{-3}$ at an electric field ranging to $495\,kV\,cm^{-1}$ [101]. Effective lowering of the dimension of polar nano-regions of the ceramics through combining high-entropy and local lattice deformation as well as ion disorder could lead to obtain high η. Furthermore, the addition of these three ions is intended to enhance the E_b and W_{rec} of ceramics because of the broad bandgap of the corresponding oxides of Al^{3+}, Zr^{4+}, and Nb^{5+}. A high-entropy dielectric material has been produced in the study by Zuo et al. with the formula $(Sr_{0.6}Bi_{0.2}Na_{0.2})$ $(Ti_{1-x}Zr_{x/2}Al_{x/4}Nb_{x/4})O_3$ (SBNTZAN, x = 0, 0.15, 0.20, 0.25, 0.30) that achieves a significant η of ~88.52% and a W_{rec} of ~$4.46\,J\,cm^{-3}$ [102]. Recently, Huan et al., in order to increase the W_{rec} and η together, added $Sr(Ti_{0.5}Zr_{0.5})O_3$ and $Bi(Zn_{0.5}(Ta_{0.75}Nb_{0.25})_{0.5})O_3$ as high-entropy parts to the $(K_{0.44}La_{0.02})Na_{0.5}NbO_3$ system. Eventually, they succeeded in attaining a remarkable energy storing capability i.e., a high η of 87.1%, a W_{rec} of $11.14\,J\,cm^{-3}$, and a significantly high E_b of $750\,kV\,cm^{-1}$ [103].

8.1.6 Supercapacitor

SCs, which are identified by their quick charge as well as discharge rates, superb power density, superior cycle stability, broad operating temperature range, and environmental friendliness, are a potential development in energy storage technology. Research on metal oxides, metal hydroxides, and metal sulfides—mostly as positive electrodes—has significantly increased in the context of materials (electrodes) practiced in energy storage devices. The built-in potential of these materials to develop and advance the next phase of powerful energy storage systems has garnered interest extensively. The large configurational entropy of HEMs makes it possible to integrate several elements into a single crystal lattice, offering exciting opportunities in material development and performance improvement [104]. Sreenivasulu et al. assessed the electrochemical activity of an SP-HEO- $(Al_{0.2}Co_{0.2}Cr_{0.2}Mn_{0.2}Ni_{0.2})_3O_4$ that is nanocrystalline in nature. The maximum specific capacity of $318\,mA\,h\,g^{-1}$ is delivered at a current density of $1\,A\,g^{-1}$, as demonstrated by the material [25]. Li et al. used polyacrylamide gel method for making $(Co_{0.2}Cr_{0.2}Fe_{0.2}Mn_{0.2}Ni_{0.2})_3O_4$ nanopowders in order to reduce excessive grain coarsening and aggregation of particles during annealing. An HEO-activated carbon as an asymmetric SC was also designed, and it produced a power density of 746 W kg^{-1} and an energy density of 24 Wh kg^{-1} [83]. A three-dimensional HEO composite $(CoCrFeMnNi)_3O_4$@Ni, with porous nature electrodes prepared by Wei et al., demonstrated specific capacitances of 942 F g^{-1} at 1 A g^{-1} current density while maintained 60% of its specific capacity when the current density reached to 5 A g^{-1}. could Moreover, the noted values for power density and energy density for HEO@Ni//AC@NF were 825 W kg^{-1} and ~37 Wh kg^{-1} and could retain a capacitance of ~90% for 10k cycles [105].

References

1 Larcher, D. and Tarascon, J.M. (2015). Towards greener and more sustainable batteries for electrical energy storage. *Nat Chem* 7 (1): 19–29.

2 Yeh, J.W., Chen, S.K., Lin, S.J. et al. (2004). Nanostructured high-entropy alloys with multiple principal elements: novel alloy design concepts and outcomes. *Adv Eng Mater* 6 (5): 299–303.

3 Cantor, B., Chang, I.T.H., Knight, P., and Vincent, A.J.B. (2004). Microstructural development in equiatomic multicomponent alloys. *Mater Sci Eng A* 375–377: 213–218.

4 Yeh, J.W. (2013). Alloy design strategies and future trends in high-entropy alloys. *JOM* 65 (12): 1759–1771.

5 Amiri, A. and Shahbazian-Yassar, R. (2021). Recent progress of high-entropy materials for energy storage and conversion. *J Mater Chem A* 9 (2): 782–823.

6 Panda, P.K., Huang, H.Y., Dash, P. et al. (2025). Liquid-phase microwave synthesis of platinum-based high-entropy alloy catalysts on carbon supports for electrochemical hydrogen adsorption/desorption and oxygen evolution/reduction reactions. *Int J Hydrog Energy* 111: 536–545.

7 Gu, S., Huang, H.Y., Lin, Y.C. et al. (2025). High-entropy alloys with low platinum content: novel catalysts for oxygen reduction and evolution reactions. *Emerg Mater* 1–14.

8 Sarkar, A., Breitung, B., and Hahn, H. (2020). High entropy oxides: The role of entropy, enthalpy and synergy. *Scr Mater* 187: 43–48.

9 Miracle, D.B. and Senkov, O.N. (2017). A critical review of high entropy alloys and related concepts. *Acta Mater* 122: 448–511.

10 Sarkar, A., Wang, Q., Schiele, A. et al. (2019). High-entropy oxides: fundamental aspects and electrochemical properties. *Adv Mater* 31 (26): 1806236.

11 Spiridigliozzi, L., Ferone, C., Cioffi, R., and Dell'Agli, G. (2021). A simple and effective predictor to design novel fluorite-structured high entropy oxides (HEOs). *Acta Mater* 202: 181–189.

12 Tseng, K.P., Yang, Q., McCormack, S.J., and Kriven, W.M. (2020). High-entropy, phase-constrained, lanthanide sesquioxide. *J Am Ceram Soc* 103 (1): 569–576.

13 Pressley, L.A., Torrejon, A., Phelan, W.A., and McQueen, T.M. (2020). Discovery and single crystal growth of high entropy pyrochlores. *Inorg Chem* 59 (23): 17251–17258.

14 Song, T., Chen, L., Gastol, D. et al. (2022). High-voltage stabilization of O_3-type layered oxide for sodium-ion batteries by simultaneous tin dual modification. *Chem Mater* 34 (9): 4153–4165.

15 Tomboc, G.M., Zhang, X., Choi, S. et al. (2022). Stabilization, characterization, and electrochemical applications of high-entropy oxides: critical assessment of crystal phase–properties relationship. *Adv Funct Mater* 32 (43): 2205142.

16 Sen, S., Palabathuni, M., Ryan, K.M., and Singh, S. (2024). High entropy oxides: mapping the landscape from fundamentals to future vistas. *ACS Energy Lett* 9 (8): 3694–3718.

17 Li, Y., Qiu, Y., Yao, G. et al. (2025). Dual-phase (AlMgCoNiCuZn)O high-entropy oxide embedded in graphite nanosheets with superior lithium storage capability. *J Energy Storage.* 115: 115993.

18 Chen, L., Deng, S., Liu, H. et al. (2022). Giant energy-storage density with ultrahigh efficiency in lead-free relaxors via high-entropy design. *Nat Commun.* 13 (1): 3089.

19 Talluri, B., Aparna, M.L., Sreenivasulu, N. et al. (2021). High entropy spinel metal oxide $(CoCrFeMnNi)_3O_4$ nanoparticles as a high-performance supercapacitor electrode material. *J Energy Storage* 42: 103004.

20 Nguyen, V.T., Sari, F.N.I., Saputro, B.W., and Ting, J.M. (2023). Structural and defect modulations of co-precipitation synthesized high-entropy Prussian blue analogue nanocubes via Cu/Zn co-doping for enhanced electrochemical performance. *J Mater Chem A.* 11 (36): 19483–19495.

21 Ren, Y., Cai, J., and Pitsch, H. (2021). Theoretical single-droplet model for particle formation in flame spray pyrolysis. *Energy Fuel* 35 (2): 1750–1759.

22 Bian, W., Li, H., Zhao, Z. et al. (2023). Entropy stabilization effect and oxygen vacancy in spinel high-entropy oxide promoting sodium ion storage. *Electrochim Acta.* 447: 142157.

23 Patra, J., Nguyen, T.X., Panda, A. et al. (2025). Yolk-shell-structured $(CrMnFeCoNi)_3O_4$ high-entropy oxide anode for high-performance lithium-ion batteries. *Electrochim Acta* 533: 146545.

24 Hota, A., Das, J.K., Mohammed, A. et al. (2025). Temperature-mediated phase control in high entropy transition metal oxides for hybrid supercapacitors. *N J Chem* 49 (19): 7824–7832.

25 Sreenivasulu, N., Kumar, U.N., Madhav, K.M.V.V. et al. (2022). Structural and electrochemical investigations on nanocrystalline high entropy spinel oxides for battery-like supercapacitor applications. *ChemistrySelect* 7 (5): e202104015.

26 Mostafaei, J., Çoruh, A., Asgari, A. et al. (2024). Supercapacitor application of Sr-doped $LaCo_{0.2}Fe_{0.8}O_3$ perovskite oxide: Investigation of structural and electrochemical properties. *J Energy Storage* 93: 112394.

27 Niculescu, G.E., Bejger, G.R., Barber, J.P. et al. (2025). Local structure maturation in high entropy oxide (Mg,Co,Ni,Cu,Zn)1-(Cr,Mn)O thin films. *Journal of the American Ceramic Society.* 108 (2): e20171.

28 Zhang, D., Xu, S., Li, T. et al. (2023 Feb 24). High-Entropy oxides prepared by Dealloying method for supercapacitors. *ACS Appl Eng Mater.* 1 (2): 780–789.

29 Pei, L., Kang, J., He, C. et al. (2025). Designing high reversibility free-standing hierarchical porous high-entropy oxide as anodes for advanced Lithium-Ion batteries. *Adv Funct Mater.* 35 (16): 2422809.

30 Ning, Y., Pu, Y., Zhang, Q. et al. (2024). Improved energy storage capacity of high-entropy ferroelectric perovskite ceramic via flash sintering. *J Power Sources* 618: 235205.

31 Zhou, S., Pu, Y., Zhang, X. et al. (2022). High energy density, temperature stable lead-free ceramics by introducing high entropy perovskite oxide. *Chem Eng J* 427: 131684.

32 Zhang, L., Li, H., Zhang, X. et al. (2025). Ultra-rapid synthesis of high-entropy MAX phases and their derivative MXenes for battery electrodes. *Angew Chem Int Ed* 137 (6): e202418538.

33 Ma, Z., Yang, Q., Meng, B. et al. (2024). Dielectric and energy storage properties of $(Ba_{0.2}Sr_{0.2}Na_{0.2}Ca_{0.2}La_{0.2})(Zr_xTi_{1-x})O_3$ high-entropy ceramics and thin films. *Int J Appl Ceram Technol* 21 (5): 3422–3434.

34 Chen, B., Zhang, X.L., Zhang, W.B. et al. (2024). Electrochemical pseudocapacitance performance of high entropy carbide (FeCoCrMnNi)C film by magnetron sputtering technology. *Ceram Int* 50 (13, Part A): 23302–23314.

35 Yang, Z., Huang, R., Zhang, J. et al. (2024). Engineering multi-ion doping by entropy for high energy storage density with high efficiency in amorphous thin film. *Ceram Int* 50 (19, Part B): 36375–36382.

36 He, Q., Li, J., Liu, W. et al. (2024). High entropy oxides for electrochemical energy storage and conversion: A critical review. *J Power Sources* 619: 235207.

37 Ji, X., Lei, P., Chen, J. et al. (2025). Synthesis and irradiation behavior of $(Gd_{0.2}Sm_{0.2}Dy_{0.2}Er_{0.2}Yb_{0.2})_2Ti_2O_7$ high-entropy pyrochlore waste forms consolidated by spark plasma sintering. *Vacumm* 232: 113893.

38 Tyagi, P.K. and Jha, S.K. (2024). Flash sintering improves the densification and mechanical properties of high entropy oxides used in anodic applications. *Materialia* 33: 101967.

39 Manchón-Gordón, A.F., Lobo-Llamas, C., Molina-Molina, S. et al. (2024). Influence of the atmosphere on the formation of high-entropy oxides within the Co-Cu-Fe-Mg-Mn-Ni-O system via reactive flash sintering. *Ceram Int* 50 (21, Part B): 42276–42285.

40 Xu, C., He, J., Chen, Z. et al. (2024). Two-step reactive flash sintering of high entropy pyrochlore: Densification, grain growth, and microstructural homogeneity. *J Am Ceram Soc* 107 (2): 797–802.

41 Manchón-Gordón, A.F., Almanza-Vergara, G.E., Molina-Molina, S. et al. (2024). Structural, Mössbauer and magnetic study of $(Mn_{0.2}Co_{0.2}Ni_{0.2}Cu_{0.2}X_{0.2})Fe_2O_4$ (X = Fe,Mg) spinel high-entropy oxides fabricated via reactive flash sintering. *J Eur Ceram Soc* 44 (14): 116686.

42 Brandt, T.G., Tuokkola, A.R., Yu, M., and Laine, R.M. (2023). Liquid-feed flame spray pyrolysis enabled synthesis of Co- and Cr-free, high-entropy spinel oxides as Li-ion anodes. *Chem Eng J* 474: 145495.

43 Luo, L., Ju, J., Wu, Y. et al. (2025). Lattice-strain engineering of high-entropy-oxide nanoparticles: regulation by flame spray pyrolysis with ultrafast quenching. *Adv Mater* 37 (10): 2418856.

44 Kamecki, B., Karczewski, J., Cempura, G. et al. (2023). Evaluation of structural and electrical properties of multicomponent spinel oxide thin films deposited via spray pyrolysis technique. *Mater Charact* 203: 113097.

45 Luo, L., Ju, J., Xi, M. et al. (2024). The micron-droplet-confined continuous-flow synthesis of freestanding high-entropy-alloy nanoparticles by flame spray pyrolysis. *Small* 20 (36): 2401360.

46 Wang, Y., Liu, J., Cao, H. et al. (2025). Facile synthesis of porous high-entropy perovskite nanoparticles through MOF gel method for solid-state supercapacitor application. *Chem Eng J* 509: 161246.

47 Yin, Y., Zhang, W.B., Zhang, X.L. et al. (2023). Low-dimensional high entropy oxide $(FeCoCrMnNi)_3O_4$ for supercapacitor applications. *Dalton Trans* 52 (26): 9005–9016.

48 Chillal, K., Halder, A., Jha, S.K. et al. (2025). Electrochemical investigation of synthesized $(Mg_{0.21}Cr_{0.21}Mn_{0.21}Fe_{0.21}Cu_{0.16})_3O_4$ high entropy oxide for supercapacitor electrode material. *Ceram Int* 51 (3): 3890–3898.

49 Gupta, A.K., Singh, A., Kumari, P. et al. (2024). Effect of synthesis routes on electrochemical properties as supercapacitor electrode for novel spinel (CuNiFeMnCo)O high entropy oxide. *Energy Storage* 6 (1): e538.

50 Wang, D., Duan, C., He, H. et al. (2023 Sep). Microwave solvothermal synthesis of Component-Tunable High-Entropy oxides as High-Efficient and stable electrocatalysts for oxygen evolution reaction. *Journal of Colloid and Interface Science.* 15 (646): 89–97.

51 Li, Y., Chen, Z., Liu, J. et al. (2023 Dec 1). Novel high entropy oxide as anode for high performance lithium-ion capacitors. *Ceramics International.* 49 (23, Part B): 38439–38447.

52 Wang, X., Song, Y., Fan, Y. et al. (2025 Jan). Accelerating the screening of high-entropy Co-Mn-Fe-Ni-Zn-O gradient films for highly stable NTC materials through co-sputtering combinatorial synthesis. *Applied Surface Science.* 15 (679): 161306.

53 Naveed, A.K., Akhavan, B., Zheng, Z. et al. (2021 Jul). Nanostructured AlCoCrCu0.5FeNi high entropy oxide (HEO) thin films fabricated using reactive magnetron sputtering. *Applied Surface Science.* 1 (553): 149491.

54 Zenkin, S., Bulakh, V., Uglov, V. et al. (2024 Sep 15). Influence of the elemental composition on the structural and mechanical characteristics of entropy-stabilized HfZrCeYMgAlO thin films prepared by reactive magnetron sputtering. *Ceramics International.* 50 (18, Part A): 33093–33097.

55 Guo, H., Wang, X., Dupuy, A.D. et al. (2022 Jan 1). Growth of nanoporous high-entropy oxide thin films by pulsed laser deposition. *Journal of Materials Research* 37 (1): 124–135.

56 Wang, K., Nishio, K., Horiba, K. et al. (2022 Feb 2). Synthesis of high-entropy layered oxide epitaxial thin films: $LiCr_{1/6}Mn_{1/6}Fe_{1/6}Co_{1/6}Ni_{1/6}Cu_{1/6}O_2$. *Crystal Growth & Design* 22 (2): 1116–1122.

57 Momma, K. and Izumi, F. (2011 Dec 1). VESTA 3 for three-dimensional visualization of crystal, volumetric and morphology data. *J Appl Cryst.* 44 (6): 1272–1276.

58 Pasupathi, A., Perumal, M., Narayanamoorthi, E. et al. (2024 Aug 1). Electrochemical charge storage performance of (Mn, Ni, Mo, Co, Fe)$_3$O$_4$ high entropy oxide nanoparticles produced via thermal plasma route. *Ceramics International.* 50 (15): 26740–26749.

59 Pasupathi, A. and Subramaniam, Y. (2024 Mar 21). A novel strategy for rapid synthesis of nanostructured high-entropy metal oxides through thermal plasma for supercapacitor applications. *Energy Fuels* 38 (6): 5534–5544.

60 Amarnath, P., Madhu, R., Praveen, K. et al. (2023 Jun 12). Phase-pure high-entropy spinel oxide (Ni,Fe,Mn,Cu,Zn)$_3$O$_4$ via thermal plasma: A promising electrocatalyst for oxygen evolution reaction. *ACS Appl Energy Mater.* 6 (11): 5899–5911.

61 Pasupathi, A., Madhu, R., Kundu, S., and Subramaniam, Y. (2024 Sep). Effect of carbon infusion on water splitting performance of $(Ni_{0.2}Co_{0.2}Cr_{0.2}Mn_{0.2}V_{0.2})_3O_4$ high entropy oxides nanoparticle synthesized via thermal plasma. *Electrochimica Acta.* 1 (497): 144621.

62 An, Y., Tian, Y., Wei, C. et al. (2021 Apr). Dealloying: An effective method for scalable fabrication of 0D, 1D, 2D, 3D materials and its application in energy storage. *Nano Today.* 1 (37): 101094.

63 Jin, Z., Lyu, J., Zhao, Y.L. et al. (2021 Mar 9). Top–Down synthesis of noble metal particles on high-Entropy oxide supports for electrocatalysis. *Chem Mater.* 33 (5): 1771–1780.

64 Zhang, M., Zhang, L., Liaw, P.K. et al. (2018 Oct 1). Effect of Nb content on thermal stability, mechanical and corrosion behaviors of hypoeutectic CoCrFeNiNbχ high-entropy alloys. *Journal of Materials Research.* 33 (19): 3276–3286.

65 Lin, Y., Xu, W., Gao, Z. et al. (2024). Self-supporting high-entropy Co-Cr-Fe-Ni-Nb oxide electrocatalyst with nanoporous structure for oxygen evolution reaction. *Chem Eng J.* 489: 151233.

66 Kirsch, A., Bøjesen, E.D., Lefeld, N. et al. (2023). High-entropy oxides in the mullite-type structure. *Chem Mater* 35 (20): 8664–8674.

67 Song, B., Yang, Y., Rabbani, M. et al. (2020). In situ oxidation studies of high-entropy alloy nanoparticles. *ACS Nano* 14 (11): 15131–15143.

68 Su, L., Ren, J., Lu, T. et al. (2023). Deciphering structural origins of highly reversible lithium storage in high entropy oxides with in situ transmission electron microscopy. *Adv Mater* 35 (19): 2205751.

69 Liu, D., Guo, P., Pan, H., and Wu, R. (2024). Emerging high-entropy compounds for electrochemical energy storage and conversion. *Prog Mater Sci* 145: 101300.

70 Wang, K., Hua, W., Huang, X. et al. (2023). Synergy of cations in high entropy oxide lithium-ion battery anode. *Nat Commun* 14 (1): 1487.

71 Xiao, B., Wu, G., Wang, T. et al. (2022). High-entropy oxides as advanced anode materials for long-life lithium-ion batteries. *Nano Energy* 95: 106962.

72 Ren, R., Xiong, Y., Xu, Z. et al. (2024). Fast synthesis of high-entropy oxides for lithium-ion storage. *Chem Eng J* 479: 147896.

73 Qiu, N., Chen, H., Yang, Z. et al. (2019). A high entropy oxide ($Mg_{0.2}Co_{0.2}Ni_{0.2}Cu_{0.2}Zn_{0.2}O$) with superior lithium storage performance. *J Alloys Compd* 777: 767–774.

74 Lu, Y., Kang, Q., Dong, F. et al. (2025). Metalloid phosphorus induces tunable defect engineering in high entropy oxide toward advanced lithium-ion batteries. *Adv Funct Mater* 35 (3): 2413782.

75 Patra, J., Nguyen, T.X., Tsai, C.C. et al. (2022). Effects of elemental modulation on phase purity and electrochemical properties of Co-free high-entropy spinel oxide anodes for lithium-ion batteries. *Adv Funct Mater* 32 (17): 2110992.

76 Ji, S., Cui, Y., Lu, T., and Xi, J. (2025). High-entropy oxide heterostructure-boosted bidirectional electrocatalysis in lithium–sulfur batteries. *ACS Appl Mater Interfaces* 17 (14): 21084–21096.

77 Gui, X., Xiang, Z., Ren, T. et al. (2025). High-entropy doped P'2 Mn-based layered oxide with superior stability and high capacity for sodium-ion batteries. *Adv Mater* 37 (20): 2417008.

78 Yue, J., Xiong, F., Shadike, Z. et al. (2025). A layer-structured high entropy oxide with highly reversible Fe^{3+}/Fe^{4+} redox as advanced cathode material for sodium ion batteries. *J Power Sources* 627: 235735.

79 Wang, H., Gao, X., Zhang, S. et al. (2023). High-entropy Na-deficient layered oxides for sodium-ion batteries. *ACS Nano* 17 (13): 12530–12543.

80 Bai, M., Qiao, W., Mei, J. et al. (2024). High-performance energy storage in $BaTiO_3$-based oxide ceramics achieved by high-entropy engineering. *J Alloys Compd* 970: 172671.

81 Zhi Wei, L., Chen, Z.-H., and Xu, J.J. (2022). Enhanced energy storage performance of $BaTi_{0.97}Ca_{0.03}O_{2.97}$-based ceramics by doping high-entropy perovskite oxide. *J AlloysCompd* 922: 166179.

82 Zhu, X., Xiong, S., Zhu, G. et al. (2024). Dielectric properties and excellent energy storage density under low electric fields for high entropy relaxor ferroelectric $(Li_{0.2}Ca_{0.2}Sr_{0.2}Ba_{0.2}La_{0.2})TiO_3$ ceramic. *J Alloys Compd* 984: 173987.

83 Li, L., Ji, P., Geng, C. et al. (2023). Facile synthesis of high-entropy $(Co_{0.2}Cr_{0.2}Fe_{0.2}Mn_{0.2}Ni_{0.2})_3O_4$ nanopowders and their electrochemical properties as supercapacitor electrode. *J Energy Storage* 73: 109182.

84 Wang, Q., Sarkar, A., Li, Z. et al. (2019). High entropy oxides as anode material for Li-ion battery applications: a practical approach. *Electrochem Commun* 100: 121–125.

85 Hou, S., Su, L., Wang, S. et al. (2024). Unlocking the origins of highly reversible lithium storage and stable cycling in a spinel high-entropy oxide anode for lithium-ion batteries. *Adv Funct Mater* 34 (4): 2307923.

86 Ghigna, P., Airoldi, L., Fracchia, M. et al. (2020). Lithiation mechanism in high-entropy oxides as anode materials for Li-ion batteries: an operando XAS study. *ACS Appl Mater Interfaces* 12 (45): 50344–50354.

87 Liu, T., Yu, L., Liu, J. et al. (2021). Understanding Co roles towards developing Co-free Ni-rich cathodes for rechargeable batteries. *Nat Energy* 6 (3): 277–286.

88 Yang, X., Wang, H., Song, Y. et al. (2022). Low-temperature synthesis of a porous high-entropy transition-metal oxide as an anode for high-performance lithium-ion batteries. *ACS Appl Mater Interfaces* 14 (23): 26873–26881.

89 Gao, S., He, H., Zhai, F. et al. (2025). Hollow graphene microsphere in situ deposited with rock-salt-type high-entropy oxide nanoparticles for enhanced rate performance as anodes in lithium-ion batteries. *ACS Appl Energy Mater* 8 (3): 1896–1907.

90 Hong, C., Tao, R., Tan, S. et al. (2025). In situ cyclized polyacrylonitrile coating: key to stabilizing porous high-entropy oxide anodes for high-performance lithium-ion batteries. *Adv Funct Mater* 35 (2): 2412177.

91 Kim, J., Yang, S., Zhong, Y. et al. (2025). High-entropy Li-rich layered oxide cathode for Li-ion batteries. *J Power Sources* 628: 235915.

92 Huang C, Luo J, R. Mansley Z, Kingan A, Campos AR, Wang Z, et al. Manganese-rich high entropy oxides for lithium-ion batteries: materials design approaches to address voltage fade. *Journal of Materials Chemistry A.* 2024;12(38):26253–65.

93 Zheng, Q., Ren, Z., Zhang, Y. et al. (2023). Surface phase conversion in a high-entropy layered oxide cathode material. *ACS Appl Mater Interfaces* 15 (3): 4643–4651.

94 Liang, P., Qi, K., Chen, S. et al. (2024). Low-electronegativity cationic high-entropy doping to trigger stable anion redox activity for high-Ni Co-free layered cathodes in Li-ion batteries. *Angew Chem Int Ed* 63 (10): e202318186.

95 Yuan, K., Tu, T., Shen, C. et al. (2022). Self-ball milling strategy to construct high-entropy oxide coated $LiNi_{0.8}Co_{0.1}Mn_{0.1}O_2$ with enhanced electrochemical performance. *J Adv Ceram* 11 (6): 882–892.

96 Zhao, C., Ding, F., Lu, Y. et al. (2020). High-entropy layered oxide cathodes for sodium-ion batteries. *Angew Chem Int Ed* 59 (1): 264–269.

97 Yao, L., Zou, P., Wang, C. et al. (2022). High-entropy and superstructure-stabilized layered oxide cathodes for sodium-ion batteries. *Adv Energy Mater* 12 (41): 2201989.

98 Tian, L., Zhang, Z., Liu, S. et al. (2022). High-entropy spinel oxide nanofibers as catalytic sulfur hosts promise the high gravimetric and volumetric capacities for lithium–sulfur batteries. *Energy Environ Mater* 5 (2): 645–654.

99 Raza, H., Cheng, J., Lin, C. et al. (2023). High-entropy stabilized oxides derived via a low-temperature template route for high-performance lithium-sulfur batteries. *EcoMat* 5 (4): e12324.

100 Chung, S.H., Wu, Y.H., Tseng, Y.H. et al. (2023). High entropy oxide $(CrMnFeNiMg)_3O_4$ with large compositional space shows long-term stability as cathode in lithium-sulfur batteries. *ChemSusChem* 16 (8): e202300135.

101 Wang, Z., Kang, R., Zhang, L. et al. (2023). Ultrahigh energy-storage capacity achieved in $(Bi_{0.5}Na_{0.5})TiO_3$-based high-entropy dielectric capacitors with linear-like polarization response. *Chem Eng J* 474: 145506.

102 Zuo, C., Yang, S., Cao, Z. et al. (2023). Combustion synthesis of high-performance high-entropy dielectric ceramics for energy storage applications. *Ceram Int* 49 (15): 25486–25494.

103 Huan, Y., Wu, L., Xu, L. et al. (2025). Superior energy-storage density and ultra-high efficiency in KNN-based ferroelectric ceramics *via* high-entropy design. *J Mater* 11 (1): 100862.

104 Bao, W., Shen, H., Zhang, Y. et al. (2024). High-entropy oxides for energy storage and conversion. *J Mat Chem A* 12 (35): 23179–23201.

105 Wei, L.Z., Shi, S.J., Guo, M.Y. et al. (2025). Self-supporting high-entropy-oxide@layered-Ni electrode for high-performance supercapacitors. *J Energy Storage* 110: 115351.

106 Krawczyk, P.A., Jurczyszyn, M., Pawlak, J. et al. (2020 Oct 27). High-entropy perovskites as multifunctional metal oxide semiconductors: synthesis and characterization of $(Gd_{0.2}Nd_{0.2}La_{0.2}Sm_{0.2}Y_{0.2})CoO_3$. *ACS Appl Electron Mater.* 2 (10): 3211–3220.

9

Noble Metal-Based High-Entropy Alloys for Energy Storage Applications

Parul Devi

Johannes Gutenberg University Mainz, Institute of Physics, Mainz 55128, Germany

9.1 Introduction

In 2004, high-entropy alloys (HEAs) were introduced first time which have five or more than five elements for obtaining stable system. To confer desirable properties on materials, alloying term has been used for a long time. A relatively small addition of the second element to the primary element is known as alloying. Recently, a new alloying strategy has come into the picture that involves combining different principal elements on the higher concentration side to create novel materials, known as HEAs [1–4].

Since their recent discovery, HEA acts as a new level of alloy complexes that have fascinated remarkable experimentation attentiveness. A key point is the reliance on high-entropy stabilization alongside the inclusion of several cations (typically five or more) in near-equiatomic proportions, which often leads to the formation of a single-phase solid solution by suppressing the usual enthalpy-driven phase segregation [5–7] (Figure 9.1) [6].

This chapter addresses the extensive compositional span that continues to be investigated and outlines productive paths for identifying the different regions within the space in which HEAs with different, probably fascinating properties may be skulking. HEA originally stemmed from the concept of high-entropy. Entropy is a thermodynamic hypothesis that refers to randomness [8]. HEAs incorporate five or more metallic and non-metallic constituents in roughly equiatomic ratio, in which each element is composed between 5% and 35%.

HEAs are extremely hard and heat-resistant, and their application in catalysis and energy storage makes them promising candidates and demonstrates exciting magnetic properties. To probe the structural disorder and local structure, an elemental-specific technique is needed to scan around each element. Interatomic interactions among different elements can explain the macroscopic properties. Recently, a model system called Cantor alloy has been used to study the local and macroscopic-scale high-entropy effect.

High-Entropy Materials for Energy Storage Devices, First Edition.
Edited by Chien-Te Hsieh, Pradeep Kumar Panda, and Arpan Kumar Nayak.

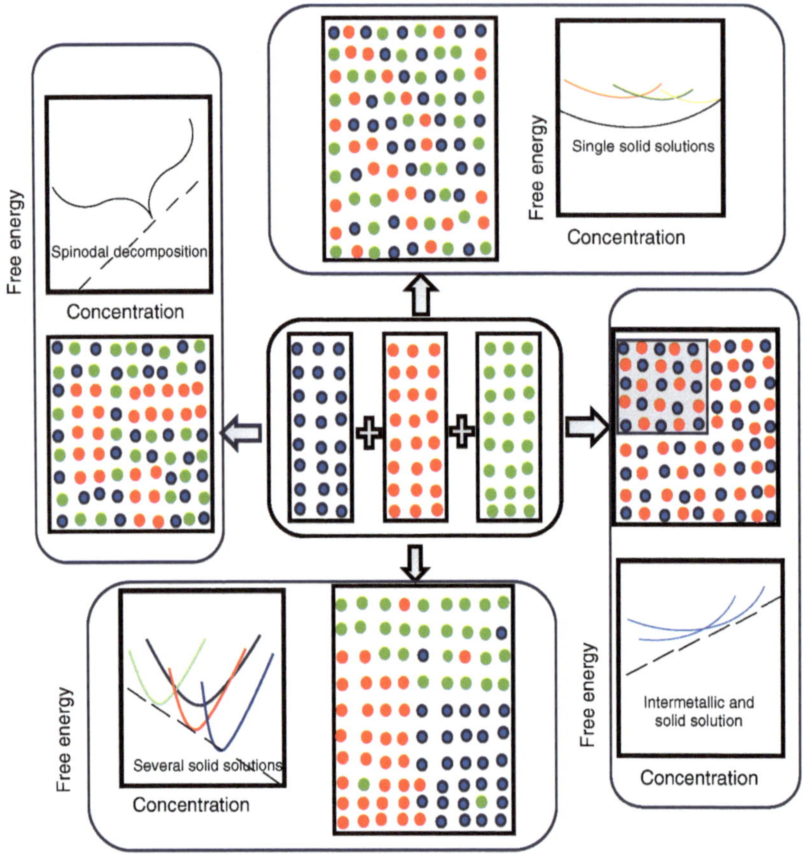

Figure 9.1 Multiple or single solid-solution formation or intermetallic compound precipitation in a solid solution. Some possible reactions occur with three different constituents, as indicated by red, green, and blue. These constituents are mixed in balanced proportions to form different compositions.

To form a stable structure, the Gibbs free energy concept is followed, which is explained by the Gibbs–Helmholtz equation:

$$\Delta G_{mix} = \Delta H_{mix} - T\Delta S_{mix} \tag{9.1}$$

where, ΔG_{mix} represents Gibbs free energy, ΔH_{mix} shows mixing enthalpy, T denotes the absolute temperature, and ΔS_{mix} is the mixing entropy [9–11].

The Hume-Rothery rule explains that the parent-phase crystal structure does not alter after the solute atom is incorporated into a multicomponent solid solution.

Alloys that keep high principal constituents, specifically $n = 5$, contribute to the free energy entropy at the temperature of melting, which is similar to the enthalpy of strong intermetallic compound formation.

$$\Omega = T_m \Delta S_{mix}/\Delta|H_{mix}| \tag{9.2}$$

where the melting temperature is denoted by T_m.

The relative stabilities of the intermetallic compound and solid solution depend on whether $-T\Delta S_{mix}$ is more negative than the enthalpy of formation. In an ideal case, the entropy of mixing is given by

$$\Delta S_{mix} = -R\sum_{i}^{N} c_i \ln c_i \tag{9.3}$$

where the ideal gas constant is denoted by R, and c_i defines the mole number of component i. It gives values of $1.39R$, $1.61R$, and $1.79R$ for an equiatomic alloy which contains 4, 5, and 6 constituents, individually [9–15].

Different entropy values ought to be pre-owned suitably to form and adjust the microstructure, which can deviate from an equiatomic amalgamation. Different microstructure occurs in these three different entropy alloys. Due to non-equiatomic coalescence, polyphase microstructure happens in low-entropy alloys [15, 16, 19]. Medium-entropy alloys also show multiphase microstructure by encompassing equiatomic and non-equiatomic constituents. The inclusion of a principal constituent increases the entropy. Table 9.1 indicates the entropy variation for different entropy alloys reliant upon remarkable constituent addition.

HEA formation follows the Hume-Rothery rule, which is established on different quantifiable variables like electronegativity, difference in atomic size, valence, Young's modulus, density, crystal structure, and boiling point [17–19]. First group elements except H are soft metals and shiny, with a high level of reactivity. During metal sintering and mechanical alloying, these are utilized as inert gases in HEAs. The second group of elements is low-density metals, which produce lightweight HEAs. Groups 3–12 are transition metal elements which have strong electrical and thermal conductivity. They are corrosion-resistant, which blocks corrosion from close-by elements. Inert noble gases preserve the least possible quantity of reaction [20–24].

Non-reactive Lanthanides are F-block elements which have more binding force, which increases the strength and hardness. Their application is to strengthen the heat deformation behavior of HEAs. Actinides are radioactive. Alloying these elements to HEAs enhances individual element performance and strengthens the solid mixture. Different HEAs arising due to the different groups of the periodic table are shown in Figure 9.2 [17].

Simply, the magnetic HEAs can be defined based on their magnetic properties. Some magnetic HEAs exhibit promising low-temperature magnetic properties

Table 9.1 Various entropy values of alloys.

Low-entropy alloy	High-entropy alloy	Medium-entropy alloy
$\Delta S \leq 1.0R$	$1.0R \leq \Delta S \leq 1.5R$	$\Delta S \geq 1.5R$
One or two principal elements	Two to four principal elements	Minimum five principal elements
Solid solutions	Intermetallic compound to construct	Single-phase solid solution

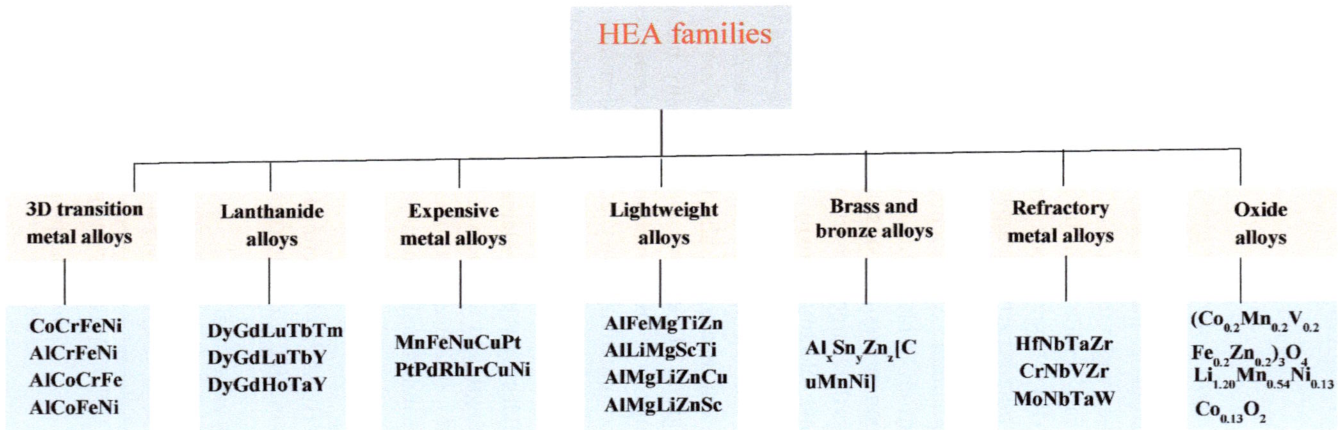

Figure 9.2 Block diagram of HEAs from the different groups of the periodic table [17].

but are non-magnetic at room temperature (RT). A higher value of saturation magnetization, coercivity, and a large exchange bias can be achieved by tuning the alloy composition, which alters the crystal structure of the phases in the resulting alloy. A small substitution of non-magnetic metal results in an increase in exchange interaction and magnetic properties. This happens because alloying results in a new phase and a change in the crystal structure that influences magnetic properties [24–27].

9.2 Synthesis Methods

Flash-Thermal Shock Synthesis: This innovative method uniformly mixes the carbon source, accompanied by metal salt precursors, as regards high temperatures. Generally, more than 2,000 K is used for burning the carbon source, which gives thermal shock. This will reduce the metal salts to metallic atoms. At high temperatures, these particles construct a solid solution which produces HEAs alongside rapid cooling 10^5 K s^{-1}. Different three-dimensional (3D) transition metallic elements, such as Fe, Co, Ni, Pt, Au, Cu, Pd, and Sn, were used to synthesize nanoparticles of HEAs, which validates the viability of this method [28–30].

Uniform distribution and rapid diffusion of metal atoms are observed in this method because of the high-temperature shock for a short duration, which leads to homogeneous compositional alloy formation. Carbon source quantity and type adjust the microstructure and alloy properties. During the heating and cooling process, precise control is required for the structure's stability and alloy formation [31–33]. To remove the impurities and residual carbon, the subsequent treatment footsteps, for example, drying and washing, create a challenge to the comprehensive implementation of the method. The challenge in this method is to achieve uniform heating over large areas, and Joule heating efficiency builds upon the carbon bearer [34].

Fast-Moving Bed Pyrolysis: Phase-segregated alloys are formed owing to the dissimilar initial depletion temperatures of the predecessors in the fast-moving bed pyrolysis method [35]. The phase separation can be prevented by diminishing the analytic nucleus dimension, which reduces the overconsumption of power consumption necessary for the evolution of nanoalloys. This method requires a carrier for the placement of mixed metal precursors, which is quickly thrust inside a high-temperature zone [36–38]. Rapid pyrolysis is used at high temperatures to break down the metal precursors at once. Different reduction potentials avoid step-by-step reduction. It is a versatile method, creating HEAs extended to γ-Al_2O_3 and zeolites, which is useful for the composition range from binary to decenary [39–41].

Two main parameters, i.e. high-temperature treatment and rapid pyrolysis, control this method. It uses graphite and carbon black substrates. Uniformly dispersed small HEA nanoparticles can be created by decomposing metal precursors. For large-scale applications, this necessitates specialized equipment, conditions, an enlightened temperature control organization, and unavoidably power expenditure correlated with high-temperature constraints [42].

Microwave Heating Synthesis: Microwave heating is an effective way as it induces thermal energy. This method is used for the rapid and facile coalescence of HEA nanoparticles. This method can generate rapidly excessive temperatures and maintain the heating throughout the procedure, which can overcome the phase separation matter in HEA nanoparticles. At 570 K, a moderately diminished graphene oxide film in Ar is used as a reference to a dummy reactant. To achieve uniform temperature distribution and a temperature up to 1,850 K within seconds, this method utilizes the uniform heat conduction properties of reduced graphene oxide and microwave absorption [43, 44].

In this method, uniformity and stability of alloy particles are maintained because of the rapid cooling process, which is ~10^4 K s^{-1}. It enables the formation of HEA PtPdFeCoNi with a particle size of ~12 nm from the direct solidification of liquid metals. This technique is validated for various carbon-based materials, for instance, 3D carbonized wood (c-wood) and one-dimensional (1D) carbon nanofibers (CNFs). Different shapes and sizes of HEA nanoparticles can be formed by this method. It is an environmentally friendly, efficient, controllable, and scalable method for the amalgamation of HEA nanoparticles compared to the radiant heating procedure [45] (Figure 9.3).

Laser Scanning Ablation Technology: Laser scanning ablation technology has been used extensively in recent years. It belongs to the physical vapor deposition procedure, which is dependent on laser technology. In the technique mentioned above, a high-energy laser beam evaporates the material from the target, creating plasma composed of ions, clusters, and hot atoms [46–48]. Through the cooling and condensation process, HEA nanoparticles are formed. Recently, this technique has been successfully able to create HEA nanoparticles at atmospheric pressure and room temperature.

To ablate metal precursors, a precise focus of a 5-ns laser pulse is used to create micrometer-sized regions. This technique shows a precise remarkability, varied composition synthesis of alloys on different substrates such as metallic, carbon, and polymeric materials [48–51]. Even though this method has various achievements, it is also confronted with challenges in practical implementation, which

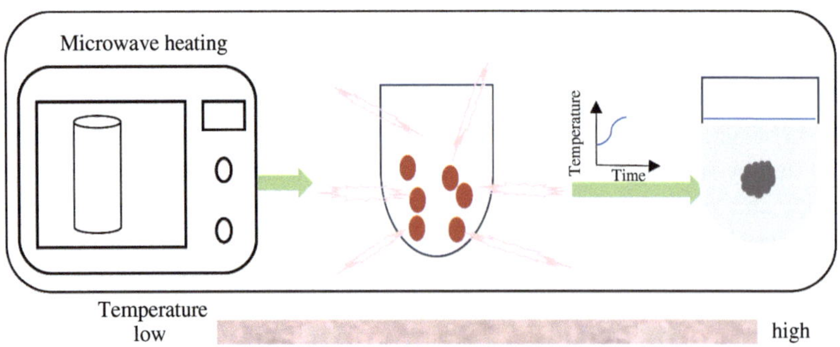

Figure 9.3 Microwave heating process for HEA nanoparticle.

include the need to avoid restrictions enforced by the equipment's pre-owned and over-ablation [52, 53].

Solvothermal Method: This method requires a suitable solvent for immersing the metal precursor and a reductant, for instance triethylene glycol, which is mixed in the company of a protective agent (e.g. PVP [polyvinyl pyrrolidone]). It does not require a high temperature or substrate. Under controlled synthesis (temperature and stirring) circumstances, metal ions are shortened to atomic scattering factors to coalesce, commencing to form HEA particles. By optimizing key reaction parameters, temperature, surfactant, reaction time, and temperature-reducing agent concentration, one can control the size, shape, and composition of HEAs.

This method shows the successful synthesis of PtPdIrRhRu using the low-temperature solvothermal procedure. It overcomes the limitation of the traditional solution-based chemical synthesis method. The average grain size can be changed by varying the enthalpy of melting or the melting point of the constituent metals. A small grain size up to 1.68 nm can be achieved by the synergistic approach. Reducing the particle size increases the surface area, and so the higher their catalytic activity. This procedure permits the precise emergence of different shapes of HEAs. These are 0D nanoparticles, 1D nanowires, 2D ultra-thin nanorings, and 3D nano-dendrites. Recently, HEAs of PtRuRhCoNi, PtPdRuRhCu, and FeCoNiMoW were successfully synthesized [54, 55]. The solvothermal process is shown in Figure 9.4 [56].

Dealloying Synthesis: The dealloying method is a top-down synthesis method that falls inside the domain of corrosion scheme. Due to its simplicity, it is widely pre-owned in the synthesis of high-disorder alloys. This method involves

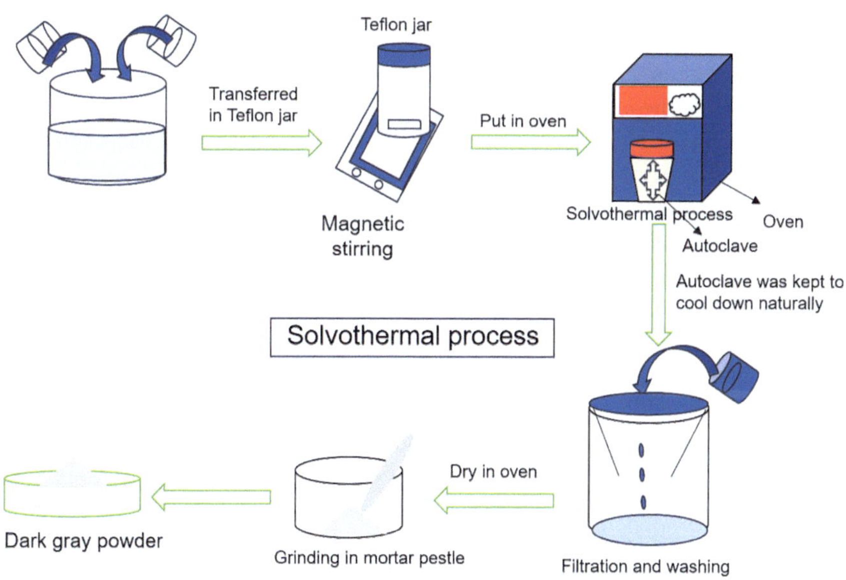

Figure 9.4 Solvothermal process for the formation and separation of HEAs from the solution [56].

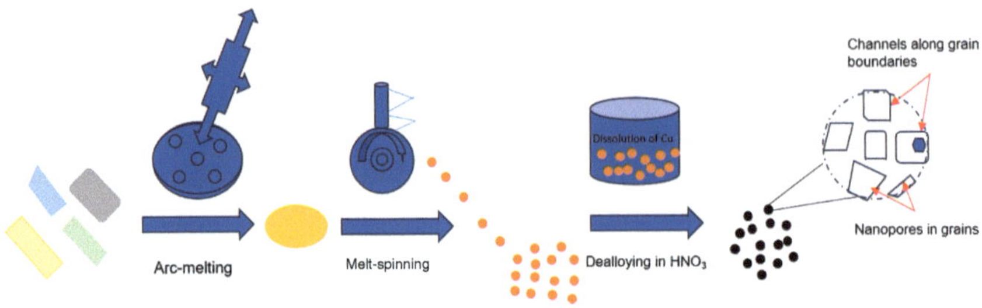

Figure 9.5 Dealloying synthesis showing the formation of quasi-HEA microspheres.

chemically dissolving specific elements to produce nanoporous HEA materials within the synthesized HEAs. Melting casting is an energy-intensive method that requires the preparation of high-entropy metal precursors that still rely on standard techniques. AlNiCuPtPdAu, which is termed a HEA nanoporous, exhibits notable CO oxidation activity and high-temperature stability. This is achieved by combining dealloying and rapid solidification methods to synthesize a multifaceted nanostructured alloy [43–46].

Employed in zinc-air batteries, these deliver an extensive specific capacity of 800 mA h g^{-1} and an open-circuit potential of 1.55 V. By combining different metal elements, a countless number of HEAs were prepared using dealloying methods, which were utilized as catalysts for the hydrogen evolution reaction (HER) and the oxygen evolution reaction (OER) [57]. Each step of the dealloying synthesis process is shown in Figure 9.5 [58].

Arc/Induction Melting: It is a primary method that is used for casting of HEAs, which ensures screening of rapid alloys with high-cost effectiveness, minimum material input, and a very high cooling rate. The required parts of the arc or induction melting are a copper stage, a turbo pump to get the vacuum, water to cool down the process while the melting of the ingot takes place, and an electric power supply. All of them should be inside a chamber. It also requires an inert gas, such as Ar, to remove the oxygen and dust inside the chamber.

First, the copper stage is cleaned with the help of acetone and ethanol. A piece of Ti is also placed inside the chamber at one of the copper stages, because it may contain more than one part. When the melting takes place, a small piece of Ti gets the excess amount of oxygen and prevents the chamber from getting oxidized. The melting takes place with the help of a Titanium tip, which is lowered down to touch the materials. This melted ingot is turned over and melted more than three times to get a uniform ingot. Afterwards, the electric power and water supply are turned off. This technique melts the ingot up to a temperature range of 3,000 °C. Lastly, the melted ingot was taken out of the chamber [55] (Figure 9.6) [59].

Additive Manufacturing: It is a rapid technique to fabricate HEAs, which can produce graded component compositionally from the layer-by-layer deposition. The variation of the feedstock powder feed rate ratio can tune the layer composition. Additive manufacturing (AM) includes the synthesis of magnetic

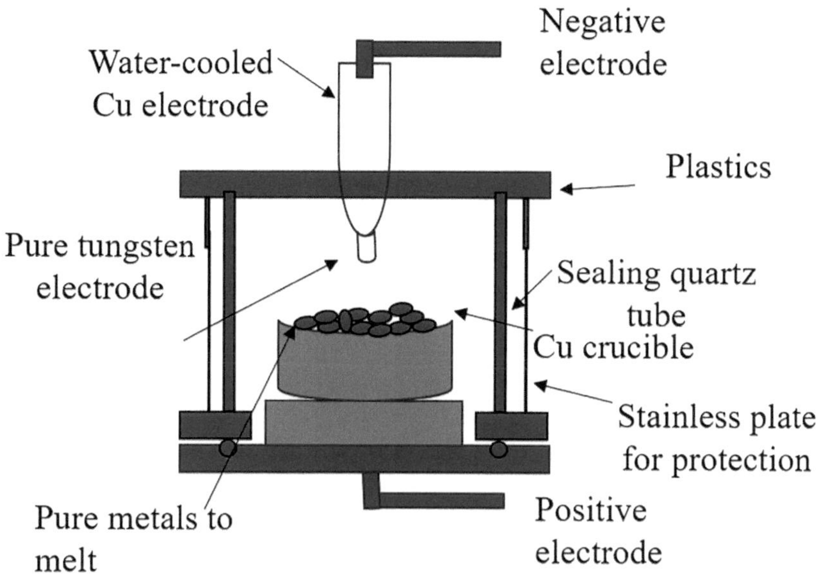

Figure 9.6 Arc melting method for synthesis of HEAs.

compositionally graded alloys such as $NiFe_x$, $CoFe_x$, $TiCr_x$, $AlCo_xCr_{1-x}FeNi$, $Al_xCoCrFeNi$, Fe–Si–B–Cu–Nb, and $(Ti\text{-}6Al\text{-}4V)_{1-x}(Fe\text{-}36Ni)_x$ [57, 60, 61].

AM for metal powders can be produced using different methods, such as mechanical alloying, chemical processes, plasma, gas, water atomization, and electrolysis. Several AM techniques have been used to produce metallic components. Among them, Laser-Engineered Net Shaping (LENS) is the most suitable technique for developing an alloy library. Three different routes allow for developing this library.

A single compositionally graded sample is produced by continuously adjusting the relative powder feed rates during the AQ process. Continuous or discrete composition gradient can happen.

Each sample with uniform composition by taking multiple samples in one run. During processing, thermal gradient tuning can be used for the microstructural gradient samples.

The LENS technique was used to conduct a combinatorial assessment of magnetic HEA $FeCo_xNiAlCr_{1-x}$ $(0<x<1)$, which is compositionally graded based on the composition–structure–magnetic property relationship. One end contains the composition of FeCoNiAl comprised of equiaxed B2 grains and it exhibits an early stage of phase separation into two different Fe-Co and Ni-Al rich regions within the B2 grains. The other end of the sample possessing FeNiAlCr composition exhibits the grains with pronounced spinodal decomposition resulting in the microstructure B2 + BCC [62].

Mechanical Milling: HEA particles can be produced from mechanical milling. The advantages of this technique are as follows:

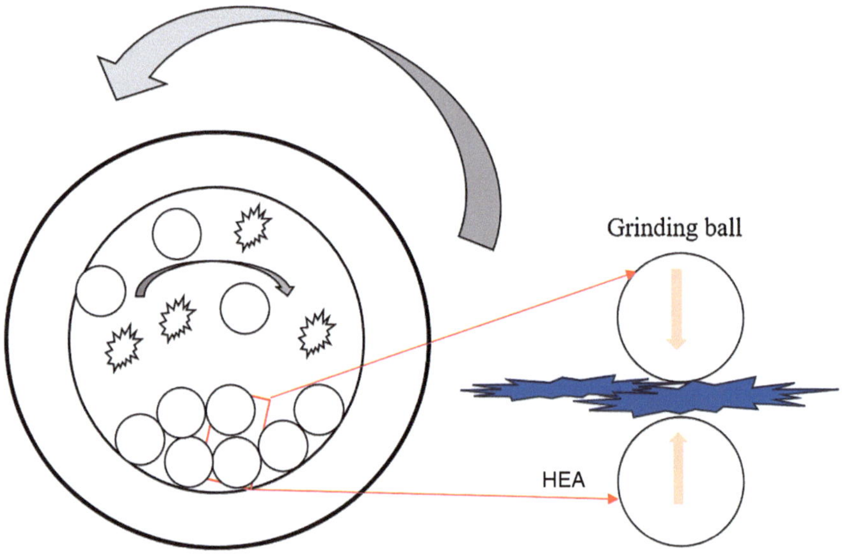

Figure 9.7 Schematic of the mechanical milling process for HEAs.

(a) Solid-solubility limit can be enlarged, (b) Grain size can be obtained in the nanometer range, (c) Crystalline, quasi-crystalline, and amorphous phases can be produced, and (d) A key limitation of chemical techniques is that they are difficult to implement for alloy synthesis, particularly when precise stoichiometric control is required. This process is used for flattening, welding, fracturing, and rewelding of the powder. The mixture of different elements and milling balls is filled into a rotating reaction chamber.

After the optimization of time and speed, alloying can be obtained by repeated welding and fracturing. The optimization of different variables, (a) mill and milling container time, (b) milling time and speed, (c) ratio of ball to powder, (d) temperature and atmosphere during milling, is essential because of the complex process of alloying. All these variables are connected, e.g. the milling time depends on the dimension of the grinding approach, milling variety, and ball-to-powder proportion (Figure 9.7) [63].

Diffusion Multiples: The diffusion multiple procedure was used for the acceleration of alloy development, such as Fe-Co-Ni-Cr-Mn, Fe-Co-Ni-Cr-Al, Fe-Co-Ni-Cr, Fe-Co-Ni-Cr-Cu, and Fe-Co-Ni-Cr-Al-Ti. To establish the composition-dependent interdiffusion coefficients in HEA FCC Fe-Co-Ni-Cr-Al at three different temperatures, 1,273, 1,323, and 1,373 K, a numerical inverse procedure consolidated with the different diffusion approaches was used.

Multiple diffusion layers, which were composed of HEA of Fe-Co-Cr-Ni, were arranged as six dissimilar pseudotrivial diffusion pairs. Heat treatment was given to these diffusion couples above 1,000 K for more than 11 hours. Afterwards, elemental concentration was dignified at numerous distances beyond each duo. The constituents that were progressed as sine-wave concentration gradients in the vicinity of the boundary suggest that the diffusion of these constituents had happened.

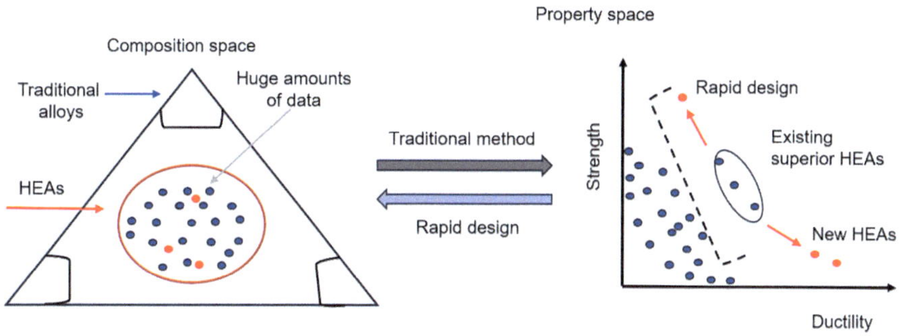

Figure 9.8 A schematic shows the synthesis by rapid alloy prototyping for the new HEAs from the existing superior HEAs [30].

Rapid Alloy Prototyping: This method was initially used to display steels Fe-30Mn-1.2C-xAl using Springer and Raabe's combinatorial approach. Now, it is used to screen HEAs. This method necessit7ates bulk casting of semi-continuous, decontamination, rolling, and sample extraction for depiction (Figure 9.8) [30].

9.3 Entropy Enhancement of HEAs

A scheme has been introduced to enhance the entropy of HEAs. Modification of solvent and solute may enhance the entropy of HEA (Figure 9.9) [64].

There are three different possible ideal cases after solvent and solute modification. These are single-phased, one high-entropy second phase, and several near-binary second phases (Figure 9.10) [64].

Entropy-enhanced alloys show improved properties over the conventional alloys. The properties such as the composition, phase, and microstructure of the modified alloys can be very different. The modified alloy could be single-phased, one high-entropy second phase, and several near-binary second phases. The diverse structure and phase of the modified alloy give considerable freedom for

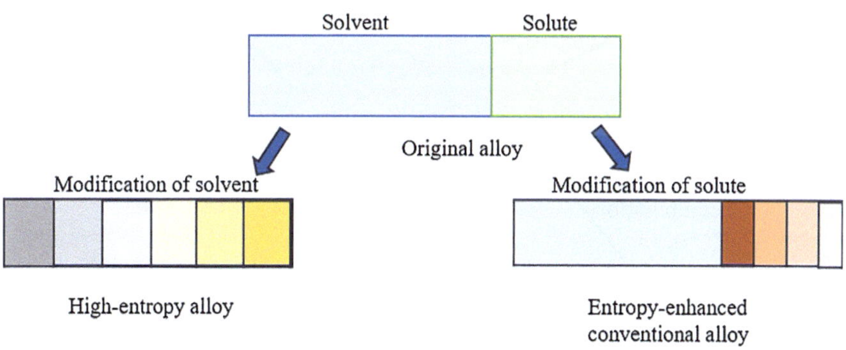

Figure 9.9 Each color represents one element and the width represents the concentration of that element [64].

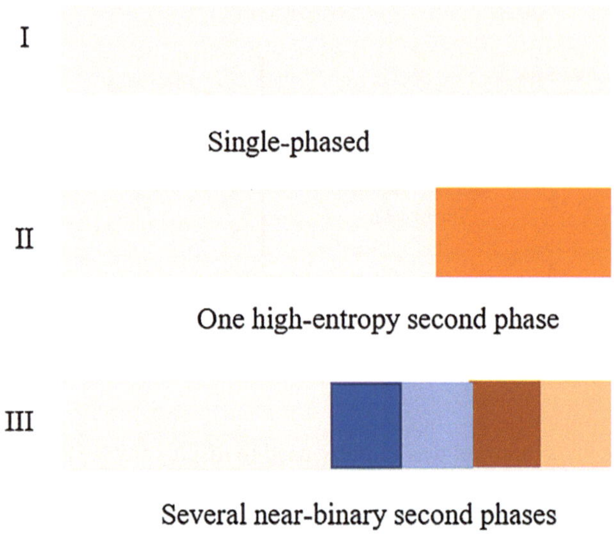

Figure 9.10 Three different cases after solute modification: (I) single-phased, (II) one high-entropy second phase, and (III) several near-binary second phases. Each color represents one phase in the alloy and the width represents the volume fraction [64].

adjusting the properties and making the strategy useful for the better properties of conventional alloy systems.

9.4 Application of HEAs

HEAs exhibit four different core effects from traditional alloys, which are the high-entropy effect, lattice distortion, sluggish diffusion, and cocktail effect. These

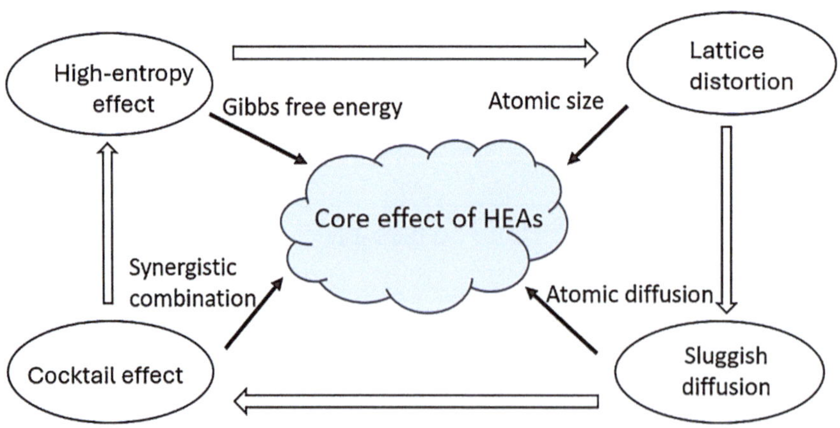

Figure 9.11 A schematic shows the four core effects of HEAs [17].

four core effects have been shown in Figure 9.11 [17]. The formation and stabilization of a single solid solution is observed from the high-entropy effect. Gibbs free energy is the main parameter for the high-entropy effect. Crystal structure and atomic size differences between the constituent elements result in significant lattice distortion, which may alter the material properties. The phase transformation in the solid solution systems is slowed down by sluggish diffusion. The synergistic enhancement of properties, which refers to the cocktail effect, may result from the interaction of the different elements. These core effects play a vital role in HEAs in various applications [1–5].

Recent years have shown us that the world has been facing a lot of challenges of the energy crisis and climate warming. This situation arises because of the following reasons. A lot of fossil fuels have been used for the advancement of society and the economic field of activity, which causes significant surroundings pollution. In addition to that, a high quantity of energy supply is needed to ensure national profitable evolution and security, which leads to energy depletion.

These heat-resistant radiation HEAs could be used as different coatings for reactors and avionics in extreme conditions. In water splitting, they can be used as catalysts and for renewable energy applications. Their macroscopic properties, such as catalytic activity, resistance against irradiation, and mechanical hardness, make them interesting and better performers than conventional alloys. Due to more element numbers, these properties are related to the local configuration number, which is huge billions.

As a consequence of that, there has been a great acceleration in the development of the shift from fossil fuels to renewable energy and the energy revolution process. The development of efficient energy conversion, storage, clean, and utilization techniques has become a hot topic of current research. Currently, green energy resources, for instance wind, tidal, and solar energy are limited by geographical constraints. In this chapter, we focus on three different applications of HEAs: lithium-ion batteries, dielectric capacitors, and supercapacitors [41–45, 65].

Lithium-Ion Batteries: Both the cathode and the anode can be replaced with HEAs in lithium-ion batteries, showing an excellent performance. Tables 9.2 and 9.3 represent the different oxides of HEAs that can replace the conventional materials in the cathode and anode of lithium-ion batteries for better performance. The stabilization of entropy confers structural stability, which makes them robust for the charging and discharging process in comparison to the traditional electrode materials [57].

The energy density as well as specific capacity can be enhanced after the establishment of electroactive cations Ni and Fe with steep redox potentials. Rock salt and spinel-type HEOs are the most promising electrodes in lithium-ion batteries because of the excessive operating voltages in comparison to the graphite-based materials. The electrochemically active elements Ni and Co in the lattice enable the HEO to perform redox reactions during the charging and discharging process. Mg^{2+} and Zn^{2+} are the electrochemically active inert cations that play a critical role in the stabilization of HEO structure during the charging and discharging process, and they lead to the excellent cyclic performance [42–44].

Table 9.2 Potential candidates for cathodes from HEAs.

HEAs	Cycle life	Initial and final capacity (mA h g^{-1}) (approximate)	Current density	References
$Li_x(CoCuMgNiZn)OF_x$	300	168 and 120	C/10	Wang et al. (2019)
Li(MnCoCrTiNb)	20	307 and 225	20 mA g^{-1}	Lun et al. (2021)
(CoMgCuNiZn)O	200	1,191 and 664	C/10	Wang et al. (2019)
$Li_{0.8}Na_{0.2}(NaCoMnAlFe)O_2$	20	80 and 62	C/10	Wang et al. (2020)
$Li(Ni_{0.2}Co_{0.2}Mn_{0.2}Ti_{0.2}Fe_{0.2})O_2$	50	85 and 67	C/10	Sturman et al. (2021)
$LiCr_{1/6}$ $MnFe_{1/6}$ $Co_{1/6}$ $Ni_{1/6}Cu_{1/6}$ O_2	40	40 and 65	5 μA cm^{-2}	Wang et al. (2022a)
$NaNi_{0.12}Cu_{0.12}$ $Mg_{0.12}Fe_{0.15}$	200	110 and 87	C/2	Zhao et al. (2020)
$Co_{0.15}Mn_{0.1}Ti_{0.1}$ $Sn_{0.1}Sb_{0.04}O_2$	20	80 and 62	20	Lun et al. (2021)
$Na_{0.67}Ni_{0.165}Cu_{0.1}$ $Zn_{0.05}Fe_{0.03}Mn_{0.65}O_2$	200	105 and 77.49	5C	Zhang et al. (2022)
$Na_{0.75}Mn_{0.5625}Ni_{0.25}$ $Cu_{0.0625}Mg_{0.0625}$ $Ti_{0.0625}O_2$	100	108 and 97	5C	Chen et al. (2022)
$NaNi_{0.12}Cu_{0.12}Mg_{0.12}$ $Fe_{0.15}Co_{0.15}Mn_{0.1}Ti_{0.1}$ $Sn_{0.1}Sb_{0.04}O_2$	500	110 and 83	3C	Zhao et al. (2020)
$NaNi_{1/4}$ $Co_{1/4}$ $Fe_{1/4}$ $Mn_{1/8}Ti_{1/8}O_2$	100	128 and 97	2C	Yue et al. (2015)
$NaFe_{0.2}Co_{0.2}Ni_{0.2}$ $Ti_{0.2}Sn_{0.1}Li_{0.1}O_2$	100	130 and 72	2C	Tian et al. (2022)
$NaCu_{0.1}Ni_{0.3}Fe_{0.2}$ $Mn_{0.2}Ti_{0.2}O_2$	100	130 and 87	5C	Ghosh et al. (2024)
$NaNi_{0.25}Mg_{0.05}Cu_{0.1}$ $Fe_{0.2}Mn_{0.2}Ti_{0.1}Sn_{0.1}O_2$	500	130 and 75	5C	Ding et al. (2022) [66]
$Na_{2/3}Li_{1/6}Fe_{1/6}Co_{1/6}Ni_{1/6}$ $Mn_{1/3}O_2$	90	171 and 89	10C	Yao et al. [45]

Table 9.3 Potential candidates for the anode from HEAs.

HEAs	Cycle life	Initial and final capacity ($mA\,h\,g^{-1}$)	Current density ($A\,g^{-1}$)	References
$(CoCrFeMnNi)_3O_4$	50	1,133 and 980	0.1	Zhao et al. (2021)
MgCoNiZnCuO-GO	1,000	460 and 460	1.0	Guo et al. (2022)
$(MgCoNiZi)_{0.65}Li_{0.35}O$	100	680 and 610	1.0	Lökçü et al. (2020)
$Mg_{0.2}Co_{0.2}Ni_{0.2}Cu_{0.2}Zn_{0.2}O$	300	1,500 and 935	0.1	Qiu et al. (2019)
$(Al_{0.2}CoCrFeMnNi)_{0.58}O_{4-\delta}$	500	2,234 and 536	0.2	Xiang et al. (2021)
$(FeCoNiCrMn)_3O_4$	300	660 and 402	0.5	Duan et al. (2021)
$(Co_{0.2}Cr_{0.2}Fe_{0.2}Mn_{0.2}Ni_{0.2})_3O_4$	50	980 and 400	0.1	Zhao et al. (2021)
$(FeNiCrMnZn)_3O_4$	185	386 and 200	0.5	Xiao et al. (2021)
$(FeCoNiCrMnXLi)_3O_4$ (X = Cu, Mg, Zn)	100	522 and 100	0.5	Duan et al. (2021)
(MgCoNiCuZn)O	100	550 and 400	0.1	Wang et al. (2021b)
Co-free $(CrMnFeNiCu)_3O_4$	50	755 and 500	0.1	Nguyen et al. (2022)
$(Mg_{0.2}Co_{0.2}Ni_{0.2}Cu_{0.2}Zn_{0.2})O$	200	950 and 450	0.1	Guo et al. (2022)
$(MnFeCoNiCuZn)_3O_{4-x}$	100	600 and 400	0.2	Dong et al. (2022)
$(CoNiZnXMnLi)_3O_4$ (X = Fe, Cr)	100	260 and 200	0.3	Tian et al. (2022)
$Ti_{1.1}V_{0.7}Cr_xNb_{1.0}Ta_{0.6}C_3Tz$	1,000	40 and 30	1	Etman et al. (2022)

The elemental interaction in lithium-ion batteries could result in additional composite behavior in comparison to the established metal oxide, which has only one or two metallic cations, after the integration of numerous metallic cations in entropy-balanced one-phase high-entropy oxides.

Dielectric Capacitor: Dielectric capacitors work on the concept of dipoles, where the electric energy can be stored and released by applying an electric field. The main challenge in the dielectric capacitor is still the low-energy density. Different routes have been developed to overcome the low-energy density issue, including interface design, defect control, and engineering of domains. Recently, a new HEO-based $BaTiO_3$ dielectric capacitor gained attention in 2019. The energy density in this dielectric capacitor is still low. That was assumed because of the rough design and unclear mechanism. Afterward, high-entropy-based ceramics and dielectric capacitors were reported, such as (K, Na)NbO_3, by introducing several elements like Li, Bi, Ba, Sc, Zr, Hf, and Ta [54, 55, 57].

These additives tailor the phase transition temperature, which results in high-entropy-based dielectric capacitors. Dielectric capacitor-based energy storage was found to have higher density in (K, Na)NbO_3 HEO after the substitution of these elements. There can be grain refinement and partially amorphous phase formation, which limits the application of HEAs in dielectric capacitors. Moreover, to enhance the performance optimization of HEAs, it is necessary to improve and recognize the endowment of each constituent to the altogether accomplishment upgrade. The recent development of artificial intelligence and machine learning can help to handout the endowment of each constituent and the upgrading of the functional performance of HEAs in dielectric capacitors [39–41].

Supercapacitor: Electrodes in conventional supercapacitors can be replaced by the HEAs, which provide higher efficiency in the application of fuel cells. These HEAs can be deposited on CNFs. For example FeNiCoMnMg deposited on CNFs hybrid electrode delivers particular power density and capacitance of $21.7\,Wh\,g^{-1}$ and $203\,F\,g^{-1}$, individually. The particular capacitance value can be increased by the incorporation of a metal oxide thin sheet on the outer exterior of FeNiCoCuSn HEA. The assembly HEA-NP/HCPCs/MO nanostructured displays the particular capacitance value of $495.4\,F\,g^{-1}$ in 1 M KOH and 94.7% of capacity retention. Different potassium-based high-entropy analogs $K(MgMnCoNiCu)Fe(CN)_6$, $K(MgMnFeCoCu)Fe(CN)_6$, $K(MgMnFeNiCu)Fe(CN)_6$, $K(MgMnFeCoCu)Fe(CN)_6$, and $K(MgMnFeCoNi)\ Fe(CN)_6$ were used as electrodes in supercapacitors. Among them, $K(MgMnFeNiCu)$-$Fe(CN)_6$ exhibits the highest electrochemical performance, which delivers a higher value of specific capacitance at the same rate [42–46].

9.5 Summary and Outlook

The future demands controlled synthesis, which is aimed at clarifying the relationship between different configurations of the target system with various element compositions and the replacement of different elements in several materials to satisfy the requirements of a large number of applications. HEAs offer various advantages in comparison to conventional alloys. Large composition space in HEAs results in essential characteristics such as thermal stability, corrosion resistance, and physical possessions, which are prerequisites for the future generation of electric machines. High electrical resistivity and low eddy current loss offer their use in rotating electric machines and power electronics applications.

Moreover, magnetic HEAs have magnetic domain walls that can move smoothly in a single-phase HEA under a magnetic field. It results in the reduction of the development of magnetic HEA-based components for the high-yielding AC magnetic assessments. Further evolution in this field is required, despite the better performance of HEAs in collaborative endeavors from every direction.

Owing to the involvement of five or more constituents in the formation of HEAs, there is an immiscibility among all the elements, which restricts the single-phase formation and obstructs their potentiality to exhibit their characteristic possessions. The established techniques rely on high temperature and pressure means to communicate these oppositions. So, it is significant to find other low-temperature coalescence techniques for HEAs that can involve a broad range of different constituents.

Several scrutinizes have been done to explore the morphology of HEAs, which focus on scale reduction at the nanometer scale and preparing different nanoparticles with regular morphology. HEAs have a lattice distortion effect, which shows that morphology control is not straightforward for HEAs. Thus, it is required to look over the arrangement and configuration relationship to produce HEAs with adjustable morphologies.

During the large-scale preparation process, HEAs synthesis necessitates using different high-quality elements, producing in high-energy and high-cost utilization. As a consequence of that, it is very crucial to study cost-effective methods for the synthesis of HEAs. One possibility could be to use natural mineral material for the formation of high-entropy electrode material. Another can be to use recycled waste electrode material to fully leverage the synergistic effects required for ultimately achieving high performance.

The synthesis of HEAs and their application in catalytic and energy storage performance can be assessed through theoretical calculations, which can be a valuable tool. Few theoretical studies have already been done on the synthesis of HEAs. However, there are limited reports on the synthesis of high-entropy oxides and sulfides. Therefore, further theoretical calculations and models are required to synthesize HEAs, which will bring a breakthrough in the advancement of HEAs [60–62, 66, 67].

The functional performance of HEAs can be improved by developing their properties related to the weight reduction, enhancing the magnetic and invar performance, weldability, stress corrosion resistance, harvesting capability of thermoelectric energy, their soft and hard magnetic features, superconductivity, electrical, magnetic and electro-migration resistivity, bioresorption, biocompatibility, and their suitability as cathode and anode; materials for batteries. An additional pathway is opened by HEAs, which are open to the unexpected properties. The functional structure–property relationship of HEAs with different compositions can be improved [68–70].

To explore the compositional space in HEAs, it is enormous to depend only on experiments. The computer skills, especially CALPHAD, have become extremely important to screen out the unsuitable alloys, which can increase the speed of development by several orders. The thermodynamic database of HEAs is still far from complete. CALPHAD accuracy can vary significantly depending on the composition. Therefore, it is urgently needed to develop a more multicomponent database [62].

The practical design of intermetallic phases of HEAs is also an important concern. Although, CALPHAD technique has many advantages in terms of software, database, and speed, it is costly. The accuracy of CALPHAD is also a major issue. Therefore, to design the practically intermetallic HEAs, experimental results are necessary. Current understanding of the selection criteria and stability of high-entropy alloy (HEA) designs remains limited. A general criterion for the selection of intermetallic phases includes preliminary results and certain phases, such as Laves and sigma phases [66–69].

High-entropy intermetallics are just conventional binary/ternary intermetallics which have a crystal structure similar to the binary intermetallic phases. These intermetallic phases are composed of many principal elements. The composition of these intermetallics is not different from a quinary equimolar alloy, as well as conventional binary/ternary Laves phases. A key limitation of chemical techniques is the difficulty of alloy synthesis with precise stoichiometric control [70].

References

1 Deng, Y., Tasan, C.C., Pradeep, K.G. et al. (2015). Design of a twinning-induced plasticity high entropy alloy. *Acta Mater* 94: 124–133.

2 Raabe, D., Herbig, M., Sandlöbes, S. et al. (2014). Grain boundary segregation engineering in metallic alloys: a pathway to the design of interfaces. *Curr Open Solid State Mater Sci* 18: 253–261.

3 Kürnsteiner, P., Wilms, M.B., Weisheit, A. et al. (2020). High strength Damascus steel by additive manufacturing. *Nature* 582: 515–519.

4 Lei, Z., Liu, X., Wu, Y. et al. (2018). Enhanced strength and ductility in a high-entropy alloy via ordered oxygen complexes. *Nature* 563: 546–550.

5 Yao, M.J., Pradeep, K.G., Tasan, C.C., and Raabe, D. (2014). A novel single phase, non-equiatomic FeMnNiCoCr high-entropy alloy with exceptional phase stability and ductility. *Scr Mater* 72–73: 5–8.

6 George, E.P., Raabe, D., and Ritchie, R.O. (2019). High entropy alloys. *Nat Rev Mater* 4 (8): 515–534.

7 Chaudhary, V., Chaudhary, R., Banerjee, R., and Ramanujan, R.V. (2021). Accelerated and conventional development of magnetic high entropy alloys. *Mater Today* 49: 231–252.

8 Chaudhary, V., Soni, V., Gwalani, B. et al. (2020). Influence of non-magnetic Cu on enhancing the low temperature magnetic properties and Curie temperature of FeCoNiCrCu (x) high entropy alloys. *Scr Mater* 182: 99–103.

9 Gorsse, S., Hachet, D., Godet, S. et al. (2025). Contributions and role of B2 nanoparticles in the exceptional strength/ductility trade-offs of $Al_{0.3}CoCrFeMnNi$ high entropy alloy. *Mater Res Lett* 1–9.

10 Senkov, O.N., Kube, S.A., Drazin, J.W. et al. (2024). Thermal and elastic properties of an A2/B2 refractory high entropy superalloy and its constituent phases. *Metall Mater Trans A* 55: 2704–2717.

11 Radhakrishnan, M., Sharma, S., Palaniappan, S. et al. (2024). Influence of thermal conductivity on evolution of grain morphology during laser-based directed energy deposition of CoCrxFeNi high entropy alloys. *Addit Manufact* 92: 104387.

12 Soni, V., Chesetti, A., Sharma, A. et al. (2025). Some insights into the high temperature phase stability of the BCC+ B2 microstructure in aluminum containing refractory high entropy alloys. *J Mater Sci* 60: 1675–1688.

13 Dasari, S., Sharma, A., Gwalani, B. et al. (2024). Coupled transmission electron microscopy and atom probe tomography reveals an interesting pathway to nanolamellar microstructures in high entropy alloys. *Microsc Microanal* 30: ozae044. 652.

14 Nartu, M.S.K.K.Y., Jha, S., Chesetti, A. et al. (2023). Microstructure and temperature dependent indentation response of additively manufactured precipitation-strengthened $Al_{0.3}Ti_{0.2}Co_{0.7}CrFeNi_{1.7}$ high entropy alloy. *Addit Manufact High Temp Energy Syst Harvest Mater* 75: 5046–5054.

15 Chen, S., Liu, P., Pei, Q. et al. (2023). Ideal plasticity and shape memory of nanolamellar high-entropy alloys. *Sci Adv* 9: eadi5817.

16 Sharma, A., Dasari, S., Soni, V. et al. (2023). B2 to ordered omega transformation during isothermal annealing of refractory high entropy alloys: implications for high temperature phase stability. *J Alloys Compd* 953: 170065.

17 Odetola, P.I., Babalola, B.J., Afolabi, A.E. et al. (2024). Exploring high entropy alloys: A review on thermodynamic design and computational modeling strategies for advanced materials applications. *Heliyon* 10: e39660.

18 Dasari, S., Sharma, A., Soni, V. et al. (2023). Crystallographic and compositional evolution of ordered B2 and disordered BCC phases during isothermal annealing of refractory high-entropy alloys. *Microsc Microanal* 29: 303–313.

19 Yun, D., Chae, H., Lee, T. et al. (2022). Stress contribution of B2 phase in $Al_{0.7}CoCrFeNi$ eutectic high entropy alloy. *J Alloys Compd* 918: 165673.

20 Nartu, M.S.K.K.Y., Chesetti, A., Dasari, S. et al. (2022). Engineering heterogeneous microstructures in additively manufactured high entropy alloys for high strength and strain hardenability. *Mater Sci Eng A* 849: 143505.

21 Liu, X., Banerjee, R., Vitos, L., and Wang, Y. (2022). Metastable high entropy alloys. *Appl Phys Lett* 120: 120401.

22 Srimark, K., Dasari, S., Sharma, A. et al. (2021). Hierarchical phase evolution in a lamellar $Al_{0.7}CoCrFeNi$ high entropy alloy involving competing metastable and stable phases. *Scr Mater* 204: 114137.

23 Dasari, S., Sharma, A., Byers, T.A. et al. (2021). Proton irradiation induced chemical ordering in an $Al_{0.3}CoFeNi$ high entropy alloy. *Appl Phys Lett* 119 (2021): 161907.

24 Gwalani, B., Dasari, S., Sharma, A. et al. (2021). High density of strong yet deformable intermetallic nanorods leads to an excellent room temperature strength-ductility combination in a high entropy alloy. *Acta Mater* 219: 117234.

25 Jagetia, A., Nartu, M.S.K.K.Y., Dasari, S. et al. (2021). Ordering-mediated local nano-clustering results in unusually large Hall-Petch strengthening coefficients in high entropy alloys. *Mater Res Lett* 9: 213–222.

26 Chen, S., Aitken, Z.H., Pattamatta, S. et al. (2021). Chemical-affinity disparity and exclusivity drive atomic segregation, short-range ordering, and cluster formation in high-entropy alloys. *Acta Mater* 206: 116638.

27 Mugale, M., Karki, S., Choudhari, A. et al. (2025). High strength-ductility combination in low-density dual phase high entropy alloys. *J Alloys Compd* 1010 (2025): 177762.

28 Kadhim, D.F., Koricherla, M.V., and Scharf, T.W. (2023). Room and elevated temperature sliding friction and wear behavior of $Al_{0.3}CoFeCrNi$ and $Al_{0.3}CuFeCrNi_2$ high entropy alloys. *Crystals* 13: 609.

29 Gwalani, B., Escobar, J., Song, M. et al. (2024). Mechanisms for high creep resistance in alumina forming austenitic (AFA) alloys. *Acta Mater* 263: 119494.

30 Garg, M., Grewal, H.S., Sharma, R.K. et al. (2023). Circumventing strength-ductility paradox in high entropy alloys through deformation processing. *J Alloys Compd* 933: 167750.

31 Adamo, C.G., Srivastava, A., Legese, S.S. et al. (2024). Integrated first principles and experimental investigation of thermoelectric transport in Zr/Ti half-Heusler-type high entropy alloys. *Energy Technol* 12: 2400061.

32 Mohanty, G.C., Gowda, C.C., Gakhad, P. et al. (2024). Enhanced energy density of high entropy alloy (Fe-Co-Ni-Cu-Mn) and green graphene hybrid supercapacitor. *Energy Storage* 6: e668.

33 Adamo, C.G., Srivastava, A., Legese, S.S. et al. (2024). Thermoelectric transport of a novel Zr-based half-Heusler high-entropy alloy. *Energy Technol* 12: 2301119.

34 Mohanty, G.C., Gowda, C.C., Gakhad, P. et al. (2024). Effect of Zn/Mn on the supercapacitor behavior of high-entropy FeCoNiCrZn/Mn alloy. *Energy Adv* 3: 2972–2985.

35 Luo, H., Li, Z., Mingers, A.M., and Raabe, D. (2018). Corrosion behavior of an equiatomic CoCrFeMnNi high-entropy alloy compared with 304 stainless steel in sulfuric acid solution. *Corros Sci* 134: 131–139.

36 Li, Z., Tasan, C.C., Springer, H. et al. (2017). Interstitial atoms enable joint twinning and transformation induced plasticity in strong and ductile high-entropy alloys. *Sci Rep* 7: 40704.

37 Shi, P., Li, R., Li, Y. et al. (2021). Hierarchical crack buffering triples ductility in eutectic herringbone high-entropy alloys. *Science* 373: 912–918.

38 Su, J., Raabe, D., and Li, Z. (2019). Hierarchical microstructure design to tune the mechanical behavior of an interstitial TRIP-TWIP high-entropy alloy. *Acta Mater* 163: 40–54.

39 Mohanty, S., Palit, M., Pradeep, K.G., and Biswas, K. (2025). Evolution of microstructure, elemental segregation and magnetic order in $Ti_{20}Co_{20}Cu_{20}Fe_{20}Ni_{20}$ high entropy alloy. *J Alloys Compd* 1010: 178089.

40 Sonkusare, R., Gurao, N.P., Biswas, K. et al. (2023). Micro-mechanisms of deformation and strengthening during high pressure torsion of CoCuFeMnNi high entropy alloy. *Materialia* 32: 101916.

41 Wu, C., Gong, Y., Liu, C. et al. (2025). Hydrogen accommodation and its role in lattice symmetry in a TiNbZr medium-entropy alloy. *Acta Mater* 288: 120852.

42 Shkodich, N.F., Smoliarova, T., Ali, H. et al. (2025). Effect of high energy ball milling, heat treatment and spark plasma sintering on structure, composition, thermal stability and magnetism in $CoCrFeNiGa_x$ ($x = 0.5$; 1) high entropy alloys. *Acta Mater* 284: 120569.

43 Sulumar, P.T., Kloenne, Z., Das, S.M. et al. (2024). Mechanisms of grain boundary α precipitation in the metastable β-titanium Ti-5Al-5Mo-5V-3Cr. *Acta Mater* 272: 119955.

44 Buttard, M., Freixes, M.L., Josserond, C. et al. (2023). Ageing response and strengthening mechanisms in a new Al-Mn-Ni-Cu-Zr alloy designed for laser powder bed fusion. *Acta Mater* 259: 119271.

45 Prithiv, T.S., Kloenne, Z., Li, D. et al. (2022). Grain boundary segregation and its implications regarding the formation of the grain boundary α phase in the metastable β-Titanium Ti–5Al–5Mo–5V–3Cr alloy. *Scr Mater* 207: 114320.

46 He, J., Makineni, S.K., Lu, W. et al. (2020). On the formation of hierarchical microstructure in a Mo-doped NiCoCr medium-entropy alloy with enhanced strength-ductility synergy. *Scr Mater* 175: 1–6.

47 Picak, S., Liu, J., Hayrettin, C. et al. (2019). Anomalous work hardening behavior of Fe40Mn40Cr10Co10 high entropy alloy single crystals deformed by twinning and slip. *Acta Mater* 181: 555–569.

48 Xie, K.Y., Yadav, D., Hackett, B.L. et al. (2024). Weak strain-rate sensitivity of hardness in the VCoNi equi-atomic medium entropy alloy. *Mater Sci Eng A* 913: 147091.

49 Kumari, P., Gupta, A.K., Mishra, R.K. et al. (2022). A comprehensive review: recent progress on magnetic high entropy alloys and oxides. *J Magn Magn Mater* 554: 169142.

50 Gou, J., Liu, G., Yang, T. et al. (2025). A high-entropy alloy showing gigapascal superelastic stress and nearly temperature-independent modulus. *Nat Commun* 16: 1227.

51 Cao, G., Liang, J., Guo, Z. et al. (2023). Liquid metal for high-entropy alloy nanoparticles synthesis. *Nature* 619: 73–77.

52 Yao, Y., Dong, Q., Brozena, A. et al. (2022). High entropy nanoparticles: synthesis-structure-property relationships and data-driven discovery. *Science* 376: cabn 3103.

53 Su, X. and Sun, Y. (2024). Synthesis of metallic high-entropy alloy nanoparticles. *Chem Soc Rev* 53: 4400–4433.

54 Khan, J.B., Panda, P.K., Yang, P.C. et al. (2024). Microwave synthesis of high-entropy alloy catalysts on graphene oxide sheets for oxygen reduction and evolution reactions. *Int J Hydrog Energy* 53: 999–1008.

55 Tahir, S., Shkodich, N., Eggert, B. et al. (2024). Synthesis of high entropy alloy nanoparticles by pulsed laser ablation in liquids: influence of target preparation on stoichiometry and productivity. *Chem Nano Mat* 10: e202400064.

56 Meng, Z., Zhou, J., Qin, C., and Wang, Z. (2025). Dealloying of quasi-high entropy alloys: fabrication of porous noble metals/metal oxides. *Metals* 15: 114.

57 Krishna, S.A., Noble, N., Radhika, N., and Saleh, B. (2024). A comprehensive review on advances in high entropy alloys: fabrication and surface modification methods, properties, applications, and future prospects. *J Manuf Process* 109: 583–606.

58 Xie, Q., Yu, Y.J., Zhang, H.J. et al. (2025). Structural engineering of high-entropy alloys for advanced electrocatalytic reactions. *Tungsten* 7: 1–31.

59 Zhang, Y., Zuo, T.T., Tang, Z. et al. (2014). "Microstructures and properties of high-entropy alloys" *Progress in Materials Science*. 61: 1–93.

60 Wang, X., Guo, W., and Fu, Y. (2021). High-entropy alloys: emerging materials for advanced functional applications. *J Mat Chem A* 9: 663–701.

61 Yin, X. and Xu, S. (2018). Properties and preparation of high entropy alloys. *MATEC Web Conf* 142: 03003.

62 George, E.P., Curtin, W.A., and Tasan, C.C. (2020). High entropy alloys: a focused review of mechanical properties and deformation mechanisms. *Acta Mater* 188: 435–474.

63 Suryanarayana, C. (2001). Mechanical alloying and milling. *Progress in Materials Science* 46: 1–184.

64 Tsai, M.-H. (2016). Three Strategies for the Design of Advanced High-Entropy Alloys. *Entropy* 18: 252.

65 Wu, C., Aota, L.S., Rao, J. et al. (2025). Hydrogen-assisted spinodal decomposition in a TiNbZrHfTa complex concentrated alloy. *Acta Mater* 285: 120707.

66 Ding, Q., Zhang, Y., Chen, X. et al. (2019). Tuning element distribution, structure and properties by composition in high-entropy alloys. *Nature* 574: 223–227.

67 Kamal, M.V., Ragunath, S., Reddy, M.H.S. et al. (2024). Recent advancements in lightweight high entropy alloys – a comprehensive review. *Int J Lightweig Mater Manuf* 7: 699–720.

68 Gu, S., Huang, H.-Y., Lin, Y.-C. et al. (2025). High-entropy alloys with low platinum content: novel catalysts for oxygen reduction and evolution reactions. *Emerg Mater* 1–14.

69 Panda, P.K., Huang, H.-Y., Dash, P. et al. (2025). Liquid-phase microwave synthesis of platinum-based high-entropy alloy catalysts on carbon supports for electrochemical hydrogen adsorption/desorption and oxygen evolution/reduction reactions. *Int J Hydrog Energy* 111: 536–545.

70 Liu, K.-C., Panda, P.K., Mallick, B.C. et al. (2024). Solid-phase microwave synthesis of high-entropy graphene quantum dots as metal-free electrochemical catalysts. *Appl Surf Sci* 648: 159061.

71 Wang, Q., Sarkar, A., Wang, D. et al. (2019). Multi-anionic and -cationic compounds: new high entropy materials for advanced Li-ion batteries. *Energy & Environmental Science* 12: 2433–2442. https://doi.org/10.1039/C9EE00368A.

72 Lun, Z., Ouyang, B., Kwon, D.H. et al. (2021). Cation-Disordered Rocksalt-Type High-Entropy Cathodes for Li-Ion Batteries. *Nature Materials* 20: 214–221.

73 Wang, Q., Sarkar, A., Li, Z. et al. (2019). High-entropy oxides as anode material for Li-ion battery applications. *Electrochemistry Communications.*

74 Wang, J., Zhang, J., Wu, T. et al. (2020). Lithium-containing layered high-entropy oxide structures. *Scientific Reports* 10: 18430. https://doi.org/10.1038/s41598-020-75134-1.

75 Sturman, J.W., Yim, C.-H., Baranova, E.A., and Abu-Lebdeh, Y. (2021). Communication — Design of $LiNi_{0.2}Mn_{0.2}Co_{0.2}Fe_{0.2}Ti_{0.2}O_2$ as a High- Entropy Cathode for Lithium-Ion Batteries Guided by Machine Learning. *Journal of The Electrochemical Society* 168 (5): 050541. https://doi.org/10.1149/1945-7111/ac00f4.

76 Wang, K. et al. (2022). Epitaxial high-entropy oxide thin films for Liion battery cathodes, *Advanced Materials*, 2022a, DOI: https://doi.org/10.1016/j.elecom.2019.02.001.

77 Zhao, C., Ding, F., Lu, Y. et al. (2020). High-Entropy Layered Oxide Cathodes for Sodium-Ion Batteries. *Angewandte Chemie International Edition* 59 (1): 264–269. https://doi.org/10.1002/ anie.201912171.

78 Kang, W., Yu, D.Y.W., Lee, P.-K. et al. (2016). P2-Type NaxCu0.15Ni0. 20Mn0.65O2 Cathodes with High Voltage for High-Power and Long-Life Sodium-Ion Batteries. *ACS Applied Materials & Interfaces* 8 (46): 31661–31668. https://doi.org/10.1021/acsami.6b10841.

79 Zhao, C., Ding, F., Lu, Y. et al. (2020). High-Entropy Layered Oxide Cathodes for Sodium-Ion Batteries. *Angew. Chem. Int. Ed.* 59 (1): 264–269. https://doi.org/10.1002/anie.201912171.

80 Yue, J. L., Zhou, Y. N., Shi, S. Q., Huang, X. Q., Yang, X. Q., et al. New O3-type layered oxide cathode $Na(NiCoFeTi)_{1/4}O_2$ for sodium-ion batteries, 2015, DOE/OSTI Technical Report. DOE Report ID: BNL-108557-2015-J&A.

81 Walczak, K., Plewa, A., Ghica, C. et al. (2022). NaMn0.2Fe0.2Co0.2Ni0.2Ti0.2O2 high-entropy layered oxide – experimental and theoretical evidence of high electrochemical performance in sodium batteries. *Energy Storage Materials* 47: https://doi.org/10.1016/j.ensm.2022.02.038.

82 Ghosh, A., Hegde, R., and Senguttuvan, P. (2024). A high-entropy O3-$Na_{1.0}Li_{0.1}Ni_{0.3}Fe_{0.1}Mn_{0.2}Ti_{0.2}O_2$ cathode with reversible phase transitions and superior electrochemical performances for sodium-ion batteries. *Journal of Materials Chemistry A* 12: 14583–14594. https://doi.org/10.1039/D4TA01137F.

83 Tian, K., Walczak, K., Zając, W., and Molenda, J. et al., "NaMn0.2Fe0.2Co0.2Ni0.2Ti0.2O2 high-entropy layered oxide: experimental and theoretical evidence of high electrochemical performance in sodium batteries, (2022), 358745649 NaMn02Fe02Co02Ni02Ti02O2 high-entropy layered oxide experimental and theoretical evidence of high electrochemical performance in sodium batteries (available at research gate). References for table 9.3.

84 Zhao, C., Ding, F., Lu, Y. et al. (2021). High-entropy spinel oxides as high-performance anode materials for lithium-ion batteries. *Advanced Energy Materials* 11: 2003012.

85 Guo, S., Wang, Y., Zhang, Y. et al. (2022). Graphene-oxide-modified high-entropy oxide anodes with enhanced lithium storage. *Journal of Materials Chemistry A* 10: 12145–12154.

86 Lökçü, E., Toparli, C., Varol, T., and Akdoğan, E. (2020). Lithium-containing high-entropy oxides as anode materials for Li-ion batteries. *Ceramics International* 46: 17172–17180.

87 Qiu, N., Chen, H., Yang, Z. et al. (2019). High-entropy oxide anode materials for lithium-ion batteries. *Electrochimica Acta* 317: 251–258.

88 Xiang, H., Xing, Y., Dai, F. et al. (2021). Oxygen-deficient high-entropy spinel oxides for high-capacity lithium storage. *Nano Energy* 85: 105977.

89 Duan, C., Liu, X., Wang, R., and Zhang, Y. (2021). Spinel high-entropy oxides as advanced anodes for lithium-ion batteries. *Journal of Power Sources* 484: 229309.

90 Zhao, C., Ding, F., and Hu, Y.-S. (2021). Entropy-stabilized spinel oxides for lithium storage. *Advanced Functional Materials* 31: 2101013.

91 Xiao, Y., Wang, Z., Chen, L., and Zhang, J. (2021). Zinc-containing high-entropy spinel oxides for lithium-ion batteries. *Journal of Alloys and Compounds* 867: 159082.

92 Duan, C., Wang, R., and Liu, X. (2021). Lithium-doped highentropy spinel oxides as advanced anode materials. *Electrochimica Acta* 391: 138947.

93 Wang, Q., Sarkar, A., Li, Z. et al. (2021). Hahn, H, "High-entropy oxides: Anode materials for lithium-ion batteries". *Electrochemistry Communications* 123: 106946.

94 Nguyen, T.T., Tran, H.N., and Pham, T.D. (2022). Cobalt-free high-entropy spinel oxides for lithium-ion batteries. *Journal of Energy Chemistry* 67: 403–411.
95 Guo, S., Sun, X., and Wang, Y. (2022). Electrochemical properties of equiatomic high-entropy oxide anodes. *Ceramics International* 48: 16590–16598.
96 Dong, Q., Hong, M., Gao, J., and Cui, M. (2022). Defectengineered high-entropy spinel oxides for lithium storage. *Energy Storage Materials* 47: 512–520.
97 Tian, K., Walczak, K., Zając, W., and Molenda, J. (2022). Lithium-containing high-entropy spinel oxides as anodes for Li-ion batteries. *Journal of Power Sources* 522: 231025.
98 Etman, A.S., Rosen, J., and Barsoum, M.W. (2022). High-entropy MXenes: synthesis and electrochemical energy storage. *Advanced Functional Materials* 32: 2109322.

10

Noble-Metal-Free High-Entropy Alloys for Energy Storage Applications

Yukti Setia, Nikita Bhatt, Maneesh Kumar, and Malaya K. Sahoo

Rajiv Gandhi Institute of Petroleum Technology, Department of Sciences & Humanities, Jais, Uttar Pradesh 229304, India

10.1 Introduction

The escalating worldwide need for energy and the exhaustion of fossil fuel resources have rendered the advancement of innovative materials for energy conversion and storage a prime concern. Over the past few decades, HEAs have been widely utilized in an extensive array of applications, like gas sensors, hydrogen production, SCs, and lithium-ion batteries (LIBs) [1, 2]. Alloy development has evolved from conventional single-metal materials to intricate multi-element systems. Traditional alloy design primarily relied on a base metal, supplemented by trace amounts of additional elements to improve specific characteristics. However, the limited electrochemical stability, mechanical strength, as well as resistance to corrosion of such materials spurred research into multi-principal element alloys, ultimately leading to the idea of HEAs [2]. In the year 2004, two research groups [3, 4] independently introduced a novel alloying approach, proposing singular-phase solid solution alloys composed of five or more elements in nearly equiatomic compositions. Since then, over 5,000 studies [5] have explored HEAs, initially focusing on their mechanical strength and thermal stability. More recently, interest in the potential uses of HEAs in energy storage and electrocatalysis has grown [6, 7]. As the need for efficient and sustainable energy storage devices continues to grow, materials with superior electrochemical activity, durability, and affordability are needed. HEAs can be broadly classified into noble metal based [8, 9] and NMF alloys, both of which hold significant potential for energy storage and conversion applications. While noble-metal-based HEAs exhibit remarkable catalytic activity and stability, their large-scale application is constrained by high costs and limited availability. In contrast, NMF-HEAs offer a sustainable and cost-effective alternative while maintaining high catalytic efficiency and electrochemical performance [10, 11]. The absence of noble metals in these alloys reduces reliance on expensive and scarce elements, making them a more viable option for commercial applications (Figure 10.1).

High-Entropy Materials for Energy Storage Devices, First Edition.
Edited by Chien-Te Hsieh, Pradeep Kumar Panda, and Arpan Kumar Nayak.

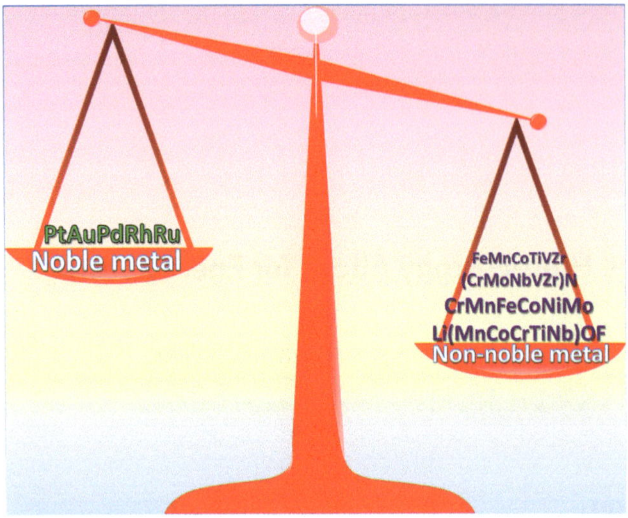

Figure 10.1 Representation of noble metal HEAs and NMF-HEAs.

NMF-HEAs are ideal for energy storage devices and electrocatalytic reactions because of their exceptional corrosion resistance, structural stability, and tunable electronic properties [12]. The need for novel materials that provide improved performance metrics and financial sustainability is highlighted by these difficulties. NMF-HEAs perform better than traditional materials because of their exceptional structural stability, which is a result of the combined effects of several metal elements. They are especially useful for energy storage applications because of this special property, which produces distinct adsorption energy distributions and enhanced reaction kinetics [13, 14]. In SCs, where their large surface areas and high electrical conductivity improve energy storage capabilities, NMF-HEAs exhibit great promise. High-power applications benefit greatly from the exceptional capacitance retention and enhanced charge–discharge efficiency of nanostructured HEAs, especially those that contain transition metals [2].

Further, NMF-HEAs have been thoroughly examined as cathode and anode materials for batteries that use zinc and lithium ions. Their entropy-stabilized structures facilitate lithium-ion diffusion, resulting in improved cycle life and rate capabilities. Additionally, the presence of multiple metallic elements enables customized redox activity, enhancing charge storage mechanisms. Because of these characteristics, HEAs are positioned as strong substitutes for traditional noble metal catalysts and are essential in creating long-term energy storage solutions. NMF-HEAs' synthesis, scalability, and long-term stability still need to be optimized for commercial applications, despite these encouraging developments. NMF-HEAs have the potential to revolutionize the energy material sector and provide high-performing, environmentally friendly solutions for advanced energy conversion and storage technologies if these challenges are overcome.

10.2 Fundamentals of HEAs

HEAs are a special kind of materials made up of several primary elements, usually five or more, taken together in roughly equimolar amounts in order to create a solid-solution structure, and each major element's concentration generally ranges between 5 and 35 atomic percent (at.%). To stabilize solid solutions and prevent the formation of detrimental intermetallic phases, Yeh and colleagues [4] hypothesized that increasing the number of elements in an alloy enhances the configurational entropy of mixing, thereby counteracting enthalpic contributions. High-entropy nanoparticles (NPs) are formed when entropy and enthalpy reach equilibrium, as described by the equation:

$$\Delta G_{mix} = \Delta H_{mix} - T\Delta S_{mix} \tag{10.1}$$

where T represents the absolute temperature, and ΔG_{mix}, ΔH_{mix}, and ΔS_{mix} represent changes in Gibbs free energy, mixing enthalpy, and configurational entropy, respectively. The synthesis and stability of HEAs are generally more favorable at high temperatures, as ΔG_{mix} is more likely to be negative under such conditions. The maximum configurational entropy of an ideal solid solution is given by [15]:

$$\Delta S_{conf} = -R\sum_{i=1}^{n} c_i \, Inc_i \tag{10.2}$$

where c_i represents the molar fraction of each component, R represents gas constant, and n represents total number of elements in the HEA, $S_{mix} = R \ln 5 = 1.61R$ [16].

10.2.1 Core Effects of HEAs

The distinct microstructures and characteristics of HEAs, such as slow diffusion, the cocktail effect, lattice distortion, and the high-entropy effect, set them apart from traditional alloys (Figure 10.2a) [17]. By affecting their behavior and performance, these effects collectively give HEAs exceptional electrocatalytic qualities. When atomic size differences in HEAs cause lattice distortion, local strain fields and defects are created, which increases catalytic activity. The cocktail effect maximizes the electrical and geometrical forms of HEAs that increase catalytic performance through the synergistic interactions with several components. When five or more elements are combined in approximately equal proportions, a strong high-entropy effect emerges, counteracting mixing enthalpy and stabilizing solid solutions, which, in turn, enhances catalytic durability in electrochemical reactions. Additionally, sluggish diffusion in HEAs results from differences in the potential energies and diffusion rates of various elemental atoms. This effect reduces atomic mobility, thereby improving structural stability, particularly in extreme environments.

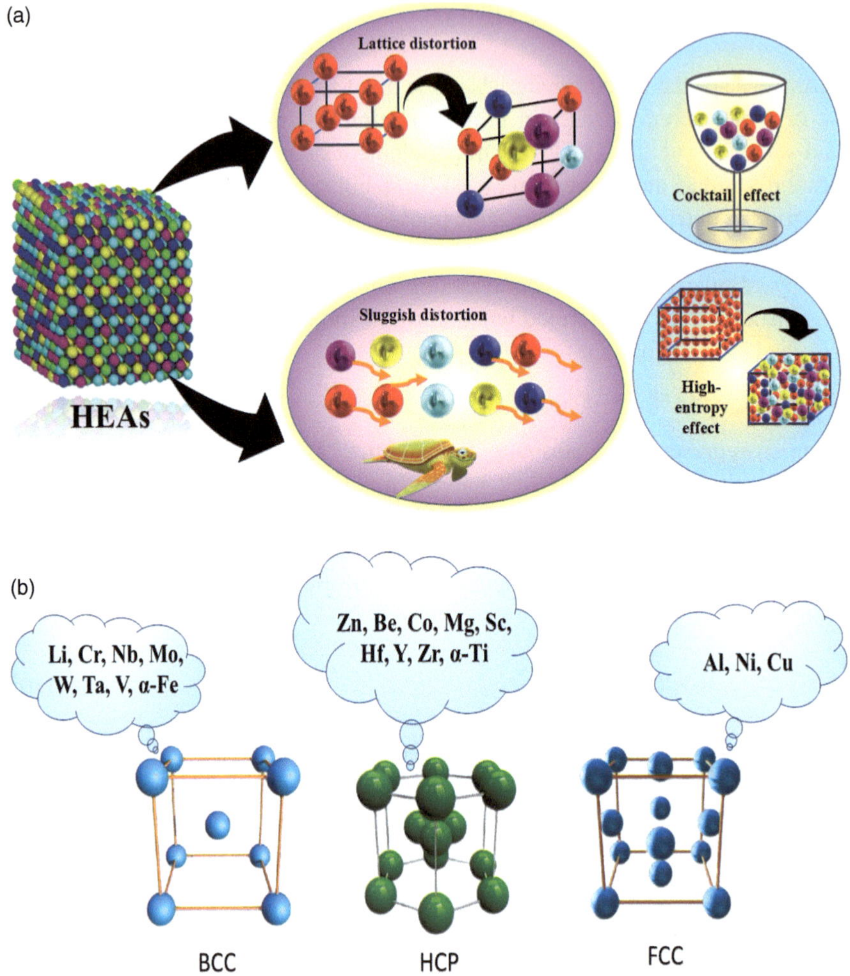

Figure 10.2 (a) A schematic illustration of HEA properties and (b) crystal structure of HEAs.

10.2.2 Crystal Structure of HEAs

The effects of high mixing entropy in HEAs promote the development of simple solid solutions, typically adopting face-centered cubic (FCC), body-centered cubic (BCC), or hexagonal close-packed (HCP) structures, as shown (Figure 10.2b) [18, 19]. Among these, the FCC lattice has been the most extensively studied. The foremost equiatomic HEAs to crystallize as a solid FCC solution in a single phase was the CrMnFeCoNi (Cantor) alloy. However, it was later discovered that below approximately 800 °C, this alloy breaks down into metallic (BCC-Cr) and intermetallic (L10-NiMn and B2-FeCo) phases [20]. It remains uncertain whether this decomposition is driven by faster diffusion pathways or the presence of favorable nucleation sites at grain boundaries. At room temperature, the metastable FCC solid-solution state can be preserved at typical cooling rates following heat treatments above 800 °C.

10.2.3 Synthesis of HEAs

To fabricate HEAs with a singular solid-solution phase along with a uniform elemental arrangement, various chemical and physical synthesis techniques can be employed to regulate thermodynamic and kinetic factors. The choice of synthesis method and processing conditions plays an important role in controlling size, phase structure, and elemental arrangement of HEAs. HEAs with singular phase are effectively produced using a variety of chemical as well as physical methods. High-energy ball milling and arc melting are two reasonably easy mechanical alloying processes that can produce HEAs in greater quantities and with higher particle sizes [21–24]. A prevalent technique for generating nanoporous HEAs involves the dealloying of a physically alloyed multiple-component alloy source. Wet-chemical techniques, including the sol-gel method, solvothermal synthesis, and ultrasonication-assisted wet chemistry, have proven effective for producing nanoscale HEAs [25, 26]. Additionally, other novel synthesis techniques, such

Table 10.1 Synthesis method of NMF-HEAs.

Energy storage application	NMF-HEAs	Physical Form	Synthesis Method	References
Li-ion battery	$[(NaBi)_{0.2}(LiLa)_{0.2}(CeK)_{0.2}Ca_{0.2}Sr_{0.2}]TiO_3$	Bulk	Ball milling and calcination	[22]
	$(MgCoNiZn)_{1-x}Li_xO$ ($x = 0.05, 0.15, 0.25$, and 0.35)	Bulk	Ball milling and calcination	[23]
	$(FeMnNiCoCr)S_2$	Bulk	High-energy ball milling	[24]
	$(Co_{0.2}Cu_{0.2}Mg_{0.2}Ni_{0.2}Zn_{0.2})O$	Nanoparticle	Flame spray pyrolysis, nebulized spray pyrolysis, reverse co-precipitation	[28]
	$Fe_{0.24}Co_{0.26}Ni_{0.10}Cu_{0.15}Mn_{0.25}$	Nanoparticle	Reflow treatment, calcination	[29]
Na-ion Battery	$Na_3VAl_{0.2}Cr_{0.2}Fe_{0.2}In_{0.2}Ga_{0.2}(PO_4)_3$	Crystal	Sol–gel method	[25]
	$Na_3V_{1.9}(Ca,Mg,Al,Cr,Mn)_{0.1}(PO_4)_2F_3$	Crystal	Sol–gel method, solid-state reaction	[30]
	$Na_x(FeMnNiCuCo)[Fe(CN)_6]$	Crystal	Co-precipitation	[31]
Supercapacitors	FeNiCoMnMg	Nanoparticle	Carbothermal shock	[27]
	(TiNbTaZrHf)C	Nanoparticle	Electrochemical	[32]
	AlCoCrFeNi	Nanoporous alloy	Selective dissolution	[33]

as sputtering deposition, fast-moving bed pyrolysis, and carbothermal shock, are used to yield fine-sized HEA NPs, although with relatively nominal yields [27]. Table 10.1 summarizes the synthesis method used for NMF-HEAs, as reported in the literature.

10.3 Applications of NMF-HEAs in Energy Storage Devices

10.3.1 Supercapacitors

SCs have emerged as one of the demanding energy storage technologies to address the energy and environmental concerns. Their key advantages include high-power density, ultrafast charge–discharge rates, high energy conversion efficiency, and long cycle life [34–36]. SCs bridge the performance gap between batteries and electrolytic capacitors. Most SCs operate based on either the electric double-layer formation principle, where charged layers form on the electrode's active surface area, or redox reactions, where charge is stored through oxidation and reduction processes. The materials used for electrodes must have a large surface area and impressive electrical conductivity while maintaining essential mechanical properties [37]. Numerous substances have been investigated for use in SCs, including chalcogenides, conducting polymers, metal oxides, and carbon derivatives. HEAs have, nevertheless, recently drawn a lot of interest for use in charge storage applications. In particular, NMF-HEAs offer an additional advantage by reducing costs, making them feasible for large-scale energy storage. These NMF-HEAs, formed by combining multiple metallic elements, exhibited exceptional electrical and magnetic properties along with outstanding thermomechanical stability at the nanoscale [34]. A summary of reported NMF-HEAs for SCs applications is given in Table 10.2. Nanoscale duplex-phase HEAs are considered a promising option

Table 10.2 NMF-HEAs for application in SCs.

Alloy composition	Specific capacitance ($F\,g^{-1}$)	Displayed potential (V)	Energy density ($Wh\,kg^{-1}$)	Power density ($W\,kg^{-1}$)	Electrolyte	References
FeCoNiCuMn	388	−0.3 to 0.4	33.4	1,700	3 M KOH	[10]
$FeCrCoMnNiAl_{0.75}$	961	0.8	NA	NA	0.5 M Na_2SO_4	[38]
FeCoNiCuZn	325.17	0.4	23.82	325	3 M KOH	[39]
FeNiCoMnMg	203	0.8	21.7	NA	6 M KOH	[27]
CoCrNiFeMn	386.66	−0.1 to 0.5	21.0	307	3 M KOH	[40]
FeCoNiCrZn	~556	−0.2 to 0.4	34.8	800	3 M KOH	[41]
FeCoNiCrMn with green carbon composite	450	1.3	33.5	1,800	3 M KOH	[42]

for creating nanoporous structures through strategically dissolving less noble phases. Continuously dispersed nanoframes with consistent outlines can produce a predictable porosity structure due to preferential dissolution of specific phases, driven by the electrochemical stability differences between FCC and BCC phases in Cantor HEAs (an equiatomic alloy of Co, Cr, Fe, Mn, and Ni). It is anticipated that porous HEA substrates will perform better as pseudocapacitors when a functional oxide layer, like MnO_2, is added. This approach involved selectively dissolving the duplex-phase Cantor $FeCrCoMnNiAl_{0.75}$ HEA, followed by pulse electrodeposition of MnO_2 layers to create metallic frameworks (MnO_2/MF). Because of its large specific surface area along with good electrical conductivity, MnO_2-decorated MnO_2/MF composite electrodes achieved a remarkable specific capacitance. The cyclic voltammetry (CV) curves of MnO_2/MF (Figure 10.3a) retained their symmetrically quasi-rectangular shape, indicating optimal capacitive behavior and outstanding high-rate performance. The capacitive efficiency contribution, calculated from CV curves, increased from 53.0% to 89.9% as the scan rate increased (Figure 10.3b), reaching 89.9% at $400\,mV\,s^{-1}$. At $5\,A\,g^{-1}$, MnO_2/MF electrodes had shown a maximum specific capacitance of $961\,F\,g^{-1}$ (Figure 10.3c). Even with increasing scan rates and current densities, MnO_2/MF electrodes maintained high specific capacitance (Figure 10.3d). This improved capacitance was assigned to the rapid ion/electron transfer between the active MnO_2 and the electrolyte [38]. In 2023, Tiwari et al. developed an HEA containing FeCoNiCuZn by the induction melting method, followed by ball milling to reduce its dimensions [39]. This HEA was utilized as a positive electrode for SC applications in both single-cell and asymmetric devices. The absence of a broad redox peak in the CV curves (Figure 10.3e) indicated that the HEA exhibited capacitive-like characteristics. However, its deviation from a purely rectangular CV suggested that the FeCoNiCuZn HEA was pseudocapacitive or quasi-ideal capacitive. To determine whether charge storage in the FeCoNiCuZn HEA electrode was adsorption controlled (surface controlled) or diffusion controlled, the following equation was applied:

$$I = av^b \tag{10.3}$$

A plot of log (I) vs. log (v) (Figure 10.3f) yielded a slope ("b" value) of approximately 0.66. This value indicated the presence of both a surface-controlled fast Faradaic reaction on the electrode's surface and a slower diffusion-controlled pseudocapacitive redox reaction involving K^+ ions within the FeCoNiCuZn HEA. The ability of these ions to access both inner and outer active sites enhanced the material's specific capacitance. Furthermore, the total charge stored in the electrode, considering both the surface-controlled charge (Q_s) and the slow diffusion-controlled charge (Q_d), was calculated using the following equation [43]:

$$Q(t) = Q_s + Q_d \tag{10.4}$$

The values of Q_s and Q_d were determined by plotting Q (t) against the inverse square root of the sweep rate using Eq. (10.5) (Figure 10.3g).

$$Q(\upsilon) = Q_s + Q_d = Q_{capacitive} + K_1\upsilon^{-1/2} \tag{10.5}$$

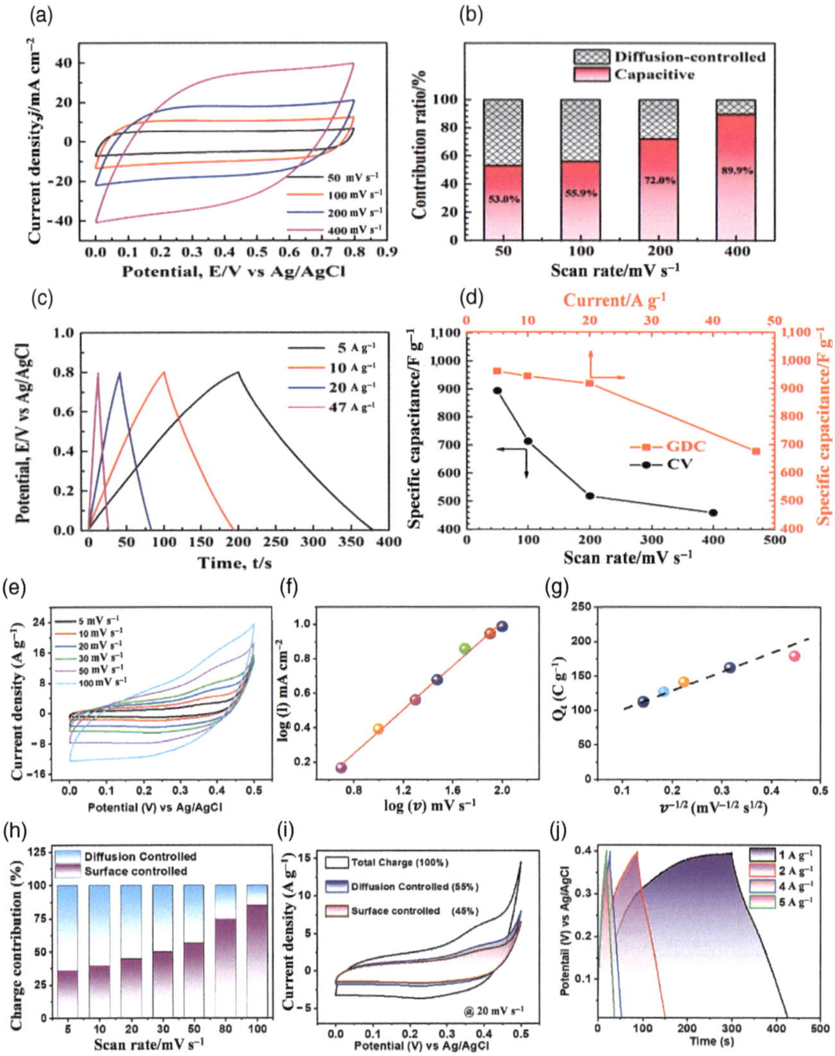

Figure 10.3 (a) CV curves, (b) contribution ratios, (c) GCD curves, and (d) the dependency of specific capacitance on scan rates and GDC currents of the MnO_2-decorated metallic frameworks. *Source:* Reference [38] / with permission of Elsevier, (e) CV curves, (f) log (*I*) vs. log (*v*), (g) total charge vs. scan rate inverse square root, (h) each scan rate contribution from Q_s and Q_d, (i) surface-controlled and diffusion-controlled contributions, (j) GCD curves at various current density. *Source:* Reference [39] / with permission of Elsevier.

The relative contributions of Q_s and Q_d indicated that charge storage was dominated by diffusion-controlled processes at smaller scan rates. This suggested that the noble-metal-free FeCoNiCuZn HEA possessed a highly porous structure with faradaic redox-active sites. Its multivalent quinary ion mesoporous channels facilitated efficient ionic transport with minimal resistance. The fraction of each process contributing to the total charge at different scan rates is shown

(Figure 10.3h). At a scan rate of 1 A g^{-1}, the total charge response comprised 55% diffusion-controlled and 45% surface-controlled contributions (Figure 10.3i). The FeCoNiCuZn HEA exhibited a gravimetric capacitance of 325.17 F g^{-1} at 1 A g^{-1} in a single-electrode system (Figure 10.3j). When used in an asymmetric device, the FeCoNiCuZn//AC device generated a power density of 325 W kg^{-1} and an energy density of 23.82 Wh kg^{-1}.

Xu et al. utilized FeNiCoMnMg HEA-NPs/Aligned electrospun carbon nanofibers (ACNFs) as electrodes for SC applications, demonstrating excellent capacitive performance in terms of specific capacitance and energy density [27]. They have developed the above electrodes by using its chloride precursor solutions at concentrations having 5, 10, 20 mM and evaluated their SC performance. With a 5 mM precursor, the electrode showed the highest capacitance and the highest specific energy density of 203 F g^{-1} and 21.7 Wh kg^{-1}, respectively, among all. These findings suggested that carbonaceous NPs and HEA-NPs (high-entropy alloy NPs [HEA-NPs]) are favorable candidates for energy storage applications. Further, the homogeneous coating enabled high capacitance, constant ion exchange, and efficient charge–discharge cycles, providing durability above 85.0% over 2,000 cycles. The authors claimed that the aligned carbon nanofibers (CNFs) enhance ion transport efficiency, playing a pivotal impact in increasing the development rate of HEA-NPs. Mohanty et al. synthesized a bulk CoCrNiFeMn HEA using a simple induction melting technique, succeeded by ball milling to get NPs for potential use in device-grade SC applications [40]. The reduction in particle size resulted in an increased surface area while maintaining a stable crystalline phase with an FCC structure. When used as an electrode in a 3-electrode system, the material showed maximum specific capacitance of 386.66 F g^{-1}@5 mV s^{-1}. Thus, a 3.0 M KOH was used to develop an asymmetric liquid-state CoCrNiFeMn HEA/activated carbon device, which achieved energy and power densities of 21 Wh kg^{-1} and 307 W kg^{-1}, respectively. The same research group also investigated the role of Zn and Mn in FeCoNiCrZn/Mn HEA for SC applications [41]. Their study revealed that substituting Mn with Zn led to the formation of a porous, flake-like morphology, resulting in an increased specific surface area. This structural enhancement contributed to a substantial enhancement in specific capacitance, increasing from 386 to 556 F g^{-1} at 50 mV s^{-1}. Furthermore, they formed an asymmetric liquid-state device, which exhibited a specific energy density of 34.8 Wh kg^{-1} at a power density of 800 W kg^{-1}, along with a maximum capacitance of 98 F g^{-1}@1 A g^{-1}. The substantial capacitive performance of the FeCoNiCrZn quinary HEA was assigned to redox transitions occurring during the intercalation and deintercalation processes.

10.3.2 Lithium-Ion Batteries

LIBs, that were first accessible to consumers in the 1990s, are now an essential part of portable devices like laptops, cell phones, and electric cars because of their higher volumetric and gravimetric energy densities [44]. Development of novel anode materials employing an alloying mechanism with Li^{+}, including P, Sn, Sb, Si, Bi, and Ge has been widely explored to fulfil the rising need for high energy density. Over conventional intercalation-based materials, these materials provide better theoretical gravimetric capacities (600–4,000 $mAh\,g^{-1}$) and volumetric capacities (>2,000 $mA\,h\,cm^{-3}$) [45]. However, over several cycles, the significant volumetric and structural alterations brought on by lithium insertion and extraction frequently result in a precipitous drop in specific capacity. Additionally, the complex alloying reaction mechanism introduces considerable hysteresis, which varies depending on charge and discharge rates due to its ohmic nature. To address these challenges, NMF-HEAs have been shown as potential materials for energy storage. High configurational entropy is known to stabilize the crystal structures of disordered multi-metallic systems, while multi-principal element mixing enhances structural and chemical diversity. Furthermore, the solid solution strengthening mechanism, where ions of different sizes are substituted, enhances mechanical and physical properties, including superior strength, ductility, fracture resistance, and an exceptional elastic strain limit [46]. Zhao et al. designed a high-entropy alloying-type anode material SnSbMnBiTe that demonstrates remarkable properties such as high electrical conductivity (5.25 $S\,cm^{-1}$), a tapping density of 3.6 $g\,cm^{-3}$, and high mixing entropy (1.61*R*). It also demonstrated viscoelastic characteristics, which help to achieve remarkable low-temperature tolerance, improved cycle stability, and a remarkable volumetric capacity of 2,408.4 $mA\,h\,cm^{-3}$. Additionally, at 5 $A\,g^{-1}$, it exhibits a rate performance of 1,017.4 $mA\,h\,cm^{-3}$ [47]. Additionally, the high-entropy SnSbMnBiTe alloy exhibits outstanding cycling stability, maintaining its capacity for more than 200 cycles with little degradation. Its high volumetric capacity and long-term durability are primarily due to its disordered atomic structure and progressive activation process, which introduce numerous defects and active sites for Li^+ storage, thereby enhancing overall capacity. Solid-state reactions, liquid-phase reactions, and sputtering are usually the primary methods used to produce HEAs. The liquid-phase reaction usually requires surfactants that are difficult to remove and could contaminate the final product, while the other two methods typically require high annealing temperatures and are costly. Additionally, bulk structures, NPs, or nanofibers make up the majority of the reported NMF-HEAs, which results in unstable structures and a dearth of exposed ion storage sites. Thus, designing a simple and environmentally friendly process to synthesize HEA and optimize its morphology remains a crucial challenge. In this regard, Jing et al. used a salt-template approach to create 2D FeCoNiCrMn HEA nanosheets [48]. The area of nanosheets increased to micrometers and the alloying temperature was lowered with the help of the NaCl template (Figure 10.4). At the 120th cycle, the developed FeCoNiCrMn HEA anode showed remarkable rate performance, a high Coulombic efficiency (CE) of 98%, and an exceptional reversible capacity of about 1,026.01 $mA\,h\,g^{-1}$. Detailed characterizations demonstrated that Mn acts as an active metal, providing capacity, and the Cr atom is essential for maintaining structural stability. The study underscores the significant impact of the cocktail effect in improving the energy storage capabilities of HEAs. This study represents a major advancement in the formation of alternative anode materials for energy storage devices and offers a new avenue for the design of HEAs. Using the high-entropy concept as inspiration, Wei et al. [49] objectively developed Li-active and

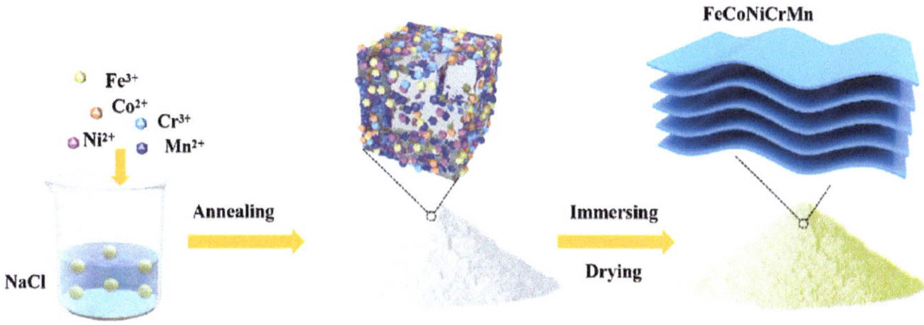

Figure 10.4 Schematic illustration of the preparation of FeCoNiCrMn HEA nanosheets. *Source:* Reference [48] / with permission of Elsevier.

highly conductive components to incorporate the basic elements into the HEA system. The successful synthesis of a particular type of atomic disorder, HEA, which is made up of GeSnSbSiFeCu-P, is significant because it provides ultra-high reversibility for LIBs along with a suitable plateau (approximately 0.5 V) and a high discharge capacity (1,448 mA h g^{-1}). The HEA/C composite demonstrates exceptional rate performances (787 mA h g^{-1} and 63% retention at 2,000 mA g^{-1}) and great cyclability (>1,600 hours operation) because of the aforesaid unique structural design (Figure 10.5). Su et al. [50] investigated the functional contributions of different components by varying the composition of the material in high-entropy amorphous thin-film (HEATF) anodes. Al, Mg, Ge, Sn, and Zn were chosen to merge with Si to create thin-film anodes using magnetron sputtering. These elements are conductive and lithium reactive. High capacity (~2,200 mA h g^{-1}) and an appropriate charging plateau (≈0.4 V) were provided by the prepared Si-based (HEATFs), where $Si_{50}(AlMgGeSn)_{12.5}$ (5- HEATF) had shown the best cycle stability and rate capability. A solid-solution reaction process with limited phase separation of the elements was found to be the cause of the improved structural stability using X-ray photoemission spectroscopy and in-situ transmission electron microscopy. Internal stress was reduced, and structural integrity was maintained by this mechanism. Moreover, the development of Sn nanocrystals immersed in the amorphous matrix increases ion diffusion electrokinetics. The results offer a fresh method for designing enhanced-energy-density thin-film LIB anodes and highlight how crucial the structural and elemental complexity of Si-based HEAs is to enhance their electrochemical performance. To improve metallic conductivity, Li et al. [51] modified stable diamond-structured Ge with P to create a tetrahedrally coordinated sphalerite structure, which was then further stabilized by adding Zn and other elements. This allowed them to study high-entropy element screening rules. Additionally, it was confirmed by both theoretical and experimental evidence that Li-storage performance increases as the atomic number increases. The sequence is $BZnGeP_3$ < $AlZnGeP_3$ < $GaZnGeP_3$ < $InZnGeP_3$. The greatest Li-ion affinity, fastest Li-ion and electronic transit, greatest Li-storage capacity and reversibility, and best mechanical integrity were all displayed by $InZnGeP_3$-based electrodes. With high-rate performance, maintaining 453 mA h g^{-1} at 20,000 mA g^{-1}, $InZnGeP_3$ demonstrated

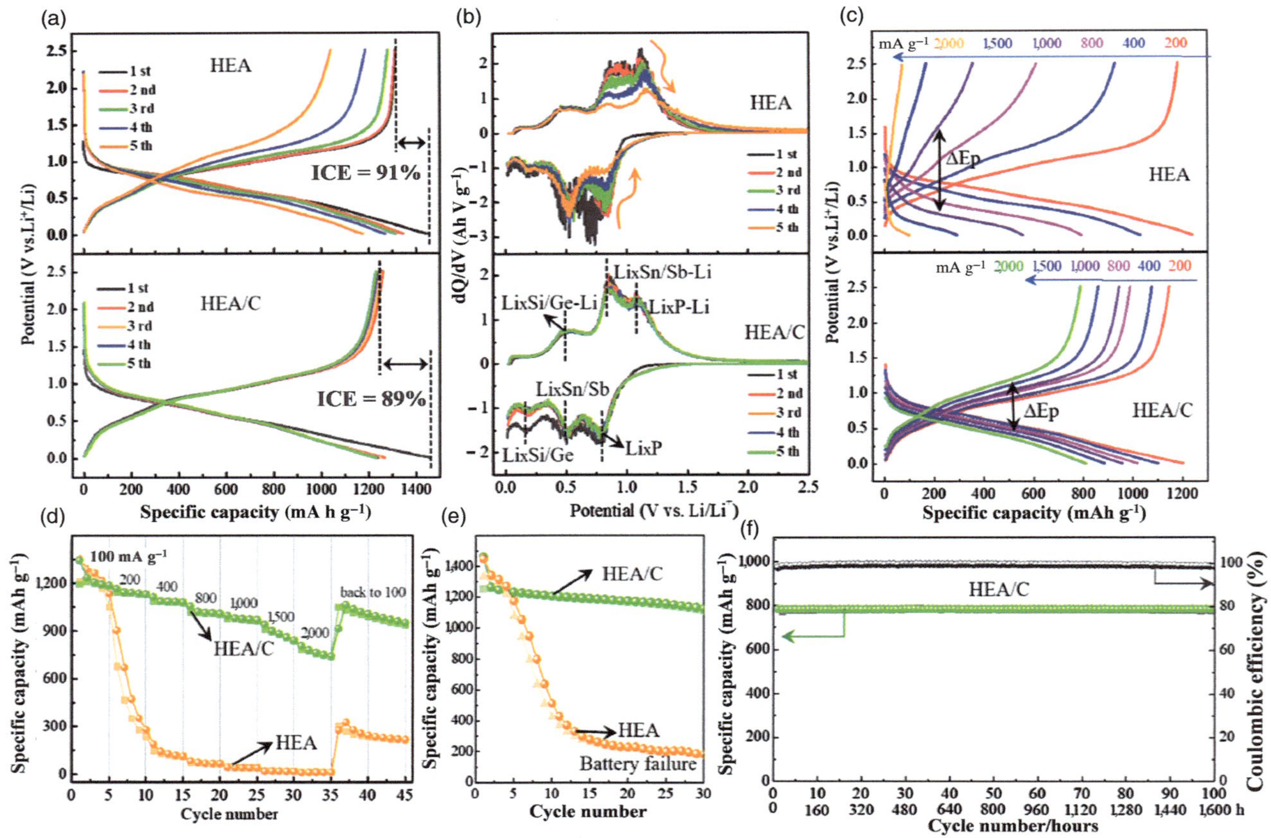

Figure 10.5 (a) Electrochemical performances of HEA and HEA/C composite, (b) comparison of the discharge/charge profiles, *dQ/dV* curves, (c) discharge/charge profiles at different current densities, (d) rate capability, and (e–f) cycle stability between HEA and HEA/C electrodes. *Source:* Reference [49] / with permission of Elsevier.

stable cycling, holding onto 1,293 mA h g^{-1}@400 mA g^{-1} after 650 cycles and 904 mA h g^{-1}@2,000 mA g^{-1} over 1,500 cycles. The Li-storage mechanisms of the $InZnGeP_3$ alloy are illustrated (Figure 10.6a). Zhu et al. [52] effectively synthesized a CuCoNiMnAl into a 2D ultra-thin nanosheet structure HEA by employing a straightforward salt-templated technique. Various annealing temperature are used to synthesize ultra-thin nanosheets. When used for lithium-ion storage, the catalyst has a reversible capacity of roughly 700 mA h g^{-1}@100 mA g^{-1} after 80 cycles. These alloys also have a remarkable rate capacity. The CuCoNiMnAl HEAs exhibited persistent electrochemical endurance with little degradation in alkaline conditions, having 390 mV@10 mA cm^{-2}. The salt-templated two-dimensional structure enhances the specific surface area while shortening the ion diffusion path, enabling the rapid insertion of additional lithium ions. This improved ion transport electrokinetics, coupled with the enhanced structural integrity and conductivity of the HEA, contributes to its overall stability. Wei et al. [53] successfully developed a $Zn_xGe_yCu_zSi_wP_2$ HEA via elemental modulation for use as an anode in LIBs. This novel HEA exhibited a good initial CE (93%), a lower volume expansion (34.5%), and an elevated Li storage capacity (>1,500 mA h g^{-1}). Its outstanding Li storage performance suggests that the HEA made of the metallic-P solid solution holds great potential for LIBs in an efficient manner. On the other hand, the HEA material's ability to maintain capacity stability beyond 1,000 cycles was not demonstrated. In this regard, Xiao et al. [54] created a novel, inexpensive, amorphous high-entropy SnSiCoCu-P alloy/graphite composite as an anode material for LIBs by replacing Li-inactive atoms with active ones. This HEA/C composite offered remarkable cycling durability with a specific capacity of 1,196.5 mA h g^{-1} after 1,000 cycles, a high initial discharge capacity exceeding 1,881 mA h g^{-1}, a strong rate capability (47.2% capacity retention@10 A g^{-1}), and nearly 100% CE. The remarkable electrocatalytic activity was assigned to the high-entropy effect, that helped in keeping structural durability during charge–discharge cycles. Additionally, the graphite coating effectively buffered volume expansion and enhanced electron transport, thereby improving the material's durability and overall lifespan.

10.3.2.1 Lithium Sulfur Batteries

Lithium-sulfur (Li-S) batteries are regarded as to be among the finest intriguing possibilities for the foreseeable future rechargeable batteries, offering an energy density 3–5 times higher than that of LIBs. During the cathodic redox reaction in Li-S batteries, lithium polysulfides (LiPSs) undergo a series of intricate transformations. Li-S batteries have a high specific capacity (1,675 mAh/g) and theoretical energy density (2,600 Wh/kg) due to the conversion of elemental sulfur (S_8) into Li_2S through a multi-electron (16 e^-) transfer process. This transformation involves a solid → liquid → solid phase transition [55]. Unfortunately, the slow redox kinetics and shuttle effect (Figure 10.6b) during the charge and discharge operations prevent Li–S batteries from increasing their energy density. The five diffusion steps for the shuttle effect are (i) generating a long chain polysulfide; (ii) separating from the sulfur host; (iii) diffusing into the electrolyte; (iv) migrating to the anode side; and (v) reacting with the lithium anode. Major difficulty in Li-S battery was insulation

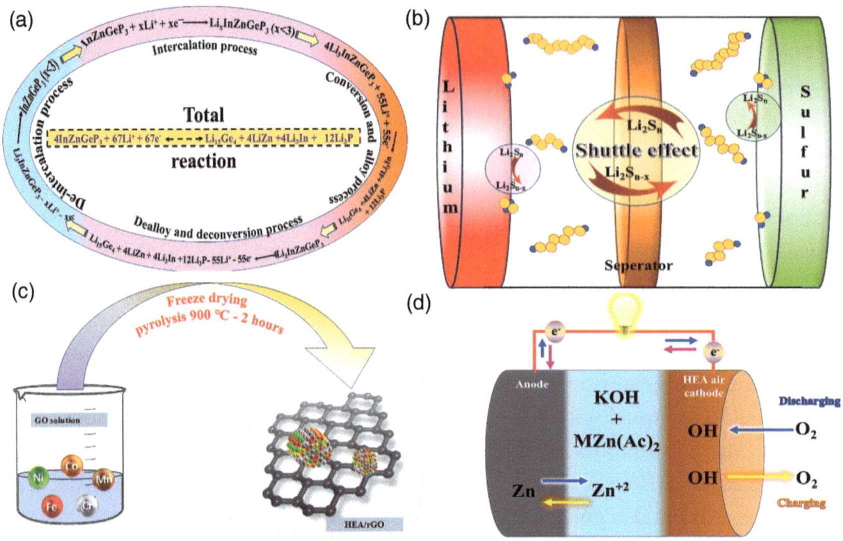

Figure 10.6 (a) The Li-storage mechanism of the $InZnGeP_3$ alloy. *Source:* Reference [51] / with permission of John Wiley & Sons, (b) the shuttle effect of polysulfide in Li-S battery, (c) a schematic representation of the HEA/rGO fabrication process, and (d) a schematic representation of the ZAB.

of sulfur and lithium sulfide, which led to a low active material utilization rate. Lithium sulfide and sulfur have densities of 1.66 and 2.06 $g\,cm^{-3}$, respectively, which results in high volume changes of 80% during the discharging–charging process [56]. Given the complexity of LiPSs conversion, it is difficult to identify a single-component catalyst capable of fully optimizing Li-S battery performance. That's where HEA comes in as versatile and tunable heterogeneous catalyst, effectively driving multi-electron and multiphase redox reactions of LiPSs, paving the way for better battery performance.

Xu et al. [57] synthesized nano-HEA to significantly improve the multi-electron and multiple-phase transformations of LiPSs in Li-S battery. Because of its high affinity for LiPSs, nano-HEA greatly reduces concentration polarization by enriching the kinetics of LiPSs adjacent to the electrode, which was 17 times greater in comparison to of the reference sample (lacking nano-HEA incorporation). Additionally, this nano-HEA catalyst significantly reduced the activation polarization, as evidenced by the greater current response, higher exchange current density, and lower Tafel slope. Consequently, Li-S batteries that are constructed with nano-HEA augmented separators exhibited exceptional capacity retention rates of 82% (0.1 C after 150 cycles in a pouch cell) and 83.3% (2 C after 500 cycles in a coin cell). Although sulfur has an extremely high theoretical specific capacity, its use in the entire cathode is reduced due to its slow kinetics and discharged products, which limit the improvement in energy density of Li-S batteries.

Wang et al. [58] used an HEA, $Fe_{0.24}Co_{0.26}Ni_{0.10}Cu_{0.15}Mn_{0.25}$, to initiate the sulfur cathode's electrochemical activity in Li-S batteries. With the complete active cathode (rather than just the sulfur element), the S/HEA-NC cathode yielded a high specific capacity of 1,079.5 mA h $g_{cathode}^{-1}$, which was equivalent to a high sulfur utilization of 89.4%. Furthermore, the sulfur cathode still attained a high discharge capacity of 868.2 mA h $g_{cathode}^{-1}$ under both ultrahigh sulfur loading (27.0 mg cm^{-2}) and lean electrolyte (3 μl mg^{-1}). This approach paved the way for further research into catalytic host materials to enhance S usage across the Li-S battery cathode.

Yao et al. [59] investigated FeCoNiCrMn HEA NPs evenly dispersed across reduced graphene oxide (HEA/rGO) by the calcination and freeze-drying method (Figure 10.6c). To enhance its charge transfer activity and the intrinsic conductivity, rGO provides a remarkable specific surface area and a well-developed porous structure that acts as a quick ion/electron transport channel. Further, well-integrated HEA NPs facilitate charge re-allocation and create well-structured electronic environment, efficiently trapping LiPSs along with catalyzing their transformation activities. The HEA/rGO-integrated separator exhibits excellent electrical conductivity, robust adsorption ability, and numerous active sites, resulting in exceptional electrochemical performance, achieving an initial capacity of 1344.6 mA h g^{-1} at 0.1 C and maintaining 782 mA h g^{-1} at 2 C. This chapter proposes an efficient technique for improving capacity and extending the cycle lifespan of Li-S batteries.

Subsequently, Wang et al. [60] used a hydrothermal process to develop 5-membered organometallic framework compounds (CoNiCuMnMo-MOF) on graphene oxide (GO) through pyrolytic reduction. It performed exceptionally well as a catalytic intercalation material for the Li-S battery separator. This was mostly because the catalyst had a diameter of around 12 nm and was widely distributed throughout rGO at a substantial density to prevent accumulation and create a more porous structure. The results of the electrochemical tests demonstrated that mentioned catalyst had a strong capacity for lithium polysulfide adsorption, efficiently speeds up Li^+ diffusion and mobility, and simultaneously possesses the catalytic use of efficiently speeding up the transformation of Li_2S_n ($4 \leq n \leq 8$) into Li_2S_2/Li_2S. At the same time, the redox process was efficiently controlled by the presence of the Cu element, leading to excellent bidirectional catalysis in HEA. An Li-S battery equipped with the catalyst with rGO modified membrane achieves an initial specific discharge capacity of 1,512.4 mA h g^{-1} at 0.2 C. Even after 1,100 cycles at 3 C, it maintains a CE exceeding 91.94%, with a minimal capacity attenuation of just 0.055%/cycle. Chen and colleagues [61] synthesized HEA NP as nitrogen-doped carbon nanotubes (FeCoNiZnCu-NCNTs) to facilitate the multiple-step redox processes in Li-S batteries. Their study revealed that the FeCoNiZn HEAs' adsorption of LiPS was increased by the addition of Cu, which raised the band centers of Fe, Co, and Ni. This optimized electronic structure significantly improved the bidirectional catalytic activity of FeCoNiZnCu-NCNTs compared to FeCoNiZn-NCNTs, thereby improving the overall electrochemical activity of Li-S batteries. In comparison to FeCoNiZn-NCNTs, the Li-S batteries containing Cu element demonstrated enhanced rate performance, achieving 679.7 mA h g^{-1}@2 C, along with improved cycling stability, exhibiting a low-capacity decay of just 0.072%/cycle over 600 cycles at 1 C^{-1}, compared to FeCoNiZn-NCNTs. Through element selection and electron structure modification, the study proved the extravagance of HEA in catalyzing redox reactions in Li-S batteries. A very effective electrocatalyst (CNT/HEA-NC) was created by Ma et al. [62] using HEA NPs embellished with conductive networks of carbon nanotubes (CNTs) and nitrogen-doped carbon (NC). The CNT provides a conductive network that facilitated rapid electron and ion transport, while the NC exhibiting robust sulfophilic characteristics successfully absorbed LiPSs, alleviating the shuttle effect. Additionally, the highly catalytic HEA NPs accelerated the bidirectional conversion of LiPSs within the complex nanostructure. Li–S batteries with separators modified with CNT/HEA-NC demonstrated exceptional stability, owing to the advantages of hierarchical confinement, delivering a discharge capacity of 692.0 mA h g^{-1} after 300 cycles, with a minimal degradation rate of just 0.03%/cycle. Furthermore, the cell maintained an impressive capacity of 521.1 mA h g^{-1} at 5 C. This research introduces an innovative approach for the construction of consolidated versatile electrocatalysts to enhance high-energy-density Li-S batteries.

10.3.2.2 Lithium Metal Batteries

Lithium metal is an effective high-performance anode component due to its modest electrochemical value (−3.04 V relative to the standard hydrogen electrode) and exceptional theoretical capacity of 3,860 mA h g^{-1} [63]. The three crucial characteristics for stabilizing lithium metal anodes are the presence of consistent lithiophilic active sites to ensure uniform Li^+ distribution [29], stable mechanical properties to prevent lithium dendrite growth [64], and anti-corrosion properties to prevent adverse reactions with the electrolyte [65, 66]. Using magnetron sputtering at room temperature, Zheng et al. [67] successfully developed HEA interphase on a lithium metal surface to study the utilization of HEAs in lithium metal batteries. The numerous lithiophilic sites present in HEA efficiently control the distribution of Li^+ flux and promote Li^+ deposition. Furthermore, AZCNF's (Al, Zn, Co, Ni, and Fe) magnetic characteristics produce a horizontal Lorentz force, which successfully prevented Li dendrite formation. Additionally, the high mechanical strength and anti-corrosion qualities helped to create a very stable anode/electrolyte interphase, which hinders the proliferation of lithium dendrites and the corrosion that goes along with them. Thus, for more than 4,000 hours at 2 mA cm^{-2}, lithium anodes treated with AZCNF exhibit remarkable stability in Li^+ stripping/plating cycling. The lithium iron phosphate (LFP) complete battery had shown an average CE of 99.7%. Thus, through high-entropy manipulation, the HEA interphases on Li metal surfaces provided adjustable management of ion accumulation behavior, bringing up new possibilities for durable Li metal batteries. Using a high-temperature redox technique, Wang et al. [68] applied lithophilic CuInNiSnCd HEA-NPs to the surface of 3D carbon hosts. By reducing the anode electrode's non-uniform current density, the HEA/CF 3D framework efficiently mitigates the volumetric growth of metallic lithium. The above catalyst minimizes the nucleation potential of lithium and enhances the lithophilicity of the carbon substrate. At 2 mA cm^{-2}/10 mA h cm^{-2}, the mentioned framework half-cell maintained a significant CE of around 89% for 800 hours. Furthermore, the Li@HEA/CF-based symmetrical cell exhibited stable performance for over 3,000 hours, demonstrating minimal hysteresis voltage at 10 mA cm^{-2}/10 mA h cm^{-2}. Additionally, after 160 cycles at 1 C, the full cell demonstrated constant cycling behavior with a CE of nearly 99% when paired with an NCM-811 cathode. This study provides valuable insights for enhancing lithium battery performance and supports the utilization of HEAs in advanced energy storage systems.

10.3.3 Zinc Air Batteries

Zinc air batteries (ZAB)s are among the most advanced energy storage technologies for their substantial energy density, ecological sustainability, and outstanding reliability. Primary ZABs are already extensively utilized in applications such as networking devices and hearing aids. However, a number of technical concerns pertaining to the zinc anode, electrolyte, and air cathode limit the useful applications of rechargeable ZABs. In the discharge–charge cycles, the lethargic kinetics of the charging oxygen evolution reaction (OER) and discharging oxygen reduction reaction (ORR) at the air cathode severely restricted the electrocatalytic activity of rechargeable ZABs [69]. This can be addressed by modifying the crystalline phases, surface atom coordination, and engineering compositions of HEAs, which offer countless design options for catalyst advancements. To this end, Fang et al. [70] synthesized NP-HEA/HEO catalyst using noble-metal-free AlFeCoNiCr. When used in aqueous ZABs, it showed a high specific capacity of approximately 800 mA h g^{-1}. Additionally, in solid-state ZABs, it exhibited a small discharge–charge potential gap of 0.76 V at 2 mA cm^{-2} with cycling stability for 60 hours. He et al. [71] used a low-temperature solution-based way to synthesize a NMF CrMnFeCoNi HEA. This HEA demonstrated exceptional OER activity, 265 mV at j value of 10 mA cm^{-2}, significantly outperforming the conventional RuO_2-based commercial catalyst. When employed as the cathode in rechargeable ZABs, the CrMnFeCoNi HEA achieved an open-circuit voltage of 1.489 V, attaining 836 mA h g^{-1} at 8 mA cm^2. Additionally, the HEA-ZAB exhibited remarkable durability, sustaining continuous charge–discharge cycling for over 240 hours at 8 mA cm^{-2} and 400 hours at 5 mA cm^{-2}.

Madan et al. [72] described a nanocrystalline HEA composed of Cu–Co–Mn–Ni–Fe (CCMNF) that was made via a casting-cum-cryomilling process. The electrochemical redox processes were improved by this multicomponent HEA because it contained numerous catalytically active sites with various functions. During testing for the air electrode catalyst of ZAB (Figure 10.6d), the HEA demonstrated stable charge–discharge performance. The ZAB utilizing the CCMNF catalyst achieved an energy density of 481.5 W kg^{-1}, while its specific capacity in an aqueous system measured at 688.6 mA h g^{-1}. Cao et al. [73] successfully deposited $Fe_{12}Ni_{23}Cr_{10}Co_{55-x}Mn_x$ HEA NPs onto CNTs using a liquid-phase reduction method followed by high-temperature sintering. Their research explored how variations in the Co: Mn influenced the design of the catalyst, ultimately impacting the performance of ZABs. The findings indicate that when the Co/Mn ratio is optimized at 30/25, the $Fe_{12}Ni_{23}Cr_{10}Co_{30}Mn_{25}$/CNT catalyst exhibits exceptional electrochemical performance, achieving a remarkable specific capacity of 760 mA h g^{-1} and an energy density of 865.5 Wh kg^{-1} in a liquid ZAB, along with remarkable

Table 10.3 List of NMF-HEAs used in battery applications.

Type	Alloy composition	Specific capacity ($mA\,h\,g^{-1}$)	Cycling performance	Advantages	Challenges	References
Li-ion battery	SnSbMnBiTe	—	Stable over 1,200 cycles	Excellent structural and electrochemical reversibility	Suppressed volume expansion	[47]
	FeCoNiCrMn	1,026.01	Stable over 500 cycles	High structural stability, enhanced energy storage performance	—	[48]
	GeSnSbSiFeCu-P	1,448	Stable Over 1,600 hours	Unique structural design and high cycling stability	Complexity in synthesis	[49]
	CuCoNiMnAl	700	Stable over 80 cycles	Exhibit long-term durability in electrochemical performance	—	[52]
	$Zn_5Ge_5Cu_5Si_5P_2$	1,204	Stable over 80 cycles	High-entropy stabilization, superior cyclability, and rate performances	Complexity in designing and synthesis	[53]
Zn-ion battery	CrMnFeCoNi	836	Stable over 1,200 cycles	Electrochemical stability in a harsh environment	Structure instability due to lattice distortions	[71]
	CuCoMnNiFe	—	Stable over 1,045 cycles	High stability and durability	Electrochemical interface instability	[72]
	FeNiCrCoMn	760	Stable Over 256 hours	Excellent bifunctional catalytic performance	Corrosion and degradation in a harsh environment	[73]
	$Fe_6Ni_{20}Co_2Mn_2Cu_{1.5}$@rGO	800	Stable over 300 hours	High-energy efficiency and good structural stability	Complexity in nanoparticle synthesis	[74]
	FeCoNiMnV/N-CNTs	—	Stable over 1,698 cycles	Enhance energy storage capabilities and improved charge transfer	Metal leaching and structural degradation	[75]
	FeCoNiCuMo	809	Stable over 1,576 cycles	Excellent mechanical stability	Material optimization for large scale	[76]
	FeCoNiMoW	857	Stable over 1,950 cycles	High energy density and power performance	Tuning of the active site for optimal performance	[77]

cycle stability exceeding 256 hours. Gao et al. [74], on the other hand, used a Joule heating technique to manufacture a multilayer rGO enclosing NMF-HEA (Fe_6Ni_{20}-$Co_2Mn_2Cu_{1.5}$@rGO). The rGO layer plays a crucial role in preventing NP aggregation, ensuring their uniform dispersion within the substrate. Additionally, its high crystallinity provides effective protection against particle detachment in harsh conditions, thereby enhancing the catalytic activity and stability of the HEA. As a result of the rGO encapsulation, Fe_6Ni_{20}-$Co_2Mn_2Cu_{1.5}$@rGO-based rechargeable ZABs demonstrated outstanding electrocatalytic activity, achieving a high specific capacity of 800 mA h $g_{Zn}{}^{-1}$, a peak power density of 154.612 mW cm^{-2} with the stability of 300 hours at 10 mA cm^{-2}. Wu et al. [75] utilized a one-step pyrolysis process at 800 °C to create an NMF FeCoNiMnV HEA supported by FeCoNiMnV HEA/N-CNTs. They optimized by adjusting the temperature of the pyrolysis process in the control groups. Leveraging the superior ORR catalytic properties as well as durability, the FeCoNiMnV HEA/N-CNTs catalyst was utilized as the cathode material for ZAB assembly to assess its practicality. For comparison, a hybrid Pt/C + RuO_2-based battery was also constructed. The assembled battery demonstrated a higher power density (185.12 mW cm^{-2}) and a higher open-circuit voltage (1.482 V).

Lin et al. [76] developed a cost-effective bifunctional FeCoNiCuMo HEA catalyst using a basic single-step pulse electrodeposition method and used it as a cathode in ZABs. The catalyst demonstrated impressive performance, achieving a high peak power density of 298 mW cm^{-2}@412 mA cm^{-2}, along with exceptional cycling stability over 1,576 cycles (788 hours) at 10 mA cm^{-2}, with only a slight voltage gap increase of 6.2%. They were also successful in producing a flexible ZAB by altering the electrolyte and incorporating CoNi-layered double hydroxide nanowire arrays into poly(vinyl alcohol) gel. This change increased the battery's performance as well as lifespan by increasing the peak power density to 102.9 mW cm^{-2} and extending the operational stability to 58 hours at 5 mA cm^{-2}. Furthermore, even after being bent repeatedly, the flexible ZAB maintained a constant voltage gap, demonstrating remarkable mechanical stability. A summary of reported NMF HEAs used in batteries for energy storage is presented in Table 10.3.

10.4 Conclusions and Perspectives

Besides being low-cost and eco-friendly alternatives for structural materials, energy storage, and catalysis, NMF-HEAs are a very emerging area in materials science. They enhance electrocatalytic applications, such as fuel cells, CO_2 reduction, and hydrogen evolution, through the generation of diverse atomic structures by means of enhanced stability and activity. Structural applications have also been found due to their mechanical characteristics including hardness, strength, and toughness of these alloys. In particular, the enhancement of phase stability and microstructure evolution will lead them to a deeper function. They are also promising candidates as alternatives in severe conditions, like high-temperature reactors and aerospace applications, because of their corrosion resistance and thermal stability. This chapter highlights the recent progress on energy storage applications from a few selected HEAs, in particular as battery and SC electrodes (adjustable compositions can contribute to prolonged lifetime, effectiveness, and charge–discharge rates). In addition, HEAs can be further improved as electrodes of batteries and SCs by introducing carbon-based materials such as graphene, CNTs, or porous carbon, which can improve electrical conductivity, surface area, and structural stability. Better cycling stability, increased capacitance, and quicker charge transfer are made possible by this synergy. With their adaptability and sustainability, NMF-HEAs have the potential to completely transform a number of industries as research advances.

Author Contributions

Yukti Setia and Nikita Bhatt contributed equally to this chapter.

References

1 Yan, X., Zhou, Y., and Wang, S. (2025). Nano-high entropy materials in electrocatalysis. *Adv Funct Mater* 35 (2): 2413115.

2 Ma, Y., Ma, Y., Wang, Q. et al. (2021). High-entropy energy materials: challenges and new opportunities. *Energy Environ Sci* 14 (5): 2883–2905.

3 Cantor, B., Chang, I.T., Knight, P., and Vincent, A.J. (2004). Microstructural development in equiatomic multicomponent alloys. *Mater Sci Eng A* 375: 213–218.

4 Yeh, J.W., Chen, S.K., Lin, S.J. et al. (2004). Nanostructured high-entropy alloys with multiple principal elements: novel alloy design concepts and outcomes. *Adv Eng Mater* 6 (5): 299–303.

5 Zhang, Y., Wang, D., and Wang, S. (2022). High-entropy alloys for electrocatalysis: design, characterization, and applications. *Small* 18 (7): 2104339.

6 Kamaruddin, H., Jianghong, Z., Yu, L. et al. (2024). A review of noble metal-free high entropy alloys for water splitting applications. *J Mater Chem A* 12 (17): 9933–9961.

7 Li, H., Lai, J., Li, Z., and Wang, L. (2021). Multi-sites electrocatalysis in high-entropy alloys. *Adv Funct Mater* 31 (47): 2106715.

8 Liu, M., Zhang, Z., Okejiri, F. et al. (2019). Entropy-maximized synthesis of multimetallic nanoparticle catalysts via a ultrasonication-assisted wet chemistry method under ambient conditions. *Adv Mater Interfaces* 6 (7): 1900015.

9 Panda, P.K., Huang, H.Y., Dash, P. et al. (2025). Liquid-phase microwave synthesis of platinum-based high-entropy alloy catalysts on carbon supports for electrochemical hydrogen adsorption/desorption and oxygen evolution/reduction reactions. *Int J Hydrog Energy* 111: 536–545.

10 Mohanty, G.C., Chowde Gowda, C., Gakhad, P. et al. (2024). Enhanced energy density of high entropy alloy (Fe-Co-Ni-Cu-Mn) and green graphene hybrid supercapacitor. *Energy Storage* 6 (4): e668.

11 Bib Khan, J., Kumar Panda, P., Dash, P., and Hsieh, C.T. (2025). Microwave-assisted synthesis of platinum-free high-entropy alloy catalysts on reduced graphene oxide sheets for enhanced oxygen reduction and evolution reactions. *Chem Eur J* 31 (11): e202403863.

12 Zeng, K., Zhang, J., Gao, W. et al. (2022). Surface-decorated high-entropy alloy catalysts with significantly boosted activity and stability. *Adv Funct Mater* 32 (33): 2204643.

13 Wang, J., Wang, Y., Lu, X. et al. (2024). Ultra-sleek high entropy alloy tights: realizing superior cyclability for anode-free battery. *Adv Mater* 36 (11): 2308257.

14 Amiri, A. and Shahbazian-Yassar, R. (2021). Recent progress of high-entropy materials for energy storage and conversion. *J Mater Chem A* 9 (2): 782–823.

15 Han, C., Fang, Q., Shi, Y. et al. (2020). Recent advances on high-entropy alloys for 3D printing. *Adv Mater* 32 (26): 1903855.

16 Zhang, X., Ren, K., Liu, Y. et al. (2024). Progress on entropy production engineering for electrochemical catalysis. *Acta Phys -Chim Sin* 40 (7): 2307057.

17 Sun, L., Wen, K., Li, G. et al. (2024). High-entropy alloys in catalysis: progress, challenges, and prospects. *ACS Mater Au* 4 (6): 547–556.

18 Hussain, I., Lamiel, C., Ahmad, M. et al. (2021). High entropy alloys as electrode material for supercapacitors: a review. *J Energy Storage* 44: 103405.

19 Zhao, Y.J., Qiao, J.W., Ma, S.G. et al. (2016). A hexagonal close-packed high-entropy alloy: the effect of entropy. *Mater Des* 96: 10–15.

20 George, E.P., Raabe, D., and Ritchie, R.O. (2019). High-entropy alloys. *Nat Rev Mater* 515–534.

21 Broge, N.L., Bondesgaard, M., Søndergaard-Pedersen, F. et al. (2020). Autocatalytic formation of high-entropy alloy NPs. *Angew Chem Int Ed* 132 (49): 22104–22108.

22 Yan, J., Wang, D., Zhang, X. et al. (2020). A high-entropy perovskite titanate lithium-ion battery anode. *J Mater Sci* 55: 6942–6951.

23 Lokcu, E., Toparli, C., and Anik, M. (2020). Electrochemical performance of $(MgCoNiZn)_{1-x}Li_xO$ high-entropy oxides in lithium-ion batteries. *ACS Appl Mater Interfaces* 12 (21): 23860–23866.

24 Lin, L., Wang, K., Sarkar, A. et al. (2022). High-entropy sulfides as electrode materials for Li-ion batteries. *Adv Energy Mater* 12 (8): 2103090.

25 Li, M., Sun, C., Ni, Q. et al. (2023). High entropy enabling the reversible redox reaction of V^{4+}/V^{5+} couple in NASICON-type sodium ion cathode. *Adv Energy Mater* 13 (12): 2203971.

26 Gu, Z.Y., Guo, J.Z., Cao, J.M. et al. (2022). An advanced high-entropy fluorophosphate cathode for sodium-ion batteries with increased working voltage and energy density. *Adv Mater* 34 (14): 2110108.

27 Xu, X., Du, Y., Wang, C. et al. (2020). High-entropy alloy NPs on aligned electronspun carbon nanofibers for supercapacitors. *J Alloys Compd* 822: 153642.

28 Sarkar, A., Velasco, L., Wang, D.I. et al. (2018). High entropy oxides for reversible energy storage. *Nat Commun* 9 (1): 3400.

29 Wang, Z., Ge, H., Liu, S. et al. (2023). High-entropy alloys to activate the sulfur cathode for lithium–sulfur batteries. *Energy Environ Mater* 6 (3): e12358.

30 Fu, W., Li, B., Wang, P. et al. (2024). A high-entropy carbon-coated $Na_3V_{1.9}(Mg, Cr, Al, Mo, Nb)_{0.1}(PO_4)_2F_3$ cathode for superior performance sodium-ion batteries. *Ceram Int* 50 (9): 16166–16171.

31 Ma, Y., Ma, Y., Dreyer, S.L. et al. (2021). High-entropy metal–organic frameworks for highly reversible sodium storage. *Adv Mater* 33 (34): 2101342.

32 Sure, J., Sri Maha Vishnu, D., Kim, H.K., and Schwandt, C. (2020). Facile electrochemical synthesis of nanoscale (TiNbTaZrHf) C high-entropy carbide powder. *Angew Chem Int Ed* 132 (29): 11928–11933.

33 Kong, K., Hyun, J., Kim, Y. et al. (2019). Nanoporous structure synthesized by selective phase dissolution of AlCoCrFeNi high entropy alloy and its electrochemical properties as supercapacitor electrode. *J Power Sources* 437: 226927.

34 Dissanayake, K. and Kularatna-Abeywardana, D. (2024). A review of supercapacitors: materials, technology, challenges, and renewable energy applications. *J Energy Storage* 96: 112563.

35 Kumar, Y.A., Roy, N., Ramachandran, T. et al. (2024). Shaping the future of energy: the rise of supercapacitors progress in the last five years. *J Energy Storage* 98: 113040.

36 Shuja, A., Khan, H.R., Murtaza, I. et al. (2024). Supercapacitors for energy storage applications: materials, devices and future directions: a comprehensive review. *J Alloys Compd* 176924.

37 Ghosh, A., Kaur, S., Verma, G. et al. (2024). Enhanced performance of laser-induced graphene supercapacitors via integration with candle-soot NPs. *ACS Appl Mater Interfaces* 16 (31): 40313–40325.

38 Yuan, Y., Xu, Z., Han, P. et al. (2021). MnO_2-decorated metallic framework supercapacitors fabricated from duplex-phase $FeCrCoMnNiAl_{0.75}$ Cantor high entropy alloy precursors through selective phase dissolution. *J Alloys Compd* 870: 159523.

39 Mohanty, G.C., Gowda, C.C., Gakhad, P. et al. (2023). Iron-cobalt-nickel-copper-zinc (FeCoNiCuZn) high entropy alloy as positive electrode for high specific capacitance supercapacitor. *Electrochim Acta* 470: 143272.

40 Mohanty, G.C., Gowda, C.C., Gakhad, P. et al. (2023). High energy density liquid state asymmetric supercapacitor devices using Co–Cr–Ni–Fe–Mn high entropy alloy. *Adv Mater* 4 (17): 3839–3852.

41 Mohanty, G.C., Gowda, C.C., Gakhad, P. et al. (2024). Effect of Zn/Mn on the supercapacitor behavior of high-entropy FeCoNiCrZn/Mn alloy. *Energy Adv* 3 (12): 2972–2985.

42 Mohanty, G.C. and Verma, A. (2025). Supercapacitive study of a high entropy alloy and green carbon composite electrode with a wide potential window (1.3 V) and its application towards an aqueous symmetric device. *N J Chem* 49 (2): 521–529.

43 Gogotsi, Y. and Penner, R.M. (2018). Energy storage in nanomaterials—capacitive, pseudocapacitive, or battery-like? *ACS Nano* 12 (3): 2081–2083.

44 Sang, P., Chen, Q., Wang, D.Y. et al. (2023). Organosulfur materials for rechargeable batteries: structure, mechanism, and application. *Chem Rev* 123 (4): 1262–1326.

45 Peng, M., Shin, K., Jiang, L. et al. (2022). Alloy-type anodes for high-performance rechargeable batteries. *Angew Chem Int Ed* 134 (33): e202206770.

46 Han, X., Wu, G., Zhao, S. et al. (2023). Nanoscale high-entropy alloy for electrocatalysis. *Matter* 6 (6): 1717–1751.

47 Zhao, J., Wei, Z., Chen, N. et al. (2024). High-entropy alloy anodes for low-strain and high-volumetric lithium-ion storage at ambient and subzero temperatures. *Energy Storage Mater* 65: 103127.

48 Li, J., Xu, W., Yin, W. et al. (2025). Rational design of large-scale high-entropy alloy nanosheets anode with excellent lithium storage performance. *Mater Chem Phys* 130450.

49 Wei, Y., Liu, X., Yao, R. et al. (2023). Embedding the high entropy alloy NPs into carbon matrix toward high performance Li-ion batteries. *J Alloys Compd* 938: 168610.

50 Su, Y., Lei, X., Chen, W. et al. (2024). Si-based all-lithium-reactive high-entropy alloy for thin-film lithium-ion battery anode. *Chem Eng J* 500: 157197.

51 Li, X., Wang, J.H., Yang, L. et al. (2024). Element screening engineering for high-entropy alloy anodes: achieving fast and robust Li-storage with optimal working potential. *Adv Mater* 36 (48): 2409278.

52 Zhu, X., Li, J., Ali, R.N., and Song, M. (2024). Two-dimensional ultra-thin CuCoNiMnAl high-entropy alloy nanosheets for lithium-ion storage and oxygen evolution reaction. *J Power Sources* 610: 234738.

53 Wei, Y., Yao, R., Liu, X. et al. (2023). Understanding the configurational entropy evolution in metal-phosphorus solid solution for highly reversible Li-ion batteries. *Adv Sci* 10 (9): 2300271.

54 Xiao, W., Xu, H., Pei, Y. et al. (2025). High entropy alloy/graphite composite as a new strategy to enhance the performance of lithium-ion batteries. *Inorg Chem Commun* 26: 114594.

55 Peng, L., Wei, Z., Wan, C. et al. (2020). A fundamental look at electrocatalytic sulfur reduction reaction. *Nat Catal* 3 (9): 762–770.

56 Ren, W., Ma, W., Zhang, S., and Tang, B. (2019). Recent advances in shuttle effect inhibition for lithium sulfur batteries. *Energy Storage Mater* 23: 707–732.

57 Xu, H., Hu, R., Zhang, Y. et al. (2021). Nano high-entropy alloy with strong affinity driving fast polysulfide conversion towards stable lithium sulfur batteries. *Energy Storage Mater* 43: 212–220.

58 Zhuang, H., Zhang, T., Xiao, H. et al. (2023). 3D Free-standing carbon nanofibers modified by lithiophilic metals enabling dendrite-free anodes for Li metal batteries. *Energy Environ Mater* 6 (3): e12470.

59 Yao, Y., Zhao, Z., Niu, R. et al. (2024). Rationally designed FeCoNiCrMn high-entropy alloys NPs hybridized rGO modified separator for highly efficient lithium-sulfur batteries. *Chem Phys Lett* 839: 141124.

60 Wang, X., Lin, P., Wu, C. et al. (2024). Superior-performance lithium–sulfur batteries: a face-centered-cubic-structure high-entropy alloy improves the bidirectional catalytic conversion of polysulfides/sulfides. *Inorg Chem Front* 11 (19): 6425–6437.

61 Chen, L., Wu, D., Li, X. et al. (2024). Modifying the electron structure of an FeCoNiZnCu high-entropy alloy with the introduction of Cu to facilitate the catalytic effect in lithium sulfur batteries. *J Mater Chem A* 12 (28): 17651–17662.

62 Ma, Y., Ren, Y., Sun, D. et al. (2024). High entropy alloy NPs dual-decorated with nitrogen-doped carbon and carbon nanotubes as promising electrocatalysts for lithium–sulfur batteries. *J Mater Sci Technol* 188: 98–104.

63 Wang, Y., Li, M., Yang, F. et al. (2023). Developing artificial solid-state interphase for Li metal electrodes: recent advances and perspective. *Energy Mater Dev* 1 (1): 9370005.

64 Wei, P., Wang, H., Yang, M. et al. (2024). Relocatable hollow multishelled structure-based membrane enables dendrite-free lithium deposition for ultra-stable lithium metal batteries. *Adv Energy Mater* 14 (22): 2400108.

65 Li, H., Zhang, F., Wei, W. et al. (2023). Promoting air stability of Li anode via an artificial organic/inorganic hybrid layer for dendrite-free lithium batteries. *Adv Energy Mater* 13 (28): 2301023.

66 Jin, C., Huang, Y., Li, L. et al. (2023). A corrosion inhibiting layer to tackle the irreversible lithium loss in lithium metal batteries. *Nat Commun* 14 (1): 8269.

67 Zheng, L., Lv, R., Luo, C. et al. (2024). Amorphous high-entropy alloy interphase for stable lithium metal batteries. *Adv Energy Mater* 14 (40): 2402042.

68 Wang, J., Yang, P., Wang, Y., and Wang, S. (2024). High entropy alloy NPs decorated carbon-based electrode as interfacial Li-ion localized accelerators for anode-free lithium metal batteries. *Carbon* 228: 119432.

69 Lee, J.S., Tai Kim, S., Cao, R. et al. (2011). Metal–air batteries with high energy density: Li–air versus Zn–air. *Adv Energy Mater* 1 (1): 34–50.

70 Fang, G., Gao, J., Lv, J. et al. (2020). Multi-component nanoporous alloy/(oxy) hydroxide for bifunctional oxygen electrocatalysis and rechargeable Zn-air batteries. *Appl Catal B Environ* 268: 118431.

71 He, R., Yang, L., Zhang, Y. et al. (2023). A CrMnFeCoNi high entropy alloy boosting oxygen evolution/reduction reactions and zinc-air battery performance. *Energy Storage Mater* 58: 287–298.

72 Madan, C., Jha, S.R., Katiyar, N.K. et al. (2023). Understanding the evolution of catalytically active multi-metal sites in a bifunctional high-entropy alloy electrocatalyst for zinc–air battery application. *Energy Adv* 2 (12): 2055–2068.

73 Cao, X., Gao, Y., Wang, Z. et al. (2023). FeNiCrCoMn high-entropy alloy NPs loaded on carbon nanotubes as bifunctional oxygen catalysts for rechargeable zinc-air batteries. *ACS Appl Mater Interfaces* 15 (27): 32365–32375.

74 Gao, L., Zhong, X., Li, Z. et al. (2024). A multi-layer reduced graphene oxide catalyst encapsulating a high-entropy alloy for rechargeable zinc-air batteries. *Chem Commun* 60 (10): 1269–1272.

75 Wu, D.H., Haq, M.U., Zhang, L. et al. (2024). Noble metal-free FeCoNiMnV high entropy alloy anchored on N-doped carbon nanotubes with prominent activity

and durability for oxygen reduction and zinc–air batteries. *J Colloid Interface Sci* 662: 149–159.

76 Lin, T.Y., Yen, F.Y., Ting, Y.C. et al. (2024). Non-precious bifunctional high entropy alloy catalyst and layered double hydroxide enhanced gel electrolyte based rechargeable flexible zinc-air batteries. *Chem Eng J* 488: 151093.

77 He, R., Yang, L., Zhang, Y. et al. (2023). A 3d-4d-5d high entropy alloy as a bifunctional oxygen catalyst for robust aqueous zinc–air batteries. *Adv Mater* 35 (46): 2303719.

11

Metal-Free High-Entropy Materials for Energy Storage Applications

Jnanranjan Panda[1], *Dipanwita Das*[2], *Subhashree Mohanty*[1], *and Sumit Majumder*[3]

[1] *Sri Sri University, Faculty of Science, Cuttack, Orissa, 754006, India*
[2] *Jadavpur University, Department of Physics, Kolkata, West Bengal, 700032, India*
[3] *KTH Royal Institute of Technology, Department of Applied Physics, School of Engineering Science, Stockholm, 114 19, Sweden*

11.1 Introduction

In the near future, the usable green energy generated from renewable sources is anticipated to replace traditional fossil fuels because of the finite supply and environmental concerns associated with the latter [1, 2]. At the current rate of extraction, global reserves of coal, natural gas, and oil are projected to last 113, 55, and 53 years, respectively [3]. However, since renewable energy sources are intermittent, it is essential to develop sustainable and energy-efficient electrical energy storage (EES) systems, such as supercapacitors (SCs), dielectric capacitors, batteries, and fuel cells, to address both the energy crisis and environmental challenges [3]. Among various EES systems, SCs and batteries have been commonly used in electric vehicles and modern-day electronic devices such as laptops, video cameras, televisions, and mobile phones.

SCs are EES systems that store energy through both Faradic and non-Faradic processes and offer various advantages such as rapid charging, high power density, and large capacity, due to their fast ion adsorption or desorption mechanism [3, 4]. However, their limited energy density hinders their large-scale development. Although various carbon-based materials, oxides, and chalcogenides have been studied for this purpose, SCs still struggle to meet the requirements of high energy density along with long cycle life. The incorporation of conducting polymers into carbon-based materials can be used to enhance the electrocatalytic performance of SCs [5]. However, limitations such as slow charging times, low power density, high cost, and safety concerns restrict the application of batteries in EES devices [3, 6]. Additionally, structural deterioration of battery electrodes during the repeated charging/discharging process leads to a significant decrease in electrocatalytic performance by further limiting their practical use in EES devices. To address

High-Entropy Materials for Energy Storage Devices, First Edition.
Edited by Chien-Te Hsieh, Pradeep Kumar Panda, and Arpan Kumar Nayak.

these issues, it is essential to develop high-quality EES materials with superior electrochemical properties.

In this context, high-entropy materials (HEMs) not only expand the range of possible compositions for structural and functional materials but also introduce new ideas and approaches for material design due to their unique entropy effect [3, 7, 8]. HEMs represent an innovative class of materials characterized by including five or more primary elements in nearly equiatomic proportions, which results in advanced configurational entropy [3, 9]. These materials are distinguished by their unique crystalline structures and remarkable properties. HEMs were first reported as high-entropy alloys (HEAs) by Cantor and Yeh et al. in 2004 through two independent studies [10, 11]. Their notable strength-to-weight ratios, fracture toughness, thermal stability, and corrosion resistance have garnered considerable interest within the field of materials science research, particularly when compared to traditional alloys. The higher configurational entropy arising from the existence of multiple elements in similar ratios promotes the development of simple solid solutions with highly symmetric crystal structures, i.e. face-centered cubic (FCC), body-centered cubic (BCC), and hexagonal close-packed (HCP) configurations. The formation of HEMs is achieved by the Gibbs free energy equation: $\Delta G_{mix} = \Delta H_{mix} - T\Delta S_{mix}$. The distinctive properties of HEMs can be attributable to four fundamental effects, i.e. high-entropy (HE), sluggish diffusion, lattice distortion, and the cocktail effect, which together contribute to their enhanced stability and performance [12, 13]. A schematic diagram of four core effects of HEMs is presented in Figure 11.1.

Numerous HEMs such as alloys [5, 6], oxides [15], chalcogenides [16], diborides [17], carbides [18], nitrides [19], and phosphides [20] have been reported in the literature for their possible applications in energy storage, catalysis, and hydrogen storage. However, most of the reported HEMs still rely on non-noble metal (e.g. Fe, Co, Ni, Mn, Mg) or noble metal (e.g. Pt, Pd, Ru, Ir) components and dominate the electrocatalyst energy market due to their favorable electronic structure and remarkable electrocatalytic capabilities [21]. Despite their numerous advantages, the use of these metal-based electrocatalysts is limited due to their high cost and natural resource scarcity, which drastically hinders their lab-scale and industrial applications. Therefore, over the past two decades, tremendous efforts have been directed to the search for low-cost, stable, and highly efficient metal-free energy storage materials with superior electrochemical performance.

In this context, metal-free HE systems consisting of carbon-based materials and organic polymers can be considered a promising choice for energy storage applications. Feng et. al. introduced the concept of meso-entropy carbon to understand the relationship between several materials based on carbon such as graphene, fullerene, and carbon nanotubes (CNTs) [22]. For example, HE graphene/carbon quantum dots (GQDs) can be prepared by doping with multiple heteroatoms such as boron (B), nitrogen (N), sulfur (S), and phosphorus (P) to improve the electrochemical performance of GQDs [23]. However, sparse attention has been given to metal-free HEMs, and very few research groups have studied their electrochemical energy storage performance. This unique composition

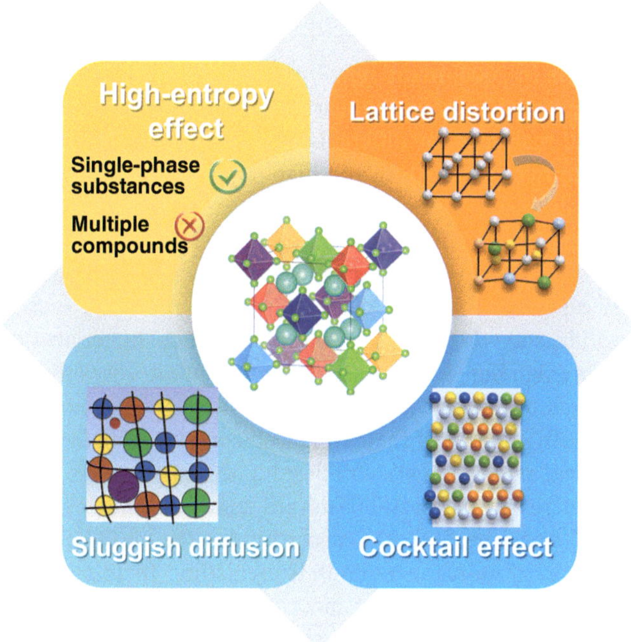

Figure 11.1 Four core effects of HEMs *Source:* Reproduced from Ref. [15] with permission of The Royal Society of Chemistry.

renders them more sustainable and economically viable alternatives to catalysts based on noble metals and non-noble HEMs, offering a unique combination of sustainability, performance, and stability. Metal-free HEMs offer a more environmentally friendly and cost-effective solution for a range of applications by eliminating the reliance on costly and limited noble metals, especially in the renewable energy sector. Their tailored electronic properties, high surface areas, and uniform pore structures enhance their effectiveness in EES devices. When used as freestanding electrodes, they eliminate the need for binders, leading to improved electron transfer kinetics and more abundant active sites. Furthermore, their excellent thermal stability, corrosion resistance, and ability to accommodate volume changes during charge/discharge cycles provide them with enhanced long-term performance and cycle stability. These distinctive advantages position metal-free HEMs as key contenders in the development of next-generation energy storage systems, offering a sustainable path toward more efficient and durable energy solutions. In this chapter, we first introduce the concept of HEMs and then classify the different types of HEMs. Subsequently, we elaborate on recent advanced synthesis methods for preparing HEMs and their characterization methods. Next, we provide an overview of recent metal-free HEMs used in various energy storage devices, including SCs, dielectric capacitors, and different types of batteries, while also highlighting the challenges and future perspectives in this field.

11.2 Classification of HEMs

After the discovery of HEMs in 2004, many new branches of HEMs have subsequently been developed. Figure 11.2a shows the yearwise developmental milestones of HEMs used as electrode materials for different types of batteries [15]. HEMs are broadly categorized into two types: HEAs and high-entropy ceramics (HECs). HECs are further divided into two types: single-anion ceramics and coexisting anion–cation ceramics, based on their ionic composition [3, 7]. High-entropy oxides (HEOs) [14], chalcogenides [16], borides [17], carbides [18], and nitrides [19] fall under the first category of single-anion HECs, while materials such as phosphates [24] and fluoride oxides [5] belong to polyanionic HECs. Materials such as HE MXenes [25] and Prussian blue [26] fall under the second category of HECs. Recently, HEMs have received substantial attention in various areas of EES technologies, as shown in Figure 11.1b, because of their enhanced configurational entropy, along with robust mechanical and chemical stability at higher temperatures.

As the global landscape increasingly transitions toward renewable energy sources and the electrification of transportation, the necessity for the most efficient, durable, and sustainable energy storage solutions is escalating. In this context, metal-free HEMs provide several benefits that meet these demands. First, their superior phase stability, stemming from their high mixing entropy, guarantees durability and reliable performance across numerous charge/discharge cycles. This stability is essential for applications such as grid energy storage and electric vehicles, where long-term dependability is critical. Second, the varied local environments generated by the multiple elements enhance their electrochemical performance, often resulting in increased capacity and improved rate capability. This feature is particularly advantageous in high-power applications that require swift energy storage and release, and these materials pave the path for additional sustainable and potentially cost-effective energy storage ways. This characteristic is in harmony with global initiatives aimed at developing technologies that are both environmentally sustainable and economically feasible over the long term. These types of HEMs are stable, are earth-abundant, and have a reduced environmental impact in contrast to their metal-based counterparts. The development of advanced production techniques is necessary to optimize the performance of metal-free HEAs to be used in energy storage applications. Metal-free HEMs can be classified into two categories, as explained in the following sections.

11.2.1 HE Carbonaceous Materials (Multielemental Doped Carbon Networks)

Here, HE carbon networks can be synthesized by doping of multiple heteroatoms such as B, N, S, P, and O to improve the electrochemical performance of carbonaceous materials [23, 27]. These heteroatom-doped, multicomponent carbonaceous materials show enhanced electrical conductivity and charge storage capacity, making them suitable for applications in various EES devices such as SCs and different types of batteries. However, very few studies are available in this particular field, and it is still in its infancy; more focus should be devoted to this area.

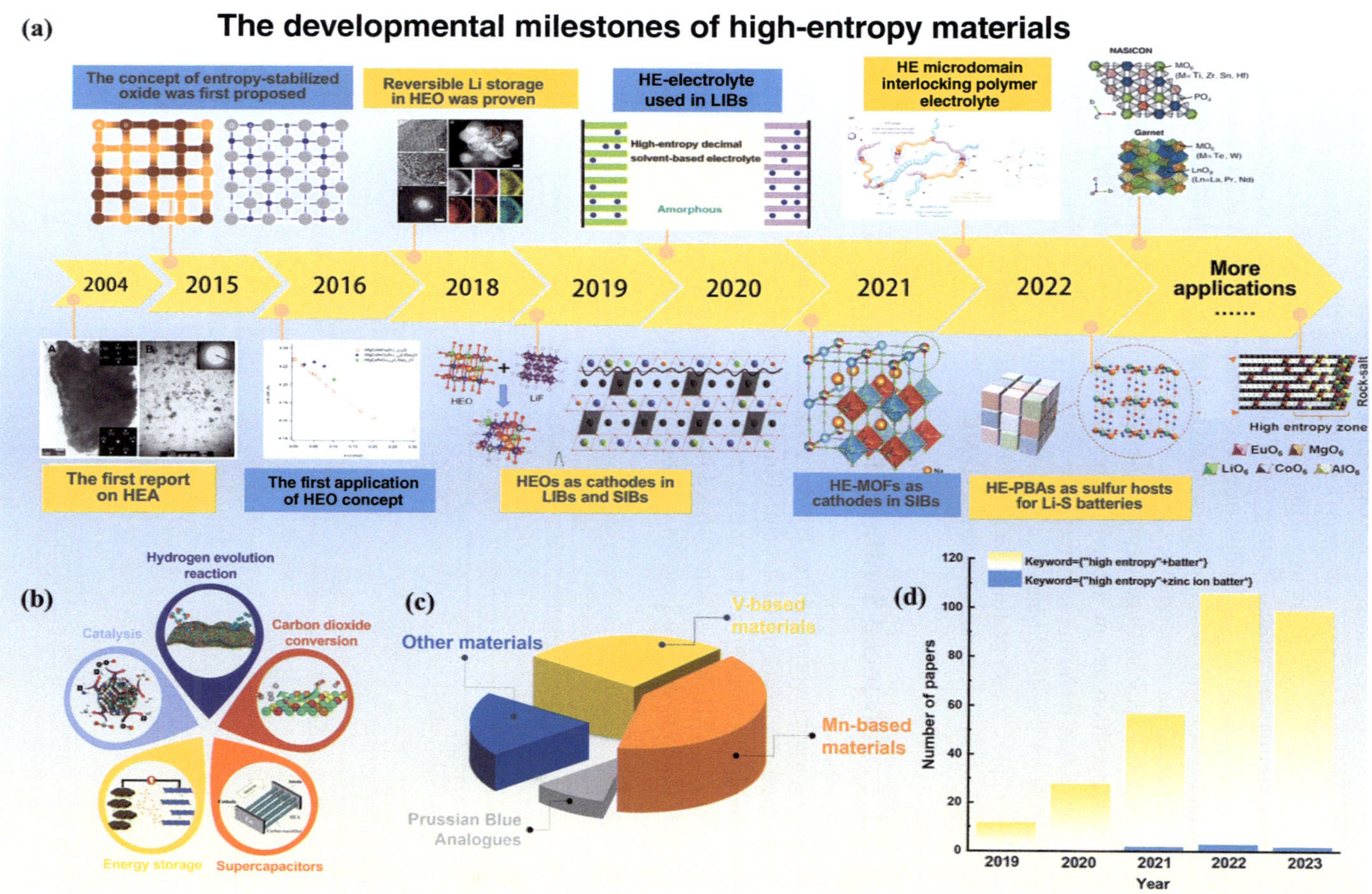

Figure 11.2 Developmental milestones of HEMs. *Source:* Reproduced from Ref. [15] / with permission of The Royal Society of Chemistry.

11.2.2 High-Entropy Polymers

High-entropy polymers (HEPs) can be designed with multiple organic monomeric units by incorporating nitrogen-rich and sulfur-rich components to enhance the electrochemical performance of the synthesized material [28, 29]. These HEPs combine the unique features of nitrogen and sulfur with conductive polymers, improving electrical conductivity, stability, and charge storage capacity. As a result, HEPs are suitable for a wide range of applications in EES devices such as SCs, batteries, and other electronic components.

Nonetheless, challenges remain, including complexities in synthesis, difficulties in composition design, scalability concerns, and the necessity for improved characterization techniques. In response to these challenges, researchers are concentrating on the development of high-throughput experimental and computational techniques, incorporating machine learning and artificial intelligence methodologies and exploring nanostructured HEMs.

11.3 Advanced Synthesis Methods of HEMs

The fabrication of metal-free HEMs necessitates advanced techniques to ensure the desired composition and structural integrity. Various synthesis methods have been developed and optimized for this purpose. Various preparation processes used in the synthesis of HEMs are summarized in detail in Figure 11.3, which provides a concise review of these techniques. The bottom-up strategy, which involves assembling

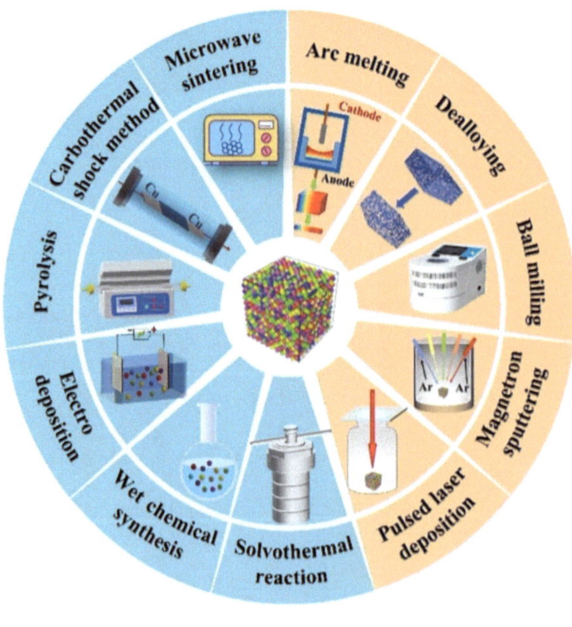

Figure 11.3 Schematic presentation of the synthesis techniques. *Source:* Reference [30] / American Chemical Society / CC BY 4.0.

Table 11.1 A brief overview of different synthesis methods.

Synthesis route	Equipment	Reaction condition	Duration	Size distribution	References
Solvothermal/ hydrothermal	Hydrothermal reactor and furnace	~200 °C and high pressure, inert atmosphere/air	Few hours to a day	Good	[30, 31]
Low-temperature oil-phase	Reactor made of stainless steel, oil bath	100–250 °C	Several hours to a day	Good	[33]
Electro-deposition	Electrochemical setup	Room temperature, air	Few minutes to several hours	Good	[34]
Pyrolysis	Tube furnace	High temperature, inert atmosphere	Few seconds	Good	[38]
Carbon thermal shock method	Current pulse device	Greater than 1,000 K and inert atmosphere	Millisecond	Good	[30, 40]
Microwave	Microwave reactor	Less than 2,000 K	Few seconds	Good	[43]
Vacuum melting technology	Vacuum furnace	1,000–3,000 °C and vacuum	Few minutes to several hours	Good	[44]
Dealloying method	Smelting furnace	500–700 °C and 1 atm, both inert and air atmospheres	Less than two hours and greater than five hours	Good	[46, 47]
Ball milling	Ball milling unit	Inert atmosphere	Few minutes to several hours	Good	[49]
Magnetron sputtering	UHV deposition system	UHV and room temperature	Few minutes to hours	Good	[30]
Pulsed laser deposition	UHV deposition system and pulsed laser	UHV, 300–1,000 °C	Few minutes to hours	Good	[50]

Source: References [30, 57].

nanostructured HEMs gradually from their molecular precursors (shown by the blue zone), and the top-down strategy, which entails transforming bulk materials into nanoscale forms (shown by the orange zone), are the two primary methods for synthesizing HEMs [30]. A brief overview of all the reported synthesis methods is shown in Table 11.1.

11.3.1 Bottom-up Approaches

These synthesis techniques mainly pertain to producing HEMs by reducing metal salt precursors, which are derived from different wet-chemical synthesis routes. In this context, we outline six techniques for the bottom-up synthesis of HEMs.

11.3.1.1 Solvothermal/Hydrothermal Synthesis Method

The solvothermal synthesis process is a conventional wet lab technique for preparing different types of nanostructured HEMs with optimized electrochemical features [30]. In this approach, various metal salts are dispersed in an aqueous solution or organic solvent under moderate to high pressure and temperature (180–250 °C), allowing for the exact regulation of particle size and morphology, which is critical for performance in energy storage applications. GQD can be prepared using this method [31], and the synthesis process is a straightforward, suitable, and eco-friendly technique for batch preparation of HEMs. Jiang et al. developed a HE metal hydroxymethoxy material, designated as FeCoNiMgCr(OH)(OCH_3), through a single-step solvothermal process. This synthesis aimed to explore the efficiency of HEMs for electrochemical water splitting applications [32].

11.3.1.2 Low-Temperature Oil-Phase Method

This is another wet-chemical synthesis route, which closely resembles the previously mentioned solvothermal technique [30]. In this process, an organic solvent is used to dissolve metallic precursors, and the reaction is allowed to proceed for a specified amount of time at a suitable low temperature. This low-temperature synthesis enables precise control over critical parameters such as size, morphology, and crystallinity of the materials, which are essential for energy storage applications. While this technique is straightforward and energy-efficient, it may not be applicable for metal precursors that are insoluble. Nanosized FeNiCrCoMn HEAs loaded on CNTs were effectively synthesized by Gong et al. using this method, suggesting that these materials could be a viable choice for zinc–air battery electrode [33].

11.3.1.3 Electro-Deposition Method

The electro-deposition method is another well-known approach for synthesizing HEMs, and the whole process is carried out using a potentiated three-electrode system [34]. The three-electrode setup consists of a platinum electrode acting as the counterelectrode, a reference electrode (Hg/HgO), and a substrate acting as the working electrode (such as glassy carbon, nickel foam, or carbon cloth). This approach facilitates both anodic and cathodic exfoliation through the use of appropriate electrolytes. It provides user-friendly operation and precise control at low temperatures, employing techniques such as potentiostatic, galvanostatic, potentiodynamic, cyclic voltammetry, and cathodic plasma. The desired particle size and edge state of the prepared materials strongly depend on various parameters, including solvent, supporting electrolyte, deposition current, deposition time, and the type of counterion used [35, 36]. Unlike other synthesis methods, major morphological changes do not occur during doping. Although this method offers several advantages, the low yield produced during the synthesis process limits its potential for scaling up. Zhang et al. synthesized CoCrFeNi HEA using this method and examined how graphene oxide (GO) affects the microstructural properties of the resulting HEAs [37]. Their findings indicated that incorporating GO enhances the quality of the thin film and decreases surface roughness.

11.3.1.4 Pyrolysis Method

Another effective and well-established method for synthesizing HEMs is spray pyrolysis, where a solution containing precursor elements is atomized into fine droplets and introduced into a high-temperature reactor under an inert atmosphere [38]. This approach is especially beneficial for producing fine, uniform particles with a precisely controlled composition. The straightforward process makes it easier to produce HEA materials in batches. Although it employs basic equipment, there are ongoing challenges related to achieving swift heating and cooling processes. Nevertheless, the introduction of fast-moving bed pyrolysis by Gao et al. in 2020 has effectively mitigated these issues [39]. The outcome is a uniform and small-sized HEA, which is produced by rapidly pyrolyzing precursors containing mixed metal salts at high temperatures.

11.3.1.5 Carbon Thermal Shock Method

Carbon thermal shock (CTS) method, a recent invention, involves quickly heating a precursor combination to extremely high temperatures for short durations [30, 40]. This technology stands out for its speed and capacity to produce materials with unique characteristics that are usually not achievable using traditional synthesis techniques. The CTS method has proven to be an effective approach for synthesizing multicomponent HEAs that exhibit enhanced electrocatalytic performance. The process involves the swift thermal heating of metal salt precursors combined with carbon materials, followed by exposure to extreme temperatures (approximately 2,000 K) and subsequent cooling in an inert atmosphere (Ar/H_2). This method produces homogeneous particles with a consistent temperature distribution, high reaction temperatures, quick heating and cooling rates, and incredibly quick reaction times. By preventing the production of secondary phases, the instantaneous high temperature makes it easier to synthesize single-phase nanosized HEAs. Thus, there is no chance of phase separation when producing quinary, senary, septenary, and even octonary HEAs with this technique.

Nanosized senary HEO FeNiCoCrMnV, with a particle size of 10 nm, was successfully synthesized by Abdelhafiz and coworkers using the CTS method. For catalytic applications, they emphasized the material's prolonged stability and remarkable electrocatalytic activity [41]. FeNiCoMnMg and FeNiCoMuCu are two different nanosized quinary HEAs that were synthesized by Xu et al. using electrospun carbon nanofibers (CNFs) following the CTS approach in a different work [42]. These authors reported the exceptional mechanical, electrical, and electrochemical qualities of these alloys, making them appropriate for electrochemical energy storage in SCs. Despite being restricted to conductive substrates such as carbon-based materials, this technique works quite well for producing evenly sized nanosized HEAs on a wide scale. However, it necessitates significant energy input to reach high temperatures within a short timeframe.

11.3.1.6 Microwave Sintering Method

This approach resembles the CTS method, which employs microwave radiation for rapid and uniform heating of precursor materials [30]. By efficiently absorbing energy through the carbon carrier, this method synthesizes HEAs by combining metal precursors with a carbon carrier and exposing them to a microwave environment. This approach necessitates a heat-absorbing carrier material, in contrast to the CTS method. The synthesis process is notably efficient, providing excellent uniformity and ease of implementation. Hsieh and his team successfully synthesized HE GQD with a particle size of 5 nm using a solid-state microwave method [23]. Their process involved the pyrolysis of two distinct carbon sources, fructose and citric acid, as well as four dopant precursors at 180 °C using pulsed microwave irradiation to produce HE GQDs. In another study, Khan and coworkers developed a platinum-free HEA catalyst, designated as $Cr_{18.2}Zr_{13.5}Ti_{5.3}Co_{17.8}Ni_{8.4}Cu_{8.5}Fe_{28.3}$ and $Al_{15.3}Mn_{18.2}Ti_{3.7}Co_{23.2}Ni_{5.9}Fe_{27.8}$, on reduced GO sheets at low temperatures using this method and assessed their electrocatalytic performance against pure platinum catalysts [43].

11.3.2 Top-down Approaches

These synthesis techniques pertain to the physical technique used in the synthesis of HEAs by reducing substantial material blocks to the nanoscale. In this context, we outline five methods for the synthesis of HEMs.

11.3.2.1 Vacuum Melting Technology

This approach represents a traditional technique for the fabrication of HEAs, employing both arc melting and induction melting processes[44]. It is particularly effective for metals characterized by high melting points, resulting in high-purity alloys, and is widely used in conventional metal processing. The smelting of sample occurs within an inert atmosphere, leading to the formation of alloy ingots. Additionally, magnetic stirring is used to achieve a homogeneous composition. Subsequently, these ingots are placed into molds and shaped into appropriate sheets through wire-cut machining. The final alloys are often made in a block form, with a limited specific surface area. Güler and coworkers synthesized AlCoCrFeNiTiX-based HEAs by adjusting the stoichiometry of titanium using this method, followed by heat treatment at various temperatures of 500, 700, and 1,000 °C [45].

11.3.2.2 Dealloying Method

The dealloying method includes the specific removal of one or more components from multicomponent HEMs using a chemical or electrochemical process resulting in the formation of a porous alloy [46, 47]. The resulting porous nanostructure provides numerous advantages, including an enhanced electrochemical surface area, which is the most essential characteristic for performance optimization in energy storage applications. This approach has been extensively employed in materials science research for the synthesis of HEAs and has demonstrated a significant level of consistency across various electrocatalytic applications. Jin and coworkers developed Al-based nanosized AlNiCoIrMo HEAs and evaluated their electrocatalytic performance in an acidic environment [48]. In addition, the as-synthesized nanocatalyst demonstrated superior cyclic stability after 7,000 cyclic voltammetry cycles.

11.3.2.3 High-Energy Ball Mill Method

This technique, commonly referred to as mechanical alloying, utilizes a high-energy ball mill along with grinding balls to synthesize sustained impacts, collisions, and repeated cold welding and fracture events within the powder [30]. HEAs are produced by the diffusion of atoms within the powdered particles. This approach is frequently integrated with temperature sintering and melting processes to fabricate HEAs. Lei et al. successfully synthesized micron-sized silicon-based HEAs with enhanced structural stability using this method, noting that the synthesized alloy could be used as an anode for lithium-ion batteries (LIBs) [49].

11.3.2.4 Magnetron Sputtering Method

Magnetron sputtering is a physical vapor deposition (PVD) method that employs ionized argon to energize a solid metal target, subsequently transforming the metal into a vapor phase that deposits onto a substrate, resulting in a thin film characterized by HE features [30]. This approach utilizes multiple target materials containing the desired elemental components to fabricate HEAs.

11.3.2.5 Pulsed Laser Deposition Method

Pulsed laser deposition (PLD) is another PVD technique in which a high-powered strong pulsed laser is focused on a target material in a vacuum chamber, which partially ionizes the target material and deposits it on a substrate [50]. The deposition can be assisted by a flow of oxygen or other background gases or occur under ultra-high vacuum conditions. Although this method is very fast and efficient for producing HEA thin films, it may not be suitable for large-scale preparation.

Cropper et al. produced nanosized AlCrFeCoNiCu HEA thin films on borosilicate glass substrates using this method and reported that the thin film deposited at room temperature showed a mixed BCC and FCC crystal structure [50]. However, they showed that increasing the deposition temperature reduced the XRD peak intensities of thin films.

11.4 Characterization Techniques

The structural, microstructural, and morphological characteristics, elemental composition, and chemical state of the prepared HEMs are examined through various material characterization techniques. phase purity and crystallographic information of the HEMs are recorded by powder X-ray diffraction pattern and selected area diffraction. The morphological, elemental composition, particle size, and microstructural information of the HEMs was acquired by field emission electron microscopy, energy-dispersive X-ray spectroscopy, high-resolution transmission electron microscopy, and high-angle annular dark-field scanning transmission electron microscopy. Additionally, lattice planes, defects, and dislocations are identified by atomic-scale HAADF-STEM and fast Fourier transform analysis. The HADDF micrograph and elemental mappings of nanosized FeNiCoMnMg HEA on electrospun CNFs are presented in Figure 11.4 [42].

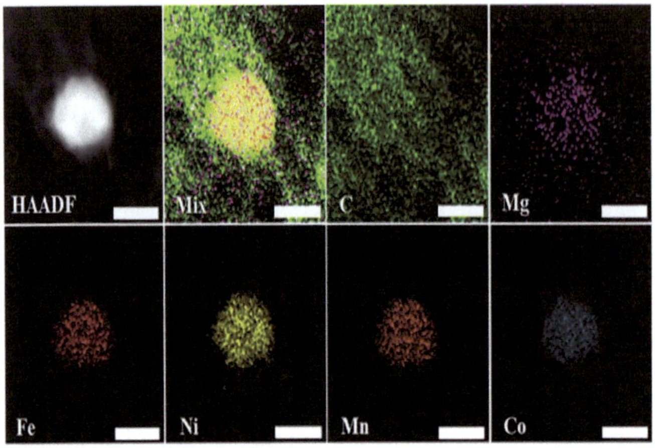

Figure 11.4 (a) The HADDF micrograph and elemental mappings of nanosized FeNiCoMnMg HEA on CNFs. (The scale bars of the images are 20 nm.) *Source:* Reproduced with permission from Ref. [42] / Elsevier.

The chemical state of the constituent elements of HEMs is analyzed using X-ray photoelectron spectroscopy and X-ray absorption spectroscopy. Compositional analysis is carried out using inductively coupled plasma mass spectroscopy and inductively coupled plasma-optical emission spectroscopy, and the concentration of the constituent elements is investigated.

11.5 Application of HEMs in Energy Storage Systems

Initially, HEMs were employed as catalysts or metal coatings for chemical reactions. Later, researchers studied the electrochemical performance of HEMs for use as electrode materials in various types of EES devices such as SCs, dielectric capacitors, and batteries. All of these EES devices consist of two electrodes and an electrolyte. SCs are considered promising EES devices due to their higher power densities and longer cycling life (>10^6 cycles) compared to batteries and dielectric capacitors. SCs are categorized into two types: electrostatic double-layer capacitors (EDLCs), which are based on non-Faradaic charge transfer, where charge accumulates electrostatically at the interface between the electrode and electrolyte [3].

The other type is the pseudocapacitor, which works on a fast surface redox reaction mechanism [3, 12]. Since no chemical reactions occur, the charging mechanism in EDLCs relies solely on electrostatic attraction, which allows them to deliver high power density and an unlimited life cycle with very low energy density. On the other hand, pseudocapacitors show high energy density with inferior cycle life. Therefore, several trials have been devoted to developing innovative electrode materials with better cycling life and improved energy density. In the case of batteries, chemical energy arises from the redox or intercalation process that converts the chemical energy to electric potential [12]. The electrode with the higher potential is selected as the cathode, while the electrode with the lower potential is selected as the anode. Here, both the electrodes not only work as charge storage mediums but also actively participate in the intercalation process or redox reactions. Therefore, batteries function as a closed system, where EES and conversion happen within the electrode materials. Batteries have an intermediate power density compared to SCs, whereas SCs exhibit high power density and low energy density. Therefore, designing a hybrid EES mechanism that will provide both high power and energy densities simultaneously, compared to batteries and EDLC alone, is highly appealing [12]. In this context, HEMs will perform as active electrode materials for the hybrid EES system. Reshuffling of multiple component elements within these HEMs provides remarkable mechanical strength, superior ionic conductivity, and outstanding thermal stability, making them highly desirable for this purpose.

11.5.1 HEMs for SCs

HEMs have demonstrated exceptional efficacy as electrode materials for SCs, characterized by prolonged cycling stability and significant capacity retention. The varied local environments fostered by the HE architecture may enhance the electrochemical performance of SC materials. Lal and coworkers developed a cost-effective catalyst by combining CNTs with FeNiCoCrAl HEO and evaluated its electrochemical behavior in both half-cell and full-cell configurations using a hydrogel electrolyte composed of polyvinyl alcohol/H_2SO_4 [51]. The cyclic voltammetry (CV) plot of the full cycle showed a quasi-rectangular shape as well as pseudocapacitive behavior, as shown in Figure 11.5a–c. The nanocomposite exhibited a remarkable capacitance retention of 100%, alongside a Coulombic efficiency reaching up to 15,000 when subjected to a current density of $5\,A\,g^{-1}$. Additionally, a high specific capacitance value of 286.0 F/g, an increased energy density of $217\,Wh\,kg^{-1}$, and an impressive power density of $24{,}521\,W\,kg^{-1}$ were recorded at a scan rate of $10\,mV\,s^{-1}$.

In a separate study, Xu et al. synthesized a nanosized FeNiCoMnMg HE alloy on CNFs through the CTS synthesis method using different precursor concentrations (5, 10, and 20 mM). The electrochemical characteristics of the as-synthesized composite sample were examined within a cell employing a 6 M KOH electrolyte solution [42]. The material synthesized from the lowest precursor concentration depicted a superior performance with a capacitance value of $203\,F\,g^{-1}$ and a specific energy density value of $21.7\,Wh\,kg^{-1}$. Moreover, the composite showed 89.2% capacitance retention with a lifespan of 2,000 cycles.

Kong et al. developed a nanoporous core–shell architecture utilizing an equiatomic FeNiCoCrAl HEA through the process of selective phase dissolution and used this material as a binder-free electrode for SC evaluations. The resulting HEA demonstrated an impressive volumetric capacitance of $700\,F/cm^3$ and a superior cycling stability exceeding 3,000 cycles, as shown in Figure 11.5d–f [52].

11.5.2 HEMs for Dielectric Capacitors

In today's electrical and electronic power systems, dielectric capacitors are crucial due to their ability to charge and discharge at ultrafast rates and their substantial

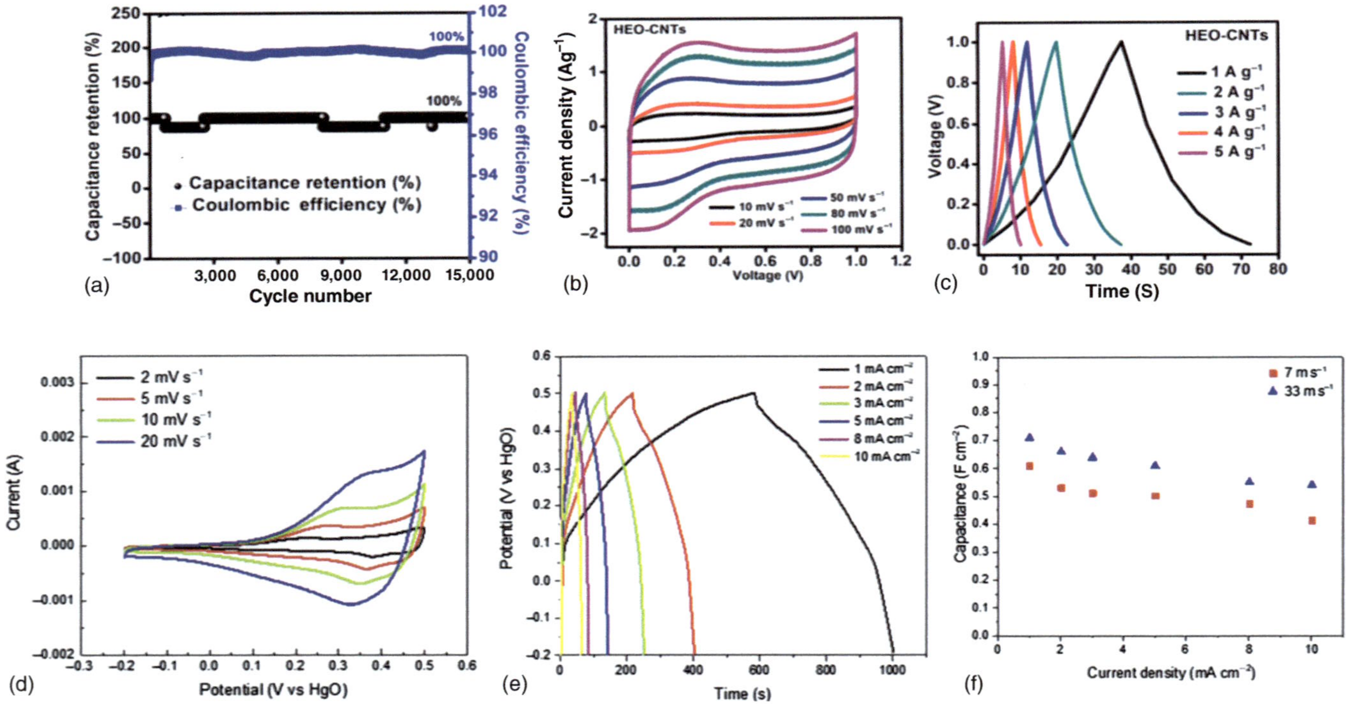

Figure 11.5 (a–c) HEO CNT-based nanocomposite: (a) Retention capacity and Coulombic efficiency at a current density of 5 A g^{-1} CV, (b) CV plot, and (c) GCD plot. *Source:* Reproduced from Ref. [51] / with permission of American Chemical Society. (d–f) Nanosized porous FeNiCoCrAl HEA: (d) CV plot, (e) GCD curves, and (f) areal specific capacitance as a function of current density. *Source:* Reproduced from Ref. [52] / with permission of Elsevier.

power densities. However, the enhancement of energy densities and efficiency poses a significant challenge for the advancement of next-generation applications that necessitate miniaturization and the incorporation of cost-effective, advanced technologies.

In this context, HEMs provide an innovative approach to enhancing dielectric energy performance due to their properties such as high configurational entropy, sluggish diffusion kinetics, and synergistic effect due to the presence of multicomponent elements [53]. Lattice distortion improves the polarization capabilities of the material by reducing the occurrence of dielectric breakdown while improving the stability of the capacitor. HEM-based systems demonstrate high dielectric constant, low dielectric loss, and excellent breakdown strength, and several attempts have been made to explore dielectric capacitors using HEMs since the first report on an entropy-stabilized oxide by Rost et al. in 2015. However, very few HEM-based systems have been reported that improve the permittivity and other dielectric properties while reducing dielectric loss [14].

Dou et al. developed HE-induced nanofibers using enhanced polymer nanocomposites and evaluated their energy storage performance [53]. They fabricated ceramic-induced nanofibers with a stable $Bi_2Ti_2O_7$ phase using the electrospinning method to improve the energy storage performance of the polyetherimide (PEI) composite. The polymer nanocomposite showed enhanced energy storage capabilities across a temperature spectrum of 25–150 °C, achieving a remarkable energy density of 6.46 J cm^{-3} at an electric field of 590 MV m^{-1} at 150 °C, exceeding the energy density of pure PEI by more than twofold. Furthermore, the temperature- and frequency-dependent dielectric performance of the synthesized composite was evaluated to explore its dielectric stability. It is known that breakdown strength (E_b) and dielectric constant (ε_r) play a crucial role in enhancing energy density. Their findings showed that the dielectric constant increased from 3.3 to 4.8 at a frequency of 100 Hz for the polymer composite compared to the pure polymer. Similarly, a high breakdown strength was observed for the composite material, as shown in Figure 11.6.

Yang and coworkers prepared a HE-stabilized $Bi_2Ti_2O_7$-based dielectric film with a thickness of 620 nm using the chemical solution deposition method, followed by fast thermal annealing [54]. Under normal conditions, $Bi_2Ti_2O_7$-based systems are thermodynamically unstable. However, their HE counterparts can remain stable by preventing decomposition and overcoming solubility limits. They reported a high energy density value of 182 J cm^{-3} with 78% efficiency at an electric field of 6.35 MV cm^{-1}. They reported that the HE strategy improves the breakdown field by reducing polarization hysteresis and the prepared films could be a promising choice for the development of high-quality dielectric capacitors.

In another study, Qian and coworkers prepared $Bi_{0.5}Na_{0.5}TiO_3$-based HE thin films by establishing a relationship between entropy, symmetry, polarization, and temperature [55]. They reported a high energy density value of 103 J cm^{-3} with 79% efficiency at an electric field of 4,143 kV cm^{-1} and extended the operating temperature range of the film to 3,320 °C. Their research findings suggested an

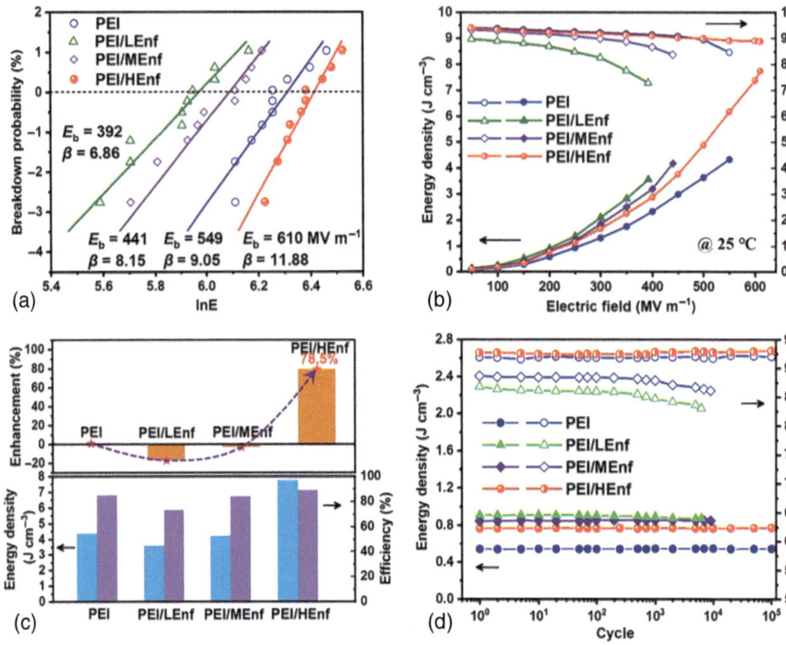

Figure 11.6 (a) Weibull distribution analysis; (b) discharge energy density (Ue) and efficiency; (c) enhancement in discharge energy density (Ue, top) relative to pristine PEI and the energy storage performance of PEI composites (bottom) with varying nanofillers; and (d) charging/discharging reliability of the films under an applied electric field of 200 MV m^{-1}. *Source:* Reproduced from Ref. [53] / with permission of John Wiley & Sons.

innovative method for the development of high-temperature dielectric energy storage capacitors.

A significant advancement in the field of dielectric ceramics has been realized through the application of a HE strategy aimed at engineering "local polymorphic distortion" within relaxor ferroelectrics. As a result, an increase in relative energy density with a reduction in dielectric loss has been achieved. The HE strategy has contributed to improved breakdown electric fields and a delay in polarization saturation. These materials demonstrate exceptional stability across frequency and temperature, which is essential for their practical applications.

11.5.3 HEMs for LIBs

The endeavor of the sustainable energy approach has shown pivotal advancement in EES technologies. Recently, LIBs are considered torchbearers for their prolonged life cycle, high efficiency, and presence of memory effect. Nevertheless, the theoretical energy storage potential of the electrode material is inadequate, failing to meet the escalating energy requirements of human society. Therefore, several efforts have been made to enhance the energy density of the electrode material utilized in different types of batteries.

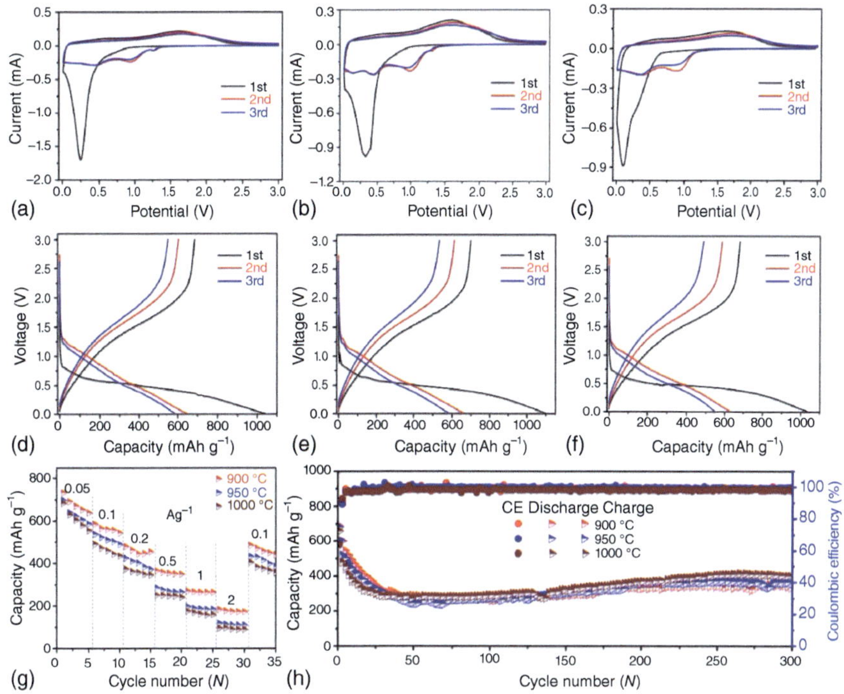

Figure 11.7 Electrochemical performance of $(FeCoNiCrMn)_3O_4$: (a–c) CV curves; (d–f) GCD curves; (g) rate capabilities; (h) cycle performance study of HEO-900, HEO-950, and HEO-1000. *Source:* Reproduced from Ref. [56] / with permission of Elsevier.

A spinel-type HEO $(FeCoNiCrMn)_3O_4$ was developed by Qi and his colleagues using a solid-state reaction technique, and its suitability as an anode electrode for LIBs was assessed [56]. They explored the influence of different synthesis temperatures, including 900, 950, and 1,000 °C, on the structural properties and lithium storage performance of the resulting material (Figure 11.7). Their results showed that the as-prepared HEO maintained a pure phase up to 900 °C. Although all three samples showed excellent cycling stability, HEO-900 with commercial mass loading demonstrated a higher charge/discharge capacity (1,034/680 $mA\,h\,g^{-1}$) and better rate capability (182 $mA\,h\,g^{-1}$ at 2 $A\,g^{-1}$) compared to the other two samples. This excellent electrochemical performance can be attributable to the synergistic effects of different cations with varying radii and high configurational entropy. The peak at 0.25 V in the cathodic process is related to the gradual reduction of transition metal oxides, which occurs from M^{4+}/M^{3+} to M^{2+} and subsequently from M^{2+} to M^0, as well as the formation of a stable solid electrolyte interface layer [57].

This report indicates the redox nature of the material, suggesting that it functions as a battery-type electrode material. The non-crystalline structure of HEO is beneficial for developing an anode that exhibits outstanding lithium storage capabilities. Furthermore, the defects present in amorphous HEO enhance the number of active sites available for Li+ storage and facilitate adequate diffusion pathways for Li+ within the electrode.

In another work, Sarkar and coworkers developed a micron-sized ($Co_{0.2}Cu_{0.2}Mg_{0.2}Ni_{0.2}Zn_{0.2}$)O HEO and evaluated it as an anode material for LIBs [58]. They showed reversible lithium storage features of HEOs by introducing the HE stabilization concept. The as-synthesized HEO showed a capacity value of 590 mA h g^{-1} with a Coulombic efficiency higher than 99% after being cycled 500 times. The increase in the capacity value during cycling is a typical electrochemical behavior of conversion-type materials, which can be ascribed to the activation process of the large-particle electrode. The medium-entropy oxide after the removal of cobalt ions resulted in a significant decay in the capacity value after just a few cycles, suggesting that changes in elemental composition can significantly affect cycling stability. In another study, Ting and coworkers prepared a surfactant-assisted spinel $(CrMnFeCoNi)_3O_{4\,HEO}$ using the hydrothermal synthesis route and evaluated its performance by using it as a LIB anode material [59]. Their results showed that the HEO exhibited a specific capacity of 1,235 mA h g^{-1} with 90% retention capacity after 200 cycles.

So far, most of the HEM-based LIB anode materials contain cobalt, which improves conductivity and reversible capacity. However, cobalt is costly, and most of the world's cobalt reserves are located in the Democratic Republic of Congo, a country with political instability, raising significant geopolitical concerns [12]. Consequently, the innovation of cobalt-free anodes has become a pivotal aspect of research efforts. In this context, Chang et al. developed a cobalt-free $(CrMnFeNiCu)_3O_4$ HEO material and evaluated the lithium storage performance of the as-synthesized sample by using it as an anode material for LIBs [60].

In addition to anode materials, HEMs have been gaining huge attention for their use as cathode materials in LIBs. Lun et al. developed a cation-disordered rock-salt-type HEO for LIBs, where the cathode compound consists of six different transition metal elements. This material exhibited 40% higher capacity as compared to conventional compounds [61]. The incorporation of multiple transition metal elements into the lattice site suppresses short-range ordering and significantly enhances lithium transport. A comparison of different types of LIB electrodes is presented in Table 11.2.

11.5.4 HEMs for Sodium-Ion Batteries (SIBs)

HEMs are also employed as SIBs, and the storage principles of SIBs are closely similar to those of LIBs, with the advantages of abundant resources, low cost, safety, etc. [62]. However, the production of a large amount of sodium ions causes a severe expansion in volume, which is highly effective for short life cycles involving low energy density. Zhao et al. [63] synthesized an O_3-type $NaNi_{0.12}Cu_{0.12}Mg_{0.12}Fe_{0.15}Co_{0.15}Mn_{0.1}Ti_{0.1}Sn_{0.1}Sb_{0.04}O_2$ layered oxide with an outstanding retention of capacity of 83% at 3C along with 500 cycles (>99% of Coulombic efficiency) with highly reversible phase transition in between O_3 and P_3 structures. Katarzyna et al. [64] demonstrated the layered oxide material $NaMn_{0.2}Fe_{0.2}Co_{0.2}Ni_{0.2}Ti_{0.2}O_2$ with an average discharging capacity of 180 mA h g^{-1} at C/10 with the appearance of the P_3 phase, which is associated with sodium-ion vacancies and redox

Table 11.2 HEM-based electrode materials for LIB electrodes

HEMs	Synthesis method	Battery/ electrode	Key points	References
$(FeCoNiCrMn)_3O_4$	Calcination	LIB/anode	Columbic efficiency, 61.6%, higher capacity (discharge/charge 1,034/680 mA h g^{-1}), better rate capability (182 mA h g^{-1} at 2 A g^{-1})	[56]
$(Co_{0.2}Cu_{0.2}Mg_{0.2}Ni_{0.2}Zn_{0.2})O$	Spray pyrolysis	LIB/anode	Capacity 590 mA h g^{-1} after 500 cycles at 100 mA g^{-1}	[58]
$(CrMnFeCoNi)_3O_4$	Hydrothermal	LIB/anode	Specific capacity of 1,235 mA h g^{-1} with 90% retention capacity after 200 cycles	[59]
$Li_{1.3}Mn^{2+}{}_{0.1}Co^{2+}{}_{0.1}Mn^{3+}{}_{0.1}Cr^{3+}{}_{0.1}Ti_{0.1}Nb_{0.2}O_{1.7}F_{0.3}$	Solid-state method	LIB/cathode	Specific capacity 307 mA h g^{-1} at a low rate (20 mA g^{-1}), and retains more than 170 mA h g^{-1} when cycling at a high rate of 2,000 mA g^{-1}	[61]

reactions of Mn, Fe, Co, and Ni. The stabilized entropy structure formed by various transition metals has a great influence in providing a layered structure and enhancing stability. Moreover, the analysis of multimetals and sodium-deficient layered oxide cathodes, for instance $Na_{0.83}Li_{0.1}Ni_{0.25}Co_{0.2}Mn_xTi_xSn_{0.45-2x}O_{2-\delta}$, featured the active role of conversional entropy in extensively improving the performance of cathode materials [65]. By controlling inert cation species, Mn^{4+}, Ti^{4+}, and Sn^{4+}, researchers accomplished higher sodium cation transport kinetics, electrochemical reversibility, and a long life cycle, which leads to higher energy density. Besides all these, the issue involved the P_2 layered structure of transition metals, i.e. mobility of sodium ions in the transition metal layers while charging or discharging. HEMs are also involved in reducing phase transition difficulties faced by P2-type layered Na_xTMO_2 structures. Yang et al. [66] developed P_2-type $Na_{0.6}(Ti_{0.2}Mn_{0.2}Co_{0.2}Ni_{0.2}Ru_{0.2})$ O_2 for sodium storage applications, and the intermediate phase OP_4 of P_2 and O_2 appeared instead of a common P_2 and O_2 phase transition, achieving an improved mobility of sodium ions in storage mechanism reactions.

11.5.5 HEMs for Lithium-Sulfur Batteries (LSBs)

LSBs are another torchbearer in the EES technology due to their theoretically strong charge capacity, energy density, low cost, and biofriendly attributes. However, capacity deterioration occurs during the CV process due to the uncontrolled shuttling effect of liquid-state polysulfides (LIPS). Therefore, several efforts have been made to develop a suitable electrocatalyst to mitigate the shuttle effect, including surface engineering of the cathode material [67]. In this context, Zhao and coworkers designed (MgCoNiCuZn)O using the concept of HE and evaluated the performance of this HEO with the KB carbon cathode [68]. The initial capacities of the cells featuring KB/S and HEO/KB/S cathodes were reported to be 1,131 $mA\,h\,g^{-1}$ and 1,191 $mA\,h\,g^{-1}$, respectively. However, the cells containing HEO retained a high capacity value of 664 $mA\,h\,g^{-1}$ even after 200 cycles, whereas the capacity of the cell with KB/S declined rapidly after the same number of cycles. It was observed that high configurational entropy facilitates the confinement of LIPS and enhances the redox reaction at the cathode by improving the cycling stability.

11.5.6 HEMs for Zn-Ion Batteries (ZIBs)

ZIBs are another alternative for sustainable and efficient energy storage devices because of their environmental safety and low-cost availability. In ZIBs, divalent zinc is directly used as the battery anode, which provides more charge transfer in the battery operation. However, the performance of the ZIBs is hindered by the limited cyclability, specific capacity, and recharge ability of efficient cathode materials. The cathode of ZIBs is mainly based on tunnel and layered-type materials such as MnO_2, V_2O_5, and PBAs, which helps intercalate and deintercalate Zn^{2+} in charging and discharging, respectively. In a limited number of cycles, the cathode structure collapses, which restricts the overall performance of the battery. To mitigate these issues, HEMs have been used as an alternative cathode material, which is beneficial because of the multiple compositional and structural features that boost the host structure in the charging and discharging of materials. Xu et al. [69] reported a HE carbon-coated NASICON-type $Na_3V_2(PO_4)_2F_3$ (NVPF@C) cathode, which shows an exceptional cyclic stability of 6,000 cycles and a capacity density of 78.2 $mA\,h\,g^{-1}$ at 50 °C. The electrochemical performance of the as-synthesized sample is shown in Figure 11.8. Similarly, Wu et al. [70] developed a HEO cathode material $(Co_{0.2}Cr_{0.2}Fe_{0.2}Mn_{0.2}Ni_{0.2})_3O_4$, which in incorporation with ZIBs showed a high specific capacity (283 $mA\,h\,g^{-1}$ at 0.1 $A\,g^{-1}$) and an excellent cycle stability of 66.61% capacity retention after 90,000 cycles at 2 $A\,g^{-1}$.

11.6 Advantages and Challenges

Metal-free HEMs show several benefits for EES applications:

- *Reduced Dependence on Conventional Metals*: The absence of metals in these materials mitigates concerns regarding the depletion and ecological consequences associated with metal extraction.
- *Tailorable Properties*: The variety of compositions enables precise adjustments to material properties, thereby enhancing energy storage efficiency.
- *Potential for Enhanced Stability*: HE arrangements may contribute to improved structural integrity throughout charge/discharge cycles.

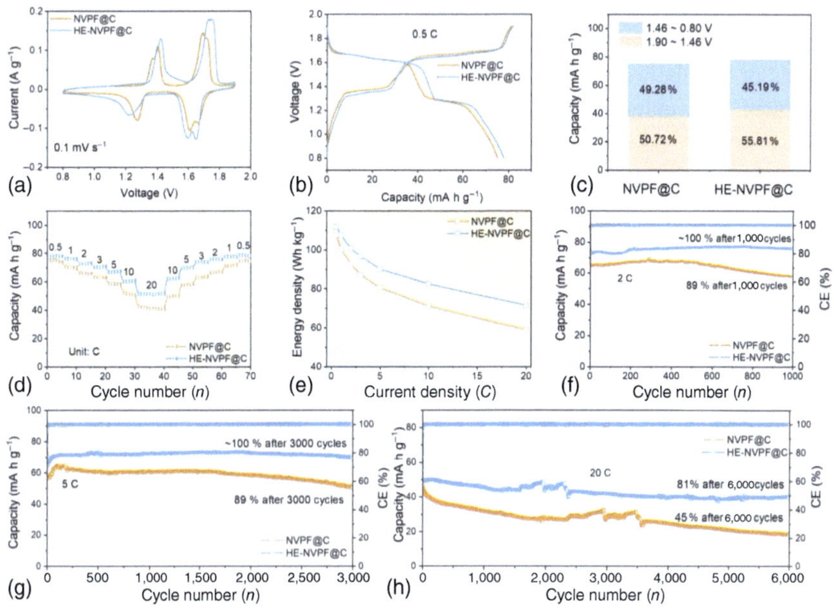

Figure 11.8 (a) CV, (b) GCD, (c) discharge capacity, (d) rate capability, and (e) energy density at different current densities. (f–h) Cycling stability at different temperatures of NVPF@C and HE-NVPF@C. *Source:* Reproduced from Ref. [70] / with permission of Elsevier

Nonetheless, there are obstacles to the advancement of metal-free HEMs in energy storage:

- *Insufficient Research*: In contrast to their metal-containing counterparts, metal-free HEMs have garnered less scholarly attention, resulting in a scarcity of studies and practical applications.
- *Performance Enhancement*: Attaining competitive energy storage capabilities without metals necessitates meticulous design and optimization of material formulations.

11.7 Conclusions

In summary, metal-free HEMs signify an intriguing and promising frontier in energy storage research, as well as producing clean energy while maintaining sustainability goals by replacing conventional metal-based systems. Their distinctive properties, which stem from intricate compositions and high configurational entropy, present the potential for substantial progress in SCs, dielectric capacitors, and possibly other energy storage technologies. Their tunable electrical properties provide a rich platform for tailoring electrochemical properties in the specific field of applications in research or industry. Although challenges persist in synthesis, comprehension, and optimization, ongoing research and development in this domain is poised to unveil new opportunities for more efficient, sustainable, and high-performance energy storage solutions by keeping aside toxicity, high cost, and weight. Moreover, in the scientific design of next-generation energy systems, via synergistic combinations or collaborations with other improved technologies, toolkits, electrolytes, substrates, and materials, HEMs may unlock excellent capabilities to store and manage energy. As we strive to expand the horizons of materials science, i.e. including experimental modeling and computational approaches, metal-free HEMs may overcome the existing barriers and will be crucial across these disciplines in meeting the increasing global demand for advanced energy storage technologies, thereby contributing to a more sustainable and energy-efficient future.

11.8 Future Perspectives

The future of metal-free HEMs in EES devices is highly promising as these materials have the potential to combine sustainability with high performance. Researchers may discover materials with customized ionic and electronic conductivity, improved durability, and reduced reliance on costly metals by applying entropy stabilization concepts to non-metallic systems, such as HEOs, carbides, or heteroatom-doped GQDs. These technological advancements could revolutionize applications such as solid-state batteries, SCs, and dielectric capacitors, offering sustainable alternatives to traditional metal-based systems. However, challenges remain, particularly in developing scalable synthesis methods with cost-effective techniques and understanding the relationships between entropy and the properties of HEMs. Advances in combinatorial chemistry, machine learning, and atomic-scale characterization will likely accelerate this frontier, paving the way for next-generation energy technologies.

References

1 Larcher, D. and Tarascon, J.M. (2015). Towards greener and more sustainable batteries for electrical energy storage. *Nat Chem* 7: 19–29.

2 Jiang, L., Zou, F., Qiao, Y., and Huang, Y. (2022). Patent analysis for generating the technology landscape and competition situation of renewable energy. *J Clean Prod* 378 (10): 134264.

3 Ma, J. and Huang, C. (2023). High entropy energy storage materials: synthesis and application. *J Energy Storage* 66: 107419.

4 Hussain, I., Lamiel, C., Ahmad, M. et al. (2021). High entropy alloys as electrode material for supercapacitors: a review. *J Energy Storage* 44: 103405.

5 Wang, Q., Yan, J., and Fan, Z. (2016). Carbon materials for high volumetric performance supercapacitors: design, progress, challenges and opportunities. *Energy Environ Sci* 9: 729–762.

6 Wu, F., Liu, M., Li, Y. et al. (2021). High-mass-loading electrodes for advanced secondary batteries and supercapacitors. *Electro Chem Energy Rev* 4 (2): 382–446.

7 Oses, C., Toher, C., and Curtarolo, S. (2020). High-entropy ceramics. *Nat Rev Mater* 5: 295–309.

8 Gao, M.C., Miracle, D.B., Maurice, D. et al. (2018). High-entropy functional materials. *J Mater Res* 33: 3138–3155.

9 Miracle, D.B. and Senkov, O.N. (2017). A critical review of high entropy alloys and related concepts. *Acta Mater* 122: 448–511.

10 Cantor, B., Chang, I.T.H., Knight, P., and Vincent, A.J.B. (2004). Microstructural development in equi atomic multicomponent alloys. *Mater Sci Eng A* 375–377: 213–218.

11 Yeh, J.W., Chen, S.K., Lin, S.J. et al. (2004). Nanostructured high-entropy alloys with multiple principal elements: novel alloy design concepts and outcomes. *Adv Eng Mater* 6 (5): 299–303.

12 Qu, J., Buckingham, M.A., and Lewis, D.J. (2023). High-entropy materials for electrochemical energy storage devices. *Energy Adv* 2: 1565–1590.

13 Amiri, A. and Yassar, R.S. (2021). Recent progress of high-entropy materials for energy storage and conversion. *J Mater Chem A* 9: 782–823.

14 Rost, C.M., Sachet, E., Borman, T. et al. (2015). Entropy-stabilized oxides. *Nat Commun* 6: 8485.

15 Han, X., Hu, Z., Lv, M. et al. (2024). High-entropy materials for aqueous zinc metal batteries. *Energy Environ Sci* 18: 97–129.

16 Nguyen, T.X.Y.H., Lin, C., and Ting, J.M. (2021). Self-reconstruction of sulfate-containing high entropy sulfide for exceptionally high-performance oxygen evolution reaction electrocatalyst. *Adv Funct Mater* 31: 2106229.

17 Gild, J., Zhang, Y., Harrington, T. et al. (2016). High-entropy metal diborides: a new class of high-entropy materials and a new type of ultrahigh temperature ceramics. *Sci Rep* 6: 37946.

18 Sarker, P., Harrington, T., Toher, C. et al. (2018). High-entropy high-hardness metal carbides discovered by entropy descriptors. *Nat Commun* 9: 4980.

19 Jin, T., Sang, X., Unocic, R.R. et al. (2018). Mechanochemical-assisted synthesis of high-entropy metal nitride via a soft urea strategy. *Adv Mater* 30: 1707512.

20 Zhao, X., Xue, Z., Chen, W. et al. (2020). Eutectic synthesis of high-entropy metal phosphides for electrocatalytic water splitting. *Chem Sus Chem* 13: 2038–2042.

21 Kamaruddin, H., Jianghong, Z., Yu, L. et al. (2024). A review of noble metal-free high entropy alloys for water splitting applications. *J Mater Chem A* 12: 9933–9961.

22 Feng, B. and Zhuang, X. (2019). The philosophy of carbon: meso entropy materials. *Faraday Discuss* 1–3.

23 Chien, L.K., Panda, P.K., Mallick, B.C. et al. (2024). Solid-phase microwave synthesis of high-entropy graphene quantum dots as metal-free electrochemical catalysts. *Appl Surf Sci* 648: 159061.

24 Qiao, H., Wang, X., Dong, Q. et al. (2021). A high-entropy phosphate catalyst for oxygen evolution reaction. *Nano Energy* 86: 106029.

25 Nemani, S.K., Zhang, B., Wyatt, B.C. et al. (2021). High-entropy 2D carbide MXenes: $TiVNbMoC_3$ and $TiVCrMoC_3$. *ACS Nano* 15 (8): 12815–12825.

26 Peng, J., Zhang, B., Hua, W. et al. (2023). A disordered Rubik's cube- inspired framework for sodium-ion batteries with ultralong cycle lifespan. *Angew Chem Int Ed* 62: e2022158.

27 Panda, P.K., Huang, H.Y., Dash, P. et al. (2025). Liquid-phase microwave synthesis of platinum-based high-entropy alloy catalysts on carbon supports for electrochemical hydrogen adsorption/desorption and oxygen evolution/reduction reactions. *Int J Hydrog Energy* 111: 536–545.

28 Qian, X., Han, D., Zheng, L. et al. (2021). High-entropy polymer produces a giant electrocaloric effect at low fields. *Nature* 600: 664–669.

29 Lei, L., Liu, J.Y., and Zhang, H.Y. (2023). Polymeric nitrogen: a review of experimental synthesis method, structure properties and lattice dynamic characterization from large scientific facilities and extreme spectroscopy perspectives. *Energ Mater Front* 4: 158–168.

30 Zhang, X., Liu, Y., Zhao, X. et al. (2024). Recent advances and perspectives of high-entropy alloys as electrocatalysts for metal-air batteries. *Energy Fuel* 38: 19236–19252.

31 Tian, R., Zhong, S., Wu, J. et al. (2016). Solvothermal method to prepare graphene quantum dots by hydrogen peroxide. *Opt Mater* 60: 204–208.

32 Jiang, Q., Gu, J., Zhang, L. et al. (2022). A highly efficient high-entropy metal hydroxy-methylate electrocatalyst for oxygen evolution reaction. *Chem Eng J* 453: 139510.

33 Cao, X., Gao, Y., Wang, Z. et al. (2023). FeNiCrCoMn high-entropy alloy nanoparticles loaded on carbon nanotubes as bifunctional oxygen catalysts for rechargeable zinc-air batteries. *ACS Appl Mater Interfaces* 15 (27): 32365–32375.

34 Zhang, H.M., Zhang, S.F., Zuo, L.H. et al. (2024). Recent advances of high-entropy electrocatalysts for water electrolysis by electro-deposition technology: a short review. *Rare Metals* 43 (6): 2371–2390.

35 Nechvoglod, O., Ostovari, M.A., and Pratskova, S. (2025). A review on high-entropy alloys coatings fabricated by electro-deposition: the correlation between composition, properties and processing parameters. *JOM* 77: 1005–1028.

36 Zhang, D., Yu, H., Xue, S. et al. (2024). Effect of graphene oxide on CoCrFeNi high-entropy alloy coatings prepared by the wetelectro-deposition method. *Surf Coat Technol* 485: 130919.

37 Zhao, J., Wang, Z., Fang, X. et al. (2023). Fast joule heating synthesis of NiCoFeCrMo high-entropy alloy embedded in graphene for water oxidation. *J Alloys Compd* 966: 171535.

38 Raj, G., Nandan, R., Kumar, K. et al. (2023). High entropy alloying strategy for accomplishing quintuple-nanoparticles grafted carbon towards exceptional high-performance overall seawater splitting. *Mater Horiz* 10 (11): 5032–5044.

39 Gao, S., Hao, S., Huang, Z. et al. (2020). Synthesis of high-entropy alloy nanoparticles on supports by the fast-moving bed pyrolysis. *Nat Commun* 11 (1): 2016.

40 Song, J.Y., Kim, C., Kim, M. et al. (2021). Generation of high-density nanoparticles in the carbothermal shock method. *Sci Adv* 7: eabk2984.

41 Abdelhafiz, A., Tanvir, A.N.M., Zeng, M. et al. (2023). Pulsed light synthesis of high entropy nano-catalysts with enhanced catalytic activity and prolonged stability for oxygen evolution reaction. *Adv Sci* 10: 2300426.

42 Xu, X., Du, Y., Wang, C. et al. (2020). High-entropy alloy nanoparticles on aligned electronspun carbon nanofibers for supercapacitors. *J Alloys Compd* 822: 153642.

43 Khan, B.J., Panda, P.K., Dash, P., and Hsieh, C.T. (2025). Microwave-assisted synthesis of platinum-free high-entropy alloy catalysts on reduced graphene oxide sheets for enhanced oxygen reduction and evolution reactions. *Chem – A Eur J* 31 (11): e202403863.

44 Yi, L., Xiao, S., Wei, Y. et al. (2023). Free standing high-entropy alloy plate for efficient water oxidation catalysis: structure/composition evolution and implication of high valence metals. *Chem Eng J* 469: 144015.

45 Güler, S., Alkan, E.D., and Alkan, M. (2022). Vacuum arc melted and heat treated AlCoCrFeNiTiX based high-entropy alloys: thermodynamic and microstructural investigations. *J Alloys Compd* 903: 163901.

46 Liu, H., Qin, H., Kang, J. et al. (2022). A freestanding nanoporous NiCoFeMoMn high-entropy alloy as an efficient electrocatalyst for rapid water splitting. *Chem Eng J* 435: 134898.

47 Cai, Z.X., Goou, H., Ito, Y. et al. (2021). Nanoporous ultra-high-entropy alloys containing fourteen elements for water splitting electrocatalysis. *Chem Sci* 12 (34): 11306–11315.

48 Jin, Z., Lyu, J., Zhao, Y.L. et al. (2021). Top-Down synthesis of noble metal particles on high-entropy oxide supports for electrocatalysis. *Chem Mater* 33 (5): 1771–1780.

49 Lei, X., Wang, Y., Wang, J. et al. (2023). Si-based high-entropy anode for lithium-ion batteries. *Small Methods* 8: 2300754.

50 Cropper, M.D. (2018). Thin films of AlCrFeCoNiCu high-entropy alloy by pulsed laser deposition. *Appl Surf Sci* 455: 153–159.

51 Lal, M.S. and Sundara, R. (2019). High entropy oxides—a cost effective catalyst for the growth of high yield carbon nanotubes and their energy applications. *ACS Appl Mater Interfaces* 11 (34): 30846–30857.

52 Kong, K., Hyun, J., Kim, Y. et al. (2019). Nanoporous structure synthesized by selective phase dissolution of AlCoCrFeNi high entropy alloy and its electrochemical properties as supercapacitor electrode. *J Power Sources* 437: 226927.

53 Dou, L., Yang, B., Lan, S. et al. (2023). High-entropy-nanofibers enhanced polymer nanocomposites for high-performance energy storage. *Adv Energy Mater* 13: 220392.

54 Yang, B., Zhang, Y., and Pan, H. (2022). High-entropy enhanced capacitive energy storage. *Nat Mater* 21: 1074–1080.

55 Qian, J., Ge, G., Yu, Z. et al. (2025). High temperature stable capacitive energy storage up to 320 °C in high-entropy dielectric thin film. *Nano Energy* 135: 110651.

56 Wang D, Jiang S, Duan C, Mao J, Dong Y, et. al. High-entropy chemistry stabilizing spinel oxide $(CoNiZnXMnLi)_3O_4$ (X = Fe,Cr) for high performance anode of Li-ion batteries. *J Alloys Compd* 2020;844: 156158.

57 Bolar, S., Ito, Y., and Fujita, T. (2024). Future prospects of high-entropy alloys as next generation industrial electrode materials. *Chem Sci* 15: 8664–8772.

58 Sarkar, A., Velasco, L., Wang, D. et al. (2018). High entropy oxides for reversible energy storage. *Nat Commun* 9: 3400.

59 Nguyen, T.X., Patra, J., Chang, J.K., and Ting, J.M. (2020). High entropy spinel oxide nanoparticles for superior lithiation–delithiation performance. *J Mater Chem A* 8: 18963–18973.

60 Luo, X.F., Patra, J., Chuang, W.T. et al. (2022). Charge–discharge mechanism of high-entropy Co-Free spinel oxide toward Li+ storage examined using operando quick-scanning X-ray absorption spectroscopy. *Adv Sci* 9: 2201219.

61 Lun, Z., Ouyang, B., Kwon, D.H. et al. (2021). Cation-disordered rocksalt-type high-entropy cathodes for Li-ion batteries. *Nat Mater* 20: 214–221.

62 Wang, P., Xi, B., Zhang, Z. et al. (2021). Atomic tungsten on graphene with unique coordination enabling kinetically boosted lithium–sulfur batteries. *Angew Chem Int Ed* 60: 15563–15571.

63 Zhao, C., Ding, F., Lu, Y. et al. (2022). High-entropy layered oxide cathodes for sodium-ion batteries. *Angew Chem Int Ed* 59 (1): 264269.

64 Walczak, K., Plewa, A., Ghica, C. et al. (2022). $NaMn_{0.2}Fe_{0.2}Co_{0.2}Ni_{0.2}Ti_{0.2}O_2$ high-entropy layered oxide – experimental and theoretical evidence of high electrochemical performance in sodium batteries. *Energy Storage Mater* 47: 500–514.

65 Wang, H., Gao, X., Zhang, S. et al. (2023). High-entropy Na-deficient layered oxides for sodium-ion batteries. *ACS Nano* 17: 12530–12543.

66 Yang, L., Chen, C., Xiong, S. et al. (2021). Multiprincipal component P_2-$Na_{0.6}$ $(Ti_{0.2}Mn_{0.2}Co_{0.2}Ni_{0.2}Ru_{0.2})O_2$ as a high-rate cathode for sodium-ion catteries. *J Am Chem Soc* 1 (1): 98–107.

67 Zhou, Z., Chen, Z., Lv, H. et al. (2023). High-entropy nanoparticle constructed porous honeycomb as a 3D sulfur Host for lithium polysulfide adsorption and catalytic conversion in Li-S batteries. *J Mater Chem A* 11: 5883–5894.

68 Zheng, Y., Yi, Y., Fan, M. et al. (2019). A high-entropy metal oxide as chemical anchor of polysulfide for lithium-sulfur batteries. *Energy Storage Mater* 23: 678–683.

69 Xu J, Galib M, Wu Z, Tao L, Shao Y, Zhang Y, Guo X, et. al. High-entropy strategy to suppress volumetric strain and enhance diffusion rate of $Na_3V_2(PO_4)2F_3$ cathode for durable and high-areal-capacity zinc-ion battery pouch cells, *Nano Energy* 2024;132:110373.

70 Wu, Z., Lv, Q., Wei, S. et al. (2025). Novel high-entropy oxides with ultra-long cycle life as cathode materials for aqueous zinc-ion batteries. *J Power Sources* 634: 236488.

12

Metal-Doped High-Entropy Materials for Energy Storage Applications

Barkha Rani, Sourav Ghosh, A. Deepak, and M. Suresh Kumar

Department of Electronics and Communication Engineering, Saveetha School of Engineering, Saveetha Institute of Medical and Technical Sciences, Chennai, Tamil Nadu, 602105, India

12.1 Introduction

Alloys are typically composed of two or more principal constituents. Additionally, trace amounts of other constituents are required to achieve the desired structural and functional properties. However, conventional materials cannot meet the demand of today's advanced industries such as aerospace, nuclear power, and defense in terms of performance, leading to the emergence of high-entropy materials (HEMs), particularly high-entropy alloys (HEAs). HEAs are composed of five or more elements with near-equimolar amounts, and each element makes a significant contribution to the behavior and hence properties of HEAs. These materials were proposed independently by Yeh and Cantor in 2004 and have since attracted extensive interest. High structural stability, entropy-stabilized phases, and compositional tailoring are the key advantages of HEAs [1, 2]. Scheme 12.1a,b depicts various classes of HEMs and their contribution to next-generation energy storage and other advanced applications, respectively [3].

12.1.1 Fundamental Effects of HEMs

Four fundamental effects of HEMs are illustrated in Figure 12.1 [4]. Random solid solution phases are stabilized by high-entropy effects. This stabilization is due to the reduction in Gibbs free energy and promotes structural uniformity. The severe lattice distortion effect is a consequence of atomic size differences, which is known to strengthen the material by introducing structural irregularities. Higher structural irregularities improve the hardness and mechanical performance of the material [5]. The sluggish diffusion effect inhibits atomic diffusion during solidification, which increases high-temperature stability and leads to nanostructured formations. Finally, the cocktail effect implies that HEA properties arise from complex synergistic interactions among constituent elements [6]. Due to these four distinct features, HEAs outperform conventional alloys in applications such as radiation

High-Entropy Materials for Energy Storage Devices, First Edition.
Edited by Chien-Te Hsieh, Pradeep Kumar Panda, and Arpan Kumar Nayak.

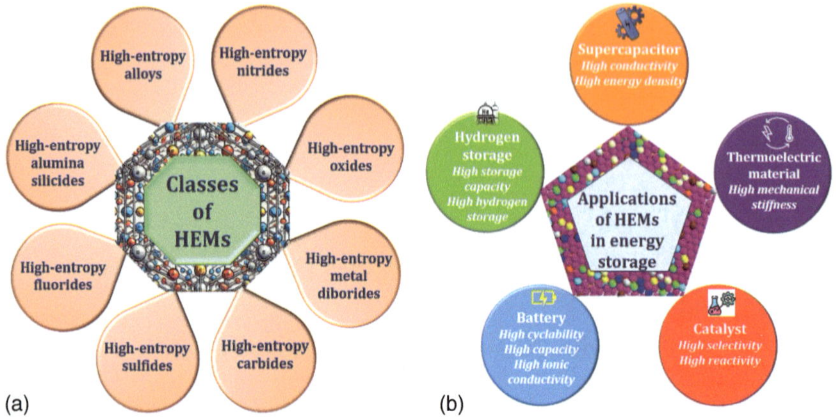

Scheme 12.1 (a) Different classes of HEMs and (b) their applications in the area of energy storage.

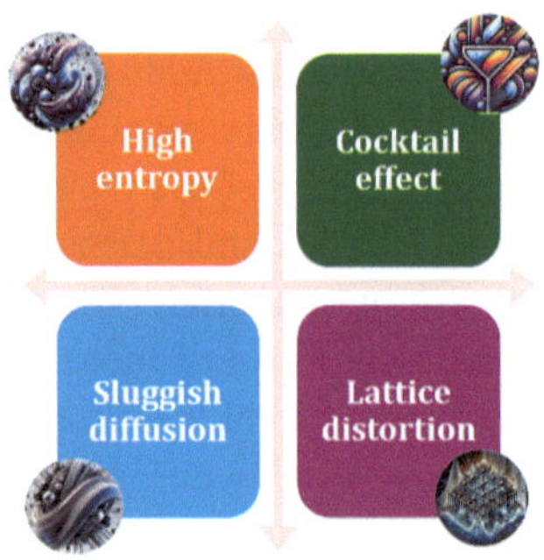

Figure 12.1 Core effects in HEMs and advantages.

resistance, wear and corrosion protection, fuel cells, and energy storage [7, 8]. This performance enhancement is crucial from an engineering standpoint.

12.1.2 Energy Storage Applications of M-HEMs

In this chapter, the discussion is focused on electrochemical energy storage. Advanced materials with superior electrochemical performance are in demand for cutting-edge energy storage technologies [8]. Conventional electrodes/electrolyte materials employed in supercapacitors (SCs) and batteries often suffer from structural instability, poor ionic conductivity, and low cycle life. Metal-doped HEMs (M-HEMs) provide high-entropy stabilization, which suppresses phase separation and improves electrochemical stability, thus addressing the above-mentioned challenges. Additionally, metal doping enhances the versatility of M-HEMs even more by enhancing conductivity, redox activity, and mechanical robustness [9].

Now, in the context of SCs, M-HEMs are researched extensively. SCs are known for their rapid charging ability, high cyclability, and power density. The charge storage mechanism of SCs can be of two major categories: (i) electrical double-layer capacitance (EDLC) and (ii) pseudocapacitive. M-HEMs, particularly high-entropy oxides (HEOs), exhibit a combination of both mechanisms. However, M-HEMs usually offer high specific capacitance because of their dominant pseudocapacitive behavior. Their rapid ion diffusivity and high capacitance retention are attributable to their distinctive electronic interactions and availability of multiple redox-active sites [10]. Transition-metal-doped HEOs such as $(Co_{0.2}Cu_{0.2}Mg_{0.2}Ni_{0.2}Zn_{0.2})O$ have demonstrated longer lifespan and enhanced charge storage capability [11]. Transition metal dopants have the ability to provide multiple oxidation states that facilitate quick and reversible charge storage, enabling enhanced electrochemical stability and prolonged cycling performance for M-HEMs [12].

Regarding battery chemistries such as lithium-ion batteries (LIBs) and lithium-sulfur batteries, M-HEMs serve as electrode materials due to their improved ionic conductivity and electrode stability [13, 14]. Structural instability is a major concern in lithium-based batteries, which leads to capacity fading and performance loss. The presence of more than one transition metal in M-HEMs results in robust electrode frameworks with high mechanical integrity. This mitigates typical battery aging mechanisms such as dendrite growth, electrode pulverization, and dissolution of polysulfide. Moreover, the high configurational entropy of M-HEMs increases Li-ion diffusion and charge transport kinetics. Thus, M-HEMs enable higher capacity retention and longer battery life. In lithium-sulfur batteries, where polysulfide dissolution and shuttle effects are pivotal disadvantages, M-HEMs ensure stability as the host matrix to efficiently encapsulate polysulfides, thereby enhancing cycle life and Coulombic efficiency [13]. The incorporation of metal dopants such as Fe, Mn, Ni, and Co further tailors its electrochemical reactivity by maximizing the charge storage efficiency even further.

Dielectric capacitors are also greatly benefited by M-HEMs [15]. They are crucial in energy storage and power conditioning applications. M-HEMs possess high permittivity, low dielectric loss, and superior thermal and electrical stability. Hence, they are suitable for high-performance energy storage applications. Engineering of the constituent materials of M-HEMs enables precise control over dielectric properties: the entropy-stabilized structure increases breakdown strength, eliminates grain boundary defects, and increases charge/discharge efficiency. Thus, M-HEMs address key limitations of conventional dielectric materials [16].

12.1.3 Synthesis and Structural Stability of M-HEMs

The tunability of material properties is greatly reliant on synthesis approaches. Hence, the synthesis of M-HEMs is crucial to their (i) electrochemical performance, (ii) structural stability, and (iii) scalability. Various synthesis techniques are demonstrated in Figure 12.2. Mechanical alloying, high-temperature sintering, solid-state microwave route, solvothermal synthesis, and chemical vapor deposition (CVD) have been employed to prepare M-HEMs, as found in the literature [17–19]. The mechanical alloying approach facilitates metastable high-entropy phase formation, whereas high-temperature sintering allows for the formation of crystalline structures with high phase stability. On the other hand, solvothermal synthesis enables the fabrication of nanostructured M-HEMs with high surface area and tailored

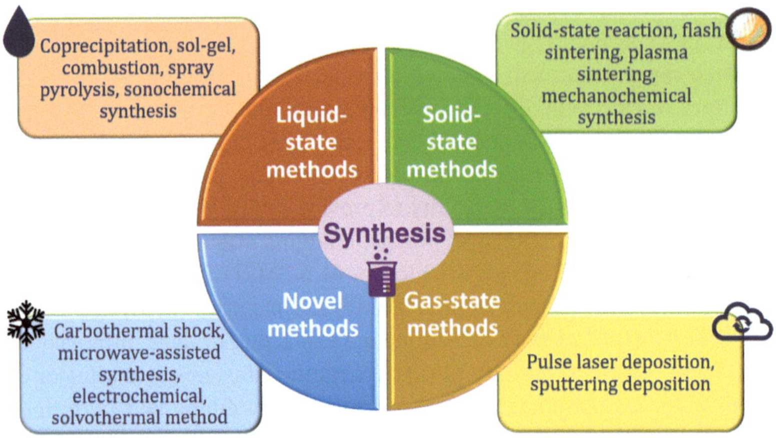

Figure 12.2 Various synthesis techniques for HEMs.

particle morphology. Lastly, CVD presents a pathway for the deposition of thin films of M-HEMs, especially for microelectronic applications. Thus, different synthesis approaches of M-HEMs can be pursued based on the targeted application.

In the modern era of scientific research, computational modeling is absolutely pivotal in the discovery and optimization of materials for engineering applications. Advanced techniques such as machine learning (ML), molecular dynamics (MD), and density functional theory (DFT) enable the reasonable prediction of material properties, electrochemical performance, and charge transport behavior. ML algorithms analyze large datasets of experimental and theoretical material properties, predicting new M-HEMs with optimized electrochemical behavior [20]. DFT calculations give insights into the electronic structure, charge transfer mechanisms, and ion diffusion kinetics, enabling rational material design. MD simulations offer significant insights into ion transport behavior at the electrode–electrolyte interface and surface interactions. Synergizing computational and experimental approaches can facilitate the development of novel M-HEMs for energy storage [21, 22].

12.2 Role of Doping in HEMs for Energy Storage

As mentioned above, doping is crucial for modulating the structural, electrochemical, and mechanical properties of M-HEMs. Dopants may modify the local atomic environment through lattice distortions and altering electronic structures. These modifications promote ion transport, which in turn improves the performance of HEAs and HEOs in electrochemical storage systems [23]. Ionic radii, valence states, and host matrix interaction are three essential factors to identify a suitable dopant. For instance, doping with small atomic-sized species such as Li, C, B, O, and N in HEAs leads to the formation of secondary ceramic phases, which results in precipitation strengthening [24]. These dopants can often be incorporated into interstitial sites. Therefore, they enhance mechanical properties via interstitial solution strengthening. The alteration of these mechanical properties directly affects the structural stability of electrode materials, which is important for the cyclability of devices under rigorous charging/discharging conditions [25].

In the present context, the inclusion of La^{3+} into spinel-structured $(CoCrFeMnNi)_3O_4$ has been reported to induce lattice distortion. The inclusion enhanced oxygen vacancies and amplified Li^+ diffusion kinetics. It stabilizes oxygen defects, which in turn improves electron transport and enhances the electrochemical activity of the material. However, doping should be optimized appropriately to obtain maximum benefit. Excessive Li^+ diffusion can cause capacity loss. Thus, an optimal doping concentration should be ensured [26].

Furthermore, heteroatom doping has been widely adopted to enhance intrinsic electrical conductivity and charge storage kinetics in HEO anodes. For example, polyaniline-coated HEOs have exhibited enhanced rate performance and extended cycling stability. Similarly, Li^+ doping in (LiCoCuMgNiZn)O anodes has been found to improve specific capacity and rate capability. This improvement is attributed to increased ionic conductivity and redox reversibility [27]. Such modifications introduce a synergistic effect that uses configurational entropy to stabilize electrochemical processes. The stabilization avoids degradation over the prolonged cycle life of devices.

Substitutional doping is another type of doping apart from interstitial doping. The substitution of B-site ions in high-entropy perovskites with elements such as Al^{3+} causes lattice distortion and defect formation. This type of doping can slow down crack growth and offer active sites for charge storage [28]. This substitutional approach has shown vast improvements by mitigating particle agglomeration and maintaining structural integrity during charge/discharge cycles [29]. In addition, rare earth element doping has shown promising effects on the structure and properties of HEMs. The distinctive 4f electronic configuration of rare earth ions enhances the redox activity. It also alters the electronic density of states, thereby facilitating charge transfer processes [30].

For instance, La-doped spinel-type HEOs demonstrate a better rate capability and higher Li^+ storage capacity due to nanosized rock salt phases and metallic nanograins. The nanoscale reduces Li^+ diffusion barriers and enhances electrode/electrolyte interface characteristics. This concludes that doping serves as a versatile strategy to tailor the mechanical, structural, and electrochemical features of HEMs. The continuous development of doping strategies in high-entropy systems has tremendous potential for next-generation storage devices.

12.3 Synthesis Methods of M-HEMs

The method of synthesis is crucial for tuning the properties of M-HEMs. Synthesis approaches such as sol-gel, wet chemical, and ball milling are common for these materials (Table 12.1). Synthesis approaches help control thermodynamic factors, enabling the formation of M-HEMs with a uniform distribution of elements in a single solid solution phase. The synthesis method significantly influences

Table 12.1 Summary and comparison of synthesis methods of HEMs or HEOs.

Parameters	Solid-state reaction	Sol-gel	Hydrothermal/ solvothermal	Coprecipitation	Spray pyrolysis
Synthesis approach	Top-down	Bottom-up	Bottom-up	Bottom-up	Bottom-up
Complexity of preparation	Simple process involving mixing and calcination	Involves multiple synthesis steps, requiring expertise	Requires precise temperature and pressure control	Complex due to multiple reaction steps and precipitation control	Requires optimized aerosol and reaction conditions
Reaction time	Long processing time due to high-temperature calcination	Moderate reaction time with additional drying and calcination steps	Short synthesis time but requires postprocessing	Short reaction time due to fast precipitation	Short synthesis cycle, ideal for rapid processing
Calcination requirement	Essential for phase formation and crystallinity	Required for phase stabilization	Optional, but may enhance crystallinity	Not required, as precipitation occurs at low temperatures	Required to obtain desired phase purity
Equipment	Requires conventional high-temperature furnaces	Requires controlled atmosphere and drying setup	Requires autoclaves and high-pressure reactors	Basic laboratory equipment is sufficient	Demands specialized aerosol generator
Cost	Economic and scalable	Costly due to expensive precursor materials	Moderate cost, requires specialized equipment	Low cost, economical for laboratory and industrial applications	Moderate cost, depends on operational scale
Scalability	Highly scalable for bulk material production	Moderate scalability but requires precise control	Difficult to scale due to equipment constraints	Limited scalability due to precipitation control challenge	Limited scalability due to process variability
Energy consumption	High due to prolonged heating cycles	Moderate due to multiple synthesis and drying steps	Moderate energy consumption for heating and pressure control	Moderate energy use, dependent on reaction conditions	High due to rapid heating and cooling requirements
Synthesis efficiency	Low efficiency due to long reaction times	Low efficiency due to precursor requirements	Moderate efficiency, dependent on reaction conditions	Moderate efficiency with controlled precipitation	High efficiency with continuous synthesis capability

Uniformity and dispersion	Moderate uniformity, limited by diffusion constraints	Moderate uniformity, affected by drying and gelation process	High uniformity due to controlled reaction kinetics	High uniformity, ensuring homogeneous material distribution	High dispersion and uniformity with fine particle formation
Particle quality	Moderate, with larger and irregular particle shapes	Moderate particle quality, depending on processing conditions	High-quality particles with controlled morphology	Poor particle quality due to agglomeration issues	Poor particle quality due to rapid quenching effects
Product purity	Moderate purity due to possible elemental segregation	Moderate purity, affected by precursor selection	High purity, allowing precise composition control	High purity due to controlled reaction environment	Moderate purity, depends on precursor uniformity
Advantages	1. Simple and widely used. 2. Produces stable phases with high crystallinity. 3. Economic and scalable.	1. Good phase control and homogeneous element distribution. 2. Useful for synthesizing fine powders and coatings. 3. Lower processing temperatures compared to solid-state methods.	1. Produces well-crystallized materials with high purity. 2. Excellent control over particle morphology and size. 3. Suitable for nanomaterial synthesis.	1. Produces high-purity materials with good composition control. 2. Cost-effective for laboratory and industrial applications. 3. Short synthesis time.	1. High efficiency and rapid synthesis. 2. Good control over dispersion and particle uniformity. 3. Continuous processing capability.
Disadvantages	1. High energy consumption due to long calcination times. 2. Limited control over particle size and morphology. 3. May lead to inhomogeneous elemental distribution.	1. High cost due to expensive precursors. 2. Requires precise control of drying and calcination. 3. Moderate scalability due to process complexity.	1. Requires high-pressure reactors and specialized equipment. 2. Difficult to scale up for industrial applications. 3. Process control is complex and time-consuming.	1. Limited scalability due to challenges in maintaining uniform precipitation. 2. Agglomeration issues can affect particle quality. 3. Requires additional heat treatment for phase formation.	1. High energy consumption due to rapid heating and cooling. 2. Limited scalability due to process variability. 3. Particle quality is affected by quenching effects.
References	[31–33]	[34]	[35]	[36, 37]	[38]

microstructural properties, phase stability, and defects. These factors can directly influence the electrochemical performance of HEMs. Various synthesis methods are discussed in detail in the following sections.

12.3.1 Solid-State Reactions

Solid-state synthesis is often utilized for preparing high-entropy ceramics and HEOs [39]. This approach involves blending metal oxides/carbonates in stoichiometric ratios. Then, the blended mixture is calcined to facilitate diffusion and phase formation. Key synthesis factors such as precursor selection, processing atmosphere, and synthesis temperature influence final material properties. Higher temperatures increase the diffusion coefficient, facilitating ion and vacancy movement. However, excessive heating can lead to oxygen deficiencies, secondary recrystallization, and increased grain size. Thus, surface area can be reduced, hindering metal (Li, Na) ion mobility. The major advantage is the ability to produce bulk materials with robust structural integrity. On the other hand, critical limitations of this synthesis technique are longer reaction times, high energy consumption, and irregularly shaped particle formation. Despite these drawbacks, materials such as $(GeSnPb)_{1/3}(SSeTe)_{1/3}$, $FeCoNiCuRuS_2$, and $AgSnSbSe_{1.5}Te_{1.5}$ demonstrate excellent thermal stability and performance when synthesized using this technique [40–42].

12.3.2 Sol-Gel Processing

Sol-gel synthesis is a low-temperature technique for producing highly homogeneous materials with controlled morphology and porosity. The process involves the formation of a colloidal sol. Then, the colloidal sol goes through gelation and subsequent thermal treatment to achieve the desired phase composition [43]. For instance, $(Co_{0.2}Cu_{0.2}Mg_{0.2}Ni_{0.2}Zn_{0.2})O$ was successfully synthesized via sol-gel combustion, reducing the synthesis temperature from 850 to 185 °C [34]. Wan et al. successfully prepared a series of Sr^{2+}-doped $La_{1-x}Sr_x(Co_{0.2}Mn_{0.2}Fe_{0.2}Ni_{0.2}Cu_{0.2})O_3$ using the sol-gel method. The prepared HEO showed a notable improvement in the performance [44]. Sr-doped $La_{0.7}Sr_{0.3}(Co,Cr,Fe,Mn,Ni)O_{3-\delta}$ was prepared via an adapted Pechini sol-gel process, yielding nanosized powders with a stable perovskite structure. The material maintains phase stability with limited strontium solubility ($x \geq 0.3$) and undergoes a phase transition at ~800 °C [45].

Chen et al. formed spinel-structured $(CrMnFeCoNi)_3O_4$ powder by glucose gelation. The molar ratio of glucose-to-metal cation (G/M) was found to influence the resultant particle size [46]. Park et al. fabricated rare earth-based HEO $(Y\text{-}La\text{-}Ce\text{-}Nd\text{-}Gd)_2O_{3+\beta}$ using the sol-gel method, achieving superior compositional uniformity [47]. This method is particularly suitable for energy storage applications due to its ability to control particle size and porosity at the molecular level, enhancing electrochemical performance in SCs and LIBs.

12.3.3 Synthesis of HEMs Using Hydrothermal and Solvothermal Processes

Solvothermal and hydrothermal processes involve aqueous or organic solvents at a certain atmosphere as temperature, pressure, pH, and precursor concentration can control the growth of crystalline nanostructures. These techniques are beneficial for synthesizing nanosized, phase-pure HEMs with controlled morphology. High-entropy sulfides, such as $(MnFeCoNiCu)S_2$ and $FeNiCoCrXS_2$ (X = Mn,Cu, Zn,Al), have been synthesized using solvothermal techniques, exhibiting superior catalytic properties [48]. Similarly, Jiang et al. synthesized (CoNiFeCuCr)Se via hydrothermal methods, achieving excellent stability and catalytic efficiency [49]. Broge et al. successfully developed a solvothermal method to synthesize phase-pure HEA nanoparticles (NPs) composed of Pt, Rh, Cu, Ni, Ir, Pd, Ru, and Co, further highlighting the potential of solvothermal synthesis [50]. Nguyen et al. prepared a HEO using a surfactant-assisted hydrothermal method, highlighting its potential for advanced electrode applications. They integrated non-equimolar Cr, Fe, Mn, Ni, and Co cations in distinct Wyckoff sites. The prepared material with flexible valence states showed improved structural stability and enhanced electrochemical performance [51].

12.3.4 Coprecipitation Method

The coprecipitation method is widely adopted for fabricating HEOs and composites due to its simplicity and scalability. This method simultaneously precipitates multiple metal precursors from a solution through pH adjustment. Then, the precipitates are treated thermally to obtain the desired phase [52]. Coprecipitation offers the significant benefit of achieving atomic-level uniformity in dopant distribution, resulting in consistent and homogeneous material characteristics [53]. Materials such as $(La_{0.2}Ce_{0.2}Nd_{0.2}Sm_{0.2}Eu_{0.2})PO_4$ have demonstrated better homogeneity when synthesized using this method [54]. Guo et al. successfully synthesized $La(CoCrFeMnNiAl_x)_{1/(5+x)}O_{3-\delta}$ ($x = 0.4$, 0.5, 0.6, and 0.7) using a coprecipitation technique followed by calcination and explored its suitability as an electrode material for SC applications [29].

12.3.5 Spray Pyrolysis

Spray pyrolysis is a rapid and efficient technique for producing thin films and fine powders. It involves atomizing a precursor solution into fine droplets and then subjecting them to thermal decomposition to produce solid particles. This method is beneficial for fabricating HEOs and other advanced materials due to its ability to produce uniform particle distributions and controlled film thicknesses [55]. HEOs such as $(Mn,Co,Fe,Ni,Cr)_3O_4$ have been synthesized via spray pyrolysis, with improved electrochemical performance due to their homogeneous structure [56]. Additionally, the flame spray pyrolysis (FSP) technique has been utilized by Phakatkar et al. to fabricate Cu-Mn-Fe-Ni-Zn HEO NPs, significantly improving their catalytic activity and energy storage properties [57, 58]. Kamecki et al. studied the fabrication of HEO thin films using low-temperature spray pyrolysis on amorphous silica (SiO_2) substrates and single-crystal sapphire (Al_2O_3). They focused on examining the structure and electrical characteristics of the $(Co,Cr,Fe,Mn,Ni)_3O_4$ compound [56].

12.3.6 Arc Melting

Arc melting is a metallurgical process where an electric arc is generated between electrodes or between metal-melting materials and the electrode. Researchers have successfully used arc melting to create various HEAs. For instance, Liu et al. employed arc melting to produce $Fe_{20}Co_{20}Ni_{20}Cr_{20}Mn_{20}$ alloy ingots. After remelting the ingots four times, a FCC-structured FeCoNiCrMn HEA was obtained [59]. Similarly, Wu et al. successfully synthesized a refractory $Hf_{25}Nb_{25}Ti_{25}Zr_{25}$ HEA exhibiting a BCC structure through arc melting [60].

12.3.7 Mechanochemical Synthesis

Mechanochemical synthesis or high-energy ball milling is a reliable top-down approach for producing HEOs. This method involves repeated mechanical impact and shear forces to induce solid-state reactions at room temperature, enabling the formation of single-phase materials with controlled defect densities. $(Hf_{0.2}Ta_{0.2}Ti_{0.2}Nb_{0.2}Mo_{0.2})C$ and $(FeCoNiMnCr)S_1$ have been successfully synthesized using this technique [1, 61]. Furthermore, Zhao et al. synthesized $Cu_4MnFeSnGeS_8$, demonstrating exceptional cycling stability in sodium-ion batteries (SIBs) [62]. This approach is advantageous for its scalability and energy efficiency, although contamination from milling media remains a challenge. Mohanty et al. developed a positive electrode from Cu-Co-Fe-Zn-Ni HEA through induction melting, followed by ball milling for size reduction [63]. Prasad et al. synthesized FeNiCoCrAlMn HEA with 2 wt% alumina using mechanically activated synthesis, involving 20 hours of ball milling. The alumina-dispersed alloy showed an FCC phase, increased hardness, improved oxidation resistance, and reduced thermal expansion, which makes it well suited for applications in high-temperature coatings [64]. Rekha et al. fabricated a multimetal HEA NPs–graphene composite using a two-step process. High-purity graphite was subjected to mechanical milling along with Cr, Fe, Ni, Cu, and Co, followed by sonication with sodium lauryl sulfate, yielding FCC-structured NPs on graphene with enhanced corrosion resistance [65].

12.3.8 Emerging Synthesis Strategies

Apart from the above-mentioned synthesis techniques, some other advanced synthesis approaches are being researched for M-HEMs. Microwave-assisted hydrothermal synthesis, electrospinning, and atomic layer deposition are being explored to enhance the performance of M-HEMs. A notable example is the

Ni-free high-entropy rock salt material (Mg-Co-Mn-Cu-Zn)O, synthesized using microwave-assisted techniques, which shows high conductivity and environmental sustainability [53]. These emerging strategies provide greater control over material composition and morphology, supporting the development of advanced energy storage technologies.

12.4 Advantages of M-HEMs in Energy Storage Applications

Development of HEMs for energy storage has opened up new possibilities for creating highly efficient and durable electrodes. However, to further enhance their electrochemical properties, researchers have turned to metal doping, a technique where specific metal atoms are introduced into the multielement lattice. This strategic modification fine-tunes the structural, electrical, and electrochemical properties of HEMs, making them ideal candidates for advanced batteries and SCs. One of the key challenges in battery and SC materials is structural degradation over repeated charge/discharge cycles. Just as a well-balanced alloy resists wear better than a single metal, M-HEMs offer greater phase stability by reducing lattice distortions and suppressing unwanted phase separation. The high-entropy effect naturally promotes a stable solid solution, and doping further reinforces this stability, preventing the formation of secondary phases that could compromise performance. This makes M-HEMs particularly useful for long-cycle applications, such as electric vehicle batteries, where stability over thousands of cycles is paramount [1, 9, 16, 20].

Imagine a highway packed with cars: if the lanes are well designed and wide, traffic flows smoothly; but if the road is uneven or narrow, congestion slows everything down. A similar principle applies to charge transport in electrode materials. Metal doping in HEMs helps by introducing additional charge carriers, widening the "lanes" for ionic and electronic movement. For instance, doping with transition metals such as Co, Ni, or Fe can increase the density of mobile electrons, reducing electrical resistance [23, 66]. Similarly, certain dopants can create structural voids that act as fast pathways for ion diffusion, ensuring efficient charge transport and improving overall energy storage efficiency [5]. In conventional battery materials, repeated charge/discharge cycles often lead to structural fatigue, much like how a metal spring weakens after being stretched and compressed repeatedly. However, in M-HEMs, the presence of multiple elements allows for a self-adjusting lattice, where different atomic sizes work together to accommodate volume changes without causing cracks or degradation. This intrinsic adaptability helps maintain the material's integrity over prolonged use, ensuring higher capacity retention. In SCs, where rapid charge storage relies on surface reactions, metal doping enhances the electrode surface area and porosity, resulting in an increased density of active sites for storing energy.

For any energy storage device, ion transport is as important as capacity. Batteries with sluggish ion transport are like slow-charging phones—frustratingly inefficient.

M-HEMs tackle this issue by reducing charge transfer resistance, allowing faster electron flow. The incorporation of metals with different oxidation states, such as Mn, Fe, or Cr, enables rapid Faradaic reactions, leading to improved charge storage kinetics. In SCs, this translates to higher power density, ensuring quick energy release when needed—ideal for applications such as regenerative braking in electric vehicles, where rapid energy uptake and discharge are essential [67]. Another unique advantage of M-HEMs is their ability to fine-tune redox chemistry, which directly influences energy storage capacity. Just as a chef carefully selects spices to enhance a dish's flavor, scientists strategically choose metal dopants to modify oxidation states and introduce multielectron transfer processes. For instance, the presence of Mo, V, or W can lead to highly reversible redox reactions, enhancing the material's efficiency in energy storage and release [68]. This flexibility makes M-HEMs particularly attractive for next-generation battery chemistries beyond traditional lithium-ion systems.

Overall, from an environmental standpoint, the use of M-HEMs can help reduce the reliance on critical materials, such as cobalt, which is expensive and associated with supply chain challenges. By substituting rare or hazardous elements with abundant and eco-friendly alternatives, M-HEMs offer a more sustainable and economic solution for energy storage. Additionally, their longer lifespan reduces material waste, contributing to a greener and more efficient energy economy.

12.4.1 Applications in Battery Technologies

a) *LIBs*: LIBs are the leading energy storage solution, widely used in applications ranging from mobile devices to electric vehicles. However, traditional electrode materials face structural degradation, slow charge transport, and limited capacity. M-HEMs offer a compelling solution by enhancing electrode stability, increasing lithium-ion diffusion, and improving electronic conductivity. Bérardan et al. initially identified Li^+ conductivity in rock-salt-structured HEOx ($10^{-3}\,S\,cm^{-1}$) [69]. Incorporating dopants such as Mg, Ti, or Zr into the cathode material can stabilize the crystal structure, preventing undesired phase transitions and capacity fading over long-term cycling [70, 71]. Qiu et al. showed that $(Cu_{0.2}Co_{0.2}Mg_{0.2}Ni_{0.2}Zn_{0.2})O$ HEOx anodes provide superior cycling stability due to entropy stabilization. Inactive MgO further prevents metal oxide agglomeration, enhancing cycle life [72]. High-entropy perovskite oxides (HEPOs) have also been explored for LIBs, with $((BiNa)_{1/5}(LaLi)_{1/5}(CeK)_{1/5}Ca_{1/5}Sr_{1/5})TiO_3$ delivering excellent cycling performance [29].

   Imagine a bridge designed with multiple materials—some providing strength, others flexibility—allowing it to withstand repeated stress without breaking. Similarly, metal doping strengthens the lattice structure in LIB electrodes by balancing atomic forces. This prevents volume expansion and contraction, which is a primary cause of electrode failure over time. Doping with elements such as Mg, Ti, Al, or Zr introduces atomic-level reinforcement, stabilizing the cathode or anode structure and preventing the formation of cracks or dislocations during charge/discharge cycles [60, 73]. Xiang et al. found that

Al^{3+}-doped $(Al_{0.2}CoCrFeMnNi)_{0.58}O_{4-\delta}$ anodes delivered more than twice the reversible specific capacity (554 mA h g^{-1} after 500 cycles at 200 mA g^{-1}) compared to $(CoCrFeMnNi)_{0.6}O_{4-\delta}$, along with excellent rate performance (634 mA h g^{-1} at 3 A g^{-1}) [74]. Chen et al. reported $(Ni_{0.2}Co_{0.2}Mn_{0.2}Fe_{0.2}Ti_{0.2})_3O_4$ anodes achieving 560 mA h g^{-1} at 100 mA g^{-1}, with full capacity retention after 100 cycles [75]. Sarkar et al. explored the electrochemical response of transition metal-based high-entropy oxides (TM-HEO) and medium-entropy

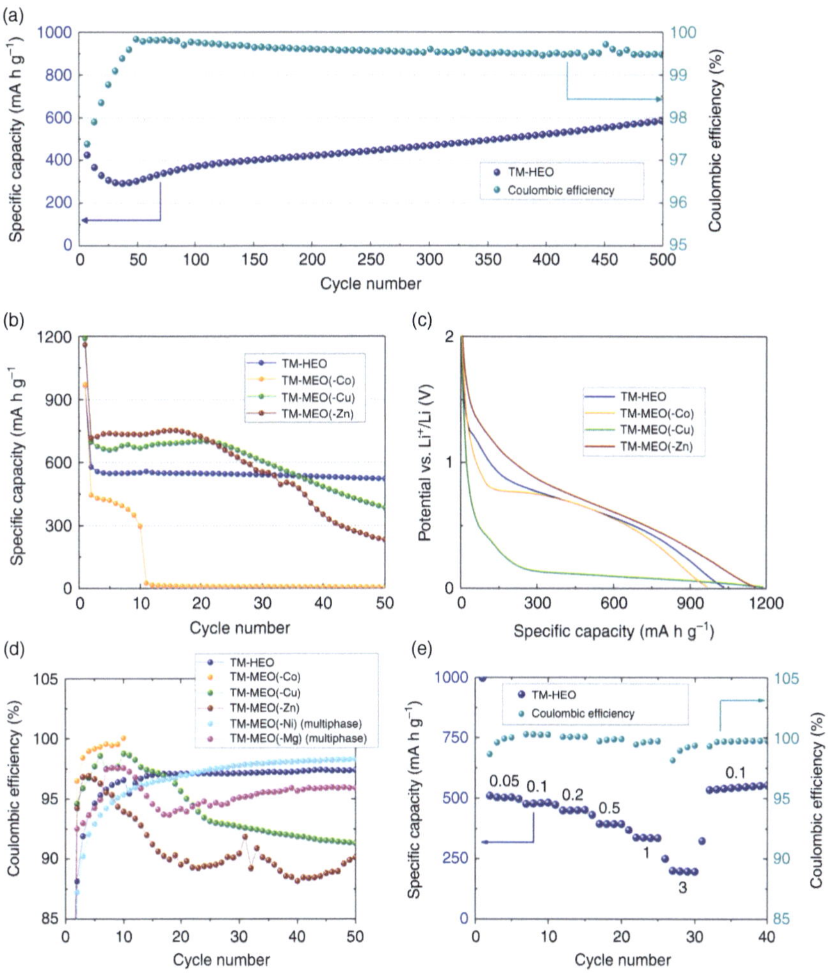

Figure 12.3 Cycling and rate performance of calcined TM-HEO: (a) Long-term cycling at 200 mA g^{-1} showing excellent capacity retention and high Coulombic efficiency. (b) Stable specific capacity at 50 mA g^{-1} indicating superior electrochemical stability compared to other compositions. (c) First-cycle discharge profiles revealing composition-dependent voltage behavior. (d) Coulombic efficiency variation with cycles at 50 mA g^{-1} illustrating their comparative cycling stability. (e) Rate performance of a high-loading (2.3 mg cm^{-2}) TM-HEO cell with current densities in A g^{-1} *Source:* Reproduced from Ref. [76] / nature communications / CC BY-4.0.

oxides (TM-MEO) to elucidate how structural complexity governs lithium-ion storage. The TM-HEO delivered markedly enhanced capacity and distinct lithiation characteristics, sustaining almost full Coulombic efficiency and stable performance at 200 mA g^{-1} (Fig. 12.3a). Even when cycled at 50 mA g^{-1}, the material preserved its structural integrity, while the Zn- and Cu-deficient analogues suffered rapid capacity fading and the Co-deficient oxide lost activity after only a few cycles (Fig. 12.3b,c). Interestingly, the Cu-free composition exhibited a lower discharge potential, indicating its suitability as an anode for primary batteries. Further evaluation of Coulombic efficiency reinforced the outstanding cycling durability of TM-HEO (Fig. 12.3d). In contrast, Ni-deficient multiphase oxides delivered much lower capacities (below 250 mAh g^{-1} after 40 cycles). Moreover, a high-loading TM-HEO electrode (2.3 mg cm^{-2}) maintained consistent rate performance across increasing current densities, highlighting its structural robustness (Fig. 12.3e). These findings clearly indicate that high configurational entropy strengthens the lattice framework and enhances charge reversibility [76].

Lithium ions should move smoothly and rapidly within the electrode for a battery to charge quickly and deliver power efficiently. However, in traditional LIB materials, some lattice structures create bottlenecks that slow down the ion transport. Metal doping creates wider and more interconnected diffusion channels, allowing lithium ions to move more freely. This is akin to adding extra lanes to a highway to reduce congestion and speed up traffic flow. By incorporating Ni, Co, or Fe, diffusion kinetics is enhanced, enabling faster charge/discharge cycles without sacrificing capacity [77].

b) *SIBs*: With concerns about lithium scarcity, SIBs have emerged as a viable alternative. However, sodium ions are larger than lithium ions, making it difficult for them to move efficiently within electrode materials. M-HEMs can be used to create open-framework structures that accommodate sodium-ion diffusion more effectively [78].

Since sodium ions require larger diffusion channels than lithium ions, metal doping helps by creating an open-framework structure that provides low-resistance pathways. Just as well-planned urban infrastructure allows for smooth vehicle movement, an optimized electrode lattice with strategic metal dopants (such as Fe, Ti, or Sn) enables fast sodium-ion transport. One of the biggest limitations of SIBs is their poor long-term stability, as sodium insertion and extraction often cause structural breakdown. Doping with metals such as Mg and Zn strengthen the material, reducing volume expansion and preventing premature degradation [79]. This ensures that SIBs can withstand thousands of cycles while maintaining high capacity, making them ideal for grid storage applications where longevity is crucial. SIBs can potentially be cheaper and more environmentally friendly than LIBs as they eliminate the need for scarce lithium and cobalt resources. M-HEMs enhance the viability of SIBs by enabling the use of earth-abundant elements such as iron and manganese, reducing the reliance on expensive and geopolitically sensitive materials.

c) *Solid-State Batteries (SSBs)*: These batteries are poised to shape the future of energy storage, offering greater energy density, enhanced safety, and extended service life than conventional liquid electrolyte systems. M-HEMs hold significant promise in the development of solid electrolytes as they can deliver high ionic conductivity and minimize interfacial resistance, facilitating efficient ion movement. The ability of HEMs to accommodate multiple elements enables the precise tuning of electrolyte characteristics, supporting the advancement of next-generation battery systems.

One of the key obstacles in SSBs is the slow movement of ions through the solid electrolyte. Unlike liquid electrolytes, where ions move freely, solid materials often suffer from rigid lattices that hinder ion diffusion. Metal doping creates defects and vacancies that act as fast ion transport channels, improving the electrolyte's performance. For example, doping $Li_7La_3Zr_2O_{12}$, a common solid electrolyte, with Al or Ta increases its ionic conductivity by several orders of magnitude [80]. A major issue in SSBs is the formation of high-resistance interfaces between the electrode and electrolyte, which slows down charge transfer. M-HEMs can modify surface chemistry to create seamless, high-contact interfaces, reducing energy loss. This is akin to using lubricants to improve mechanical efficiency in engines—by ensuring a smooth interface, the battery operates more efficiently. SSBs operate under higher pressures and temperatures than conventional LIBs, requiring materials that can withstand mechanical stress. Metal doping enhances the toughness and flexibility of electrode materials, making them resistant to cracking or delamination. This ensures that SSBs maintain stable performance over extended periods, making them a promising candidate for electric vehicles, aerospace, and high-energy-density storage systems.

A key benefit of M-HEMs is their high specific capacity. Traditional graphite anodes in LIBs offer a capacity of 372 mA h g^{-1}, while the capacity of HEM anodes can exceed 900–1,500 mA h g^{-1}, depending on the composition [71]. M-HEMs improve redox activity by introducing additional charge storage sites, resulting in improved rate capability and increased capacity. During charge/discharge cycles, conventional electrode materials experience mechanical stress and volume expansion, leading to structural degradation and capacity loss. However, M-HEMs have a self-adjusting lattice structure, allowing them to accommodate ion insertion and extraction without significant volume changes. For instance, entropy-stabilized oxides such as $Co_{0.2}Cu_{0.2}Mg_{0.2}Ni_{0.2}Zn_{0.2}O$ exhibit long cycle life, retaining 80% of their initial capacity after 1,000 cycles [81].

Introducing metal dopants in HEMs helps create percolating diffusion pathways, reducing charge transfer resistance and enhancing ion mobility. This is particularly useful in SSBs, where M-HEMs can act as superionic conductors, achieving ionic conductivities comparable to those of liquid electrolytes while improving safety. By reducing the reliance on cobalt and nickel, HEM cathodes address supply chain concerns and reduce battery production costs. Additionally, these materials allow for the incorporation of earth-abundant elements such as iron, manganese, and titanium, making battery manufacturing more sustainable. Graphite has traditionally served as the primary anode material in LIBs, but its limited energy storage capacity has driven the search for alternatives. M-HEMs present a promising alternative, offering higher lithium storage capacity, improved cycling stability, and enhanced conductivity. Certain compositions, such as entropy-stabilized transition metal oxides, exhibit superior lithium-ion diffusion properties, reducing charge/discharge times and improving the overall battery performance. Additionally, silicon, a high-capacity anode material, suffers from extreme volume expansion during lithiation. Doping silicon anodes with HEMs helps mitigate this issue by providing a stabilizing framework that prevents excessive expansion and structural collapse [82].

In the case of cathodes, HEMs improve energy density and long-term stability. Traditional lithium cobalt oxide and nickel manganese cobalt cathodes suffer from capacity fade and structural instability due to repeated lithium-ion intercalation and deintercalation. By replacing these conventional cathodes with high-entropy compositions such as $Li(Mn,Fe,Co,Cr,Ti)O_2$, researchers have observed significantly improved cycling performance and reduced reliance on expensive cobalt [83].

SSBs deliver superior energy density and safety over conventional LIBs and require advanced electrolytes with high ionic conductivity. High-entropy ceramic electrolytes have demonstrated promising results in this field, with enhanced lithium-ion transport properties and improved thermal stability. Sulfide-based high-entropy electrolytes have been explored for both Na-ion and Li-air batteries, showing significant performance improvements over conventional solid electrolytes. Despite their advantages, M-HEMs face several challenges. Producing these materials involves intricate processes that demand accurate regulation of annealing temperatures, reaction parameters, and atomic arrangements to ensure uniform phase formation. Additionally, characterization of HEMs remains challenging due to their complex phase interactions, which require advanced techniques, including high-resolution transmission electron microscopy (HRTEM) and synchrotron-based X-ray diffraction (XRD). Scalability is another concern as large-scale production of HEMs remains more expensive than conventional battery materials. However, advances in scalable synthesis techniques such as ball milling, sol-gel processing, and coprecipitation are making industrial applications more feasible.

M-HEMs represent a paradigm shift in battery technology, providing enhanced energy density, superior stability, and better ionic conductivity compared to traditional materials. Their ability to tune electrochemical properties through composition design makes them highly versatile for future energy storage applications. Additional details regarding the application of HEMs in energy storage can be found in Table 12.2.

12.4.2 Applications in SCs

12.4.2.1 Enhancing Electrode Performance

SCs rely on fast charge storage mechanisms, and their performance is heavily dependent on the electrode material's conductivity, surface area, and electrochemical stability. Metal doping improves these aspects by modifying the electronic structure, leading to better charge transfer kinetics. For example, doping carbon-based

Table 12.2 Various M-HEMs as battery electrodes.

Composition	Cycles	Initial capacity (mAh g^{-1})	Final capacity (mAh g^{-1})	Current density/ rate (A g^{-1})	References
$Mg_{0.2}Co_{0.2}Ni_{0.2}Cu_{0.2}Zn_{0.2}O$	300	~1,500	~935	0.1	[72]
$(Al_{0.2}CoCrFeMnNi)_{0.58}O_{4-\delta}$	500	2,234	536	0.2	[74]
$(MgCoNiZn)_{0.65}Li_{0.35}O$	100	925	610	1	[84]
$Co_{0.2}Mg_{0.2}Ni_{0.2}Cu_{0.2}Zn_{0.2}O$	300	1,325	417	1	[85]
$Li_{0.8}Na_{0.2}(NaCoMnAlFe)O_2$	20	80	62	—	[86]
$Co_{0.15}Mn_{0.1}Ti_{0.1}Sn_{0.1}Sb_{0.04}O_2$ $NaNi_{0.12}Cu_{0.12}Mg_{0.12}Fe_{0.15}$	200	110	87	0.5C	[87]
$La_{0.8}Sr_{0.2}(Cr_{0.2}Mn_{0.2}Fe_{0.2}Co_{0.2}Ni_{0.2})O_3$/Sulfur/PVDF/CNF	500	1,038	714	1C	[88]
$(Cu_{0.7}Ni_{0.6}Fe_{0.6}Sn_{0.5}Mn_{0.4})O_4$/CNF	400	907	435	1C	[89]
$(Ni_{0.2}Co_{0.2}Cu_{0.2}Mg_{0.2}Zn_{0.2})O$/ Sulfur/Ketjen Black	800	1,244	784	0.5C	[90]
$Li_{1.8}(FeCoZnCrMn)_3O_x$	300	388	484	0.5	[91]
$Zn_{0.5}Co_{0.5}Mn_{0.5}Fe_{0.5}Al_{0.5}Mg_{0.5}O_4$	5,000	950	233	2	[77]
$[(FeCoNiMn)_{0.88}Na_{0.12}]_3O_4$	300	470	90	1	[92]
$Li_{0.1}(LiLaCaSrBa)Ti_{0.9}Al_{0.1}O_3$	100	59	57	0.1	[93]
$[(Bi,Na)_{1/5}(La, Li)_{1/5}(Ce, K)_{1/5}Ca_{1/5}Sr_{1/5}]TiO_3$	300	84	115	1	[29]
$Zn_{0.5}Ge_{0.5}Cu_{0.5}Si_{0.5}P_2$	100	1,581	701	3.2	[94]
$NaNi_{0.12}Cu_{0.12}Mg_{0.12}Fe_{0.15}Co_{0.15}Mn_{0.1}Ti_{0.1}Sn_{0.1}Sb_{0.04}O_2$	500	110	83%	3C	[87]
$Co_{0.08}Mn_{0.08}Ni_{0.08}Fe_{1.96}Mg_{0.08}Nd_{0.01}Gd_{0.01}Sm_{0.01}Pr_{0.01}O_4$/S	300	1,146	76%	0.5C	[95]
$(Mg_{0.2}Mn_{0.2}Ni_{0.2}Co_{0.2}Zn_{0.2})Fe_2O_4$/S	500	1,368.7	63.5%	1C	[96]
$Na_3(Ti_{0.2}Cr_{0.2}Zr_{0.2}V_{0.2}Mn_{0.2})_2(PO_4)_3$	2,000	118.5	80.4%	20C	[97]
$NaFe_{0.2}Co_{0.2}Ni_{0.2}Ti_{0.2}Sn_{0.1}Li_{0.1}O_2$	100	130.6	72%	0.1C	[98]

electrodes with transition metals such as Ni or Co introduces additional pseudocapacitive redox reactions, significantly increasing the overall capacitance.

12.4.2.2 Hybrid SCs

Hybrid SCs bridge the gap between batteries and traditional capacitors by combining high energy density with fast charge/discharge rates. M-HEMs are particularly well suited for this application, as their multimetal composition allows for the simultaneous optimization of high capacity (like batteries) and high power efficiency (like capacitors). As a result, they are highly suitable for systems that require swift energy discharge, such as in electric vehicles, power infrastructure, and backup solutions for industrial operations. As researchers push the limits of the SC technology, the versatility of M-HEMs provides a platform for designing next-generation devices with ultrahigh energy densities. By incorporating novel dopants, optimizing surface structures, and exploring 3D nanostructured architectures, future SCs could achieve unprecedented performance, enabling compact and efficient energy storage for diverse applications.

Wan et al. reported that Sr incorporation significantly enhances the electrochemical performance of HEPOs. The $La_{0.9}Sr_{0.1}(Fe_{0.2}Ni_{0.2}Co_{0.2}Mn_{0.2}Cu_{0.2})O_3$ electrode attained a maximum specific capacitance of 204.33 F g^{-1} at 2 mV s^{-1}, outperforming the undoped sample. It retained 74% capacity from 1 to 5 A g^{-1} and exhibited stable cycling over 1,000 cycles [44]. Guo et al. demonstrated that HEPOs (x = 0.4–0.7) exhibited capacitances between ~259 and 35 F g^{-1} at 1 A g^{-1}, maintaining 85.01–91.25% capacity after 2,000 cycles. These properties stem from entropy stabilization and Al^{3+} doping, which prevent ion aggregation during charge/discharge cycles [28]. A Ni-Cu-Fe-Co-Zn HEA electrode was engineered by Mohanty et al. for SC applications, delivering a specific capacitance of 325.17 F g^{-1} at 1 A g^{-1} in 3 M KOH. DFT analysis confirmed enhanced chemical stability, yielding an energy density of ~23 Wh kg^{-1} at 325 W kg^{-1} [63]. Chen et al. synthesized $(CrMnFeCoNi)_3O_4$ HEO powders for SCs, achieving ~490 F g^{-1} at 0.5 A g^{-1}. Their superior behavior is attributed to high conductivity, a wide potential window, and synergistic cation effects. The HEO/AC asymmetric SC showed a 0–1.6 V window, delivering 104.1 Wh kg^{-1} at 3,991.8 W kg^{-1} [46]. Jiang et al. (Figure 12.4) demonstrated that all $KMFe(CN)_6$ samples showed reversible redox peaks with a ~40 mV separation, indicating efficient cation transfer. Oxidation peak potentials ranged between 0.51 and 0.68 V depending on elemental composition. Among them, Mg-containing $KMFe(CN)_6$ showed higher capacities (50–60 F g^{-1} at 200 mV s^{-1}) compared to $K(MnFeCoNiCu)Fe(CN)_6$, highlighting the role of Mg in enhancing performance. Additionally, single-metal PBAs were synthesized, and Mg-containing ones showed the maximum specific capacitance (175 F g^{-1} at 5 mV s^{-1}), confirming the benefits of a high-entropy structure [99].

Ongoing research is anticipated to drive new advancements in energy storage chemistry, enabling faster charging, longer-lasting, and more environmentally friendly energy storage solutions. Further information on the use of HEMs in SCs is given in Table 12.3.

12.4.3 Applications in Dielectric Capacitors

Dielectric materials are essential components in modern electronics, especially capacitors, where they influence energy storage, charge retention, and device efficiency. The dielectric properties required for capacitors vary based on application needs such as high dielectric constants; minimal energy losses are critical for capacitive energy storage, while minimal leakage currents ensure reliable information storage [108]. Metal doping in HEMs enables the fine-tuning of dielectric properties through compositional complexity, which introduces lattice distortions, strain, and oxygen vacancies that influence polarization behavior

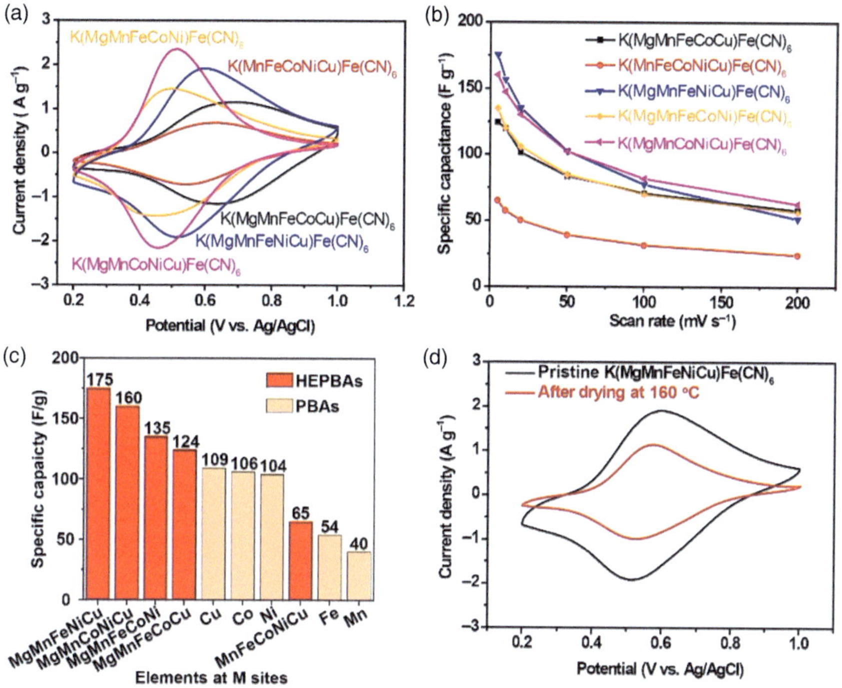

Figure 12.4 (a) CV curve of $KMFe(CN)_6$ series measured in 1 M Na_2SO_4 with a scan rate of 5 mV s^{-1}. (b) Specific capacitance of $KMFe(CN)_6$ series derived from CVs at different scan rates. (c) Comparative analysis of specific capacitance between HEPBAs and single-metal PBAs at a scan rate of 5 mV s^{-1}. (d) CVs at 5 mV s^{-1} for $K(MgMnFeNiCu)Fe(CN)_6$ in both its standard and dried forms. *Source:* Reproduced from Ref. [99] / with permission of Elsevier.

[109, 110]. For instance, Zhang et al. demonstrated that high-κ HEM ceramics, such as $(Li_{0.2}\,La_{0.2}Ba_{0.2}Sr_{0.2}Ca_{0.2})Nb_2O_{6-\delta}$, can achieve dielectric constants up to 1.1×10^6 at 100 Hz ~1,290 times higher than their conventional counterparts [111]. However, strong dielectric relaxation effects due to oxygen vacancies cause a marked frequency dependence. Beyond 10^3 Hz, the dielectric constant declines significantly, highlighting a common challenge in both HEMs and traditional dielectric materials. Nonetheless, certain HEM compositions, such as $[(Ce_{0.5}K_{0.5})(Bi_{0.5}Na_{0.5})_{0.25}Ba_{0.25}Sr_{0.25}Ca_{0.25}]_{1-x}TiO_3$, exhibit remarkable frequency stability, maintaining a high dielectric constant ($\kappa = 535$) over a broad range from 10^2 to 10^9 Hz [112].

Beyond capacitive applications, HEM dielectrics offer solutions for overcoming solubility limits and stabilizing unstable phases, as observed in the pyrochlore-based $Bi_2Ti_2O_7$ system. A composition such as $(Bi_{3.25}La_{0.75})(Ti_{3-3x}Zr_xHf_xSn_x)O_{12}$ achieves high energy densities (182 J cm^{-3} at 6.35 MV cm^{-1}) with excellent thermal stability, making it ideal for gate insulation materials [9]. Emerging fields such as printed electronics are also exploring alternatives to conventional high-κ dielectrics, such as electrolyte-based materials that improve semiconductor interfaces and reduce

Table 12.3 Various M-HEMs as SC electrodes.

Composition	Electrolyte	Stability% (cycle)	Capacitance ($F g^{-1}$)	Current density ($A g^{-1}$)/scan rate ($mV s^{-1}$)	References
$(FeCoCrMnZn)_3O_4$	1 M KOH	82.8% (1,000)	340.3	0.5	[100]
$MnO_2/FeCrCoMnNiAl_{0.75}$	0.5 M Na_2SO_4	81% (6,000)	961	5	[101]
$(Zr_{0.5}Ti_{0.5}Ce_{0.5}Hf_{0.5})O_7$	1 M Na_2SO_4	—	703	1	[102]
$La_{0.7}Bi_{0.3}Mn_{0.4}Fe_{0.3}Cu_{0.3}O_3$	6 M KOH	—	480.95	0.5	[103]
$(CrMnFeCoNi)_3O_4$	2 M KOH	76% (1,000)	239	0.5	[104]
$La_{0.9}Sr_{0.1}(Co_{0.2}Mn_{0.2}Fe_{0.2}Ni_{0.2}Cu_{0.2})O_3$	2 M KOH	100% (1,000)	204.33	2 $mV s^{-1}$	[44]
FeCoNiCuMn HEA	3 M KOH	92.3% (5,000)	194	2	[66]
$La(CoCrFeMnNiAl_x)_{1/(5+x)}O_{3-\delta}$	2 M KOH	88.61% (2,000)	353.65	1	[28]
(FeCoNiCuZn) HEA	3 M KOH	—	325.17	1	[63]
$Ti_{1.1}V_{1.2}Cr_{0.8}Nb_{1.0}Mo_{0.9}C_4T_x$	2 M KOH	—	284.6	1	[105]
$La(Mn_{0.2}Fe_{0.2}Co_{0.2}Ni_{0.2}Cu_{0.2})O_3$	3 M KOH	84.2% (10,000)	609.3	—	[10]
$(FeCoNiCuMn)_3O_4$	3 M KOH	92% (7,500)	313	5 $mV s^{-1}$	[106]
$K(MgMnFeCuNi)Fe(CN)_6$	1 M Na_2SO_4	—	175	5 $mV s^{-1}$	[99]
$(FeCoCrMnNi)_3O_4$	1 M KOH	—	332.2	0.3	[12]
$(FeCoNiCrMn)_3O_4$	2 M KOH	80.77%	639	1	[107]

supply voltage requirements—a concept that could be adapted for high-entropy systems to enhance transistor efficiency and reduce power consumption [113, 114].

12.5 Computational Modeling for M-HEM Development

Despite their potential, the vast compositional space of HEAs presents a challenge: discovering optimal compositions through traditional experimental methods is time-consuming and costly. ML has emerged as a transformative tool, enabling rapid material discovery, optimizing properties, and providing deeper insights into the structure–property relationships of HEAs. The following section discusses the role of ML in HEM research. In materials science, ML is integrated in general materials research by:

- Predicting material properties based on compositional and processing parameters.
- Reducing experimental costs by identifying promising candidates before synthesis.
- Enhancing the understanding of complex relationships between atomic structure and macroscopic properties.

ML incorporation in HEMs is a crucial aspect for realizing Materials 4.0. The Materials 4.0 paradigm aligns with Industry 4.0 through (a) data-driven material discovery, (b) predictive modeling (performance, lifecycle, etc.), and (c) intelligent design strategies (materials recommender system). Conventional experimental and computational approaches are often deemed to be slow and resource-intensive [115]. ML facilitates the conventional process by analyzing datasets, identifying hidden patterns, and making high-accuracy predictions. The prediction can be related to properties, synthesis parameters, and performance of HEMs. Hence, in the context of Industry 4.0, ML-driven HEM research enables property optimization, performance prediction, and adaptive manufacturing [116]. By using high-throughput experiments, AI-driven simulations, and self-learning algorithms, ML ensures that new materials are not only stronger and more durable but also tailored for specific industrial applications with minimal waste and cost. This seamless integration of ML into materials science paves the way for next-generation alloys, high-performance coatings, and resilient structural materials, revolutionizing industries such as aerospace, automotive, and energy storage. Vadivel et al. synthesized $BiOXCO_3$ (X = Br, Cl, I) materials via a solvothermal method to investigate their SC performance. Electrochemical analysis revealed that $BiOXCO_3$ [5], with the highest CO_3 concentration, achieved the maximum specific capacitance (645 F g^{-1} at 1 A g^{-1}), outperforming another composition. The enhanced performance is attributed to improved charge transfer, ion diffusion, and Faradaic storage. This study demonstrates how variations in anion concentration influence electrochemical behavior, offering new opportunities for advanced Bi-based HEMs in energy storage applications [117].

ML methods commonly used in HEA research include the following: (a) *supervised learning* (e.g. neural networks, decision trees) for predicting mechanical properties; (b) *unsupervised learning* (e.g. clustering techniques) for identifying hidden patterns in material datasets; and (c) *reinforcement learning* for design optimization [20].

12.5.1 ML in HEMs

Phase Prediction: HEAs' phase stability is critical in determining their properties (mechanical and thermal). ML models, trained on large datasets, can predict phase formation more efficiently compared to conventional thermodynamic calculation-based approaches. For example, researchers have successfully employed random forest and support vector machine methods to classify

HEAs into three categories: (i) single-phase solid solutions, (ii) intermetallic compounds, and (iii) multiphase systems.

Mechanical Property Optimization: ML helps optimize the mechanical properties by analyzing large datasets. Neural networks and gradient boosting algorithms have been applied to predict yield strength, hardness, and ductility based on composition and processing conditions [118].

Chen et al. utilized a computational approach to enhance the yield strength of Cr–Fe–Co–Ni HEAs through vanadium addition. An ML framework, integrating support vector regression and MD, identified cohesive energy mismatch, vacancy formation energy, and valence electron concentration as key factors influencing strength. Bayesian adaptive alloy design pinpointed $Cr_{16}V_5Fe_9Co_{35}Ni_{35}$ as optimal composition. Simulated tensile tests validated the predicted yield strength trends, aligning with experimental data as demonstrated in Figure 12.5 [119].

Composition Design and Property Prediction: The extensive range of possible compositions in HEAs presents a significant challenge in pinpointing the most optimal alloy combinations. ML techniques, particularly deep learning and generative models, offer a powerful approach to composition design. By analyzing known alloy databases, these models can generate novel compositions with tailored properties [120].

For instance, generative adversarial networks have been used to propose new HEA compositions that maximize strength while maintaining ductility [121]. Reinforcement learning algorithms further refine these predictions by iteratively adjusting compositions based on performance metrics [122].

Processing Optimization: Synthesis processes and relevant parameters such as annealing temperature, cooling rate, and strain rates are crucial for the final properties of HEAs. ML can model past experimental data to recommend optimal processing parameters for achieving desired microstructures, thus creating a materials recommender system. For example, Bayesian optimization has been particularly effective in refining processing parameters for complex HEAs [123].

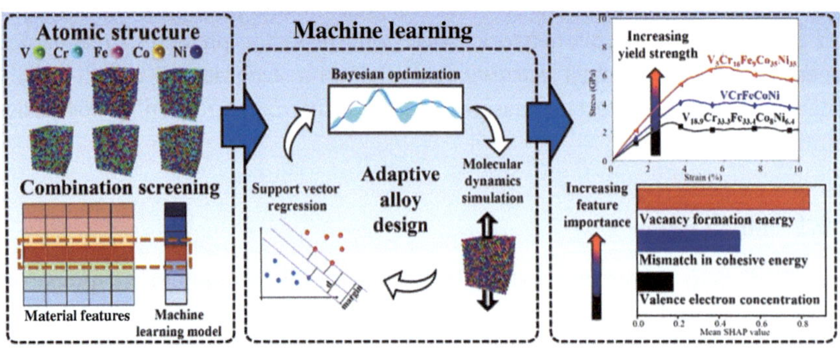

Figure 12.5 Use of ML in adaptive alloy design that resulted in increasing yield strength in HEMs. *Source:* Reproduced from Ref. [119] / with permission of Elsevier.

Despite its advantages, ML in HEA research faces several challenges: (i) data scarcity: high-quality datasets are limited, making it difficult to train accurate models; (ii) interpretability: many ML models operate "black boxes," which poses challenges in deriving clear and interpretable physical insights; (iii) extrapolation issues: ML models struggle with predicting properties for compositions far outside their training datasets [124]. Hybrid approaches that integrate physics-based simulation techniques such as physics-induced neural network algorithms can help overcome this limitation [125].

12.6 Conclusions and Future Perspectives

The emergence of M-HEMs is a paradigm shift in the field of materials science, with the potential to engineer physical and chemical properties. In electrochemical energy storage, the distinctive features of M-HEMs resolve some long-standing issues. The compositional complexity of materials causes slow diffusion kinetics for atomic rearrangement and hence structural stability in charge/discharge cycles. This is especially timely in lithium-ion, sodium-ion, and new multivalent systems, where volume change and phase transition tend to constrain electrode lifespan. Metallic dopants also improve this performance by modulating local electronic structures, defect concentrations, and redox activity—parameters directly related to charge storage capacity and rate capability.

Emerging evidence shows that the addition of transition metal dopants such as V, Mo, or Mn creates multivalent redox sites and thus enables pseudocapacitive response superimposed on diffusion-controlled storage. Hybrid mechanisms, in conjunction with the phase structures stabilized by entropy, provide a new avenue to achieve high energy and power densities simultaneously. Doping choice can also be tailored to control electronic conductivity, electrolyte compatibility, and surface reactivity—key factors that govern the performance of high-efficiency storage devices under diverse environmental conditions.

Yet the future for M-HEMs is fraught with challenges and promises. High on the list is control and comprehension of local chemical ordering at the atomic scale. While appearing disordered, many high-entropy phases possess short-range ordering that dominates macroscopic behavior such as phase stability, ion conductivity, and toughness. High-end *in situ* and operando characterization techniques—supported by theoretical modeling and ML codes—are essential to unravelling these buried structural motifs and their consequences on energy-relevant properties.

A second exciting frontier is the combination of HEMs with nanostructuring and hierarchical design. Through multiscale architectures engineered from atomic disorder to mesoscale porosity, next-generation electrodes can gain from improved ion transport, active surface area, and mechanical durability. Dopants here are not just compositional but architectural: inducing phase segregation, stabilization of interfaces, or selective crystallographic orientation favorable to target electrochemical pathways.

Scalability and sustainability are key parameters for mass adoption. Scaling up from laboratory-scale synthesis routes (e.g. carbothermal shock, arc melting) to cost-effective and scalable routes such as solution-based or sol-gel routes is necessary. Of utmost importance is the introduction of earth-abundant, nontoxic dopants to make the methods economically and environmentally sustainable.

In the future, the coupling of high-throughput computing, data-driven design, and additive manufacturing will accelerate the discovery and implementation of M-HEMs into operating energy systems. Beyond energy storage, these materials also have potential for thermoelectric, catalytic, and magnetic applications and therefore are perfect candidates for integrated energy platforms that couple generation, storage, and functional use in a single material structure.

In short, HEM-doped materials are a paradigm-breaking pathway to multifunctional material design, particularly for energy storage devices. Through the use of compositional flexibility, entropic stabilization, and adjustability via controlled doping, these materials have the potential to revolutionize performance standards in electrochemical devices. Continued research efforts driven by a deep understanding and facilitated by advanced tools will be required to unlock their full potential and transition from academic curiosity to the industrial pillar.

References

1 Ma, J. and Huang, C. (2023). High entropy energy storage materials: synthesis and application. *J Energy Storage* 66: 107419.

2 Hussain, A., Jabeen, N., Hassan, N.U. et al. (2023). High-entropic relaxor ferroelectric perovskites ceramics with A-site modification for energy storage applications. *Ceram Int* 49 (23, Part B): 39419–39427.

3 Álvarez-Montaño, V.E., Kumar, M., Sharma, S. et al. (2023). A recent look at high-entropy ceramics based on doping engineering/technology and the future scope of their novel applications. *Mater Lett* 349: 134785.

4 Zhang, Y., Zuo, T.T., Tang, Z. et al. (2014). Microstructures and properties of high-entropy alloys. *Prog Mater Sci* 61: 1–93.

5 Owen, L.R. and Jones, N.G. (2018). Lattice distortions in high-entropy alloys. *J Mater Res* 33 (19): 2954–2969.

6 Wang, L., Zhang, L., Lu, X. et al. (2023). Surprising cocktail effect in high entropy alloys on catalyzing magnesium hydride for solid-state hydrogen storage. *Chem Eng J* 465: 142766.

7 Gu, X., Guo, X.-B., Li, W.-H. et al. (2024). High-entropy materials for application: electricity, magnetism, and optics. *ACS Appl Mater Interfaces* 16 (40): 53372–53392.

8 Fu, M., Ma, X., Zhao, K. et al. (2021). High-entropy materials for energy-related applications. *iScience* 24 (3): 102177.

9 Yang, B., Zhang, Y., Pan, H. et al. (2022). High-entropy enhanced capacitive energy storage. *Nat Mater* 21 (9): 1074–1080.

10 Rani, B. and Sahu, N.K. (2020). Electrochemical properties of $CoFe_2O_4$ nanoparticles and its rGO composite for supercapacitor. *Diam Relat Mater* 108: 107978.

11 Moździerz, M., Świerczek, K., Dąbrowa, J. et al. (2022). High-entropy $Sn_{0.8}(Co_{0.2}Mg_{0.2}Mn_{0.2}Ni_{0.2}Zn_{0.2})_{2.2}O_4$ conversion-alloying anode material for Li-ion cells: altered lithium storage mechanism, activation of Mg, and origins of the improved cycling stability. *ACS Appl Mater Interfaces* 14 (37): 42057–42070.
12 Yin, Y., Zhang, W.-B., Zhang, X.-L. et al. (2023). Low-dimensional high entropy oxide $(FeCoCrMnNi)_3O_4$ for supercapacitor applications. *Dalton Trans* 52 (26): 9005–9016.
13 Gonçalves, J.M., Santos, É.A., Martins, P.R. et al. (2023). Emerging medium- and high-entropy materials as catalysts for lithium-sulfur batteries. *Energy Storage Mater* 63: 102999.
14 Santos, É.A., Fernandes, R.C., Vicentini, R. et al. (2023). On the electrochemical properties of lithium-sulfur batteries. *J Energy Storage* 71: 108203.
15 Yang, B., Liu, Y., Lan, S. et al. (2023). High-entropy design for dielectric materials: status, challenges, and beyond. *J Appl Phys* 133 (11).
16 Schweidler, S., Botros, M., Strauss, F. et al. (2024). High-entropy materials for energy and electronic applications. *Nature Rev Mater* 9 (4): 266–281.
17 Sun, Y. and Dai, S. (2024). Synthesis of high-entropy materials. *Nat Synth* 3 (12): 1457–1470.
18 Singh, A.K., Jhala, R., Patil, N. et al. (2025). Cutting-edge surface generation on super Co_6O_5 high entropy alloy using EDM process approach. *Results Eng* 26: 104604.
19 Liu, K.-C., Kumar Panda, P., Chandra Mallick, B. et al. (2024). Solid-phase microwave synthesis of high-entropy graphene quantum dots as metal-free electrochemical catalysts. *Appl Surf Sci* 648: 159061.
20 Liu, X., Zhang, J., and Pei, Z. (2023). Machine learning for high-entropy alloys: progress, challenges and opportunities. *Prog Mater Sci* 131: 101018.
21 Elkatatny, S., Abd-Elaziem, W., Sebaey, T.A. et al. (2024). Machine-learning synergy in high-entropy alloys: a review. *J Mater Res Technol* 33: 3976–3997.
22 Bansal, A., Kumar, P., Yadav, S. et al. (2023). Accelerated design of high entropy alloys by integrating high throughput calculation and machine learning. *J Alloys Compd* 960: 170543.
23 He, M.Y., Shen, Y.F., Jia, N., and Liaw, P.K. (2021). C and N doping in high-entropy alloys: a pathway to achieve desired strength-ductility synergy. *Appl Mater Today* 25: 101162.
24 Brito-Garcia, S.J., Mirza-Rosca, J.C., Jimenez-Marcos, C., and Voiculescu, I. (2023). EIS study of doped high-entropy alloy. *Meta* 13 (5): 883.
25 Zhu, C., Xu, L., Liu, M. et al. (2023). A review on improving mechanical properties of high entropy alloy: interstitial atom doping. *J Mater Res Technol* 24: 7832–7851.
26 Chen, S., Bao, M., Jia, Y. et al. (2024). Boosting high-rate Li-ion storage properties by La(III) ion doping in spinel $(Co_{0.2}Cr_{0.2}Fe_{0.2}Mn_{0.2}Ni_{0.2})_3O_4$ high-entropy oxide anode. *J Adv Ceram* 13 (6): 769–779.
27 Liu, X., Xing, Y., Xu, K. et al. (2022). Kinetically accelerated lithium storage in high-entropy (LiMgCoNiCuZn)O enabled by oxygen vacancies. *Small* 18 (18): 2200524.

28 Guo, M., Liu, Y., Zhang, F. et al. (2022). Inactive Al^{3+}-doped La(CoCrFe $MnNiAl_x)_{1/(5+x)}O_3$ high-entropy perovskite oxides as high performance supercapacitor electrodes. *J Adv Ceram* 11 (5): 742–753.

29 Yan, J., Wang, D., Zhang, X. et al. (2020). A high-entropy perovskite titanate lithium-ion battery anode. *J Mater Sci* 55 (16): 6942–6951.

30 Aslam, A., Morley, N.A., Amin, N. et al. (2021). Study of structural, optical and electrical properties of La^{3+}doped $Mg_{0.25}Ni_{0.15}Cu_{0.25}Co_{0.35}Fe_{2-x}La_xO_4$ spinel ferrites. *Phys B Condens Matter* 602: 412565.

31 Zhang, J., Yan, J., Calder, S. et al. (2019). Long-range antiferromagnetic order in a rocksalt high entropy oxide. *Chem Mater* 31 (10): 3705–3711.

32 Cheng, B., Lou, H., Sarkar, A. et al. (2019). Pressure-induced tuning of lattice distortion in a high-entropy oxide. *Commun Chem* 2 (1): 114.

33 Zhang, Z.-H., Liu, Z.-F., Lu, J.-F. et al. (2014). The sintering mechanism in spark plasma sintering – proof of the occurrence of spark discharge. *Scr Mater* 81: 56–59.

34 Gautam, A. and Ahmad, M.I. (2021). Low-temperature synthesis of five component single phase high entropy oxide. *Ceram Int* 47 (15): 22225–22228.

35 Zhao, C., Cai, W., Sun, N. et al. (2024). Facile preparation of porous high-entropy alloy FeCoNiCuMn and its OER performance. *J Phys Chem Solids* 184: 111668.

36 Sarkar, A., Djenadic, R., Usharani, N.J. et al. (2017). Nanocrystalline multicomponent entropy stabilised transition metal oxides. *J Eur Ceram Soc* 37 (2): 747–754.

37 Shaw, S.K., Gangwar, A., Sharma, A. et al. (2021). Structural and magnetic properties of nanocrystalline equi-atomic spinel high-entropy oxide $(AlCoFeMnNi)_3O_4$ synthesised by microwave assisted co-precipitation technique. *J Alloys Compd* 878: 160269.

38 Leng, J., Wang, Z., Wang, J. et al. (2019). Advances in nanostructures fabricated via spray pyrolysis and their applications in energy storage and conversion. *Chem Soc Rev* 48 (11): 3015–3072.

39 Hong, W., Chen, F., Shen, Q. et al. (2019). Microstructural evolution and mechanical properties of (Mg,Co,Ni,Cu,Zn)O high-entropy ceramics. *J Am Ceram Soc* 102 (4): 2228–2237.

40 Deng, Z., Olvera, A., Casamento, J. et al. (2020). Semiconducting high-entropy chalcogenide alloys with ambi-ionic entropy stabilization and ambipolar doping. *Chem Mater* 32 (14): 6070–6077.

41 Chang, C.-B., Lu, Y.-R., and Tuan, H.-Y. (2023). High-entropy NaCl-type metal chalcogenides as K-ion storage materials: role of the cocktail effect. *Energy Storage Mater* 59: 102770.

42 Cavin, J., Ahmadiparidari, A., Majidi, L. et al. (2021). 2D high-entropy transition metal dichalcogenides for carbon dioxide electrocatalysis. *Adv Mater* 33 (31): 2100347.

43 Wang, G., Qin, J., Feng, Y. et al. (2020). Sol-gel synthesis of spherical mesoporous high-entropy oxides. *ACS Appl Mater Interfaces* 12 (40): 45155–45164.

44 Wan, Z., Lv, J., and Zhang, Y. (2025). Sr^{2+}-doped $La_{1-x}Sr_x(Co_{0.2}Mn_{0.2}Fe_{0.2}Ni_{0.2}Cu_{0.2})O_3$ high-entropy perovskite oxides as supercapacitor electrodes. *Ionics* 31: 3671–3682.

45 Dąbrowa, J., Olszewska, A., Falkenstein, A. et al. (2020). An innovative approach to design SOFC air electrode materials: high entropy $La_{1-x}Sr_x$(Co, Cr,Fe,Mn,Ni)$O_{3-\delta}$($x = 0$, 0.1, 0.2, 0.3) perovskites synthesized by the sol–gel method. *J Mater Chem A* 8 (46): 24455–24468.

46 Chen, B., Zhang, W.-B., Yin, Y. et al. (2025). Electrochemical energy storage and rectification performance of high-entropy oxide $(CrMnFeCoNi)_3O_4$. *J Energy Storage* 109: 115228.

47 Park, T., Javadinejad, H.R., Kim, Y.-K. et al. (2022). Effect of processing route on the crystal structure and physical properties of bixbyite high-entropy oxides. *J Alloys Compd* 893: 162108.

48 Wu, L., Shen, X., Ji, Z. et al. (2023). Facile synthesis of medium-entropy metal sulfides as high-efficiency electrocatalysts toward oxygen evolution reaction. *Adv Funct Mater* 33 (3): 2208170.

49 Jiang, Z., Yuan, Y., Tan, L. et al. (2023). Self-reconstruction of (CoNiFeCuCr)Se high-entropy selenide for efficient oxygen evolution reaction. *Appl Surf Sci* 627: 157282.

50 Broge, N.L.N., Bertelsen, A.D., Søndergaard-Pedersen, F., and Iversen, B.B. (2023). Facile solvothermal synthesis of Pt-Ir-Pd-Rh-Ru-Cu-Ni-Co high-entropy alloy nanoparticles. *Chem Mater* 35 (1): 144–153.

51 Nguyen, T.X., Patra, J., Chang, J.-K., and Ting, J.-M. (2020). High entropy spinel oxide nanoparticles for superior lithiation–delithiation performance. *J Mater Chem A* 8 (36): 18963–18973.

52 Tanmathusorachai, W., Aulia, S., Rinawati, M. et al. (2024). High-entropy prussian blue analogue derived heterostructure nanoparticles as bifunctional oxygen conversion electrocatalysts for the rechargeable zinc–air battery. *ACS Appl Mater Interfaces* 16 (45): 62022–62032.

53 Biesuz, M., Spiridigliozzi, L., Dell'Agli, G. et al. (2018). Synthesis and sintering of (Mg, Co, Ni, Cu, Zn)O entropy-stabilized oxides obtained by wet chemical methods. *J Mater Sci* 53 (11): 8074–8085.

54 Zhao, Z., Chen, H., Xiang, H. et al. (2019). $(La_{0.2}Ce_{0.2}Nd_{0.2}Sm_{0.2}Eu_{0.2})PO_4$: a high-entropy rare-earth phosphate monazite ceramic with low thermal conductivity and good compatibility with Al_2O_3. *J Mater Sci Technol* 35 (12): 2892–2896.

55 Ren, J., Kumkale, V.Y., Hou, H. et al. (2025). A review of high-entropy materials with their unique applications. *Adv Compos Hybrid Mater* 8 (2): 195.

56 Kamecki, B., Karczewski, J., Cempura, G. et al. (2023). Evaluation of structural and electrical properties of multicomponent spinel oxide thin films deposited via spray pyrolysis technique. *Mater Charact* 203: 113097.

57 Dimitriou, C., Psathas, P., Solakidou, M., and Deligiannakis, Y. (2023). Advanced flame spray pyrolysis (FSP). Technologies for engineering multifunctional nanostructures and nanodevices. *Nanomaterials* 13 (23): 3006.

58 Phakatkar, A.H., Saray, M.T., Rasul, M.G. et al. (2021). Ultrafast synthesis of high entropy oxide nanoparticles by flame spray pyrolysis. *Langmuir* 37 (30): 9059–9068.

59 Liu, W.H., Wu, Y., He, J.Y. et al. (2013). Grain growth and the Hall–Petch relationship in a high-entropy FeCrNiCoMn alloy. *Scr Mater* 68 (7): 526–529.

60 Wu, Y.D., Cai, Y.H., Wang, T. et al. (2014). A refractory $Hf_{25}Nb_{25}Ti_{25}Zr_{25}$ high-entropy alloy with excellent structural stability and tensile properties. *Mater Lett* 130: 277–280.

61 Sedegov, A., Vorotilo, S., Tsybulin, V. et al. (2019). Synthesis and study of high-entropy ceramics based on the carbides of refractory metals. *IOP Conf Ser Mater Sci Eng* 558 (1): 012043.

62 Zhao, J., Zhang, Y., Chen, X. et al. (2022). Entropy-change driven highly reversible sodium storage for conversion-type sulfide. *Adv Funct Mater* 32 (45): 2206531.

63 Mohanty, G.C., Chowde Gowda, C., Gakhad, P. et al. (2023). Iron-cobalt-nickel-copper-zinc (FeCoNiCuZn) high entropy alloy as positive electrode for high specific capacitance supercapacitor. *Electrochim Acta* 470: 143272.

64 Prasad, H., Singh, S., and Panigrahi, B.B. (2017). Mechanical activated synthesis of alumina dispersed FeNiCoCrAlMn high entropy alloy. *J Alloys Compd* 692: 720–726.

65 Rekha, M.Y., Mallik, N., and Srivastava, C. (2018). First report on high entropy alloy nanoparticle decorated graphene. *Sci Rep* 8 (1): 8737.

66 Mohanty, G.C., Chowde Gowda, C., Gakhad, P. et al. (2024). Enhanced energy density of high entropy alloy (Fe-Co-Ni-Cu-Mn) and green graphene hybrid supercapacitor. *Energy Storage* 6 (4): e668.

67 R AS, Jacob G. Exploring the charge storage ability of the spinel-type high entropy oxide $(MnFeCoNiZn)_3O_4$ nanoparticles for supercapacitor applications *Chem Eng J Adv* 2025;21:100708.

68 Das, S. and Robi, P.S. (2018). Mechanical alloying of W-Mo-V-Cr-Ta high entropy alloys. *IOP Conf Ser Mater Sci Eng* 346 (1): 012047.

69 Bérardan, D., Franger, S., and Meena, A. (2016). Dragoe NJJoMCA. Room temperature lithium superionic conductivity in high entropy oxides. *J Mater Chem A* 4 (24): 9536–9541.

70 Lin, B., Wu, X., He, X. et al. (2025). High-entropy doping in NASICON cathodes: activating the V^{4+}/V^{5+} redox couple and inducing a reversible single solid-solution phase reaction for advanced sodium ion batteries. *J Colloid Interface Sci* 690: 137299.

71 Ritter, T.G., Pappu, S., and Shahbazian-Yassar, R. (2024). High-entropy materials for lithium batteries. *Batteries* 10 (3): 96.

72 Qiu, N., Chen, H., Yang, Z. et al. (2019). A high entropy oxide $(Mg_{0.2}Co_{0.2}Ni_{0.2}Cu_{0.2}Zn_{0.2}O)$ with superior lithium storage performance. *J Alloys Compd* 777: 767–774.

73 Brahmanandan, S., Nair, S., and Santhanagopalan, D. (2023). High-performance Zr-doped P3-type $Na_{0.67}Ni_{0.33}Mn_{0.67}O_2$ cathode for na-ion battery applications. *Crystals* 13 (9): 1339.

74 Xiang, H.-Z., Xie, H.-X., Chen, Y.-X. et al. (2021). Porous spinel-type $(Al_{0.2}CoCr FeMnNi)_{0.58}O_{4-\delta}$ high-entropy oxide as a novel high-performance anode material for lithium-ion batteries. *J Mater Sci* 56 (13): 8127–8142.

75 Chen, T.-Y., Wang, S.-Y., Kuo, C.-H. et al. (2020). In operando synchrotron X-ray studies of a novel spinel $(Ni_{0.2}Co_{0.2}Mn_{0.2}Fe_{0.2}Ti_{0.2})_3O_4$ high-entropy oxide for energy storage applications. *J Mater Chem A* 8 (41): 21756–21770.

76 Sarkar, A., Velasco L., Wang D. et al. (2018). High entropy oxides for reversible energy storage. *Nat Commun* 9: 3400.

77 Li, S., Peng, Z., and Fu, X. (2023). $Zn_{0.5}Co_{0.5}Mn_{0.5}Fe_{0.5}Al_{0.5}Mg_{0.5}O_4$ high-entropy oxide with high capacity and ultra-long life for Li-ion battery anodes. *J Adv Ceram* 12 (1): 59–71.

78 Caroline, G., Nair, N., Nair, S.V. et al. (2024). High-entropy materials for sodium-ion batteries. *Next Sustain* 4: 100044.

79 Banerjee, S., Choudhary, R.B., and Ansari, S. (2024). $Na_3V_2(PO_4)_3$ derived cathode materials for sodium-ion batteries (SIBs): a review. *Future Batter* 4: 100010.

80 Eggestad, K., Selbach, S.M., and Williamson, B.A.D. (2024). Doping implications of Li solid state electrolyte $Li_7La_3Zr_2O_{12}$. *J Mater Chem A* 12 (26): 15666–15675.

81 Guo, H., Shen, J., Wang, T. et al. (2022). Design and fabrication of high-entropy oxide anchored on graphene for boosting kinetic performance and energy storage. *Ceram Int* 48 (3): 3344–3350.

82 Zhou, Y., Wang, Z., Niu, P. et al. (2025). Realizing fast-charging capability of silicon anode via ternary doping and structural disorder. *J Colloid Interface Sci* 691: 137372.

83 Rostami, H., Valio, J., Suominen, P. et al. (2024). Advancements in cathode technology, recycling strategies, and market dynamics: a comprehensive review of sodium ion batteries. *Chem Eng J* 495: 153471.

84 Lökçü, E., Toparli, Ç., and Anik, M. (2020). Electrochemical performance of $(MgCoNiZn)_{(1-x)}Li_{(x)}O$ high-entropy oxides in lithium-ion batteries. *ACS Appl Mater Interfaces* 12 (21): 23860–23866.

85 Ghigna, P., Airoldi, L., Fracchia, M. et al. (2020). Lithiation mechanism in high-entropy oxides as anode materials for Li-ion batteries: an operando XAS study. *ACS Appl Mater Interfaces* 12 (45): 50344–50354.

86 Wang, J., Cui, Y., Wang, Q. et al. (2020). Lithium containing layered high entropy oxide structures. *Sci Rep* 10 (1): 18430.

87 Zhao, C., Ding, F., Lu, Y. et al. (2020). High-entropy layered oxide cathodes for sodium-ion batteries. *Angew Chem Int Ed* 59 (1): 264–269.

88 Tian, L., Zhang, Z., Liu, S. et al. (2023). High-entropy perovskite oxide nanofibers as efficient bidirectional electrocatalyst of liquid-solid conversion processes in lithium-sulfur batteries. *Nano Energy* 106: 108037.

89 Fan, H., Si, Y., Zhang, Y. et al. (2024). Grapevine-like high entropy oxide composites boost high-performance lithium sulfur batteries as bifunctional interlayers. *Green Energy Environ* 9 (3): 565–572.

90 Raza, H., Cheng, J., Lin, C. et al. (2023). High-entropy stabilized oxides derived via a low-temperature template route for high-performance lithium-sulfur batteries. *EcoMat* 5 (4): e12324.

91 Petrovičovà, B., Xu, W., Musolino, M.G. et al. (2022). High-entropy spinel oxides produced via sol-gel and electrospinning and their evaluation as anodes in Li-ion batteries. *Appl Sci* 12 (12): 5965.

92 Bian, W., Li, H., Zhao, Z. et al. (2023). Entropy stabilization effect and oxygen vacancy in spinel high-entropy oxide promoting sodium ion storage. *Electrochim Acta* 447: 142157.

93 Yazhou, K. and Zhiren, Y. (2022). Synthesis, structure and electrochemical properties of Al doped high entropy perovskite Lix(LiLaCaSrBa)$Ti_{1-x}Al_xO_3$. *Ceram Int* 48 (4): 5035–5039.

94 Wei, Y., Yao, R., Liu, X. et al. (2023). Understanding the configurational entropy evolution in metal-phosphorus solid solution for highly reversible Li-ion batteries. *Adv Sci* 10 (9): 2300271.

95 Chatterjee, A., Ganguly, D., Sundara, R., and Bhattacharya, S.S. (2023). Rare-earth doped configurational entropy stabilized high entropy spinel oxide as an efficient anchoring/catalyst functional interlayer for high-performance lithium-sulfur battery. *Batter Supercaps* 6 (7): e202300082.

96 Tian, L., Zhang, Z., Liu, S. et al. (2022). High-entropy spinel oxide nanofibers as catalytic sulfur hosts promise the high gravimetric and volumetric capacities for lithium–sulfur batteries. *Energy Environ Mater* 5 (2): 645–654.

97 Wu, B., Luxa, J., Šturala, J. et al. (2024). A room-temperature chloride-conducting metal–organic crystal $[Al(DMSO)_6]Cl_3$ for potential solid-state chloride-shuttle batteries. *Energy Environ Mater* 7 (1): e12530.

98 Tian, K., He, H., Li, X. et al. (2022). Boosting electrochemical reaction and suppressing phase transition with a high-entropy O_3-type layered oxide for sodium-ion batteries. *J Mater Chem A* 10 (28): 14943–14953.

99 Jiang, W., Wang, T., Chen, H. et al. (2021). Room temperature synthesis of high-entropy Prussian blue analogues. *Nano Energy* 79: 105464.

100 Liang, B., Ai, Y., Wang, Y. et al. (2020). Spinel-type $(FeCoCrMnZn)_3O_4$ high-entropy oxide: facile preparation and supercapacitor performance. *Materials* 13 (24).

101 Yuan, Y., Xu, Z., Han, P. et al. (2021). MnO_2-decorated metallic framework supercapacitors fabricated from duplex-phase $FeCrCoMnNiAl_{0.75}$ Cantor high entropy alloy precursors through selective phase dissolution. *J Alloys Compd* 870: 159523.

102 Tang, P., Cao, Y., Hao, L. et al. (2022). The preparation of high-performance aqueous supercapacitor with high-entropy pyrochlore-type electrode and super-concentrated electrolyte. *Ceram Int* 48 (2): 2660–2669.

103 Nan, H., Lv, S., Xu, Z. et al. (2023). Inducing the cocktail effect in yolk-shell high-entropy perovskite oxides using an electronic structural design for improved electrochemical applications. *Chem Eng J* 452: 139501.

104 Talluri, B., Aparna, M.L., Sreenivasulu, N. et al. (2021). High entropy spinel metal oxide $(CoCrFeMnNi)_3O_4$ nanoparticles as a high-performance supercapacitor electrode material. *J Energy Storage* 42: 103004.

105 Ma, W., Qiu, Z., Wang, M. et al. (2023). A novel high-entropy MXene $Ti_{1.1}V_{1.2}Cr_{0.8}Nb_{1.0}Mo_{0.9}C_4T_x$ for high-performance supercapacitor. *Scr Mater* 235: 115596.

106 Chandra Mohanty, G., Das, S., and Verma, A. (2024). Synthesis of $(FeCoNiCuMn)_3O_4$ spinel-high entropy oxide and green carbon from agricultural waste for supercapacitor application. *RSC Adv* 14 (46): 33830–33842.

107 Zhang, D., Xu, S., Li, T. et al. (2023). High-entropy oxides prepared by dealloying method for supercapacitors. *ACS Appl Eng Mater* 1 (2): 780–789.

108 Bersuker, G., Zeitzoff, P., Brown, G., and Huff, H.R. (2004). Dielectrics for future transistors. *Mater Today* 7 (1): 26–33.

109 Kim, L., Jung, D., Kim, J. et al. (2003). Strain manipulation in $BaTiO_3/SrTiO_3$ artificial lattice toward high dielectric constant and its nonlinearity. *Appl Phys Lett* 82 (13): 2118–2120.

110 Sharma, Y., Lee, M.-C., Pitike, K.C. et al. (2022). High entropy oxide relaxor ferroelectrics. *ACS Appl Mater Interfaces* 14 (9): 11962–11970.

111 Zhang, X., Li, J., Ni, B. et al. (2022). Dielectric properties of novel high-entropy $(La_{0.2}Li_{0.2}Ba_{0.2}Sr_{0.2}Ca_{0.2})Nb_2O_{6-\delta}$ tungsten bronze ceramics. *J Mater Sci* 57 (33): 15901–15912.

112 Zhou, J., Li, P., Zhang, X. et al. (2022). Effect of configurational entropy on dielectric properties of high-entropy perovskite oxides $(Ce_{0.5},K_{0.5})_x[(Bi_{0.5},Na_{0.5})_{0.25}Ba_{0.25}Sr_{0.25}Ca_{0.25}]_{1-x}TiO_3$. *J Mater Sci Mater Electron* 33 (26): 20721–20730.

113 Kim, S.H., Hong, K., Xie, W. et al. (2013). Electrolyte-gated transistors for organic and printed electronics. *Adv Mater* 25 (13): 1822–1846.

114 Feng, X., Marques, G.C., Rasheed, F. et al. (2019). Nonquasi-static capacitance modeling and characterization for printed inorganic electrolyte-gated transistors in logic gates. *IEEE Trans Electron Dev* 66 (12): 5272–5277.

115 Jose, R. and Ramakrishna, S. (2018). Materials 4.0: materials big data enabled materials discovery. *Appl Mater Today* 10: 127–132.

116 Mishra, A.R., Rani, P., Alshamrani, A.M. et al. (2025). Assessing the industry 4.0 strategies in the automobile manufacturing firm using a combined compromise solution-based ranking method. *Appl Soft Comput* 174: 113037.

117 Vadivel, S., Dharmavarapu, V.S.H., Sujita, P. et al. (2025). A robust approach for designing a bismuth high entropy material $(BiOXCO_3)$ as a novel electrode material for supercapacitor applications. *Mater Res Bull* 184: 113225.

118 Bundela, A.S. and Rahul, M.R. (2022). Machine learning-enabled framework for the prediction of mechanical properties in new high entropy alloys. *J Alloys Compd* 908: 164578.

119 Chen, L., Jarlöv, A., Seet, H.L. et al. (2023). Exploration of V-Cr-Fe-Co-Ni high-entropy alloys with high yield strength: A combination of machine learning and molecular dynamics simulation. *Comput Mater Sci* 217: 111888.

120 Zhang, J., Cai, C., Kim, G. et al. (2022). Composition design of high-entropy alloys with deep sets learning. *NPJ Comput Mater* 8(1):89.

121 Chen, C., Zhou, H., Long, W. et al. (2023). Phase prediction for high-entropy alloys using generative adversarial network and active learning based on small datasets. *Sci China Technol Sci* 66 (12): 3615–3627.

122 Wang, X., Tran, N.-D., Zeng, S. et al. (2022). Element-wise representations with ECNet for material property prediction and applications in high-entropy alloys. *NPJ Comput Mater* 8 (1): 253.

123 Kumar, S., Pradhan, H., Shah, N. et al. (2023). Machine learning enabled processing map generation for high-entropy alloy. *Scr Mater* 234: 115543.

124 Jain, A. (2024). Machine learning in materials research: developments over the last decade and challenges for the future. *Curr Opinion Solid State Mater Sci* 33: 101189.

125 Wan, X., Li, Z., Yu, W. et al. (2023). Machine learning paves the way for high entropy compounds exploration: challenges, progress, and outlook. *Adv Mater* 4: 2305192.

13

Noble Metal-Doped High-Entropy Materials for Energy Storage Applications

Rajashree Sahoo[1], *Saswat Mohapatra*[2], *Swagat Kumar Purohit*[3], *and Arpan Kumar Nayak*[3]

[1] *Gandhi Academy of Technology and Engineering, Department of Basic Science and Humanities, Berhampur, Odisha, 761008, India*
[2] *University of Houstan, Department of Electrical and Computer Engineering, Houstan, TX, 77204, USA*
[3] *Physics Section, Department of Education in Science and Mathematics, Regional Institute of Education (NCERT), Mysuru, Mysuru, Karnataka, 570006, India*

13.1 Introduction

With the evolution of technology and the increase in advanced applications, there is an increasing demand for materials with enhanced properties and functionalities. An emerging strategy to address this demand is by increasing the chemical complexity of synthesized materials beyond conventional limits and developing a new type of substance called high-entropy materials (HEMs). These materials are composed of several elements arranged within the crystal structure in a single phase, providing a broad compositional range with a high number of possible element combinations [1]. The broad range of possibilities renders HEMs with good efficiency and greater sustainability.

High-entropy alloys (HEAs) have at least five major metal elements, and all these elements contribute to the atomic composition, ranging from 5% to 35%. They can also include minor elements in trace amounts, which allows for enhanced tunability of HEAs.

The concept of HEAs was initially introduced by Professor Ye Junwei in 2004. Later, Yeh et al. [2] discussed the same. The composition of HEAs can be defined as follows:

$$n_{major \geq 5},\ 5\ \text{at.}\ \% \leq c_i \leq 35\ \text{at.}\ \%$$

$$n_{minor \geq 0},\ c_j \leq 5\ \text{at.}\ \%$$

where n_{major} and n_{minor} represent the primary and secondary elements, respectively, while c_i and c_j denote their atomic percentages. Based on this definition, a HEA system may include elements in equal proportions, a non-equimolar mixture, or several minor elements combined. The rationale behind this is the high mixing entropy, which encourages the development of a solid solution phase.

High-Entropy Materials for Energy Storage Devices, First Edition.
Edited by Chien-Te Hsieh, Pradeep Kumar Panda, and Arpan Kumar Nayak.

The second definition is rooted in the concept of entropy, which describes HEAs as alloys with a configurational entropy greater than $1.61R$ existing in their random solution state, where R represents the gas constant [3–6].

$$\Delta s_{mix} \geq 1.61R$$

From a thermodynamics perspective, entropy is a fundamental state function that quantifies disorder within a system. A higher level of disorder corresponds to an increase in entropy. Based on Boltzmann's statistical thermodynamics principle [7], the system's entropy can be represented as follows:

$$S = k \ln W$$

where k represents Boltzmann's constant, which is related to the molar gas constant R ($R = 1.38 \times 10^{13}\ \mathrm{J\,K^{-1}}$), and W denotes the thermodynamic probability, indicating the overall figure of the microscopic state [7]. Due to the increase in the number of microscopic states, the entropy of the system also increases. Thus, entropy serves as the measure of the total number of microscopic states within a system.

The characteristics of HEMs can be adjusted by carefully choosing the elements and their compositions. Key characteristics that emerge due to high configurational entropy include localized lattice distortions, cocktail effects, a high density of defects, and enhanced crystal structure stability. Furthermore, for specific applications, it is important to focus not only on structures in a single phase but also on structures in the multiphase, which may fulfil the necessary requirements. This strategy significantly expands the possibilities for customizing the properties of HEMs, making them particularly valuable in the fields of electronics and energy. Currently, studies on HEMs as functional materials are application-driven that focus on the properties stemming from their distinctive high-entropy design. However, entropy stabilization, as previously mentioned, only occurs under specific energy conditions and can be challenging to verify.

Single-phase solid solution HEAs can adopt a variety of crystal constructions, including body-centered cubic (bcc) [8, 9], face-centered cubic (fcc), hexagonal close-packed (hcp) [10], and orthorhombic (orth) structures [1]. Four fundamental "core effects" of HEAs have been identified in catalysis: (i) the high-entropy effect influencing thermodynamics; (ii) the sluggish diffusion effect affecting kinetics; (iii) the lattice distortion effect affecting the structure; and (iv) the cocktail effect governing properties (Figure 13.1).

The high-entropy effect is considered the central concept, which is primarily used to describe the formation of a single-phase alloy composed of multiple principal elements. Based on the conventional Gibbs phase rule [4], the total number of phases present in an alloy system under constant pressure is determined as follows:

$$P = n + 1 - f$$

where P represents the total number of equilibrium phases, n represents the number of elements contained, and F represents the degree of freedom within the alloy system.

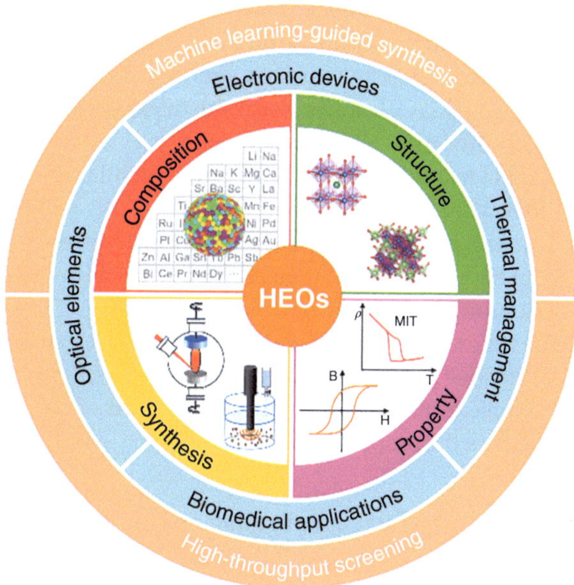

Figure 13.1 Schematic of HEMs in various applications *Source:* Reference [7] / with permission of American Chemical Society.

13.1.1 Cocktail Effect

One of the most fascinating aspects of HEMs is the so-called cocktail effect. This refers to the unique behavior that emerges when different elements are combined—producing properties that cannot be explained by looking at the individual elements alone [8]. These unusual characteristics arise from the complex interactions between the components at the microscopic level, often leading to unexpected and sometimes remarkable material behavior.

By tweaking the stoichiometry of HEMs, we can significantly change how these interactions occur. Such adjustments can alter the oxidation states or chemical surroundings of the elements involved, ultimately affecting the material's properties [9]. This means we can fine-tune the performance of a material by choosing specific elements and carefully controlling how much of each is used. In fact, the way these elements interact with each other locally plays a key role in shaping the material's overall behavior and performance.

13.1.2 Lattice Distortions

Lattice distortions have a significant impact on the behavior and properties of materials. These distortions happen when atoms or ions of dissimilar sizes, charges (valence states), or electronegativities are packed together in the same part of a crystal structure or sublattice. The greater the differences in their sizes—specifically their atomic or ionic radii—the more pronounced the distortion tends to be. In other words, when there is a big mismatch between atoms or ions involved, the crystal structure gets more strained, which can lead to significant changes in the material's properties. In HEAs such as Cantor alloy (FeCoNiCrMn), lattice distortions due to the presence of diverse elements were found to be relatively minor on average. As a result, simulations that assume undistorted lattices show great potential. However, local fluctuations can still reach up to 3%, and causes lattice distortion that impacts both lattice energy and enthalpy change (ΔH). The Madelung constant defines the potential of electrostatics in the crystal that varies when the symmetry of the unit cell is reduced due to local displacements of atoms or ions. As a result, distances between the nearer and the next nearer neighbors are no longer static; rather, they form a distribution. Research on lattice distortion effects in alloys has shown that even a 1% change in the interatomic space can lead to an approximately 12% change in interatomic forces, demonstrating a substantial impact [10]. Comparatively, a 10% change in spacing could result in a complete breakdown of spacing variations of the crystal, which may significantly influence the properties of the material.

13.1.3 Structural Stabilization

Structural stabilization through entropy often occurs when TΔS in the Gibbs–Helmholtz equation outweighs the changes in mixing enthalpy [1]. Additionally, other forms of structural stabilization play a crucial role in phase transitions, which include those observed during ion insertion in battery materials. Outside the solubility limit of binary materials, phase stabilization properties have been observed, including the emergence of structural types that are uncommon under standard conditions. These materials have complex compositions and tend to exhibit greater stability compared to simpler counterparts. While it cannot be definitively stated that such a HEM is stabilized with respect to entropy, beneficial effects are evident, even if their exact mechanisms remain unclear. The roots of these observable effects are expected to be as varied as the different systems in which stabilization can be observed. Rather than being determined by a single characteristic, it is more probable that multiple factors contribute together to the overall structural stabilization.

Doping HEMs with noble metals in energy storage applications aims to leverage the unique properties of both materials: HEMs offer tunable composition and enhanced stability, while noble metals provide high catalytic activity, leading to improved energy storage performance [1].

13.2 Outline of the Applications of HEAs in Battery Fabrication

High-energy lithium-ion batteries (LIBs) have gained a dominant position in the automotive market and are expected to retain this position for the foreseeable future. Figure 13.2 illustrates the historical development of LIBs and outlines their projected advancements over the upcoming decades [11, 12]. To achieve ranges exceeding 500 km in electric vehicles, the development of efficient high-capacity cathodes and anodes, along with innovative electrolytes, is crucial. The evolution of LIBs

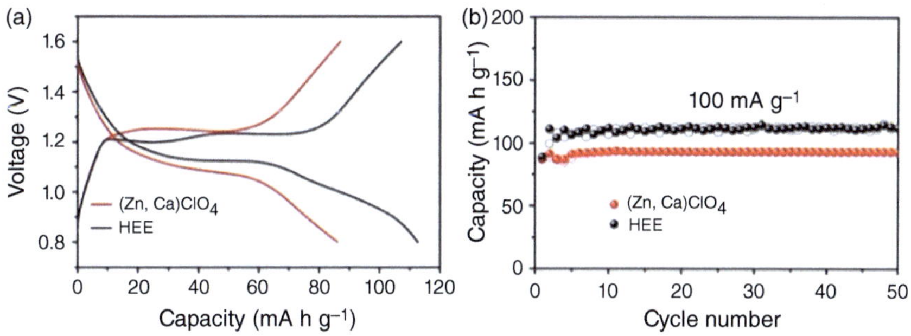

Figure 13.2 (a) Charging/discharging curve for the CCM-I_2|HEE|Zn battery; (b) cycling behaviors related to the binary electrolyte (Zn, Ca) ClO_4. *Source:* Reference [11] / with permission of American Chemical Society.

can be regarded as a revolution in energy storage and electrode materials. To date, continuous innovations in battery components, including anodes, cathodes, electrolytes, cell design, separators, and protection systems, remain essential [13–15].

According to the "Made in China 2025" development plan, specific energy density targets for LIBs have been set as follows: 300 Wh kg^{-1} by 2020, 400 Wh kg^{-1} by 2025, and 500 Wh kg^{-1} by 2030. These energy density figures refer to single cells. The primary factor limiting the energy density of LIBs is the underlying chemical system. Since chemical reactions occur within the cathode and the anode, the energy density of battery cells largely depends on advancements in these electrochemical systems. Active electrode materials play a vital role in determining the electrochemical performance and suitability of batteries for automotive applications. Their impact primarily relies on volumetric capacity, specific capacity, rate capability, discharge potential, intrinsic safety, and capacity retention.

Graphite has extensively been utilized as the negatively charged electrode (anode) of LIBs because of its stable intercalation mechanism. However, its theoretical capacity is relatively low, limited to 372 mA h g^{-1} [16]. To improve battery performance, researchers have been actively exploring transition metal oxides (TMOs) as promising materials for next-generation LIBs. These materials offer greater theoretical capacities than traditional graphite anodes [17]. Some commonly studied TMOs include CoxOy, ZnO, CuO, MnxOy, Fe_3O_4, and Fe_2O_3, which are gaining attention not only for their high capacity but also for their lower cost [18]. Among them, iron oxides stand out as particularly appealing anode materials thanks to their natural abundance, eco-friendly nature, and affordability. For example, Fe_2O_3 operates through a specific reaction mechanism and can theoretically provide an excellent specific capacity of 1,006 mA h g^{-1} [19].

Cobalt oxides, including CoO and Co_3O_4, have gained significant attention because of their excellent redox properties [19]. For instance, CoO can deliver a desirable theoretical specific capacity value of 716 mA h g^{-1} through well-defined electrochemical reactions. Manganese oxides such as MnO and Mn_2O_3 are also widely studied—not only because they are abundant in nature, but also because they offer several advantages, including low operating voltages (around 1.03 and 1.43 V), lower polarization, higher theoretical capacities (756 mA h g^{-1} for

MnO and 1,018 mA h g^{-1} for Mn_2O_3), and good environmental compatibility [19]. Furthermore, other TMOs such as NiO and CuO have also been explored as promising anode ingredients for LIBs. A summary of their expected capacities is provided in the following paragraphs.

Because of their strong and stable crystal structures and superior lithium-ion conductivity, entropy-stabilized metal oxides have become attractive alternatives to traditional graphite anodes in recent years [16]. $(Co_{0.2}Cu_{0.2}Mg_{0.2}Ni_{0.2}Zn_{0.2})O$ is a noteworthy example; it has an excellent initial discharging capacity of 980 mA h g^{-1} and adopts a rock salt structure [18]. It is interesting to note that by carefully eliminating specific components of this material, researchers can produce medium-entropy oxide anodes with specific characteristics for enhanced battery performance. The removal of one element has been shown to affect the material's overall electrochemical activity and cycle stability [20].

The increased cycling stability is ascribed to the enhancement in the structural integrity of HEMs. During cycling, the structure exhibits strong stability, where Co and Cu actively participate in the conversion reaction and Mg helps maintain the material's framework [17]. Ni and Zn contribute to improving the energy density, with Zn offering a low delithiation/lithiation potential, which results in enhanced energy density and output voltage [21, 22]. Guo et al. illustrated that alterations to high-entropy (HEOs) oxides can notably increase cycling lifespan up to 1,000 cycles with a stable capacity of 460 mA h g^{-1} at a desirable current density of 1,000 mA g^{-1} [23]. After 200 cycles, charge density decreases up to 100 mA g^{-1} and capacity surpasses 950 mA h g^{-1}. These findings underscore the outstanding structural stability of HEMs when used as anode candidates.

13.3 Synthesis

There are two primary routes to synthesize HEAs, as shown in Figure 13.3: a top-down method that breaks down bulk materials (orange zone) and a bottom-up method that reduces metal salt precursors (blue zone).

13.3.1 Top-down Approach

This approach involves physically fragmenting large material blocks to produce HEAs. In this section, we outline five different top-down methods used in HEA synthesis.

13.3.2 Vacuum Melting Methodology

It is a traditional method for synthesizing HEAs, which typically involves arc or induction melting. In this process, alloy ingots are prepared by melting the materials in an argon atmosphere, while applying magnetic stirring to promote uniform mixing. Afterward, the ingots are molded and shaped into sheets using wire-cut machining [25, 26]. This method is widely used in conventional metal materials

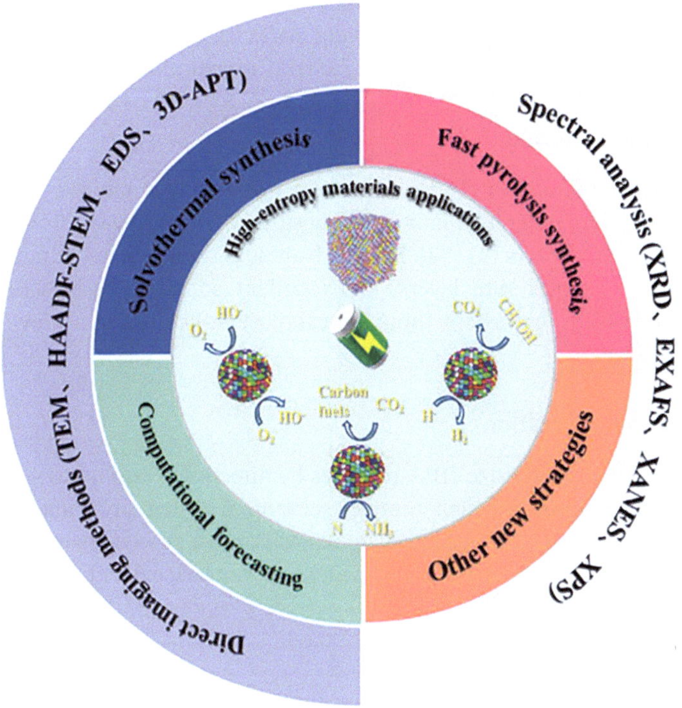

Figure 13.3 Schematic representation of HEA synthesis routes broadly categorized based on starting materials and fabrication mechanisms. *Source:* Reference [24] / with permission of American Chemical Society.

research, particularly for high-melting-point metals, as it produces high-purity alloys. However, it is an expensive process in which alloys are formed in blocks due to the restricted specific surface area.

13.3.3 Dealloying Method

This technique selectively removes or reduces specific components from an alloy by exploiting the differences in chemical activity between its elements through chemical or electrochemical processes [27, 28]. It facilitates the formation of porous alloy structures via atomic diffusion and aggregation. Although commonly used in the fabrication of porous alloys, this process often starts with producing an alloy sheet or pellet using methods including arc melting. Similarly, Jin et al. used dealloying to develop well-dispersed PtPdCuAgAu HEA clusters on the oxide of the HEM, serving as a bifunctional catalytic agent for metal-air batteries [29].

13.3.4 Ball Milling Technique

It is also called mechanical alloying method, which involves using a high-energy mill where powder particles undergo repeated impacts, collisions, cold welding, and fracturing through the action of grinding balls. Through this process, HEAs are synthesized as atoms diffused within the powder particles [30]. It is frequently combined with sintering and melting processes for HEA preparation.

13.3.5 Magnetron Sputtering

This is physical vapor deposition (PVD) technique that employs ionized argon to excite metal targets, ionizing metal ions and depositing them onto a substrate to synthesize HEA films or clusters [31, 32]. It has advantages such as a low film formation temperature, compact film layers, and excellent adhesion. It is widely used to produce HEA films or clusters, though it can pose difficulties in achieving well-dispersed particles.

13.3.6 Pulsed Laser Deposition

This technique is used to synthesize HEA particles by directing a laser onto small regions of many metal targets. The high-energy laser ionizes or partially evaporates the aimed materials, allowing the metal ions to deposit onto a substrate and form particles [33]. This method is fast and efficient, which may not be ideal for large-scale particle production.

13.3.7 Bottom-up Approach

In the bottom-up approach, HEAs are synthesized by chemically reducing metal salt precursors, using methods such as solvothermal synthesis. In the following section, we outline six different bottom-up methods for HEA preparation.

13.3.8 Solvothermal Synthesis

This method is a traditional wet chemistry technique for alloy preparation. In this method, metal salts are mixed with alcohol or water-based solvents and subjected to low-temperature heating for reduction, which results in HEA nanoparticles (NPs) deposited on the substrate [34]. This approach is simple, energy-efficient, and well suited for batch particle production. However, managing the synthesis process can be difficult, potentially causing challenges such as particle agglomeration and oxidation. Using this method, PtIrPdRhRu NPs can be successfully synthesized in a straightforward and environmentally friendly manner [35].

13.3.9 Low-Temperature Oil-Phase Technique

This technique is similar to the solvothermal method. In this method, metal salts are dissolved in the highly volatile solvents; the appropriate temperature is maintained for a specific duration. After multiple washing steps, HEAs are dried in a vacuum drying oven [36, 37]. At low temperatures, this method works better and yields uniformly sized alloy particles. Although it is straightforward and energy-efficient, it may be less effective when dealing with insoluble metal salts. Sun et al. used this method to prepare HEA nanowires [38].

13.3.10 Electrodeposition Processing

In this method, a substrate such as iron plate or nickel foam is prepared in the desired ratio, followed by the addition of a metal salt solution with a complexing agent. Then, electrodeposition is conducted in this solution using a three-electrode system [39, 40], in which the substrate is considered the working electrode simultaneously; the reference electrode is fabricated using Hg/HgO; and finally, the counterelectrode is fabricated using Pt. Electrodeposition is user-friendly and controllable at low temperatures, encompassing methods such as potentiated, galvanostatic, cyclic voltammetry (CV), cathodic plasma, pulse, and droplet-based electrodeposition [41]. The particle size of the HEA can be tailored by modifying factors such as deposition current, solution concentration, and deposition duration. This approach consumes less energy than high-temperature furnace techniques. Zhang et al. [41] successfully fabricated FeCoNiMnW HEA using pulse current electrodeposition on the carbon fibers of carbon paper. However, the uniformity of the alloy phase depends on the deposition potential of metal ions and the presence of complexing agents. Additionally, this technique requires a conductive substrate for deposition.

13.3.11 Pyrolysis Method

This is one of the extensively used methods in the synthesis of HEA NPs. The process begins with the polymerization of metal salt solutions, followed by pyrolysis as well as reduction in an Ar or H_2 environment at high temperatures. This process is simple and enables large-scale alloy material production. It requires basic processes and follows a well-known synthesis procedure; however, achieving rapid heating and cooling remains a challenge [42]. Gao et al. [43] introduced a bed pyrolysis approach to overcome these limitations; this ensures the thorough mixing of metal precursors and concurrent high-temperature decomposition, taking into account the formation of small nuclei.

13.3.12 Carbon Thermal Shock Method

This method uses Joule heating from electrical current to synthesize HEA particles [44]. Metal salt precursors are equally blended with the carbon material and subjected to rapid heating and cooling under an Ar or H_2 atmosphere. This method promotes the generation of uniform particles through consistent temperature distribution, swift heating and cooling, high reaction temperatures, and extremely short reaction durations. The intense heat instantly pyrolyzes the mixed metal salts, leading to simultaneous metal droplet formation. The brief holding time maintains an optimal concentration of the metal atoms present on the carbon fiber surface, balancing droplet splitting and fusion in order to prevent excessive particle growth. This technique is highly effective in producing HEA particles with consistent size efficiently and on a huge scale, though it is limited to conductive substrates. Compared to conventional tubular furnace pyrolysis, it has improved efficiency while reducing energy consumption.

13.3.13 Microwave Sintering Method

In this process, a metal salt is added to the carbon carrier and exposed to a microwave field, where the carbon carrier absorbs substantial energy to facilitate HEA formation. This method provides benefits such as higher efficiency and rapid heating. However, in contrast to the carbon thermal shock (CTS) method, it specifically requires a heat-absorbing carrier material for effective energy transfer [45, 46].

Top-down approaches generally require sufficient pressure, high temperatures, and inert gas conditions. On the other hand, bottom-up approaches should consider factors such as metal compatibility among different elements and reduction potential. Given the distinct characteristics of these HEA synthesis methods and the necessities of catalysts for metal-air batteries, these techniques have significant potential for producing nanoscale catalysts tailored to specific applications. Unlike other HEA applications, using HEAs as catalysts necessitates careful consideration of the intrinsic catalytic properties of the constituent metal elements and the size of the particles of HEAs.

These challenges place important limitations on how HEAs can be designed and synthesized, emphasizing the need for improved and more adaptable synthesis techniques. Several techniques, including solvothermal synthesis, low-temperature oil-phase technique, pulse laser deposition (PLD), dealloying, pyrolysis, microwave sintering, and CTS, have shown potential for synthesizing nanoscale HEA catalysts suitable for metal-air battery applications. Among these, solvothermal synthesis, microwave sintering, oil-phase processing at lower temperatures, pyrolysis, and CTS are considered more practical for real-world applications. Still, beyond just refining synthesis methods and addressing design challenges, further research is essential to boost the long-term stability of these catalysts.

13.3.14 Laser Ablation Synthesis

It is one of the synthesis methods to produce HEA NPs. In this process, a focused laser beam is used to vaporize the material located at the surface of a solid HEA target. As the material is ejected, it cools down and condenses into tiny NPs. This technique provides a straightforward and efficient method for preparing multicomponent HEAs, comparable to the CTS method. The most prominent benefit of laser ablation is that, though high-powered laser is used, it uses less energy compared to the CTS method while producing similar HEA nanocatalysts (Figure 13.4).

13.3.15 PLD Method

It is another synthesis technique that utilizes a powerful pulsed laser beam to generate HEA NPs. PLD involves the rapid vaporization of the target material inside a highly vacuum atmosphere filled with inert gas. Just like laser ablation, the efficiency of PLD is largely influenced by how well the target material absorbs the energy of the incident laser beam, the subsequent interaction between the laser and the surface, plume formation, plasma dynamics, and ultimately, the uniform

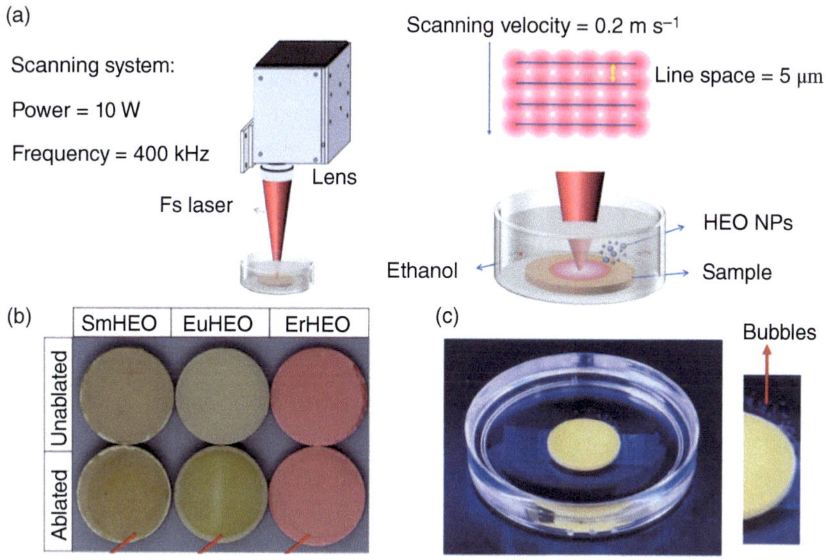

Figure 13.4 (a) Schematic of the laser ablation technique in liquids with the implemented scan strategy; (b) images of the samples SmHEO, EuHEO, and ErHEO before and after laser ablation for the formation of the colloid; (c) bubble formation during the laser ablation method. *Source:* Reference [47] / with permission of American Chemical Society.

deposition of the material onto the substrate. This method is highly efficient, and by adjusting the laser parameters, we can control the key properties of the synthesized HEA, including particle size, dispersity, composition, and morphology.

In 2018, Cropper [48] used the PLD methodology for the deposition of AlCrFeCoNiCu HEA in thin films onto a borosilicate glass substrate. The material exhibited a combination of bcc and fcc crystal structures. After deposition, when the temperature is increased, X-ray diffraction (XRD) peaks became less intense, showing changes in the material's phase structure. The thin films produced were about 35 nm thick, with fcc crystallites roughly matching the film thickness. In contrast, the size of the bcc crystallites varied depending on the temperature used during deposition. While the findings showed that PLD is an effective method for synthesizing HEA nanostructures, the study has not demonstrated how these structures perform in electrocatalytic applications.

Furthermore, in 2022, Chen et al. [49] used PLD to fabricate CrMnFeCoNi HEA thin films on nickel foam and carbon cloth substrates for oxygen evolution reaction (OER). They reported the low overpotentials of 315 and 287 mV required to reach an attainable current density of 10 mA cm^{-2} on nickel foam and carbon cloth, respectively. Additionally, the researchers used electrochemical CV over 500 cycles, which facilitated periodic metal redissolution and redeposition, contributing to enhanced stability. As a result, the HEA demonstrated excellent durability, maintaining performance for 300 hours on carbon cloth and 360 hours on nickel foam. Interestingly, these authors also used a magnetic field in order to enhance the OER kinetics by leveraging spin polarization effects.

Despite PLD's efficiency and versatility in HEA synthesis, one key limitation similar to laser ablation is the challenge of scaling up to produce large electrodes for applications in industries, which has not yet been addressed to our knowledge.

13.3.16 Laser Irradiation Synthesis

In this process, a powerful pulsed laser beam is used to rapidly heat and then cool down metallic precursors, resulting in the formation of multicomponent HEAs [50]. This technique is considered one of the most efficient methods for HEA synthesis, which operates on a millisecond timescale. Recent research has discovered the use of laser metallurgy for creating multimetallic nanomaterials applicable in various fields, such as circuit patterning [51, 52] and electrocatalytic reactions like hydrogen evolution reaction (HER) and also OER.

In 2022, Yan et al. used a powerful nanosecond-pulsed laser beam to synthesize HEAs for OER applications. They produced binary (FeCo, FeNi), ternary (FeCoNi), and quinary (MnFeCoNiZn, MnFeCoNiCu) HEA NPs using metal-organic frameworks (MOFs) as precursors. The resulting NPs exhibited a narrow size distribution. FeNi alloy-based NPs had an average size measuring nearly 7 nm and showed excellent OER activity with a brilliant potential of 282 mV at a desirable current density of 10 mA cm^{-2}, outperforming the standard RuO_2 catalyst, which required 328 mV under the same conditions. FeNi alloy exhibited the lowest resistance for charge transfer in comparison with RuO_2 and the other synthesized HEAs, which was revealed from an electrochemical impedance spectroscopy study. This study also confirmed the absence of phase separation among constituent metals. The localized reducing condition created during laser irradiation promoted HEA formation, while the rapid thermal cycling prevented phase segregation. However, despite fixing critical laser parameters such as pulse energy, wavelength, beam size, frequency, and scribing speed, the researchers did not thoroughly investigate how changes in these parameters might affect the resulting HEA morphology or their electrocatalytic performance.

Later in the same year, Lu et al. [52] fabricated a bifunctional FeCoNiCuPtIr HEA catalyst using laser irradiation, which was designed to work efficiently for both HER and OER. Their method was very simple; they just used a drop of solution of mixed metal precursors onto multiwalled carbon nanotubes (MWCNTs) and then hit it with ultrafast laser pulses. These powerful pulses broke down the metal compounds quickly and formed tiny, evenly distributed NPs across the surface of CNTs. The most interesting thing is that the laser heated the material from room temperature to around 1,600 K in just 50 ms and then cooled it just as quickly. This rapid heating and cooling helped shape the final nanostructure and facilitated the degradation of the precursor solution and subsequent HEA NP formation. By adjusting the laser exposure duration, the researchers controlled the NP size, with longer durations leading to larger particles due to increased diffusion. The entire synthesis took a millisecond without any phase separation.

The resulting HEA catalyst exhibited brilliant electrocatalytic activity, with an excellent potential of 255 mV at 10 mA cm^{-2} for OER and only 21 mV at 10 mA cm^{-2}

for HER, outperforming standard Pt/C and RuO_2 catalysts for their respective reactions. These findings show the catalyst also demonstrated stability at the current densities of 10, 20, and 30 mA cm^{-2}, maintaining performance over 20 hours of continuous catalysis. While this approach is remarkably fast and cost-effective, further studies involving larger-scale samples are necessary to determine its feasibility for industrial HEA NP production (Table 13.1).

Nguyen et al. [53] synthesized $(MnFeCoNiCr)_3O_4$ particles using a surfactant-aided hydrothermal technique. After 200 charge/discharge cycles at 500 mA g^{-1}, the material retained 90% of its capacity, delivering 1,235 mA h g^{-1}. It also exhibited an outstanding rate capability, achieving 500 mA h g^{-1} at a desirable current density of 2,000 mA g^{-1}. Wang et al. [54] manufactured single-phase spinel $(FeCoNiCrMn)_3O_4$ via a solid-state reaction performed at 900 °C. The resulting material delivered a desirable specific capacity of 1,034 mA h g^{-1} during discharge and 680 mA h g^{-1} during charge, along with an impressive rate performance, maintaining 182 mA h g^{-1} at an excellent current density of 2 A g^{-1}. To analyze how the structure evolved with temperature, *in situ* XRD at higher temperatures was utilized. Xiang et al. [55] developed spinel-structured nanocrystalline powders of $(Al_{0.2}CoCrFeMnNi)_{0.58}O_{4-\delta}$ with a high density of oxygen vacancies using a solution combustion synthesis technique. When compared to the undoped $(CoCrFeMnNi)_{0.6}O_{4-\delta}$, the Al^{3+}-doped electrode exhibited significantly enhanced electrochemical activity, delivering a capacity of 554 mA h g^{-1} at 200 mA g^{-1} after 500 cycles over twice the capacity of the undoped counterpart. It also demonstrated an admirable rate capability, maintaining 634 mA h g^{-1} at 3 A g^{-1}. ChanQin Duan developed a novel six-element HEO-coated cathode material, $Na(Fe_{0.2}Co_{0.2}Ni_{0.2}Ti_{0.2}Sn_{0.1}Li_{0.1})O_2$, using a high-temperature solid-state synthesis method [56]. Zheng et al. [57] synthesized a spinel-type HEO, specifically $(FeNiCrMnMgAl)_3O_4$, using a solution combustion technique followed by ball milling aimed at the reduction in particle size. The material, designed for LIB applications, showed a desirable capacity of 657 mA h g^{-1} (after 200 cycles) with a desirable current value of 0.2 A g^{-1} and exhibited a strong rate capability, retaining 350 mA h g^{-1} at 4 A g^{-1}. Meanwhile, Liu et al. [58] prepared a sequence of high-entropy spinel oxides in the stoichiometric form $(CrFeMnNiCo_x)_3O_4$ (where $x = 2, 3, 4$) using a solution combustion synthesis method. These materials demonstrated reversible capacities of 467.8, 574.1, and 506.2 mA h g^{-1} at 200 mA g^{-1}, respectively. In addition, they developed four novel spinel HEOs using a glycine–nitrate-based solution combustion method [58–60]. Su et al. [60] synthesized $(MgTiZnNiFe)_3O_4$ and $(CoTiZnNiFe)_3O_4$ using a solid-state combustion technique and investigated their Li storage mechanisms through *in situ* electron diffraction analysis. Sun et al. [61] synthesized a HEO with the stoichiometric composition of materials as $(Cr_{0.2}Mn_{0.2}Fe_{0.2}Co_{0.2}Ni_{0.2})_3O_4$, featuring an Fd3m spinel crystal structure, using a solid-state reaction method. This material demonstrated a high specific capacity of 560 mA h g^{-1} at an excellent current density of 100 mA g^{-1} and showed outstanding long-term stability, maintaining 100% capacity retention over 5,000 cycles. Xiao et al. [62] prepared $(FeCoNiCrMn)_3O_4$ through the oxidation of FeCoNiCrMn alloy powder. Their study highlighted that the high-entropy effect contributes to enhanced structural stability and reduced

Table 13.1 The synthesis methods, structure, role, and performance of HEOs used in energy storage applications.

Composition	Synthesis method	Structure	Application	Role	Performance	References
$(MnFeCoNiCr)_3O_4$	Surfactant-assisted hydrothermal	Spinel	LIBs	Anode	1,235 $mA\,h\,g^{-1}$ (500 $mA\,g^{-1}$)	[53]
$(FeCoNiCrMn)_3O_4$	High-temperature solid-state reaction	Spinel	LIBs	Anode	1,034/680 $mA\,h\,g^{-1}$ (500 $mA\,g^{-1}$, 1st cycle)	[54]
$(Al_{0.2}CoCrFeMnNi)_{0.58}O_4$	High-temperature solid-state reaction	Spinel	LIBs	Anode	554 $mA\,h\,g^{-1}$ (200 $mA\,g^{-1}$ after 500 cycles)	[55]
$(FeCoNiCrMnCuLi)_3O_4$	High-temperature solid-state reaction	Spinel	LIBs	Anode	1,000.1/626.6 $mA\,h\,g^{-1}$ (50 $mA\,g^{-1}$, 1st cycle)	[56]
$(FeCoNiCrMnMgLi)_3O_4$	High-temperature solid-state reaction	Spinel	LIBs	Anode	1,031/592.1 $mA\,h\,g^{-1}$ (50 $mA\,g^{-1}$, 1st cycle)	[56]
$(FeCoNiCrMnZnLi)_3O_4$	High-temperature solid-state reaction	Spinel	LIBs	Anode	1,049.9/706 $mA\,h\,g^{-1}$ (50 $mA\,g^{-1}$, 1st cycle)	[56]
$(FeNiCrMnMgAl)_3O_4$	Combustion	Spinel	LIBs	Anode	657 $mA\,h\,g^{-1}$ (200 $mA\,g^{-1}$ after 200 cycles)	[57]
$(CrFeMnNiCo_2)_3O_4$	Combustion	Spinel	LIBs	Anode	467.8 $mA\,h\,g^{-1}$ (200 $mA\,g^{-1}$)	[58]
$(CrFeMnNiCo_3)_3O_4$	Combustion	Spinel	LIBs	Anode	574.1 $mA\,h\,g^{-1}$ (200 $mA\,g^{-1}$)	[58]
$(CrFeMnNiCo_4)_3O_4$	Combustion	Spinel	LIBs	Anode	506.2 $mA\,h\,g^{-1}$ (200 $mA\,g^{-1}$)	[58]
$(CrFeMnCoMgLi)_3O_4$	Combustion	Spinel	LIBs	Anode	393.1 $mA\,h\,g^{-1}$ (100 $mA\,g^{-1}$)	[59]
$(CrFeMnCoMg)_3O_4$	Combustion	Spinel	LIBs	Anode	395.3 $mA\,h\,g^{-1}$ (100 $mA\,g^{-1}$)	[59]
$(CrFeMnCoNiLi)_3O_4$	Combustion	Spinel	LIBs	Anode	395 $mA\,h\,g^{-1}$ (100 $mA\,g^{-1}$)	[59]
$(CrFeMnCoNi)_3O_4$	Combustion	Spinel	LIBs	Anode	407.8 $mA\,h\,g^{-1}$ (100 $mA\,g^{-1}$)	[59]
$(MgTiZnNiFe)_3O_4$	Solid-state sintering	Spinel	LIBs	Anode	166.8/424.7 $mA\,h\,g^{-1}$ (100 $mA\,g^{-1}$, 1st cycle)	[60]

$(CoTiZnNiFe)_3O_4$	Solid-state sintering	Spinel	LIBs	Anode	423.9/674.7 mA h g^{-1} (100 mA g^{-1}, 1st cycle)	[60]
$(Cr_{0.2}Mn_{0.2}Fe_{0.2}Co_{0.2}Ni_{0.2})_3O_4$	Solid-state reaction	Spinel	LIBs	Anode	560 mA h g^{-1} (100 mA g^{-1})	[61]
$(FeCoNiCrMn)_3O_4$	Solid-state reaction	Spinel	LIBs	Anode	692.8 mA h g^{-1} (500 mA g^{-1} after 260 cycles)	[62]
$[(Bi,Na)_{1/5}(La,Li)_{1/5}(Ce,K)_{1/5}Ca_{1/5}Sr_{1/5}]TiO_3$	Solid-state reaction	Perovskite	LIBs	Anode	120.4 mA h g^{-1} (also 1,000 mA g^{-1} after 300 cycles)	[63]
$Na_{0.7}Mn_{0.4}Ni_{0.3}Cu_{0.1}Fe_{0.1}Ti_{0.1}O_{1.95}F_{0.1}$	Solid-state reaction	O3/P2-type	SIBs	Cathode	86.7 mA h g^{-1} (800 mA g^{-1})	[64]
$NaNi_{0.12}Cu_{0.12}Mg_{0.12}Fe_{0.15}Co_{0.15}Mn_{0.1}Ti_{0.1}Sn_{0.1}Sb_{0.04}O_2$	Solid-state reaction	O3-type	SIBs	Cathode	110 mA h g^{-1} (12 mA g^{-1})	[65]
(MgCoNiCuZn)O	Nebulized spray pyrolysis	Rock-salt	LIBs	Anode	549 mA h g^{-1} (100 mA g^{-1})	[66]
(LiMgCoNiCuZn)O	Nebulized spray pyrolysis	Rock-salt	LIBs	Anode	714 mA h g^{-1} (100 mA g^{-1})	[66]
$(Co_{0.2}Cr_{0.2}Fe_{0.2}Mn_{0.2}Ni_{0.2})_3O_4$	Sol-gel method	Spinel	SCs	Cathode	75 F g^{-1} (1 A g^{-1})	[67]
$(CrMnFeNiCu)_3O_4$	hydrothermal	Spinel	LIBs	Anode	340 mA h g^{-1} @ 2 A g^{-1}	[68]

band gap, while the spinel structure facilitates efficient ion transport pathways within the material. Yan et al. [63] synthesized a high-entropy perovskite oxide, formulated as $[(Bi, Na)_{1/5}(La, Li)_{1/5}(Ce, K)_{1/5}Ca_{1/5}Sr_{1/5}]TiO_3$, to serve as a substance for the anode of LIBs. Using the solid-state reaction method, the substance attained a capacity of 120.4 mA h g^{-1} and maintained nearly 100% capacitive retention at a current density of 1 A g^{-1} (300 cycles). The O_3-type layered structure of the HEO was found to effectively suppress intermediate phase transitions, minimize charge ordering and Na vacancy formation, and reduce structural defects such as interlayer sliding during electrochemical cycling [63]. Additionally, Zhou [64] developed a biphasic P_2/O_3, which is a HEO having an elemental composition with a specific stoichiometric ratio as $Na_{0.7}Mn_{0.4}Ni_{0.3}Cu_{0.1}Fe_{0.1}Ti_{0.1}O_{1.95}F_{0.1}$, through a solid-state synthesis process. This material demonstrated an outstanding capacity retention over a broad temperature varying from 40 to 50 °C [64]. Zhao et al. [65] introduced a novel high-entropy strategy for designing layered oxide cathodes in sodium-ion batteries. They developed an O_3-type layered oxide with the composition $NaNi_{0.12}Cu_{0.12}Mg_{0.12}Fe_{0.15}Co_{0.15}Mn_{0.1}Ti_{0.1}Sn_{0.1}Sb_{0.04}O_2$, which demonstrated enhanced cycling stability, retaining approximately 83% of its capacity after 500 cycles, and exhibited a brilliant rate capability, maintaining about 80% capacity retention with a 5.0C rate.

These authors found that the material showed a reversible shift between two structures, O_3 and P_3, during charging and discharging. Interestingly, more than 60% of the energy capacity came from the O_3 structure. They presumed that having multiple transition metals in the HEM helped it handle the changes that occur when Na ions move in and out. This offers a promising direction for improving the performance of sodium-ion batteries.

In another study, Liu and team [66] used a hydrothermal process to add helpful ions to the oxygen-deficient areas of a rock salt oxide made from Mg, Co, Ni, Cu, and Zn. This change made it more easier for ions and electrons for moving through the substance, leading to a steady discharge capacity of 549 mA h g^{-1} at 0.1 A g^{-1}. When Li was added to create Li-MO, the capacity increased even more to 714 mA h g^{-1}. These results highlight how rock salt oxides can be adjusted in both structure and makeup, making them great materials for next-generation energy storage solutions.

Ma et al. synthesized $(FeNiCrMnZn)_3O_4$ and $(FeNiCrMnMg)_3O_4$ using high-energy ball milling [68]. Both compounds exhibited single-phase spinel structure, characterized by a cubic lattice where oxygen anions form a tightly packed arrangement and metal cations occupy interstitial tetrahedral and octahedral sites. The incorporation of multiple metal elements in near-equimolar proportions leads to a stable solid solution, stabilized by high configurational entropy. These authors focused on how the substitution of Zn and Mg influences the electrochemical properties of the HEOs. They found that $(FeNiCrMnZn)_3O_4$ exhibited superior rate properties, enhanced cycle performance, and favorable electrochemical impedance spectroscopy results compared to $(FeNiCrMnMg)_3O_4$. This indicates that incorporating electrochemically active metals such as Zn can significantly boost the activity of HEOs in LIBs as anode materials.

13.3.17 Supercapacitors

Supercapacitor (SC) energy storage mechanisms are generally classified into two important categories: (i) electrostatic interactions between the electrode and electrolyte ions in electric double-layer capacitors (EDLCs) and (ii) Faradaic redox reactions at the electrode–electrolyte interface in pseudocapacitors [69]. However, EDLCs typically suffer from low energy density, while pseudocapacitors often exhibit poor cycle life. To overcome these drawbacks, there is a high need to design electrode materials that can provide both higher energy density and lasting cycling stability.

In this context, Lal et al. [70] introduced a new strategy by incorporating HEOs with CNTs. They synthesized Al-, Ni-, Fe-, Co-, and Cr-based HEOs and combined them with CNTs using chemical deposition. The resulting composite electrodes were tested in a full cell using a polyvinyl alcohol/sulfuric acid (H_2SO_4) hydrogel electrolyte. Impressively, the HEO–CNT hybrid retained 100% of its capacitance and Coulombic efficiency even after 15,000 charging/discharging cycles at $5\,A\,g^{-1}$. The encapsulation of HEOs within CNTs not only helped enhance the number of active spots by modifying the redox behavior of the surrounding carbon layers but also protected the HEOs from direct contact with the electrolyte, effectively mitigating degradation over prolonged use.

Meanwhile, Miao et al. [71] developed a series of perovskite-type HEO materials, $La(CoCrFeMnNiAl_x)_{1/(5+x)}O_{3-\delta}$ (where $x = 0.4, 0.5, 0.6, 0.7$), and evaluated their performance as SC electrodes. These materials exhibited an excellent cycling stability, retaining 85.01%, 88.61%, 86.37%, and 91.25% of their initial capacitance (for $x = 0.4$–0.7, respectively) after 2,000 cycles at an efficient current density of $1{,}000\,mA\,g^{-1}$. When the scan rate increased, ohmic resistance and polarization at the electrode–electrolyte interface also increased, causing redox peaks in CV to shift in both directions. TMO agglomerations typically decrease redox activity and can cause cracking in the crystal structure. To address this, Al^{3+}, a relatively inert ion, was introduced into $La(CoCrFeMnNi)_{0.2}O_{3-\delta}$. The addition of Al^{3+} was found to reduce TM oxide agglomeration and introduce lattice distortion, thereby enhancing the electrochemical characteristics of the material. However, excessive Al^{3+} doping hindered ion migration pathways, ultimately leading to reduced capacitance retention during extended cycling.

13.3.18 Dielectric Capacitors

Dielectric capacitors are increasingly used in pulse power applications across various sectors such as microelectronics, healthcare, and aerospace, and their advantages include high energy density, rapid charge/discharge capabilities, and strong chemical stability [72–74]. The energy storage density and efficiency of these dielectric materials can be measured using the following equations:.

$$W = \int_{0}^{Pmax} Edp \tag{13.1}$$

$$W_{rec} = \int_{P_r}^{P_{max}} Ed_p$$

$$\eta = \frac{W_{rec}}{W}$$

In Eq. (13.1), W represents the overall density for the energy storage of the substance, which includes both the energy that can be recovered, called recoverable energy (W_{rec}), and the energy lost during the process; η denotes the efficiency of energy storage; E is the strength of the applied electric field; P refers to the polarization strength; P_{max} is the polarization with the maximum value; and P_r is the remnant (residual) polarization.

To achieve higher energy storage efficiency and density, it is essential to have a low P_r, high P_{max}, and a strong breakdown field E. Minimizing P_r allows the dielectric material to store and release energy more effectively, with minimal losses. This characteristic is especially critical in applications that demand efficient energy management and rapid charge/discharge cycles, such as in advanced battery systems. Dielectric capacitors are generally divided into two categories based on how their dielectric permittivity responds to an electric field: linear and nonlinear dielectrics.

In linear dielectrics, polarization responds directly and proportionally to the applied electric field, with no delay or hysteresis. While this predictable behavior is useful, these materials have a downside: they can only store a limited amount of energy because their saturation polarization is relatively low [75].

In contrast, nonlinear dielectrics possess a more intricate relationship between electric field and polarization, often involving hysteresis. The most common examples include relaxor ferroelectrics, paraelectrics, ferroelectrics, and antiferroelectrics (including double relaxor ferroelectrics), each representing distinct hysteresis loop characteristics. Response time and operational stability depend on the choices between linear and nonlinear dielectrics. Their unique hysteresis behaviors allow for the optimization of dielectric capacitors and customized designs tailored to specific applications.

Among these, relaxor ferroelectrics are notable due to their low dielectric loss, low remnant polarization, and high saturation polarization combined with strong breakdown field strength, making them highly suitable for energy storage purposes. Furthermore, the highly disordered cation or anion configurations found in HEMs support the development of relaxor properties, offering new avenues for advanced capacitor design.

High-entropy ceramics have demonstrated remarkable potential in the field of dielectric energy storage, particularly through their ability to drastically reduce the loss tangent. For instance, the loss tangent remains significantly low (<0.015) from room temperature up to 125 °C. This performance is attributable to the formation of polar nanoregions (PNRs) with short relaxation times, a result of the substantial atomic disorder at A-sites induced by the high-entropy effect.

Pu et al. [76] developed high-entropy ceramics with the chemical composition $(Na_{0.2}Bi_{0.2}Ba_{0.2}Sr_{0.2}Ca_{0.2})TiO_3$ and investigated their energy storage characteristics. These ceramics displayed a discharge energy density of 1.02 J cm^{-3} under an electric field of 145 kV cm^{-1}, which was enhanced to 1.56 J cm^{-3} by doping with 0.5 wt% Mn^{2+} [76–78]. Incorporating cations with various ionic sizes and valences into the structure increases ion disorder, partially disturbing the long-range ferroelectric domain order. This results in nonuniform microregions and a dispersion coefficient (γ) of 1.72, further improving the relaxor behavior of the material.

Bismuth titanate-derived ceramics have gained attention of late because of their high dielectric constants, strong remnant polarization, and robust energy storage performance. However, a major disadvantage of these materials is their high loss tangent, which undermines energy efficiency. For example, traditional $(Na_{0.5}Bi_{0.5})TiO_3$ (BNT) ceramics are known for high coercive fields, low electrical resistivity, and low depolarization temperatures, all of which can limit their energy storage potential.

To address these challenges, researchers have applied high-entropy design strategies by introducing multiple cations at the A-site of BNT, which encourages the development of PNRs. These PNRs enhance the material's energy storage capability by providing higher maximum polarization (P_{max}) and lower remnant polarization (P_r), making such materials suitable for both energy storage devices and non-volatile memory applications. The disordered solid solution also disrupts macroscopic ferroelectric domains, promoting the relaxor effect.

The high-entropy strategy has also been prolonged to materials such as $BaTiO_3$ and $Bi_{1.5}ZnNb_{1.5}O_7$, where the introduction of randomly distributed doped cations leads to complex, nonuniform microstructures with enhanced energy storage properties. A summary of their energy densities, breakdown field strengths, and efficiencies is presented in Table 13.2.

In a recent development, Chen et al. developed a new high-entropy composition, $(K_{0.2}Na_{0.8})_{0.8}Li_{0.08}Ba_{0.02}Bi_{0.1}O_3$, featuring multiphase nanoclusters and randomly oriented oxygen octahedra. This composition disturbed the long-range ferroelectric order via a combination of orthorhombic, tetragonal, rhombohedral, and cubic (R–O–T–C) nanoclusters, forming small PNRs. As a result, the material achieved a high energy density of 10.06 J cm^{-3} with an expected efficiency of 90.8%. Likewise, the integration of $Bi(Li_{0.2}Y_{0.2}Mg_{0.2}Ti_{0.2}Ta_{0.2})O_3$ into a base of $(0.9Ba(Ti_{0.97}Ca_{0.03})O_3$–$0.1Bi_{0.55}Na_{0.45}TiO_3)$ resulted in a composite named (0.9BTC–0.1BNT–BLYMTT), which delivered a substantial energy density of 4.89 J cm^{-3} at 240 kV cm^{-1} with an efficiency of 91.2%.

Another innovative design introduced a random local electric field to break long-range ferroelectric order using the ceramic composition $(Bi_{0.5}Na_{0.5})(Ti_{1-2x}Fex Nbx)O_3$ (x = 1/3), synthesized from $NaNbO_3$, $(Bi_{0.5}Na_{0.5})TiO_3$, and $BiFeO_3$. This design led to the formation of small PNRs, significantly boosting P_{max}. The presence of weakly coupled PNRs, polymorphic oxygen octahedra, and an increased grain boundary density contributed to a high breakdown strength. The material transitioned into a nonpolar macroscopic state when the electric field was removed, yielding near-zero P_r. Altogether, these features enabled an exceptionally high energy density of 13.8 J cm^{-3} with a good efficiency of 82.4%.

The high-entropy method has also been successfully applied to thin films, leading to the development of $(Bi_{3.25}La_{0.75})(Ti_{3-3x}Zr_xHf_xSn_x)O_{12}$ (x = 0.4) films, which formed a stabilized pyrochlore-type $Bi_2Ti_2O_7$-based dielectric layer. By fine-tuning

Table 13.2 Comparison of energy storage performance metrics: storage density, efficiency, and dielectric strength of high-entropy dielectric materials.

System	W_{rec} (J cm^{-3})	E_b (kV cm^{-1})	η (%)	References
$Ba(Zr_{0.2}Ti_{0.2}Sn_{0.2}Hf_{0.2}Nb_{0.2})O_3$	0.55	290	—	[79]
$Ba(Zr_{0.2}Ti_{0.2}Sn_{0.2}Hf_{0.2}Ta_{0.2})O_3$	0.68	370	—	[79]
$(Bi_{0.2}Na_{0.2}K_{0.2}Ba_{0.2}Ca_{0.2})TiO_3$	0.68	129	87.5	[80]
$(Bi_{0.2}Na_{0.2}K_{0.2}La_{0.2}Sr_{0.2})TiO_3$	0.96	180	90.0	[81]
$(Bi_{0.2}Na_{0.2}K_{0.2}La_{0.2}Sr_{0.2})(Ti_{0.8}Sc_{0.2})O_3$	1.09	220	80.0	[82]
$(Bi_{0.2}Na_{0.2}Ba_{0.2}Ca_{0.2}Sr_{0.2})TiO_3$	1.37	127	72.7	[83]
$(Na_{0.2}Bi_{0.2}Ca_{0.2}Sr_{0.2}Ba_{0.2})TiO_3$	1.32	110	91.0	[84]
$(Bi_{0.2}Na_{0.2}Ba_{0.2}Ca_{0.2}Sr_{0.2})TiO_3$–0.5 wt% $MnCO_3$	1.56	160	70.2	[78]
$Bi_{1.5}Zn_{0.75}Mg_{0.25}Nb_{0.75}Ta_{0.75}O_7$	2.72	650	91.0	[85]
$0.88(Bi_{0.4}Na_{0.2}K_{0.2}Ba_{0.2})TiO_3$–$0.12Sr(Mg_{1/3}Nb_{2/3})O_3$	3.37	310	90.8	[86]
0.8 $Na_{0.5}Bi_{0.5}TiO_3$–$0.2Sr(Zr_{0.2}Sn_{0.2}Hf_{0.2}Ti_{0.2}Nb_{0.2})O_3$	3.52	283	73.4	[87]
$0.9(0.75BaTiO_3$–$0.25Na_{0.5}Bi_{0.5}TiO_3)$–$0.1Bi(Zn_{0.2}Mg_{0.2}Al_{0.2}Sn_{0.2}Zr_{0.2})O_3$	3.74	273	82.2	[88]

the atomic configuration-based entropy, these films demonstrated desirable and stable microstructural properties. This was highlighted by the presence of lattice distortion-based nanocrystalline grains with a disordered phase having amorphous-like properties.

These structural features contributed to a higher breakdown strength, significantly improving the film's durability and performance under intense electric fields. Moreover, the coexistence of disordered amorphous-like regions and distorted nanocrystals helped suppress polarization-switching hysteresis, enabling faster and more reversible polarization responses. Together, these improvements led to enhanced energy storage capabilities, with the film achieving a higher energy density of 182 J cm^{-3} and an efficiency of 78% for energy storage applications.

As shown in Table 13.2, Zhou et al. [79] investigated high-entropy perovskite oxides of the form $Ba(Zr_{0.2}Ti_{0.2}Sn_{0.2}Hf_{0.2}Me_{0.2})O_3$ (Me = Nb, Ta) and reported promising dielectric energy storage properties. The energy density storage reached up to 0.68 J cm^{-3}, with an excellent breakdown field strength of 370 kV cm^{-1} for the Ta-substituted composition. Although energy storage efficiency (η) was not explicitly stated, the materials exhibited symmetric P–E loops, suggesting moderate efficiencies. The study highlighted the potential of high-entropy ceramics for use in lead-free dielectric capacitors due to their enhanced dielectric strength and capabilities for energy storage. Liu et al. [80] studied flash-sintered high-entropy ceramics with the composition $(Bi_{0.2}Na_{0.2}K_{0.2}Ba_{0.2}Ca_{0.2})TiO_3$ and reported enhanced dielectric strength with better energy storage performance. The material showed a strong potential for energy storage, safely holding up to 0.68 J cm^{-3} and withstanding electric fields as high as 129 kV cm^{-1}. It also showed an impressive energy efficiency of 87.5%, mainly because of its neat and balanced electric response. These results show how combining high-entropy design with flash sintering can greatly improve the performance of lead-free materials used in energy storage applications.

Yang and Zheng [81] developed a special ceramic material using different elements $[(Bi_{0.2}Na_{0.2}K_{0.2}La_{0.2}Sr_{0.2})TiO_3]$, which showed excellent energy storage performance. This nanostructured material stored up to 0.96 J cm^{-3} of energy, could handle electric fields as high as 180 kV cm^{-1}, and had a high efficiency of 90%. These great results came from its tiny, well-organized structure and the way its ingredients were mixed, which helped reduce energy loss and improved how much electric field it could handle.

Later, in another study [82], Yang and Zheng added a small amount of scandium (Sc) to the material $[(Bi_{0.2}Na_{0.2}K_{0.2}La_{0.2}Sr_{0.2})(Ti_{1-x}Sc_x)O_3]$ and found even better performance. With the right amount of Sc (x = 0.2), the ceramic stored more energy—up to 1.09 J cm^{-3}—handled stronger electric fields (220 kV cm^{-1}), and still maintained a good efficiency of 80%. These improvements make it a strong option for lead-free, high-efficiency energy storage devices. The enhanced performance was attributed to the harmonious outcome of high-entropy composition and Sc-induced structural modification, which optimized the dielectric response and reduced energy loss. Sun et al. [83] investigated $(Bi_{0.2}Na_{0.2}Ba_{0.2}Ca_{0.2}Sr_{0.2})TiO_3$ high-entropy ceramics synthesized via the hydrothermal method and reported enhanced electrical properties for energy storage applications. These materials achieved an energy storage density of 1.37 J cm^{-3}, a breakdown field strength of 127 kV cm^{-1}, and an enhanced energy storage efficiency of 72.7%. The improved performance was attributed to the uniform microstructure and reduced grain size resulting from the hydrothermal process, which enhanced dielectric breakdown strength and optimized the energy storage behavior of the material. Wang et al. [84] reported high-entropy relaxor ferroelectric ceramics with simultaneously exceptional energy storage efficiency and energy storage density at low electric fields. The optimized composition, $(Na_{0.2}Bi_{0.2}Ca_{0.2}Sr_{0.2}Ba_{0.2})TiO_3$, exhibited an energy density storage of 1.32 J cm^{-3}, a breakdown field strength of 110 kV cm^{-1} and a significantly higher energy storage efficiency of 91.0%. These results highlight the usefulness of the high-entropy design in achieving superior energy storage performance at lower operating fields, making the material a strong candidate for next-generation energy storage capacitors. Fang et al. [78] examined Mn-modified high-entropy relaxor ferroelectric ceramics with the composition $(Na_{0.2}Bi_{0.2}Ca_{0.2}Sr_{0.2}Ba_{0.2})TiO_3$–0.5 wt% $MnCO_3$ and reported enhanced energy storage properties. This material achieved an excellent energy storage density of 1.56 J cm^{-3}, a breakdown field strength of 160 kV cm^{-1}, and a desirable energy storage efficiency of 70.2%. The introduction of Mn effectively suppressed domain wall motion and improved dielectric breakdown strength, taking into account the improved energy storage behavior suitable for high-performance capacitor applications. Chen et al. [85] developed pyrochlore-based high-entropy ceramics with the composition $Bi_{1.5}Zn_{0.75}Mg_{0.25}Nb_{0.75}Ta_{0.75}O_7$, targeting capacitive energy storage applications. This material demonstrated an excellent energy storage density of 2.72 J cm^{-3}, an ultrahigh breakdown field strength of 650 kV cm^{-1}, and an excellent energy storage efficiency of 91.0%. The brilliant performance was attributable to the pyrochlore structure and high-entropy effect, which together enhanced dielectric stability and reduced energy loss, making it a suitable candidate for advanced energy storage devices. Yan et al. [86] fabricated lead-free high-entropy relaxor ceramics based on the $0.88(Bi_{0.4}Na_{0.2}K_{0.2}Ba_{0.2})TiO_3$–$0.12Sr(Mg_{1/3}Nb_{2/3})O_3$ composition and achieved an excellent dielectric performance for energy storage applications. These ceramics displayed a brilliant energy storage density of 3.37 J cm^{-3}, a breakdown field strength of 310 kV cm^{-1}, and a higher energy storage efficiency of 90.8%. These outstanding properties were accountable for the harmonious effect of high-entropy design and relaxor behavior, which resulted in slim *P–E* loops and enhanced breakdown strength, making the material highly suitable for the enhanced performance of lead-free capacitors for energy storage.

Zhou et al. [87] studied NBT-based ceramics enhanced with HEOs, focusing on a specific mix: $0.8Na_{0.5}Bi_{0.5}TiO_3$–$0.2Sr(Zr_{0.2}Sn_{0.2}Hf_{0.2}Ti_{0.2}Nb_{0.2})O_3$. They found a major boost in energy storage performance. The ceramics could store 3.52 J cm^{-3} of recoverable energy, withstand electric fields up to 283 kV cm^{-1}, and reach an efficiency of 73.4%. These improvements were mostly because of the HEO, which helped stabilize the material's behavior across different temperatures and reduced energy loss. This makes it a strong choice for energy storage systems that require both good performance and thermal stability.

In a follow-up study [88], Zhou et. al synthesized another lead-free ceramic using a different formula: $0.9(0.75BaTiO_3$–$0.25Na_{0.5}Bi_{0.5}TiO_3)$–$0.1Bi(Zn_{0.2}Mg_{0.2}Al_{0.2}Sn_{0.2}Zr_{0.2})O_3$. This material performed even better, storing 3.74 J cm^{-3} of energy, handling fields up to 273 kV cm^{-1}, and reaching 82.2% efficiency. The key to this success was the use of a high-entropy perovskite structure, which improved both dielectric properties and thermal reliability. Altogether, this makes the material a strong candidate for future high-performance capacitive energy storage devices.

13.4 Summary and Future Scope

This chapter primarily explained the idea of HEMs and their four core effects. It emphasized the synthesis techniques of these materials and explained their application in batteries, dielectric capacitors, SCs, and plasmonic studies. Noble metal-doped HEMs represent a potential candidate in the energy storage technology. It is evident that noble metals such as Pt, Ru, and Pd, when incorporated into high-entropy systems, significantly improve energy storage performance due to their superior catalytic and conductive properties. Despite existing challenges in terms of synthesis, scalability, and cost, continued research and development can unlock new applications and improve the efficiency and longevity of energy storage devices.

Hence, there is a huge scope for exploring novel noble metal-doped HEM combinations for different applications, and the production of these materials should be scaled up for commercial applications. As the demand for high-performance, durable, and sustainable energy storage systems continues to grow, noble metal-doped HEMs are poised to play a major role in the evolution of clean energy technologies.

References

1 Schweidler, S., Botros, M., Strauss, F. et al. (2024 Apr). High-entropy materials for energy and electronic applications. *Nat Rev Mater* 9: 266–281.

2 Yeh, J.W., Chen, S.K., Lin, S.J. et al. (2004). Nanostructured high-entropy alloys with multiple principal elements: novel alloy design concepts and outcomes. *Adv Eng Mater* 6 (5): 299–303.

3 He, Q.F., Ding, Z.Y., Ye, Y.F., and Yang, Y. (2017). High-throughput preparation and characterization of high-entropy alloys. *JOM* 69 (11): 2092–2098.

4 Xin, Y., Li, S.H., Qian, Y.Y. et al. (2020). High-entropy materials for heterogeneous catalysis: a new frontier. *ACS Catal* 10 (19): 11280–11306.

5 Miracle, D.B. and Senkov, O.N. (2017). A critical review of high entropy alloys and related concepts. *Acta Mater* 122: 448–511.

6 Miracle, D.B., Miller, J.D., Senkov, O.N. et al. (2014). Exploration and development of high entropy alloys for structural applications. *Entropy* 16 (1): 494–525.

7 Zou, J., Tang, L., He, W., and Zhang, X. (2024). High-entropy oxides: pioneering the future of multifunctional materials. *ACS Nano* 18 (51): 34492–34530. https://doi.org/10.1021/acsnano.4c12538.

8 Zhang, X., Liu, Y., Zhao, X. et al. (2024). Recent advances and perspectives of high-entropy alloys as electrocatalysts for metal-air batteries. *Energy Fuel* 38: 19236–19252.

9 Usharani, N.J., Shringi, R., Sanghavi, H. et al. (2020). Role of size, alio-/multi-valency and non-stoichiometry in the synthesis of phase-pure high entropy oxide (Co,Cu,Mg,Na,Ni,Zn)O. *Dalton Trans* 49: 7123–7132.

10 Fultz, B. (2010). Vibrational thermodynamics of materials. *Prog Mater Sci* 55: 247–352.

11 Tian, M., Zhu, C., and Luo, K. (2024). Selective compatibility of high-entropy electrolytes for low-temperature aqueous zinc–iodine batteries. *ACS Sustain Chem Eng* 12 (52): 18577–18587. https://doi.org/10.1021/acssuschemeng.4c06557.

12 Xu, J., Cai, X., Cai, S. et al. (2023). High-energy lithium ion batteries: recent progress and a promising future in applications. *Energy Environ Mater* 6: 60–85.

13 Lipu, M.S.H., Hannan, M.A., Hussain, A. et al. (2018). A review of state-of-the-art lithium-ion battery management systems: challenges, solutions, and future directions. *J Clean Prod* 205: 115–133.

14 Hannan, M.A., Hoque, M.M., Hussain, A. et al. (2018). State-of-the-art and energy management system of lithium-ion batteries in electric vehicle applications: issues and recommendations. *IEEE Access* 6: 19362–19378.

15 Zeng, X., Li, M., Abd El-Hady, D. et al. (2019). Commercialization of lithium battery technologies for electric vehicles. *Adv Energy Mater* 9: 1900161.

16 Sturman, J.W., Baranova, E.A., and Abu-Lebdeh, Y. (2022). Review: high-entropy materials for lithium-ion battery electrodes. *Front Energy Res* 10: 1–15.

17 Shahbazian-Yassar, R. and Amiri, A. (2021). Recent progress of high entropy materials for energy storage and conversion. *J Mater Chem A* 9: 782–823.

18 Akrami, S., Edalati, P., Fuji, M., and Edalati, K. (2021). High-entropy ceramics: review of principles, production and applications. *Mater Sci Eng R Rep* 146: 100644.

19 Fang, S., Bresser, D., and Passerini, S. (2020). Transition metal oxide anodes for electrochemical energy storage in lithium- and sodium-ion batteries. *Adv Energy Mater* 10: 1902485.

20 Sarkar, A., Velasco, L., Wang, D. et al. (2018). High entropy oxides for reversible energy storage. *Nat Commun* 9: 3400.

21 Ortiz, M.G., Visintin, A., and Real, S.G. (2021). Synthesis and electrochemical properties of nickel oxide as anodes for lithium-ion batteries. *J Electroanal Chem* 883: 114875.

22 Wang, L., Zhang, G., Liu, Q., and Duan, H. (2018). Recent progress in Zn-based anodes for advanced lithium ion batteries. *Mater Chem Front* 2: 1414–1435.

23 Guo, H., Shen, J., Wang, T. et al. (2022). Design and fabrication of high-entropy oxide anchored on graphene for boosting kinetic performance and energy storage. *Ceram Int* 48: 3344–3350.

24 Chen, S., Wang, Y., Guo, P. et al. (2023). *Energy Fuel* 37 (1): 36–57. https://doi.org/10.1021/acs.energyfuels.2c03011.

25 Yi, L., Xiao, S., Wei, Y. et al. (2023). Freestanding high-entropy alloy plate for efficient water oxidation catalysis: structure/composition evolution and implication of high valence metals. *Chem Eng J* 469: 144015.

26 Suhadolnik, L., Bele, M., Smiljanić, M. et al. (2024). Enhancing oxygen evolution functionality through anodization and nitridation of compositionally complex alloy. *Mater Today Chem* 35: 101835.

27 Liu, H., Qin, H., Kang, J. et al. (2022). A freestanding nanoporous NiCoFeMoMn high-entropy alloy as an efficient electrocatalyst for rapid water splitting. *Chem Eng J* 435: 134898.

28 Cai, Z.X., Goou, H., Ito, Y. et al. (2021). Nanoporous ultra-high-entropy alloys containing fourteen elements for water splitting electrocatalysis. *Chem Sci* 12 (34): 11306–11315.

29 Jin, Z., Lyu, J., Zhao, Y.L. et al. (2021). Top-down synthesis of noble metal particles on high-entropy oxide supports for electrocatalysis. *Chem Mater* 33 (5): 1771–1780.

30 Strozi, R.B., Leiva, D.R., Huot, J. et al. (2021). Synthesis and hydrogen storage behavior of Mg-V-Al-Cr-Ni high-entropy alloys. *Int J Hydrog Energy* 46 (2): 2351–2361.

31 Yang, P., Shi, Y., Xia, T. et al. (2023). Novel self-supporting thin film electrodes of FeCoNiCrMn high entropy alloy for excellent oxygen evolution reaction. *J Alloys Compd* 938: 168582.

32 Wang, S., Xu, B., Huo, W. et al. (2022). Efficient FeCoNiCuPd thin-film electrocatalyst for alkaline oxygen and hydrogen evolution reactions. *Appl Catal B Environ* 313: 121472.

33 Huang, C.L., Lin, Y.G., Chiang, C.L. et al. (2023). Atomic scale synergistic interactions lead to breakthrough catalysts for electrocatalytic water splitting. *Appl Catal B Environ* 320: 122016.

34 Yu, H., Chen, W., Liu, Z., and Du, C. (2024). Self-supported amorphous high entropy alloy/crystalline Ni_2P heterostructure for water oxidation under alkaline condition. *Colloids Surf A Physicochem Eng Asp* 686: 133354.

35 Bertelsen, A.D., Hansen, A.R., Broge, N.L.N. et al. (2022). Composition space of PtIrPdRhRu high entropy alloy nanoparticles synthesized by solvothermal reactions. *Chem Commun* 58 (91): 12672–12675.

36 Zhan, C., Xu, Y., Bu, L. et al. (2021). Subnanometer high-entropy alloy nanowires enable remarkable hydrogen oxidation catalysis. *Nat Commun* 12 (1): 6261.

37 Kang, Y., Cretu, O., Kikkawa, J. et al. (2023). Mesoporous multimetallic nanospheres with exposed highly entropic alloy sites. *Nat Commun* 14 (1): 4182.

38 Sun, Y., Zhang, W., Zhang, Q. et al. (2023). A general approach to high-entropy metallic nanowire electrocatalysts. *Matter* 6 (1): 193–205.

39 Zhang, H.M., Zhang, S.F., Zuo, L.H. et al. (2024). Recent advances of high-entropy electrocatalysts for water electrolysis by electrodeposition technology: a short review. *Rare Metals* 43 (6): 2371–2390.

40 Bian, H., Wang, R., Zhang, K. et al. (2023). Facile electrodeposition synthesis and super performance of nano-porous Ni-Fe-Cu-Co-W high entropy alloy electrocatalyst. *Surf Coat Technol* 459: 129407.

41 Chang, S.Q., Cheng, C.C., Cheng, P.Y. et al. (2022). Pulse electrodeposited FeCoNiMnW high entropy alloys as efficient and stable bifunctional electrocatalysts for acidic water splitting. *Chem Eng J* 446: 137452.

42 Raj, G., Nandan, R., Kumar, K. et al. (2023). High entropy alloying strategy for accomplishing quintuple-nanoparticles grafted carbon towards exceptional high-performance overall seawater splitting. *Mater Horiz* 10 (11): 5032–5044.

43 Gao, S., Hao, S., Huang, Z. et al. (2020). Synthesis of high-entropy alloy nanoparticles on supports by the fast moving bed pyrolysis. *Nat Commun* 11 (1): 2016.

44 Zhao, J., Wang, Z., Fang, X. et al. (2023). Fast Joule heating synthesis of NiCoFeCrMo high-entropy alloy embedded in graphene for water oxidation. *J Alloys Compd* 966: 171535.

45 Tang, J., Xu, J.L., Ye, Z.G. et al. (2021). Microwave sintered porous CoCrFeNiMo high entropy alloy as an efficient electrocatalyst for alkaline oxygen evolution reaction. *J Mater Sci Technol* 79: 171–177.

46 Maulana, A.L., Chen, P.C., Shi, Z. et al. (2023). Understanding the structural evolution of IrFeCoNiCu high-entropy alloy nanoparticles under the acidic oxygen evolution reaction. *Nano Lett* 23 (14): 6637–6644.

47 Li, T., Tang, R., Lurun, X. et al. (2025). Femtosecond laser ablation of high-entropy ceramics $RE_2(Ce_{0.2}Zr_{0.2}Hf_{0.2}Sn_{0.2}Ti_{0.2})2O_7$ (RE = Sm, Eu, Er) in liquid for colloid synthesis. *J Phys Chem C* 129 (7): 3707–3720. https://doi.org/10.1021/acs.jpcc.4c04851.

48 Cropper, M.D. (2018). Nanostructured high-entropy materials: surface properties and applications. *Appl Surf Sci* 455: 153–159.

49 Chen, J., Ling, Y., Yu, X. et al. (2022). High entropy alloy-based anodes for lithium-ion batteries. *J Alloys Compd* 929: 167344.

50 Wang, H., Chen, L., Pang, H. et al. (2020). MOF-derived electrocatalysts for oxygen reduction, oxygen evolution and hydrogen evolution reactions. *Chem Soc Rev* 49: 1414–1448.

51 Wang, Q. and Astruc, D. (2020). State of the art and prospects in metal–organic frameworks for electrochemical energy applications. *Chem Rev* 120: 1438–1511.

52 Lu, Y., Huang, K., Cao, X. et al. (2022). High-entropy alloy nanoparticles with dual active centers for oxygen evolution reaction. *Adv Funct Mater* 32: 2110645.

53 Nguyen, T.X., Patra, J., Chang, J.K., and Ting, J.M. (2020). High entropy spinel oxide nanoparticles for superior lithiation–delithiation performance. *J Mater Chem A* 8: 18963–18973. https://doi.org/10.1039/d0ta04844e.

54 Wang, D., Jiang, S., Duan, C. et al. (2020). Spinel-structured high entropy oxide $(FeCoNiCrMn)_3O_4$ as anode towards superior lithium storage performance. *J Alloys Compd* 844: 156158. https://doi.org/10.1016/j.jallcom.2020.156158.

55 Xiang, H.Z., Xie, H.X., Chen, Y.X. et al. (2021). Porous spinel-type $(Al_{0.2}CoCrFeMnNi)_{0.58}O_{4-\delta}$ high-entropy oxide as a novel high-performance anode material for lithium-ion batteries. *J Mater Sci* 56: 8127–8142. https://doi.org/10.1007/s10853-021-05805-5.

56 Duan, C., Tian, K., Li, X. et al. (2021). New spinel high-entropy oxides $(FeCoNiCrMnXLi)_3O_4$ (X = Cu, Mg, Zn) as the anode material for lithium-ion batteries. *Ceram Int* 47: 32025–32032. https://doi.org/10.1016/j.ceramint.2021.08.091.

57 Zheng, Y., Wu, X., Lan, X., and Hu, R. (2021). A spinel $(FeNiCrMnMgAl)_3O_4$ high entropy oxide as a cycling stable anode material for Li-ion batteries. *Processes* 10: 49. https://doi.org/10.3390/pr10010049.

58 Liu, C., Bi, J., Xie, L. et al. (2023). High entropy spinel oxides $(CrFeMnNiCo_x)_3O_4$ ($x = 2, 3, 4$) nanoparticles as anode material towards electrochemical properties. *J Energy Storage* 71: 108211. https://doi.org/10.1016/j.est.2023.108211.

59 Liu, C., Bi, J., Xie, L. et al. (2023). Electrochemical properties of four novel high-entropy spinel oxides used as lithium-ion battery anodes synthesized by the glycine-nitrate SCS method. *J Mater Sci* 58: 8005–8021. https://doi.org/10.1007/s10853-023-08498-0.

60 Liu, C., Bi, J., Xie, L. et al. (2023). Preparation and electrochemical properties of two novel high entropy spinel oxides $(MgTiZnNiFe)_3O_4$ and $(CoTiZnNiFe)_3O_4$ by solid-state reaction. *Mater Today Commun* 35: 106315. https://doi.org/10.1016/j.mtcomm.2023.106315.

61 Sun, Z., Zhao, Y., Sun, C. et al. (2022). High entropy spinel-structure oxide for electrochemical application. *Chem Eng J* 431: 133448. https://doi.org/10.1016/j.cej.2021.133448.

62 Xiao, B., Wu, G., Wang, T. et al. (2022). High-entropy oxides as advanced anode materials for long-life lithium-ion batteries. *Nano Energy* 95: 106962. https://doi.org/10.1016/j.nanoen.2022.106962.

63 Yan, J., Wang, D., Zhang, X. et al. (2020). A high-entropy perovskite titanate lithium-ion battery anode. *J Mater Sci* 55: 6942–6951. https://doi.org/10.1007/s10853-020-04482-0.

64 Zhou, P., Che, Z., Liu, J. et al. (2023). High-entropy P2/O3 biphasic cathode materials for wide-temperature rechargeable sodium-ion batteries. *Energy Storage Mater* 57: 618–627. https://doi.org/10.1016/j.ensm.2023.03.007.

65 Zhao, C., Ding, F., Lu, Y. et al. (2020). High-entropy layered oxide cathodes for sodium-ion batteries. *Angew Chem Int Ed* 59: 264–269. https://doi.org/10.1002/anie.201912171.

66 Liu, X., Xing, Y., Xu, K. et al. (2022). Kinetically accelerated lithium storage in high-entropy (LiMgCoNiCuZn)O enabled by oxygen vacancies. *Small* 18: e2200524. https://doi.org/10.1002/smll.202200524.

67 Li, L., Ji, P., Geng, C. et al. (2023). Facile synthesis of high-entropy $(Co_{0.2}Cr_{0.2}Fe_{0.2}Mn_{0.2}Ni_{0.2})_3O_4$ nanopowders and their electrochemical properties as supercapacitor electrode. *J Energy Storage* 73: 109182.

68 Luo, X.F., Patra, J., Chuang, W.T. et al. (2022). Charge–discharge mechanism of high-entropy co-free spinel oxide toward Li+ storage examined using operando quick-scanning X-Ray absorption spectroscopy. *Advanced Science* 9 (21): 2201219.

69 Sharma, S. and Chand, P. (2023). Supercapacitor and electrochemical techniques: a brief review. *Results Chem* 5: 100885.

70 Sharma, S. and Chand, P. (2019). High entropy oxides—a cost-effective catalyst for the growth of high yield carbon nanotubes and their energy application. *ACS Appl Mater Interfaces* 11 (34): https://doi.org/10.1021/acsami.9b08794.

71 Guo, M., Liu, Y., Zhang, F. et al. (2022). Inactive Al^{3+}-doped $La(CoCrFeMnNiAl_x)_{1/(5+x)}O_3$ high-entropy perovskite oxides as high-performance supercapacitor electrodes. *J Adv Ceram* 11 (5): 742–753.

72 Wang, G., Lu, Z., Li, Y. et al. (2021). Electroceramics for high-energy density capacitors: current status and future perspectives. *Chem Rev* 121 (10): 6124–6172.

73 Peng, X., Pu, Y., Du, X. et al. (2021). Tailoring of ferroelectrics in (Na_2O, K_2O)-Nb_2O_5-SiO_2 glass-ceramics via control of the crystallization kinetics. *Chem Eng J* 422 (1): 130027.

74 Pan, H., Li, F., Liu, Y. et al. (2019). Ultrahigh-energy density lead-free dielectric films via polymorphic nanodomain design. *Science* 365: 578–582.

75 Chen, D., Zhu, X., Yang, X. et al. (2023). A review on structure–property relationships in dielectric ceramics using high-entropy compositional strategies. *J Am Ceram Soc* 106: 6602–6616.

76 Pu, Y.P., Zhang, Q.W., Li, R. et al. (2019). Dielectric properties and electrocaloric effect of high-entropy $(Na_{0.2}Bi_{0.2}Ba_{0.2}Sr_{0.2}Ca_{0.2})TiO_3$ ceramic. *Appl Phys Lett* 115 (22): 223901.

77 Shang, Y.C., Pu, Y.P., Zhang, Q.W. et al. (2022). Effect of configuration entropy on dielectric relaxor, ferroelectric properties of high-entropy $(NaBiBa)_x(SrCa)_{(1-3x)}/2TiO_3$ ceramics. *J Mater Sci Mater Electron* 33 (8): 5359–5369.

78 Fang, J.Q., Wang, T., Li, K. et al. (2022). Energy storage properties of Mn-modified $(Na_{0.2}Bi_{0.2}Ca_{0.2}Sr_{0.2}Ba_{0.2})TiO_3$ high-entropy relaxor-ferroelectric ceramics. *Results Phys* 38: 105617.

79 Zhou, S.Y., Pu, Y.P., Zhang, Q.W. et al. (2020). Microstructure and dielectric properties of high entropy $Ba(Zr_{0.2}Ti_{0.2}Sn_{0.2}Hf_{0.2}Me_{0.2})O_3$ perovskite oxides. *Ceram Int* 46 (6): 7430–7437.

80 Liu, J., Ren, K., Ma, C.Y. et al. (2020). Dielectric and energy storage properties of flash-sintered high-entropy $(Bi_{0.2}Na_{0.2}K_{0.2}Ba_{0.2}Ca_{0.2})TiO_3$ ceramic. *Ceram Int* 46 (12): 20576–20581.

81 Yang, W.T. and Zheng, G.P. (2021). High energy storage density and efficiency in nanostructured $(Bi_{0.2}Na_{0.2}K_{0.2}La_{0.2}Sr_{0.2})TiO_3$ high-entropy ceramics. *J Am Ceram Soc* 105 (2): 1083–1094.

82 Yang, W. and Zheng, G. (2022). Tuning the dielectric and energy storage properties of high entropy ceramics $(Bi_{0.2}Na_{0.2}K_{0.2}La_{0.2}Sr_{0.2})(Ti_{1-x}Sc_x)O_3$ by Sc-doping at B-site in perovskite structure. *J Electroceram* 49: 53–62.

83 Sun, W.Y., Zhang, F., Zhang, X. et al. (2022). Enhanced electrical properties of $(Bi_{0.2}Na_{0.2}Ba_{0.2}Ca_{0.2}Sr_{0.2})TiO_3$ high-entropy ceramics prepared by hydrothermal method. *Ceram Int* 48 (13): 19492–19500.

84 Wang, T., Li, Y.J., Zhang, X.F. et al. (2023). Simultaneous excellent energy storage density and efficiency under applied low electric field for high entropy relaxor ferroelectric ceramic. *Mater Res Bull* 157: 112024.

85 Chen, Y.Y., Qi, J.L., Zhang, M.H. et al. (2022). Pyrochlore-based high-entropy ceramics for capacitive energy storage. *J Adv Ceram* 11 (7): 1179–1185.

86 Yan, B., Chen, K.P., and An, L.N. (2023). Design and preparation of lead-free $(Bi_{0.4}Na_{0.2}K_{0.2}Ba_{0.2})TiO_3$–$Sr(Mg_{1/3}Nb_{2/3})O_3$ high-entropy relaxor ceramics for dielectric energy storage. *Chem Eng J* 453: 139921.

87 Zhou, S.Y., Pu, Y.P., Zhao, X.Y. et al. (2022). Dielectric temperature stability and energy storage performance of NBT-based ceramics by introducing high-entropy oxide. *J Am Ceram Soc* 105 (7): 4796–4804.

88 Zhou, S.Y., Pu, Y.P., Zhang, X.Q. et al. (2022). High energy density, temperature stable lead-free ceramics by introducing high entropy perovskite oxide. *Chem Eng J* 427: 131684.

14

Morphology-Dependent High-Entropy Materials for Energy Storage Applications

Pranshula Panigrahi[1,2], *Manoj Kumar Mallick*[2,3], *and Akshaya Kumar Palai*[1]

[1] *Central Institute of Petrochemicals Engineering and Technology, School for Advanced Research in Petrochemicals (SARP), LARPM, Bhubaneswar, Odisha, 754024, India*
[2] *Utkal University, Department of Chemistry, Bhubaneswar, Odisha, 751004, India*
[3] *National Dong-Hwa University, Department of Materials Science and Engineering, Hualien, 974301, Taiwan*

14.1 Introduction

High-entropy materials (HEMs) are made up of five or more elements in near-equimolar ratios, stabilized by configurational entropy [1]. In case of materials such as graphite in lithium-ion battery (LIB) anodes or cobalt oxide in supercapacitors, the whole properties are dominated by one or two elements; however, in HEMs, a unique synergy known as the "cocktail effect" is observed, where multielement interactions provide exceptional properties such as lattice distortion, enhanced thermal and chemical stability, and tunable electrochemical behavior [2]. HEMs came to light after the discovery of high-entropy alloys (HEAs) in 2004, when Cantor et al. [3] demonstrated that alloys such as CoCrFeNiMn could form single-phase structures despite their compositional complexity. Since then, research has expanded to include high-entropy oxides (HEOs) [4], sulfides [5], carbides [6], and ceramics [7], positioning HEMs as promising contenders for advanced energy storage applications, including batteries, dielectric capacitors, and supercapacitors. This can recognize the potential of HEMs to address the limitations of traditional materials, such as limited capacity or poor cycling stability, in meeting modern energy demands [8].

Scheme 14.1 illustrates the central role of morphology in determining the performance of HEMs for energy storage applications. Various synthesis techniques—such as solid-state processing, hydro/solvothermal methods, electrospinning, and sol-gel self-assembly—are employed to engineer distinct morphologies, including nanofibers, hollow structures, thin films, and porous architectures. These morphologies significantly influence key material characteristics such as surface area, ion/electron transport, mechanical stability, lattice accommodation, and electrochemical activity. These enhanced properties, in turn, enable the optimization of HEMs for a range of energy storage devices, including LIBs, sodium-ion batteries

High-Entropy Materials for Energy Storage Devices, First Edition.
Edited by Chien-Te Hsieh, Pradeep Kumar Panda, and Arpan Kumar Nayak.

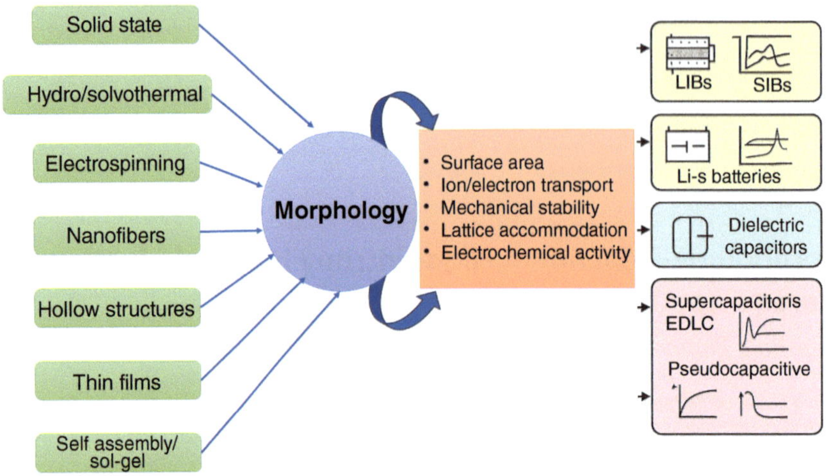

Scheme 14.1 Schematic representation showing the relationship between synthesis techniques, morphological control, and the enhanced properties of HEMs for various energy storage applications.

(SIBs), lithium-sulfur (Li-S) batteries, dielectric capacitors, and supercapacitors, operating via electric double-layer or pseudocapacitive mechanisms. The schematic emphasizes the integrated nature of synthesis, structure, and function, highlighting morphology as a bridge between material design and application performance.

Morphology—the size, shape, and structural arrangement of a material—plays a pivotal role in dictating the performance of HEMs in energy storage systems [9]. For example, nanoparticles offer high surface areas (>100 m^2 g^{-1}, measured by BET) that increase ion adsorption and redox reaction sites, which is critical for supercapacitors achieving capacitances >200 F g^{-1} [10]. Nanosheets, with their two-dimensional geometry, facilitate rapid ion and electron transport by reducing diffusion lengths, enhancing rate capability in batteries to >500 mA h g^{-1} at high C rates [11]. Porous frameworks, such as mesoporous or hollow structures, improve electrolyte accessibility and accommodate volume changes during charge/discharge cycles, boosting mechanical stability in Li-S batteries with sulfur utilization >70%, and it has been found that after electrospinning from liquid precursors, a single-phase $(Y_{0.2}Yb_{0.2}Sm_{0.2}Eu_{0.2}Er_{0.2})_2O_3$ solid solution with bixbyite structure can be formed by calcining at a temperature as low as 500 °C [12, 13]. For example, $(CrMnFeCoNi)_3O_4$ spinel nanoparticle-based supercapacitor showed a capacitance of 239 F g^{-1} and a specific energy of 24.1 Wh kg^{-1} at a current density of 0.5 A g^{-1} with 76% capacitance retention after 1,000 cycles [14]. FeCoNiCrMn HEA with a two-dimensional ultrathin nanosheet was used in LIBs. It showed a high specific capacity (1,026.01 mA h g^{-1} after 120 cycles), high rate performance, and long lifespan at 2 A g^{-1} (>500 cycles) [15]. The porous spinel-phase HEO $(Cr_{0.2}Fe_{0.2}Co_{0.2}Ni_{0.2}Zn_{0.2})3O_4$ used in LIBs delivered 1,022 mA h g^{-1} after 1,000 cycles at 1 A g^{-1} and 220 mA h g^{-1} at an ultrahigh current density of 30 A g^{-1}. The porous nanostructure not only promotes the transport

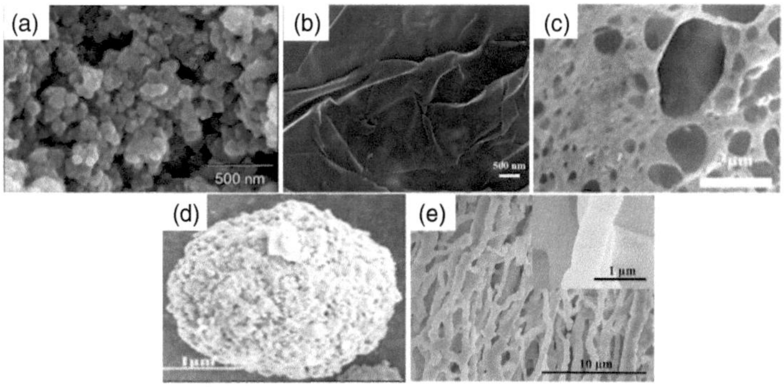

Figure 14.1 HEO nanoparticle (a) *Source:* Reference [13] / with permission of Elsevier, HEM nanosheet (b) *Source:* Reference [14] / with permission of Elsevier, porous HEO (c) *Source:* Reference [15] / American Chemical Society, HEO microsphere (d) *Source:* Reference [16] / American Chemical Society, and HEO nanofibers (e) *Source:* Reference [17] / John Wiley & Sons.

of the electrolyte but also alleviates the volume change of active materials upon cycling [16]. (MgCoNiCuZn)O–rGO hollow sphere was synthesized via a simple hydrothermal process and used in LIBs, in which it showed a capacity increase phenomenon, reaching near 1,400 mA h g^{-1} after 800 cycles at 1.0 A g^{-1} [17]. Rock salt $Mg_{0.2}Co_{0.2}Ni_{0.2}Cu_{0.2}Zn_{0.2}O$ HEO nanofiber used in LIBs delivered a reversible capacity of 365 mA h g^{-1} after 300 cycles at 2 A g^{-1} [18] (Figure 14.1).

In contrast to traditional materials, where morphology is often a secondary consideration, HEMs allow morphology to amplify their inherent multielement advantages, making it a primary design parameter [19]. This interplay between composition and structure unlocks new possibilities for optimizing charge storage efficiency, capacity, and durability, addressing challenges such as low energy density or degradation in conventional systems [20].

This chapter explores the important relationship between morphology and HEM performance in energy storage, focusing on advanced synthesis techniques, characterization methods, fundamental mechanisms, and practical applications. We examine how controlled morphologies enhance LIBs, SIBs, and Li-S batteries by improving capacity and stability, with reported values exceeding 1,000 mA h g^{-1} for LIB anodes and 300 mA h g^{-1} for SIBs [21]. In dielectric capacitors, morphology-engineered HEMs boost dielectric constants ($\varepsilon_r > 1{,}000$) and energy density (>10 J cm^{-3}), while in supercapacitors, they increase specific capacitance and power delivery (>500 F g^{-1} in pseudocapacitive HEOs) [22]. The discussion is based on advanced synthesis approaches, ranging from solid-state milling [1] to wet chemical methods [4] and emerging techniques such as atomic layer deposition (ALD) [23] that allow precise morphology control. Characterization techniques such as scanning electron microscopy (SEM), X-ray diffraction (XRD), and electrochemical impedance spectroscopy (EIS) reveal how structure plays an important role in application [24], while different mechanisms explain the impact of morphology on ion transport, lattice stability, and capacitive behavior [25].

By discussing the relationship between synthesis, structure, and application, this chapter provides comprehensive details on morphology-dependent HEMs, which help meet the increasing demands of portable electronics, electric vehicles, and grid-scale energy storage systems in a sustainable future (IEA, 2024).

14.2 Advanced Synthesis Techniques for Morphology Control

14.2.1 Solid-State Methods

Solid-state synthesis techniques, such as ball milling and spark plasma sintering (SPS), provide robust and scalable pathways to fabricate HEMs with controlled morphologies for energy storage applications. High-energy ball milling mechanically alloys multiple elements, such as Co, Cu, Fe, Ni, and Mn, into nanostructured powders, a method widely adopted since its application to HEAs by Cantor et al [3]. This process involves milling metal oxides in a planetary ball mill with grinding balls (e.g. zirconia balls, 5–10 mm diameter) at speeds of 300–600 rpm for 10–50 hours [1]. Particle size and shape depend on milling duration, balls-to-powder ratio (typically 10 : 1), and energy input, yielding nanoparticles (20–100 nm) or nanorods with aspect ratios of 3–5. To introduce porosity, templating agents such as NaCl (5–20 wt%) are added during milling and later leached out with deionized water, creating interconnected pore networks (5–50 nm pore size) that are ideal for supercapacitor electrodes or battery anodes that can interact better with the electrolyte [2]. This method's simplicity and ability to process large-scale quantities make it appealing to industries, though achieving uniform multielement distribution requires careful optimization to prevent phase segregation [26].

SPS consolidates premilled HEM powders into dense, nanostructured forms under high pressure (50–100 MPa) and temperature (800–1,200 °C). The rapid heating rate (up to 100 $°C\,min^{-1}$) and short sintering times (5–15 minutes) preserve nanoscale features, such as grains (10–50 nm) or hierarchical structures, when combined with templated precursors [27]. For example, SPS of a CoCrFeNiMn alloy powder mixed with carbon nanotubes (1–5 wt%) produces a porous, conductive matrix suitable for dielectric capacitors, with porosity controlled by nanotube burnout [28]. Morphology—whether dense pellets or porous frameworks—depends on sintering parameters and precursor design, offering flexibility for energy storage needs. However, the reliance of SPS on high-cost equipment limits its widespread adoption, a drawback highlighted by Zhang et al. [20] (Figure 14.2).

14.2.2 Wet Chemical Approaches

Wet chemical methods, including hydrothermal and solvothermal synthesis, can synthesize HEMs with precise, tunable morphologies such as nanosheets, hollow spheres, and nanocubes, which is critical for enhancing ion accessibility and charge storage capacity. Hydrothermal synthesis involves reacting metal precursors

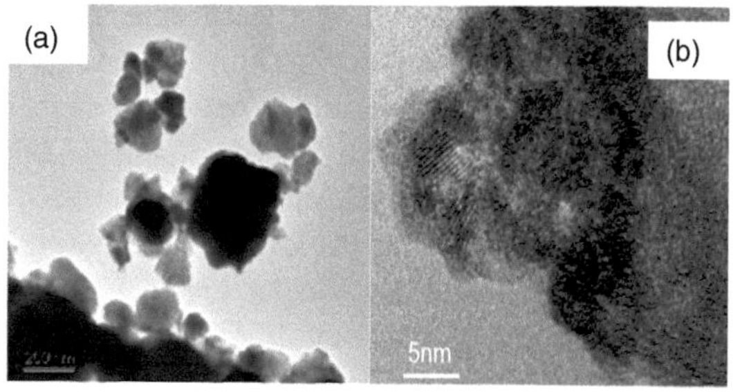

Figure 14.2 TEM of milled nanoparticles (a) *Source:* Reference [28] / with permission of Elsevier and HEO powder (b) *Source:* Reference [29] / Springer Nature / CC BY 4.0.

(e.g. nitrates or chlorides of Co, Cu, Fe, Ni, Mn in equimolar ratios) in an aqueous solution within a sealed autoclave at 150–250 °C and 10–20 bar pressure for 6–24 hours. Surfactants such as cetyltrimethylammonium bromide (CTAB, 0.1–1 mM) help in direct anisotropic growth, forming nanosheets with thicknesses of 5–20 nm and lateral dimensions up to 1 μm [13]. For instance, $CoCuFeNiMnO_x$ HEOs synthesized hydrothermally exhibit high surface areas (50–150 $m^2\ g^{-1}$, measured by BET), which is ideal for LIB anodes where capacity exceeds 1,000 $mA\,h\,g^{-1}$ (Chen et al., 2021). By adjusting pH (8–12), temperature, or precursor concentration, morphology can be fine-tuned; higher pH favors thinner, more uniform sheets due to altered nucleation kinetics, as noted by Bao et al. [18]. This method is widely accepted for its ability to control reaction conditions, though it requires postprocessing steps such as washing and drying at 80 °C to remove residual surfactant [29].

Solvothermal synthesis extends this approach by using organic solvents such as ethylene glycol or dimethylformamide (DMF) under similar conditions. The solvent's polarity and viscosity affect reaction kinetics, yielding hollow spheres (diameters 50–200 nm) or nanocubes (edge lengths 20–100 nm) over 6–24 hours [5]. For example, a solvothermal HEO with a hollow structure enhances electrolyte penetration in supercapacitors, achieving specific capacitances >300 $F\,g^{-1}$ due to increased active sites [10]. The organic medium slows down nucleation, allowing multielement precursors to self-assemble into ordered shapes, though solvent cost and disposal pose scalability challenges [8] (Figure 14.3).

14.2.3 Electrospinning and Deposition

Electrospinning and electrodeposition offer innovative routes to synthesize HEMs with fibrous or thin-film morphologies, enhancing conductivity and surface area. Electrospinning extrudes a polymer–metal precursor solution—such as equimolar Co, Ni, Fe, Cu, and Mn acetates dissolved in polyvinylidene fluoride

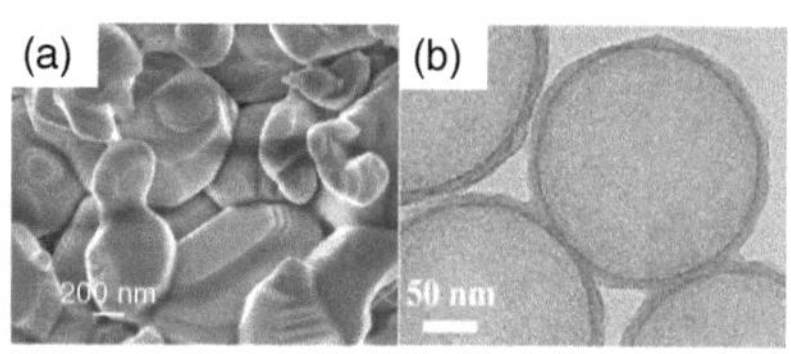

Figure 14.3 SEM of hydrothermal nanosheets (a) *Source:* Reference [30] / with permission of Elsevier and TEM of solvothermal hollow spheres (b). *Source:* Reference [31] / The Royal Society of Chemistry.

(PVDF, 10–15 wt%)—through a needle under a high-voltage field (10–20 kV) at a flow rate of 0.5–2 ml h^{-1}). The thin space of 16.3 was created on glassy carbon electrodes by electrically alloying each noble metal layer at 1,000 °C for five seconds after a carefully regulated sequential ALD technique [32]. The resulting nanofibers (50–500 nm diameter) are calcined at 400–600 °C for two to four hours to remove the polymer and crystallize the HEM. Fiber morphology depends on solution viscosity, voltage, and needle-to-collector distance (10–20 cm), producing high-aspect-ratio structures (>100 : 1) that improve electron transport in supercapacitor electrodes [20]. An HEA–carbon composite nanofiber delivers >200 F g^{-1} due to its interconnected network, though precursor stability needs to be optimized [33].

Electrodeposition deposits HEM films or porous networks onto substrates (e.g. Ni foam) from a bath of metal salts (e.g. $CoSO_4$, $NiSO_4$, 0.05–0.1 M each) at 1–5 V for 10–60 minutes [34]. Current density (1–10 mA cm^{-2}) and time control film thickness (100 nm^{-1} μm) and porosity, with higher currents yielding dendritic structures. A CoCrFeNiMn film on Ni foam enhances battery rate performance (>500 mA h g^{-1} at 5 °C) by providing short ion diffusion paths [22]. In order to achieve the anticipated electrochemical performance, this review highlights the significance of adjusting the composition, crystal structure, various phase forms, thermodynamic and kinetic characteristics, and surface morphology of HEAs and their derivatives. An approach for identifying the best electrode materials for particular electrochemical processes in an industrial setting is provided by theoretical explanations and synthesis routes for fabricating possible HEA electrode materials [35] (Figure 14.4).

14.2.4 Templating and Self-Assembly

Wang et al. (2020) presented a sol-gel synthesis of spherical mesoporous HEOs (e.g. Ni–Co–Cr–Fe–Mn oxide) with high specific surface areas (42–143 m^2 g^{-1}) and large pore sizes (5.5–8.3 nm). This method employs polyphenol as a polymerizable ligand to form metal/polyphenol-formaldehyde resin colloidal spheres, which, upon calcination, yield mesoporous HEOs. These materials exhibit potential for applications such as catalysis and DNA detection [36]. Soft templating with polymers (e.g. Pluronic F127, 0.5–2 wt%) forms micelles or layers, as in lamellar $CoCuFeNiMnO_x$ for supercapacitors (>50% pseudo capacitance) [13]. Template removal adds cost and complexity.

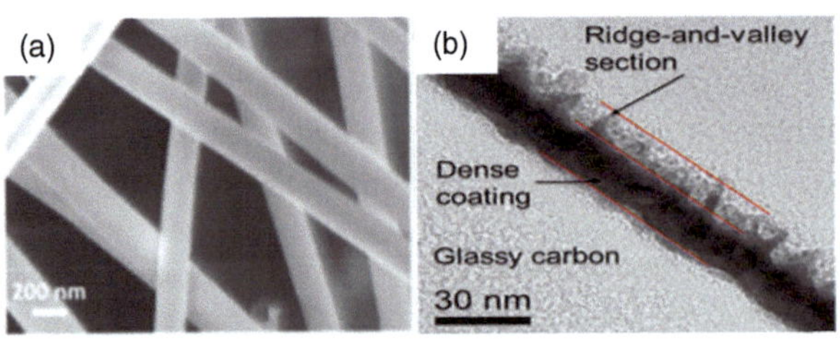

Figure 14.4 SEM of electrospun nanofibers (a) *Source:* Reference [35] / with permission of Elsevier and TEM of ALD thin film (b) *Source:* Reference [31] / The Royal Society of Chemistry / CC BY 3.0.

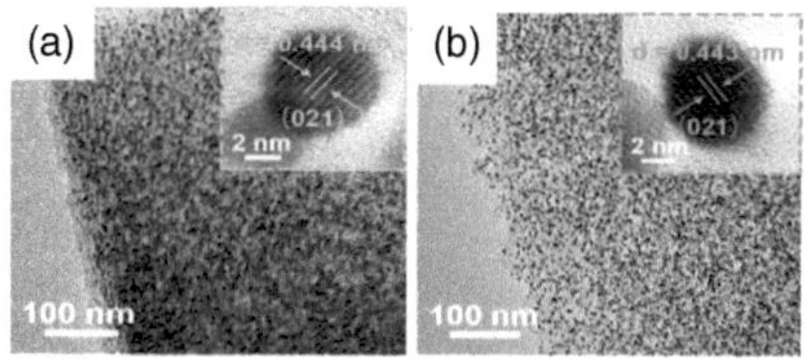

Figure 14.5 HEO via self-sacrificing method SEM (a) and TEM (b). *Source:* Zheng et al. [38] / John Wiley & Sons / CC BY 4.0.

Self-assembly spontaneously organizes HEMs into nanosheets or nanoparticles via slow cooling (100 to 25 °C over 12 hours) in ethanol, exploiting entropy stabilization [37]. Morphology depends on precursor ratios and solvent, though scaling is sensitive to conditions [14]. Using a self-sacrificing template approach and rapid Joule heating, a number of two-dimensional HEOs with a lot of active sites are created. With a low overpotential (199 mV@10 mA cm^{-1}, 266 mV@100 mA cm^{-2}), a small Tafel slope (40 mV dec^{-1}), and exceptional long-term stability (functioning at 500 mA cm^{-2} for 100 hours without appreciable decay), HEO $(FeCoNiMoRu)3O_4$ stands out among them in terms of oxygen evolution reaction (OER) performance. The nanosheet structure's huge active surface area, shorter ion transport pathway, entropy stabilization mechanism, and multielement synergism are all responsible for the flawless performance of $(FeCoNiMoRu)3O_4$, which is shown in Figure 14.5 [38].

14.2.5 Emerging Techniques

Laser ablation pulverizes an HEM target (e.g. CoCrFeNiMn) with pulses (10–100 mJ, 10 Hz) in water, producing nanoparticles (<50 nm) tuned by pulse energy [8]. ALD produces ultrathin HEM coatings (1–10 nm) using precursor pulses (e.g. metal organics), which are ideal for thin-film capacitors [23]. These methods offer precision but are limited by cost and equipment [18] (Table 14.1).

14.3 Characterization of Morphology and Properties

14.3.1 Morphological Analysis

Understanding the morphology of (HEMs) is fundamental to correlating their structure with performance in energy storage applications, and advanced imaging techniques provide critical insights into their size, shape, and structural features. SEM visualizes surface morphologies such as nanoparticles, nanorods, or and networks, offering resolutions down to 1 nm under optimal conditions [39]. For example, SEM imaging of $CoCuFeNiMnO_x$ HEOs synthesized hydrothermally reveals nanosheet arrays with thicknesses of 5–20 nm and pore diameters of 5–50 nm in templated structures, which are directly related to increased surface area (>100 m^2 g^{-1}) for supercapacitors [13]. Transmission electron microscopy (TEM) complements SEM by resolving internal structures at atomic scales (<0.2 nm resolution), such as lattice fringes (~0.25 nm) in HEO nanoparticles or the hollow interiors

Table 14.1 Preparation of morphology-controlled HEMs.

Method	Morphology	Key preparation parameter	Example material	References
Electrospinning	Nanofibers	Voltage, flow rate, polymer concentration	Ceramic nanofibers	[38]
High-energy ball milling + SPS	HEA cermets	Milling speed (300–400 rpm), SPS sintering temperature (800–1,200 °C), heating rate	Ti(C, N)-(HfZrTaNbTi) C5-HEA	[27]
Microwave-assisted hydrothermal synthesis	Coated cathode	Microwave heating time (10–20 minutes), hydrothermal temperature (~180 °C)	Li-S battery cathode	[30]
Microwave solvothermal strategy	Hollow carbon spheres	Microwave heating duration (15–30 minutes), precursor concentration	HEO/carbon composite	[31]
Sol-gel synthesis	Spherical mesoporous HEOs	Calcination temperature (500–800 °C), use of polyphenol ligand	Mesoporous HEOs	[36]
AI-driven synthesis	Varies (AI-guided)	AI-driven optimization of synthesis steps (precursor ratio, reaction time, temperature)	HEMs	[37]
Self-sacrificing template	2D HEOs	Sacrificial template degradation time, chemical bath conditions	2D HEOs	[38]
High-energy ball milling + SPS	HEA	Ball milling time and speed, sintering time and temperature	AlCrCuFeNi HEA	[28]
Solvothermal coupling	Hollow carbon spheres + HEO	Solvent choice, solvothermal time, microwave-assisted solvothermal reaction	HEO for OER	[31]
Microwave hydrothermal	Nanostructured HEO	Precise microwave power and heating time	HEOs	[29]
ALD	Noble metal HEA nanofilm	Deposition time, substrate temperature, subsurface amorphization	Hydrogen evolution catalyst	[31, 32]
Sol-gel method	Spherical mesoporous HEOs	Precursor concentration, calcination temperature, and ligand-to-metal ratio	Mesoporous spherical HEOs	[36]

Electrodeposition	Metal-selective nanostructures	Electrolyte concentration, electrode potential, electrolyte interface tuning	Cobalt and nickel electrodeposition	[35]
Ball milling + SPS	HEAs	Ball milling duration, sintering rate, alloy composition	Ti20-Al16-V16-Fe16-Ni16-Cr16 HEA	[28]
ALD	High-entropy metal nanofilm	Layer thickness, amorphization layer, ALD-controlled deposition	Noble metal HEA	[31, 32]
Joule heating synthesis	Porous carbon/HEA composite	Joule heating duration and current, porous carbon fiber structure	Ultralight electromagnetic wave absorber	[34]
Solvothermal method	2D oxides	Sacrificial template oxidation and degradation during solvothermal synthesis	2D HEOs	[38]

of solvothermally prepared spheres [40]. High-resolution TEM (HR-TEM) confirms crystallinity and elemental distribution via energy-dispersive X-ray spectroscopy (EDS), showing uniform multielement dispersion critical for HEM stability [11].

Atomic force microscopy (AFM) provides three-dimensional surface profiling, measuring nanoscale roughness and thickness with subnanometer precision [29]. For 2D HEMs such as nanosheet-based HEOs, AFM quantifies thicknesses (5–15 nm) and surface roughness (Ra ~1–3 nm), which influence ion accessibility and charge transfer in battery electrodes [22]. For instance, smoother surfaces in electrodeposited CoCrFeNiMn films correlate with lower impedance in LIBs, achieving >800 mA h g^{-1} [25]. These techniques collectively quantify how synthesis methods—e.g. hydrothermal vs. templating—yield distinct morphologies, with SEM/TEM showing shape and porosity and AFM linking surface features to electrochemical performance [8]. Their combined use ensures a comprehensive morphological analysis, which is essential for optimizing HEMs in energy storage systems. Colombo et al. [29] used SEM to investigate the surface coating of HEM-based double-coated cathodes synthesized via microwave-assisted hydrothermal synthesis. SEM confirmed a homogeneous distribution of nanoscale coatings, which is critical for enhancing the stability and cyclability of Li-S batteries [41]. Li et al. [41] used TEM to characterize layered (oxy)hydroxide nanosheets produced via a self-reliant electrolytic method. TEM images confirmed nanosheet thickness, defect sites, and crystallinity, which are key factors for enhancing electrocatalytic water oxidation efficiency [42]. Khan et al. [42] used AFM to investigate the microstructure and nanoscale roughness of $AlCoCrCu_{0.5}FeNi$ HEA thin films. AFM measurements helped correlate surface roughness with microstructure evolution and phase transformation, which directly affect corrosion resistance, wear resistance, and conductivity [43] (Figure 14.6).

14.3.2 Structural and Compositional Insights

Structural and compositional characterization of HEMs elucidates how morphology is associated with their multielement nature, influencing energy storage properties. XRD analyzes phase purity and crystallite size, which are critical for confirming the entropy-stabilized single-phase structures of HEMs [44]. For example, XRD of $CoCuFeNiMnO_x$ HEOs shows a single rock salt phase with peaks at 2θ ~43° (200 plane), and the Scherrer equation ($D = K\lambda/\beta\cos\theta$) estimates crystallite sizes of 10–50 nm, reflecting the impact of morphology on crystallinity [45]. Smaller crystallites in nanoparticles enhance surface area, facilitating better electrochemical reactions, while larger crystallites in dense SPS-processed HEMs contribute to structural stability, leading to improved cycling performance. For instance, a study on high-entropy spinel oxide nanoparticles demonstrated a capacity retention of approximately 90% after 200 cycles at a charge/discharge rate of 500 mA g^{-1}. This stability is attributed to entropy stabilization effects that maintain the crystalline oxide framework during lithiation and promote the recovery of the spinel phase during delithiation [46]. Additionally, research on Si-based HEA anodes revealed that incorporating HEA elements decreases anisotropic stress and

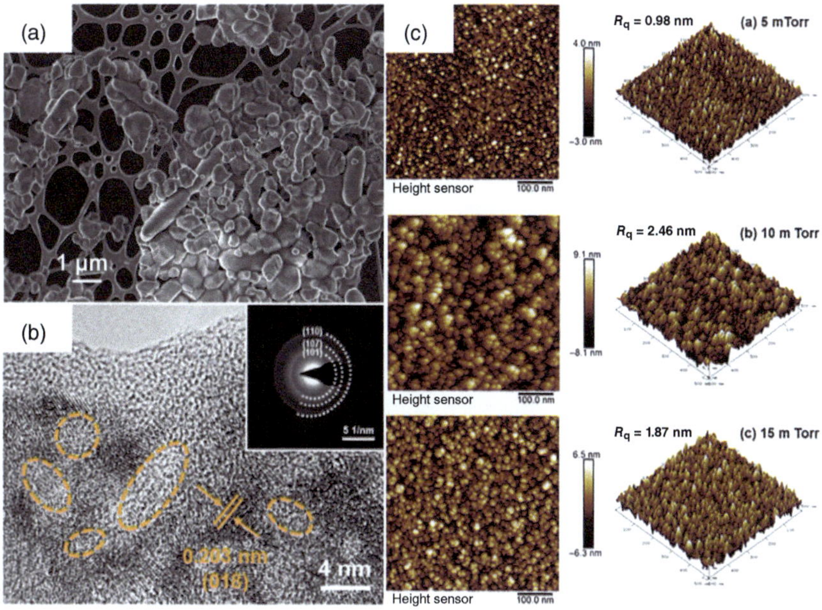

Figure 14.6 SEM of nanosheets (a) *Source:* Reference [29] / Springer Nature / CC BY 4.0, TEM of lattice fringes (b) *Source:* Reference [13] / with permission of Elsevier, and AFM for height profile (c) *Source:* Reference [42] / with permission of Elsevier.

enhances ductility, resulting in a capacity retention of 93.5% after 100 cycles. This improved stability is due to the reversible lithiation and delithiation processes between the HEA phases, which maintain structural integrity during cycling [47]. *In situ* XRD during cycling reveals lattice expansion (<5%) in porous HEMs, linking morphology to mechanical resilience, and *in situ* neutron diffraction studies on carbon-strengthened CoCrFeMnNi HEAs have demonstrated that lattice strain evolution during tensile deformation is correlated with mechanical properties, highlighting the role of lattice distortions in mechanical resilience [13].

X-ray photoelectron spectroscopy (XPS) probes surface chemistry and oxidation states, offering insights into how morphology affects electrochemical activity. For HEOs, XPS detects shifts in binding energies—e.g. Mn 2p at 641-643 eV indicating Mn^{2+}/Mn^{3+}—revealing redox-active sites enhanced by high-surface-area morphologies such as nanosheets. The detection of these shifts is crucial, as the presence of mixed oxidation states, such as Mn^{2+}/Mn^{3+}, reveals redox-active sites that are essential for electrochemical processes. High-surface-area morphologies such as nanosheets can enhance these redox-active sites by providing more exposed surface area, thereby improving electrochemical activity. This enhancement is particularly significant in applications such as batteries and catalysts, where surface reactions are critical. Furthermore, a study on quinary layered double hydroxide-derived HEOs demonstrated that calcination temperature significantly affects morphology, transitioning from nanosheets to nanospheres. This morphological change is correlated with variations in surface area and pore volume, which in turn influence the

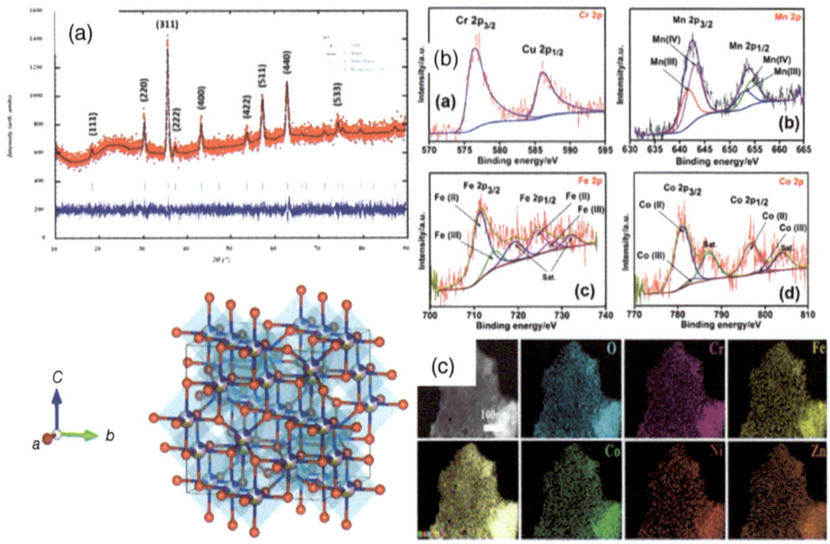

Figure 14.7 XRD pattern of HEO (a) *Source:* Adapted from Ref. [13], XPS of Mn 2p, Cr 2p, Fe 2p, and Co 2p spectra (b) *Source:* Adapted from Ref. [13], and EDS elemental map (c) *Source:* Reference [15] / American Chemical Society.

electrochemical performance of the material [48]. Depth profiling via Ar^+ sputtering shows uniform elemental distribution in electrodeposited films, contrasting with surface enrichment in milled nanoparticles, affecting battery performance [49]. XPS also quantifies oxygen vacancies, which increase with porosity and boost pseudocapacitance in supercapacitors (>300 F g^{-1}) [50]. These techniques—XRD for bulk structure and XPS for surface composition—together demonstrate how morphology modulates HEM properties, with porous structures enhancing ion storage and thin films optimizing charge transfer (Figure 14.7).

14.3.3 Electrochemical Characterization

A study synthesized $(NiMnCrCoFe)_3O_4$ HEOs with a bimodal particle size distribution through a hydrothermal method followed by annealing. This morphology led to an initial discharge capacity of 527 mA h g^{-1} at 5 A g^{-1} and a 99.8% capacity retention after 300 cycles, outperforming uniform nanoscale or microscale particles [51]. Nanocrystalline $CoNiFeCrMnO_x$ HEO thin films, prepared via sol-gel dip-coating and annealed at 400 °C, demonstrated high OER activity. The nanostructured morphology, rich in defective oxygen sites, contributed to an overpotential of 258 mV at 10 mA cm^{-2} and a low onset potential of +1.33 V vs. RHE [52]. The oxygen content in $(AlCrTaTiZr)O_x$ thin films significantly affected their resistivity. Films with higher oxygen concentrations exhibited increased resistivity, highlighting the role of oxygen vacancies in tuning electrical properties [53].

EIS is a pivotal technique for evaluating the electrochemical performance of HEMs, particularly in relation to their morphology. Nyquist plots derived from

EIS measurements typically display a semicircle at high frequencies, representing charge transfer resistance (R_{ct}) and a linear tail at low frequencies, indicative of Warburg impedance (Z_w), which is correlated with ion diffusion processes. An investigation into Ti-doped CoCrFeMoNi HEAs revealed that doping influences the microstructure and electrochemical behavior. EIS measurements demonstrated that Ti doping led to a refined microstructure, which in turn reduced the charge transfer resistance, enhancing the alloy's corrosion resistance [54]. Research on the corrosion behavior of $CoCrFeAlNi_{2.1}$ HEAs in various environmental conditions showed that factors such as chloride and sulfate ion concentrations, as well as pH levels, significantly affect the surface morphology and, consequently, the electrochemical impedance characteristics. Variations in these environmental factors led to changes in the oxide layer's integrity, affecting both R_{ct} and Z_w [55]. Investigations into high-entropy garnet electrolytes indicated that phase composition and local structural disorder substantially affect ionic conductivity. EIS spectra revealed that certain compositions exhibited superior Li^+ conductivity, emphasizing the importance of tailored phase structures in optimizing electrochemical performance. The corrosion behavior of two HEMGs, FeCoNiCrB and FeCoNiCr(BSi), was examined in this study. The total number of metalloid atoms was maintained at 20% in both situations. EIS and linear polarization resistance (LPR) measurements in a 3 wt% NaCl solution were used to examine the electrochemical behavior of these alloys [17].

14.4 Mechanisms Governing Morphology-Dependent Performance

14.4.1 Surface Area and Active Sites

The performance of HEMs in energy storage is affected by their surface area and the availability of active sites, both of which are controlled by morphology. High-surface-area morphologies, such as mesoporous structures or nanoparticles, significantly enhance ion adsorption and redox reaction sites, which are key to boosting capacitance and capacity. Understanding is crucial regarding the mechanisms by which morphology plays a key role in the performance enhancement of these materials for applications such as supercapacitors and batteries [56].

Nanostructured morphologies, such as mesoporous frameworks and nanosheets, provide a high specific surface area that enhances ion storage capacity and increases the density of electroactive sites, which is critical in energy storage applications such as supercapacitors and batteries. A study on ZnCoNiFeMg high-entropy layered hydroxides reported a surface area of $380.5\,m^2\,g^{-1}$, leading to an increased number of active sites and improved electrocatalytic performance [49]. This is attributable to the increased accessibility of electrolyte ions to the material's surface, a phenomenon well documented in porous materials. Mesoporous HEOs exhibit surface areas $>100\,m^2\,g^{-1}$ (measured by BET), which directly enhances the ion-accessible surface area, improving specific capacitance ($>300\,F\,g^{-1}$) due to higher electroactive site density [57].

14.4.2 Ion and Electron Transport

Morphology governs ion and electron transport in HEMs, directly affecting rate capability and power delivery in energy storage devices. One-dimensional (1D) nanowires and two-dimensional (2D) nanosheets reduce diffusion lengths, enhancing ion transport as described by the diffusion coefficient equation, $D = L^2/t$, where L is the diffusion length and t is time. For example, electrospun HEO nanowires (diameter 50–200 nm) exhibit diffusion coefficients $>10^{-10}$ cm^2 s^{-1}, enabling rapid Li^+ transport in LIBs and delivering >365 mA h g^{-1} at 5 °C. Similarly, hydrothermal HEO nanosheets (thickness 5–20 nm) shorten electron pathways along their planar surfaces, improving conductivity and achieving high-rate performance in SIBs (>300 mA h g^{-1} at 10 °C) [58]. Hierarchically porous HEOs facilitate rapid ion diffusion by reducing diffusion path lengths, resulting in high-rate performance (300 mA h g^{-1} at 5C) and lower charge transfer resistance.

3D hierarchical structures, such as porous networks, further enhance transport by providing interconnected channels for electrolyte penetration, reducing tortuosity. EIS measurements show lower Warburg impedance (Z_w ~0.1–0.5 Ω s$^{-1/2}$) in 1D and 2D HEMs compared to bulk counterparts (~1–2 Ω s$^{-1/2}$), reflecting faster ion diffusion. Porous nanoparticles (2–50 nm pore size) increase ion-accessible surface area, enhancing specific capacitance and minimizing Warburg diffusion impedance [59].

EIS measurements have demonstrated that 1D and 2D HEMs exhibit lower Warburg impedance ($Z_w \approx$ 0.1–0.5 Ω s$^{-1/2}$) compared to their bulk counterparts (approximately 1–2 Ω s$^{-1/2}$), indicating faster ion diffusion. For instance, a study on high-entropy metal oxide $(NiMnCrCoFe)_3O_4$ anode materials with controlled morphology reported that nanostructured forms facilitated enhanced lithium-ion transport, resulting in superior electrochemical performance [60]. This transport enhancement, driven by morphology, leverages HEMs' multielement synergy, enabling them to outperform traditional materials such as metal oxides with slower kinetics. Morphology thus acts as a critical lever for optimizing charge dynamics in energy storage systems.

14.4.3 Lattice Effects

Lattice effects in HEMs, stemming from their multielement composition and entropy stabilization, interact with morphology to enhance structural stability during energy storage operation. Lattice distortion—caused by varying atomic sizes and electronic interactions—creates a robust framework that resists degradation, a feature amplified by morphology. For instance, porous HEOs exhibit lattice expansion <5% during Li^+ intercalation, as confirmed by *in situ* XRD, due to their ability to distribute strain across multiple elements [61]. This stability enables >95% capacity retention over 1,000 cycles in LIBs, far surpassing conventional single-element oxides. The superior electrochemical performance of HEOs as anode materials in LIBs emphasizes their high capacity retention over extended cycling, which surpasses that of conventional single-element oxides [62].

Morphology enhances these effects by tailoring stress accommodation. Hollow or mesoporous HEMs buffer volume changes in Li-S batteries, trapping polysulfides and maintaining structural integrity (>70% sulfur utilization after 200 cycles). Hollow-structured HEOs effectively buffer volume changes during cycling in Li-S batteries, trap polysulfides, and maintain structural integrity, leading to high sulfur utilization and prolonged cycle life [63]. The synergy of lattice distortion and porosity reduces mechanical failure, a common issue in traditional materials such as sulfur cathodes. In supercapacitors, nanosheet HEOs leverage lattice flexibility to sustain repeated ion adsorption/desorption, supporting long-term stability (>10,000 cycles) [64]. This interplay between lattice effects and morphology underscores the potential of HEMs for durable energy storage applications.

14.4.4 Dielectric and Capacitive Behavior

Morphology influences dielectric and capacitive behavior in HEMs, which is critical for dielectric capacitors and supercapacitors. In dielectric capacitors, uniform nanoparticulate high-entropy ceramics (e.g. $Ba(Zr,Ti,Sn,Hf)O_3$) enhance the dielectric constant (ε_r > 1,000) by optimizing electric field distribution and minimizing defects [65]. Smaller particle sizes (10–50 nm) increase polarization, achieving energy densities >10 J cm^{-3} with low dielectric loss (tan $\delta < 0.01$), as measured by EIS [66]. This morphology-driven enhancement outperforms traditional ceramics such as $BaTiO_3$ (~5 J cm^{-3}) [14].

In supercapacitors, layered or nanosheet HEOs boost pseudocapacitance by exposing redox-active sites, contributing >50% of total capacitance (e.g. >500 F g^{-1} in $NiFeCoMnCuO_x$) [15]. Cyclic voltammetry (CV) analysis reveals broad redox peaks (0.2–0.6 V), reflecting fast faradaic reactions enabled by high surface area and short diffusion paths. Porous morphologies further enhance electric double-layer capacitance (EDLC) in HEA–carbon composites (>200 F g^{-1}), which is driven by ion accumulation in 2–5 nm pores [16]. Morphology thus tailors the dielectric and capacitive properties of HEMs, optimizing energy and power performance

14.5 Potential Energy Storage Applications

14.5.1 LIBs

LIBs demand high-capacity, stable anodes, and morphology-tuned HEMs outperform traditional graphite (~372 mA h g^{-1}) by leveraging nanoparticles, nanosheets, and porous structures. Yang et al. [16] synthesized porous spinel-phase $(Cr_{0.2}Fe_{0.2}Co_{0.2}Ni_{0.2}Zn_{0.2})_3O_4$ via low-temperature methods, achieving 1,022 mA h g^{-1} after 1,000 cycles at 1 A g^{-1} and 220 mA h g^{-1} at 30 A g^{-1}. The porous nanostructure enhances electrolyte transport and mitigates volume changes, improving cycling stability [16]. Li et al. [15] reported FeCoNiCrMn nanosheets, delivering 1,026.01 mA h g^{-1} after 120 cycles at 2 A g^{-1}, with 2D geometry reducing diffusion lengths for superior rate performance (>500 cycles) [15]. Gao et al. [17] developed hollow (MgCoNiCuZn)O-rGO microspheres via hydrothermal synthesis, achieving ~1,400 mA h g^{-1} after 800 cycles at 1 A g^{-1} due to increased active sites and resilience [17]. Su et al. [18] used electrospun rock salt $Mg_{0.2}Co_{0.2}Ni_{0.2}Cu_{0.2}Zn_{0.2}O$ nanofibers, yielding 365 mA h g^{-1} after 300 cycles at 2 A g^{-1}, with high-aspect-ratio fibers boosting electron transport [18]. These morphologies address the limitations of LIBs in capacity and durability, supporting applications in electric vehicles and portable devices [7].

14.5.2 SIBs

SIBs, which are viable for grid storage due to the abundance of sodium, benefit from HEMs that enhance Na^+ diffusion and stability. Gao et al. explored high-entropy layered cathodes (e.g. ZnCoNiFeMg), achieving >300 mA h g^{-1} at 10C. Nanosheet morphology reduces diffusion lengths, with high surface areas (~380.5 m^2 g^{-1}) improving ion accessibility [5]. Bao et al. observed that porous HEOs facilitate electrolyte penetration, maintaining structural integrity during cycling, a critical advantage over conventional $NaFeO_2$ [20]. These morphology-driven enhancements position HEMs as high-capacity, durable alternatives for sustainable SIBs, addressing energy density challenges [7].

14.5.3 Li-S Batteries

Li-S batteries offer a theoretical energy density of ~2,600 Wh kg^{-1}, but polysulfide shuttling and volume expansion limit their practical use [23]. Morphology-engineered HEMs mitigate these issues. Gao et al. reported hollow (MgCoNiCuZn)O spheres that trap polysulfides and buffer volume changes, achieving >70% sulfur utilization after 200 cycles [17]. Wang et al. [38] synthesized porous $(Y_{0.2}Yb_{0.2}Sm_{0.2}Eu_{0.2}Er_{0.2})_2O_3$ nanofibers via electrospinning, enhancing electrolyte access and volumetric capacity [12]. These structures outperform traditional sulfur cathodes by improving stability and sulfur utilization, making HEMs promising for high-energy Li-S systems [23].

14.5.4 Dielectric Capacitors

Dielectric capacitors require high energy density and dielectric constants, where HEMs excel through uniform morphology. Peng et al. synthesized nanoparticulate $Ba(Zr,Ti,Sn,Hf)O_3$ via advanced methods, achieving $\varepsilon_r > 1{,}000$ and >10 J cm^{-3} with low loss ($\tan \delta < 0.01$). Small particle sizes (10–50 nm) optimize polarization, surpassing $BaTiO_3$ (~5 J cm^{-3}) [29]. Xiang et al. highlighted the potential of high-entropy ceramics, observing that morphology control via ALD enhances dielectric properties [39]. These HEMs suit pulsed power and microelectronics, leveraging precise synthesis for superior performance [23].

14.5.5 Supercapacitors

Supercapacitors, which are valued for power density, utilize HEMs in EDLC and pseudocapacitive modes, with morphology dictating efficiency. Xu et al. [13] developed porous CoCrFeNiMn–carbon nanofiber composites via electrospinning, achieving >200 F g^{-1} due to high surface areas (>100 m^2 g^{-1}) for ion

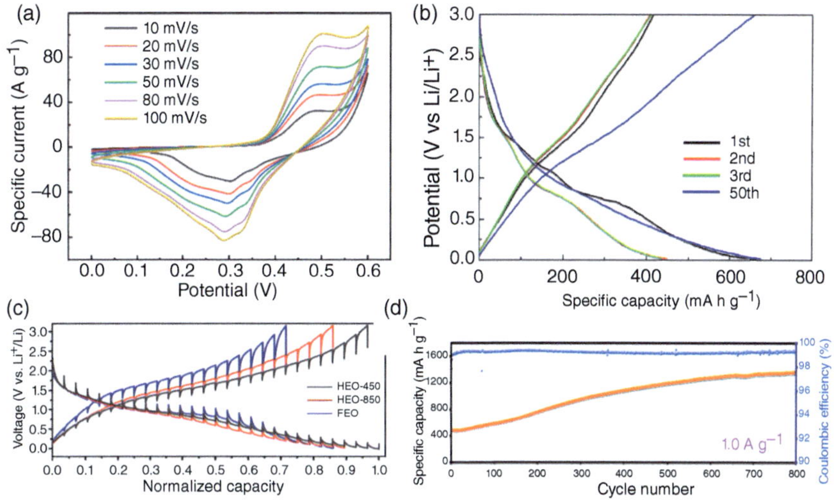

Figure 14.8 CV curve at different scan rates of HEO (a) *Source:* Reference [66] / with permission of American Chemical Society, galvanostatic charge/discharge curves of HEA in specific cycles (b) *Source:* Reference [14] / with permission of Elsevier, charge/discharge capacity curves of HEO during GITT measurements (c) *Source:* Reference [15] / American Chemical Society, and charge/discharge cycle curves of HEO–rGO composite at 1.0 A g^{-1} (d) *Source:* Reference [16] / American Chemical Society.

accumulation [13]. Talluri et al. [14] synthesized $(CrMnFeCoNi)_3O_4$ nanoparticles, delivering 239 F g^{-1} and 24.1 Wh kg^{-1} at 0.5 A g^{-1} with 76% retention after 1,000 cycles, leveraging redox-active sites [14]. Yin et al. [41] reported nanosheet $NiFeCoMnCuO_x$, exceeding 500 F g^{-1} due to enhanced pseudocapacitance from exposed sites [41]. These morphologies outperform activated carbon (~100 F g^{-1}), enhancing charge storage and stability [10].

14.5.6 Synthesis and Characterization Insights

Advanced synthesis—solid-state milling [1], hydrothermal technique [10], electrospinning [12], and ALD [32]—tailors HEM morphologies. SEM [40] and TEM [29] reveal nanoparticle size (20–100 nm) and porosity (5–50 nm), while XRD [43] confirms single-phase stability (e.g. rock salt peaks at 2θ ~43°) [18]. EIS [46] shows reduced charge transfer resistance in nanosheets, linking morphology to electrochemical performance [15]. These tools validate how structure drives function across applications [7] (Figure 14.8, Table 14.2).

14.6 Challenges and Future Perspectives

14.6.1 Synthesis Challenges

Synthesizing HEMs with controlled morphologies faces significant hurdles, particularly in achieving uniformity and scalability. The multielement composition of

Table 14.2 Applications of HEMs in energy storage.

Application	Morphology	Material	Performance metric	Advantage	References
LIBs	Porous nanoparticles	$(Cr_{0.2}Fe_{0.2}Co_{0.2}Ni_{0.2}Zn_{0.2})_3O_4$	1,022 mA h g^{-1} after 1,000 cycles at 1 A g^{-1}	Enhanced electrolyte transport, stability	[16]
	Nanosheets	FeCoNiCrMn	1,026.01 mA h g^{-1} after 120 cycles at 2 A g^{-1}	Rapid ion/electron transport	[15]
	Hollow spheres	(MgCoNiCuZn)O–rGO	~1,400 mA h g^{-1} after 800 cycles at 1 A g^{-1}	Increased active sites, resilience	[17]
	Nanofibers	$Mg_{0.2}Co_{0.2}Ni_{0.2}Cu_{0.2}Zn_{0.2}O$	365 mA h g^{-1} after 300 cycles at 2 A g^{-1}	High-aspect-ratio electron transport	[18]
SIBs	Nanosheets	ZnCoNiFeMg	>300 mA h g^{-1} at 10C	Fast Na^+ diffusion, high surface area	[5]
Li-S batteries	Hollow structures	(MgCoNiCuZn)O	>70% sulfur utilization after 200 cycles	Polysulfide trapping, volume buffering	[17]
	Porous nanofibers	$(Y_{0.2}Yb_{0.2}Sm_{0.2}Eu_{0.2}Er_{0.2})_2O_3$	High volumetric capacity	Enhanced electrolyte access	[12]
Dielectric capacitors	Nanoparticles	$Ba(Zr,Ti,Sn,Hf)O_3$	$\varepsilon_r > 1,000$, >10 J cm^{-3}	High polarization, low loss	[29]
Supercapacitors (EDLC)	Porous composite	CoCrFeNiMn–carbon nanofibers	>200 F g^{-1}	High surface area for ion accumulation	[13]
Supercapacitors (pseudocapacitive)	Nanoparticles	$(CrMnFeCoNi)_3O_4$	239 F g^{-1}, 24.1 Wh kg^{-1} at 0.5 A g^{-1}	Abundant redox sites, stability	[14]
	Nanosheets	$NiFeCoMnCuO_x$	>500 F g^{-1}	Enhanced pseudocapacitance, fast reactions	[4]

HEMs complicates phase stability, as varying atomic sizes can lead to segregation during solid-state methods such as high-energy ball milling or SPS [2]. For instance, CoCrFeNiMn alloys may form secondary phases unless milling conditions (e.g. 300–600 rpm, 10–50 hours) are optimized, affecting nanoparticle or nanorod uniformity [1]. Wet chemical methods, such as hydrothermal synthesis, provide finer control—yielding $CoCuFeNiMnO_x$ nanosheets (5–20 nm thick)—but residual surfactants and agglomeration challenge reproducibility at large scale [10]. Scalability is a persistent issue, especially for templating and advanced techniques such as ALD. Hard templating creates mesoporous HEOs with high surface area (>100 $m^2\ g^{-1}$), but template removal adds cost and complexity, limiting industrial viability [20]. ALD produces ultra-thin HEM films (1–10 nm) with precision, as observed in RhRuPtPdIr nanofilms [32], yet its high equipment cost and slow deposition rate restrict it to small-scale applications [32]. Addressing these challenges requires innovative strategies to ensure morphology control, compositional homogeneity, and cost-effectiveness, which are critical for deploying HEMs in energy storage systems such as LIBs and supercapacitors [7].

14.6.2 Characterization Gaps

Characterization of morphology-dependent HEMs struggles to keep pace with synthesis advances, particularly in capturing dynamic behavior during energy storage operations. *Ex situ* techniques such as SEM and XRD offer static insights—e.g. nanosheet thickness or crystallite size (10–50 nm) in $CoCuFeNiMnO_x$ [40, 43]—but fail to reveal real-time changes during cycling [13]. For example, porous HEOs in LIBs show <5% lattice expansion via *in situ* XRD [16], yet pore structure evolution remains poorly understood without operando methods. This limits the understanding of degradation mechanisms and the association between morphology and performance [15].

In situ techniques such as operando TEM and EIS are promising but underused. Operando TEM could monitor nanosheet collapse in supercapacitors [13], while EIS tracks ion diffusion shifts (e.g. Warburg impedance changes) in $(NiMnCrCoFe)_3O_4$ anodes [46]. However, these require specialized setups, which slows down their widespread adoption [14]. Bridging this gap demands investment in real-time tools to clarify how morphology drives the performance and longevity of HEMs [85], especially in high-rate applications such as SIBs (>300 $mA\ h\ g^{-1}$ at 10 °C) [5].

14.6.3 Future Directions

Future progress in morphology-dependent HEMs relies on integrating artificial intelligence (AI), advanced characterization, and novel applications. AI-driven synthesis, as explored by Rao et al. [37], could optimize morphology (e.g. nanosheets, porous structures) by predicting outcomes from parameters such as precursor ratios, reducing experimental overhead [37]. This approach could enhance scalable methods such as electrospinning, which are used in cost-effective batteries and supercapacitor electrodes [18]. Real-time TEM of hollow (MgCoNiCuZn)O in Li-S batteries could optimize polysulfide retention, boosting energy density [17]. ALD-deposited HEM thin films, such as RhRuPtPdIr [32], promise high ionic conductivity for solid-state batteries [32], while nanofiber HEMs (e.g. $Mg_{0.2}Co_{0.2}Ni_{0.2}Cu_{0.2}Zn_{0.2}O$) could enable flexible supercapacitors (>200 F g^{-1}) for wearables [18]. These advancements position HEMs to meet sustainable energy needs by 2030, supporting applications from grid storage to electric vehicles (IEA, 2024) [7].

14.7 Conclusion

This chapter highlights that morphology-dependent HEMs play a critical role in energy storage, which is driven by advanced synthesis, detailed characterization, and mechanistic understanding. HEMs offer unique properties such as lattice distortion and cocktail effect, which are significantly enhanced by morphology [1, 2]. Synthesis techniques, including solid-state milling [1], hydrothermal methods [10], electrospinning [12], and ALD [32], enable precise control over nanoparticles, nanosheets, and porous frameworks. For example, porous $(Cr_{0.2}Fe_{0.2}Co_{0.2}Ni_{0.2}Zn_{0.2})_3O_4$ achieves >1,000 mA h g^{-1} in LIBs [16], while nanosheet $NiFeCoMnCuO_x$ exceeds 500 F g^{-1} in supercapacitors [41]. SEM, XRD, and EIS confirm that these structures enhance surface area (>100 m^2 g^{-1}), ion transport ($D > 10^{-10}$ cm^2 s^{-1}), and stability (<5% lattice expansion) [40, 43, 46]. Across applications, HEMs outperform conventional materials: LIB anodes exceed graphite (~372 mA h g^{-1}) [15], Li-S cathodes improve sulfur utilization (>70%) [17], dielectric capacitors achieve $\varepsilon_r > 1{,}000$ [29], and supercapacitors surpass activated carbon (~100 F g^{-1}) [13]. Looking ahead, morphology-engineered HEMs are efficient in addressing global energy demands, powering sustainable technologies such as electric vehicles and grid storage [7]. Their ability to combine high energy density, power delivery, and durability makes them the cornerstone for next-generation systems, promising a cleaner, more efficient future [20].

References

1 Yeh, J.-W., Chen, S.-K., Lin, S.-J. et al. (2004). Nanostructured high-entropy alloys with multiple principal elements: novel alloy design concepts and outcomes. *Adv Eng Mater* 6 (5): 299–303.
2 Miracle, D.B. and Senkov, O.N. (2017). A critical review of high entropy alloys and related concepts. *Acta Mater* 122: 448–511.
3 Cantor, B., Chang, I.T.H., Knight, P., and Vincent, A.J.B. (2004). Microstructural development in equiatomic multicomponent alloys. *Mater Sci Eng A* 375: 213–218.
4 Rost, C.M., Sachet, E., Borman, T. et al. (2015). Entropy-stabilized oxides. *Nat Commun* 6 (1): 8485.
5 Gao, X., Zhang, X., Liu, X. et al. (2023). Recent advances for high-entropy based layered cathodes for sodium ion batteries. *Small Methods* 7 (9): 2300152.

6 Sarker, P., Harrington, T., Toher, C. et al. (2018). High-entropy high-hardness metal carbides discovered by entropy descriptors. *Nat Commun* 9 (1): 4980.

7 Ma, Y., Ma, Y., Wang, Q. et al. (2021). High-entropy energy materials: challenges and new opportunities. *Energy Environ Sci* 14 (5): 2883–2905.

8 Wang, Y., Robson, M.J., Manzotti, A., and Ciucci, F. (2023). High-entropy perovskites materials for next-generation energy applications. *Joule* 7 (5): 848–854.

9 Hang, B.T. and Anh, T.T. (2021). Controlled synthesis of various Fe_2O_3 morphologies as energy storage materials. *Sci Rep* 11 (1): 5185.

10 Zhu, Z., Zhang, Y., Kong, D. et al. (2024). A novel high entropy hydroxide electrode material for promoting energy density of supercapacitors and its efficient synthesis strategy. *Small* 20 (20): 2307754.

11 Ran, B., Li, H., Cheng, R. et al. (2024). High-entropy oxides for rechargeable batteries. *Adv Sci* 11 (25): 2401034.

12 Xing, Y., Dan, W., Fan, Y., and Li, X.'a. (2022). Low temperature synthesis of high-entropy $(Y_{0.2}Yb_{0.2}Sm_{0.2}Eu_{0.2}Er_{0.2})2O_3$ nanofibers by a novel electrospinning method. *J Mater Sci Technol* 103: 215–220.

13 Talluri, B., Aparna, M.L., Sreenivasulu, N. et al. (2021). High entropy spinel metal oxide $(CoCrFeMnNi)3O_4$ nanoparticles as a high-performance supercapacitor electrode material. *J Energy Storage* 42: 103004.

14 Li, J., Wei, X., Yin, W. et al. (2025). Rational design of large-scale high-entropy alloy nanosheets anode with excellent lithium storage performance. *Mater Chem Phys* 130450.

15 Yang, X., Wang, H., Song, Y. et al. (2022). Low-temperature synthesis of a porous high-entropy transition-metal oxide as an anode for high-performance lithium-ion batteries. *ACS Appl Mater Interfaces* 14 (23): 26873–26881.

16 Gao, S., He, H., Zhai, F. et al. (2025). Hollow graphene microsphere in situ deposited with rock-salt-type high-entropy oxide nanoparticles for enhanced rate performance as anodes in lithium-ion batteries. *ACS Appl Energy Mater*.

17 Su, J., Cao, Z., Jiang, Z. et al. (2022). High entropy oxide nanofiber by electrospun method and its application for lithium battery anode material. *Int J Appl Ceram Technol* 19 (4): 2004–2015.

18 Bao, W., Shen, H., Zhang, Y. et al. (2024). High-entropy oxides for energy storage and conversion. *J Mater Chem A* 12 (35): 23179–23201.

19 Tsai, M.-H. and Yeh, J.-W. (2014). High-entropy alloys: a critical review. *Mater Res Lett* 2 (3): 107–123.

20 Guo, R., Yang, Y., Zhao, C. et al. (2024). The role of high-entropy materials in lithium-based rechargeable batteries. *Adv Funct Mater* 34 (18): 2313168.

21 Palm, R., Moncada, A.M.B., and Gonçalves, J.M. (2024). From medium-to high-entropy hydroxides for hybrid supercapacitors: a review. *J Mater Chem A* 12: 29402–29431.

22 Raza, H. (2023). *High Entropy Materials (HEMs) for High-Performance Lithium-Sulfur Batteries*, vol. xxiii, 161.

23 Echigo, T., Aruguete, D.M., Murayama, M., and Hochella Jr, M.F. (2012). Influence of size, morphology, surface structure, and aggregation state on reductive

dissolution of hematite nanoparticles with ascorbic acid. *Geochim Cosmochim Acta* 90: 149–162.

24 Sriram, G., Hegde, G., Dhanabalan, K. et al. (2024). Recent trends in hierarchical electrode materials in supercapacitor: synthesis, electrochemical measurements, performance and their charge-storage mechanism. *J Energy Storage* 94: 112454.

25 Sun, G., Sun, L., Xie, H., and Liu, J. (2016). Electrospinning of nanofibers for energy applications. *Nanomaterials* 6 (7): 129.

26 Haché, M.J.R., Cheng, C., and Zou, Y. (2020). Nanostructured high-entropy materials. *J Mater Res* 35 (8): 1051–1075.

27 Yazdani, N., Toroghinejad, M.R., and Moazzen, P. (2024). Tribological and magnetic properties of AlCrCuFeNi high entropy alloy fabricated by MA-SPS. *Mater Chem Phys* 327: 129855.

28 Li, Z., Chen, L., Chang, F. et al. (2022). Synthesis, microstructure and properties of Ti (C, N)-(HfZrTaNbTi) C5-HEA high-entropy cermets by high-energy ball milling and spark plasma sintering. *Ceram Int* 48 (20): 30826–30837.

29 Colombo, R., Garino, N., Versaci, D. et al. (2022). Designing a double-coated cathode with high entropy oxides by microwave-assisted hydrothermal synthesis for highly stable Li–S batteries. *J Mater Sci* 57 (33): 15690–15704.

30 He, H., Kou, P., Zhang, Z. et al. (2024). Coupling high entropy oxide with hollow carbon spheres by rapid microwave solvothermal strategy for boosting oxygen evolution reaction. *J Colloid Interface Sci* 653: 179–188.

31 Zou, Y., Jing, L., Zhang, J. et al. (2024). ALD-made noble metal high entropy alloy nanofilm with sub-surface amorphization for enhanced hydrogen evolution. *J Mater Chem A* 12 (10): 5668–5678.

32 Wang, S., Liu, Q., Li, S. et al. (2024). Joule-heating-driven synthesis of a honeycomb-like porous carbon nanofiber/high entropy alloy composite as an ultralightweight electromagnetic wave absorber. *ACS Nano* 18 (6): 5040–5050.

33 Kim, K., Raymond, D., Candeago, R., and Xiao, S. (2021). Selective cobalt and nickel electrodeposition for lithium-ion battery recycling through integrated electrolyte and interface control. *Nat Commun* 12 (1): 6554.

34 Bolar, S., Ito, Y., and Fujita, T. (2024). Future prospects of high-entropy alloys as next-generation industrial electrode materials. *Chem Sci* 15 (23): 8664–8722.

35 Li, Z., Zhou, F., Baosheng, X., and Guo, D. (2022). Characterization of novel high-entropy $(La_{0.2}Nd_{0.2}Sm_{0.2}Dy_{0.2}Yb_{0.2})\ 2Zr_2O_7$ electrospun ceramic nanofibers. *Ceram Int* 48 (9): 12074–12078.

36 Gong, C., Sun, S., Zhang, Y. et al. (2019). Hierarchical nanomaterials via biomolecular self-assembly and bioinspiration for energy and environmental applications. *Nanoscale* 11 (10): 4147–4182.

37 Goldstein, J.I., Newbury, D.E., Michael, J.R. et al. (2017). *Scanning Electron Microscopy and X-Ray Microanalysis*. Springer.

38 Zheng, S., Jiuyi, H., Feng, R. et al. (2024). Ultrathin template approach to synthesize high-entropy intermetallic nanoparticles for hydrogen evolution reaction. *Small Struct* 5 (6): 2300537.

39 Binnig, G., Quate, C.F., and Gerber, C. (1986). Atomic force microscope. *Phys Rev Lett* 56 (9): 930.
40 Liu, K.-C., Panda, P.K., Mallick, B.C. et al. (2024). Solid-phase microwave synthesis of high-entropy graphene quantum dots as metal-free electrochemical catalysts. *Appl Surf Sci* 648: 159061.
41 Li, M., Xi, X., Wang, H. et al. (2022). A universal, green, and self-reliant electrolytic approach to high-entropy layered (oxy) hydroxide nanosheets for efficient electrocatalytic water oxidation. *J Colloid Interface Sci* 617: 500–510.
42 Khan, N.A., Akhavan, B., Zhou, H. et al. (2019). High entropy alloy thin films of $AlCoCrCu_{0.5}FeNi$ with controlled microstructure. *Appl Surf Sci* 495: 143560.
43 Tomboc, G.M., Zhang, X., Choi, S. et al. (2022). Stabilization, characterization, and electrochemical applications of high-entropy oxides: critical assessment of crystal phase–properties relationship. *Adv Funct Mater* 32 (43): 2205142.
44 Liu, T., Gao, Y., Bei, H., and An, K. (2018). In situ neutron diffraction study on tensile deformation behavior of carbon-strengthened CoCrFeMnNi high-entropy alloys at room and elevated temperatures. *J Mater Res* 33 (19): 3192–3203.
45 Xue, T., Wang, Y., Yang, L. et al. (2023). Development of quinary layered double hydroxide-derived high-entropy oxides for toluene catalytic removal. *Catalysts* 13 (1): 119.
46 Yu, B.-Y., Chen, Y.-Y., Wang, W.-B. et al. (2008). Depth profiling of organic films with X-ray photoelectron spectroscopy using C60+ and Ar+ co-sputtering. *Anal Chem* 80 (9): 3412–3415.
47 Pappu, S., Nanaji, K., Mandati, S. et al. (2020). Cost-effective synthesis of electrodeposited NiCo2O4 nanosheets with induced oxygen vacancies: a highly efficient electrode material for hybrid supercapacitors. *Batter Superc* 3 (11): 1209–1219.
48 Tian, X., Li, H., Rui Chang, Y. et al. (2024). Rapid, self-sacrificing template synthesis of two dimensional high-entropy oxides toward high-performance oxygen evolution. *J Mater Chem A* 12 (6): 3276–3282.
49 Wang, X.L., Jin, E.M., Sahoo, G., and Jeong, S.M. (2023). High-entropy metal oxide $(NiMnCrCoFe)3O_4$ anode materials with controlled morphology for high-performance lithium-ion batteries. *Batteries* 9 (3): 147.
50 Wu, Q., Alkemper, A., Lauterbach, S. et al. (2024). Fabrication of nanocrystalline high-entropy oxide $CoNiFeCrMnO_x$ thin film electrodes by dip-coating for oxygen evolution electrocatalysis. *Energy Adv* 3 (4): 765–773.
51 Wei, B., Jin, X., Pang, J. et al. (2022). Prediction of electrochemical impedance spectroscopy of high-entropy alloys corrosion by using gradient boosting decision tree. *Mater Today Commun* 32: 104047.
52 Panahi, S.L., Bruna, P., and Pineda, E. (2022). Effect of Si and B on the electrochemical behavior of FeCoNiCr-based high-entropy amorphous alloys. *Materials* 15 (24): 8897.
53 Modupeola, D. and Popoola, P. (2023). High entropy nanomaterials for energy storage and catalysis applications. *Front Energy Res* 11: 1149446.

54 Yan, X., Zhou, Y., and Wang, S. (2025). Nano-high entropy materials in electrocatalysis. *Adv Funct Mater* 35 (2): 2413115.
55 Liang, B., Ai, Y., Wang, Y. et al. (2020). Spinel-type $(FeCoCrMnZn)3O_4$ high-entropy oxide: facile preparation and supercapacitor performance. *Materials* 13 (24): 5798.
56 Pei, L., Kang, J., He, C. et al. (2025). Designing high reversibility free-standing hierarchical porous high-entropy oxide as anodes for advanced lithium-ion batteries. *Adv Funct Mater* 35 (16): 2422809.
57 Djenadic, R., Sarkar, A., Clemens, O. et al. (2017). Multicomponent equiatomic rare earth oxides. *Mater Res Lett* 5 (2): 102–109.
58 Zou, X., Zhang, Y.-R., Huang, Z.-P. et al. (2023). High-entropy oxides: an emerging anode material for lithium-ion batteries. *Chem Commun* 59 (91): 13535–13550.
59 Wang, Y.-Q., Wang, H.-M., Jiang, Y.-C. et al. (2023). High-entropy oxide nanofibers as catalytic host promising high volumetric capacity of sulphur-based composites for lithium–sulphur batteries. *ACS Appl Energy Mater* 6 (16): 8377–8387.
60 Jiang, Y., Du, M., Geng, P. et al. (2024). CoO/MoO3@ nitrogen-doped carbon hollow heterostructures for efficient polysulfide immobilization and enhanced ion transport in Lithium-Sulfur batteries. *J Colloid Interface Sci* 664: 617–625.
61 Xiang, H., Xing, Y., Dai, F.-z. et al. (2021). High-entropy ceramics: present status, challenges, and a look forward. *J Adv Ceram* 10: 385–441.
62 Peng, H., Tiantian, W., Liu, Z. et al. (2024). High-entropy relaxor ferroelectric ceramics for ultrahigh energy storage. *Nat Commun* 15 (1): 5232.
63 Wang, H., Zhang, J., Jiang, S. et al. (2023). TiO3-based high-entropy dielectric ceramics with ultrahigh recoverable energy density and high energy storage efficiency. *J Mater Chem A* 11: 4937–4945.
64 Yin, Y., Zhang, W.-B., Zhang, X.-L. et al. (2023). Low-dimensional high entropy oxide $(FeCoCrMnNi)3O_4$ for supercapacitor applications. *Dalton Trans* 52 (26): 9005–9009.
65 Sarkar, A., Wang, Q., Schiele, A. et al. (2019). High-entropy oxides: fundamental aspects and electrochemical properties. *Adv Mater* 31 (26): 1806236.
66 Zhang, D., Shilin, X., Li, T. et al. (2023). High-entropy oxides prepared by dealloying method for supercapacitors. *ACS Appl Eng Mater* 1 (2): 780–789.

15

Industrial Aspects of High-Entropy Materials for Energy Storage Applications

Chandan Kumar Panda[1], Subhashree Behera[2], Hyun-Suk Kim[2], and Jungseek Hwang[1]

[1] *Sungkyunkwan University, Department of Physics, 2066 Seobu-ro, Jangan-gu, Suwon 16419, Republic of Korea*
[2] *Dongguk University, Department of Energy and Materials Engineering, 30 Pildong-ro 1-gil, Jung-gu, Seoul 04620, Republic of Korea*

15.1 Introduction

Energy storage has emerged as an essential aspect of a sustainable future due to global environmental concerns and increasing energy consumption in various sectors, especially because of the growing urgency to incorporate renewable energy sources. As industrial sectors evolve rapidly to accept fluctuating renewable energy sources and to enhance grid stability, energy storage methods such as lithium-ion batteries (LIBs), sodium-ion batteries (SIBs), and supercapacitors are becoming extremely crucial. Additionally, advanced electrode materials are in huge demand because of the massively expanding electric vehicle (EV) sector to deliver improved safety characteristics, longer cycle life, and enhanced power capabilities. Conventional electrode materials, which are mostly based on graphite, layered oxides (e.g. $LiCoO_2$), and metal phosphates (e.g. $LiFePO_4$), have issues like limited capacity, structural instability, and resource scarcity; hence, to overcome these, researchers are now looking into novel alternatives [1–4] (Table 15.1). The need for large-scale energy storage solutions is also essential for the transition toward renewable energy storage systems to stabilize the grid, to control the imbalances between the supply and huge demand, and to facilitate peak shaving [9]. For example, LIBs are extensively used for their high energy density and well-matured technology. However, the components used, lithium, nickel, and cobalt, are highly expensive and have resource constraints. Researchers are now looking new and advanced materials [10]. Hence, SIBs and potassium-ion batteries have become attractive choices because of their cost-effectiveness and abundance, yet gaining high performance and durability from the available electrode materials is extremely challenging [7, 11]. Therefore, industrial electrode materials should be both affordable and environmentally friendly, also exhibiting robust stability,

High-Entropy Materials for Energy Storage Devices, First Edition.
Edited by Chien-Te Hsieh, Pradeep Kumar Panda, and Arpan Kumar Nayak.

Table 15.1 Comparison of HEMs and conventional electrode materials.

Parameter	Conventional electrodes (e.g. graphite, LFP)	HEMs (HEAs/HEOs)	References
Specific capacity ($mA\,h\,g^{-1}$)	Graphite: 150–200 $mA\,h\,g^{-1}$ LFP: ~160 $mA\,h\,g^{-1}$ Limited by intercalation chemistry	HEAs/HEOs: 300–600 $mA\,h\,g^{-1}$ Enabled by multielectron reactions and entropy-stabilized structures	[5, 6]
Rate capability	Moderate Constrained by solid-state diffusion and phase transitions	High lattice distortion and sluggish diffusion effect facilitate faster ion transport	[6, 7]
Thermal stability	Limited (especially above 100 °C) Structural degradation common in LFP and NCMs	Excellent (stable up to 300–500 °C) Due to high phase stability of entropy-stabilized systems	[5, 8]
Material cost and availability	High cost due to reliance on Co, Ni Vulnerable to supply chain instability	Cost-effective by using abundant elements such as Fe, Mn, Mg compositions	[5, 8]
Cycle life (80% retention)	Typically, 1,000–3,000 cycles Capacity fades due to structural fatigue and SEI degradation	Often >3,000–5,000 cycles Structural integrity maintained due to high-entropy stabilization	[5, 7]
Phase stability	Prone to degradation over cycles Phase transitions and segregation issues	Entropy-stabilized single-phase solid solutions Suppresses phase segregation and ensures long-term cycling	[4–6]

high ionic and electronic conductivity, minimal volume expansion during cycling, and significant energy and power densities [5]. In light of this, it is becoming increasingly necessary to develop next-generation electrode materials, such as high-entropy materials (HEMs) and their compounds, to overcome the current limitations in traditional energy storage systems and make it possible to deploy advanced batteries and supercapacitors on an industrial scale [12].

HEMs, especially high-entropy alloys (HEAs), where numerous elements are present in near-equimolar ratios, lead to a single-phase solid solution with enhanced thermodynamic stability and structural resilience [13, 14]. The inherent high disorder and complexity give rise to very high configurational entropy, which

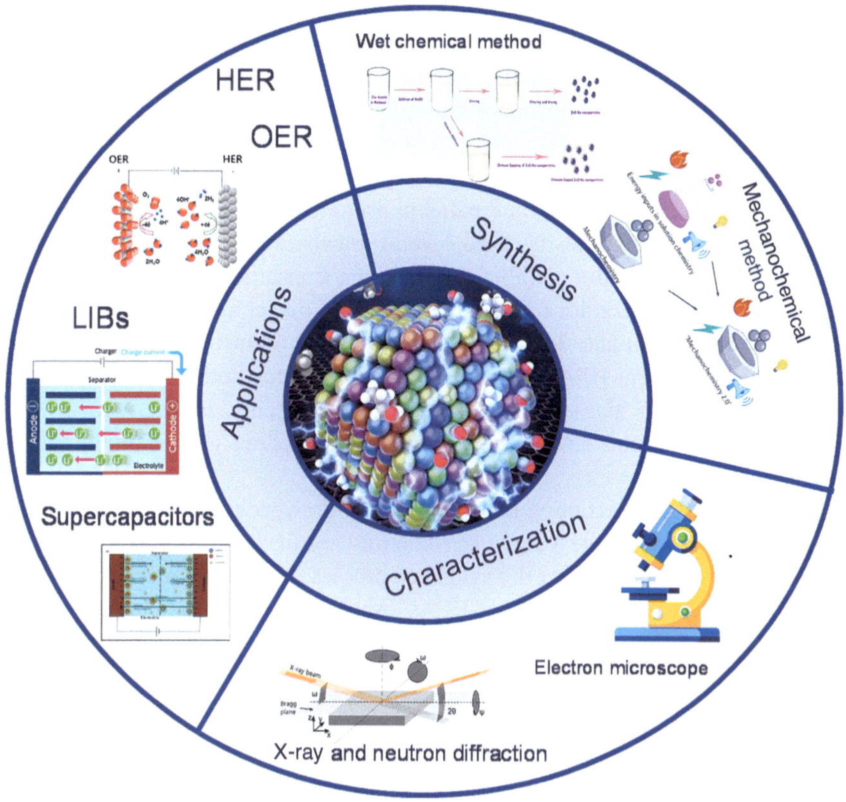

Figure 15.1 Schematic of HEMs for energy storage applications.

generates exceptional strength, outstanding corrosion resistance, and delayed atomic diffusion behavior—qualities that make HEAs suitable for energy storage purposes [9, 15]. Their remarkable electrochemical performance is one of the primary characteristics of HEAs in the energy storage sector (Figure 15.1). Usually, in conventional alloys, phase separation and secondary phase formation are observed during prolonged electrochemical cycling, but in HEAs, entropy-driven stabilization prevents these flaws [16]. As reported in recent studies, HEAs show better cycling stability, higher specific capacities, and enhanced capacity retention when compared to traditional electrode materials [1, 6]. As ceramics analogous to HEAs, cation-disordered high-entropy oxides (HEOs), for example, have exhibited promising electrochemical behaviors, delivering significantly higher reversible capacities and robust cycling stability; these characteristics make them appealing for LIB and SIB cathodes [6, 17]. Furthermore, high-entropy spinel oxides and other HEA-derived compounds, because of their distinct lattice structure, possess efficient electron and ion transport channels. Complementarily, high-entropy sulfides (HESs) offer excellent redox kinetics, multivalent charge storage capability, and high electrical conductivity, making them highly attractive for next-generation alkali-ion batteries and conversion-type electrodes [18].

Hence, HEAs show improved charge storage and rate capabilities, which is essential for next-generation solid-state batteries and supercapacitors [5, 12]. Thus, ongoing research on HEAs in energy storage systems is a revolutionary step toward overcoming the existing performance limitations, making them primary candidates for energy storage solutions [7].

Traditional electrode materials such as graphite anodes, layered transition metal oxides ($LiCoO_2$, $LiNiMnCoO_2$), and metal phosphates ($LiFePO_4$), despite dominating the energy storage industry, have several drawbacks, such as low theoretical capacities, modest cycling stability, limited rate capacities, and environmental concerns related to hazardous components such as nickel and cobalt. Thus, their scalability and sustainable long-term deployment in industrial applications are limited [9]. However, HEAs, because of their compositional flexibility, allow the incorporation of abundant and inexpensive elements, which reduces supply chain concerns and environmental impacts compared to traditional battery components [15].

In all aspects, by overcoming the fundamental drawbacks of traditional battery components, HEAs provide a strategic route forward for the energy storage sector. The following sections discuss the synthesis process, fundamental properties, and application-specific developments of HEAs, exploring their potential for industrial deployments [8, 19].

15.2 Fundamentals of HEMs

Advanced materials with more constituent elements mixed in almost equimolar proportions and having high configurational entropy to produce unique structural properties are termed HEMs. The theory of HEAs was initially proposed in 2004 by Yeh et al. and Cantor et al. [20], which is a subset of HEMs. HEAs have been proven to have strong solid solution phase stability and unique mechanical and thermodynamic characteristics, due to the very high-entropy resulting from their multielemental nature [20]. With time, the class of HEMs has been spread to HEOs, HESs, high-entropy nitrides (HENs), high-entropy carbides (HECs), and high-entropy borides, and there have been several studies on these materials for the past few years (Figure 15.2). These HEMs show enormous potential in energy storage systems, catalytic processes, and applications under challenging conditions such as high temperature, high pressure, and high levels of irradiation environments [20, 21].

15.2.1 Categorization

According to their structure and composition, HEMs are divided into different groups (Figure 15.3a,b). In HEAs, there are many elements that form a solid solution. Because of the high configurational entropy of HEAs, their mechanical strength, corrosion resistance, and high-temperature stability properties are enhanced [13, 22]. HEOs, which have several cations in one lattice structure,

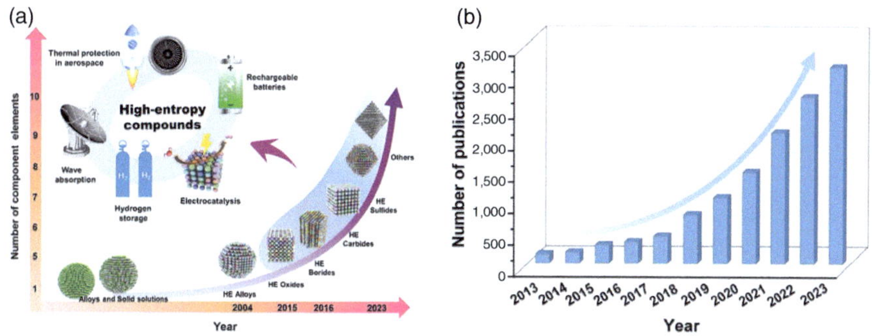

Figure 15.2 (a) Total number of publications on HEMs by year. The information was gathered on December 31, 2023, from the Web. *Source:* Reproduced with permission from Ref. [7] / Elsevier. (b) The evolution of HEMs over time.

exceptional electrochemical stability, and ionic conductivity, provide a tailor-made environment for battery- and energy storage-related applications [23]. A similar group of covalent ceramic materials called HECs, because of their exceptionally high-temperature stability and high resilience quality, is extremely useful in harsh environments such as high-pressure and aero-scope applications [23]. Hard ceramic multicomponent compounds known as HENs are suitable candidates for electronic devices and wear-resistant coating applications resulting from very high electrical and chemical resistances [11]. In the same manner, high-entropy borides, which are known for their very high mechanical strength and extremely high melting temperatures, are used in several refractory applications [24]. Additionally, HESs and high-entropy phosphates have drawn immense interest from industrial researchers, especially in energy storage and electrocatalysis applications, due to their tunable electrical performance [25]. HEMs have prolonged stability because of their ability to hinder phase segregation, which is attributable to the high configurational entropy resulting from their complex atomic arrangements [9].

15.2.2 Structural Diversity

HEMs can have several crystal structures: rock salt (FCC) is common in HEOs and HECs and offers outstanding ionic conductivity and structural stability. As we can control the ferroelectric and dielectric characteristics of perovskites, they are used in capacitor applications for energy storage. Spinels are used in LIBs as a cathode because of their excellent ionic conductivity and great cycling stability (Figure 15.3a).

15.2.3 The Four Core Effects

High-Entropy Effect: The Gibbs free energy in HEMs is reduced by the high-entropy; thus, the development of brittle intermetallic phases is suppressed (Figure 15.3c). This effect is mainly observed in HEAs and HEOs, where they can maintain single-phase structures even under tough conditions due to the stability provided by high configurational entropy [26, 27].

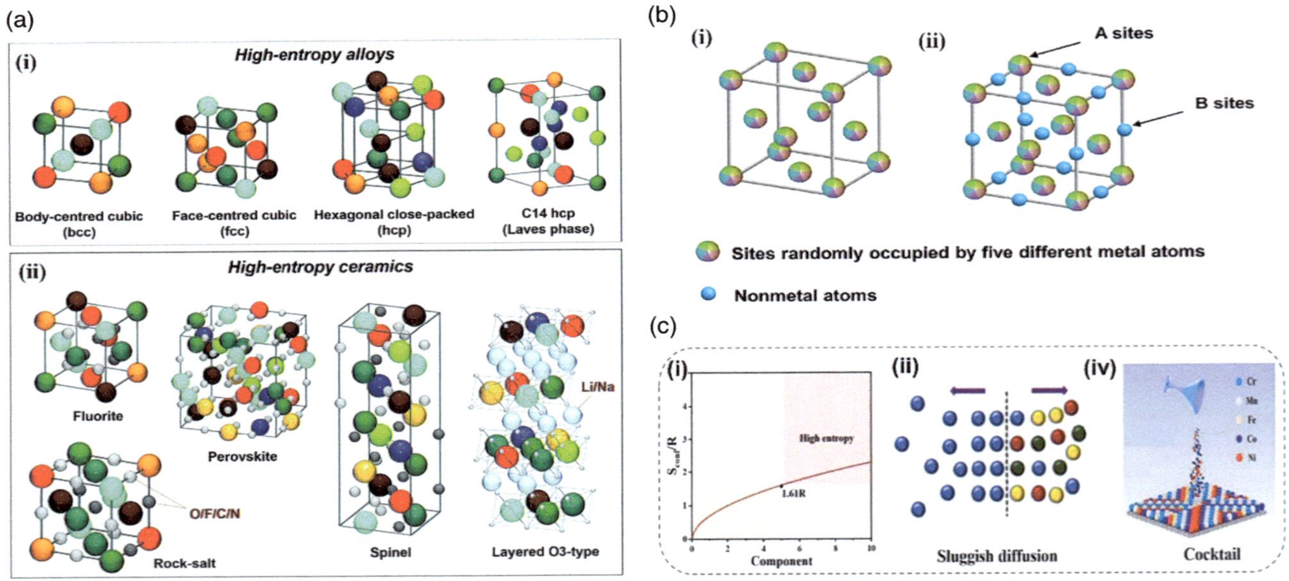

Figure 15.3 (Reproduced with permission from Ref. [9] Copyright 2021 Royal Society of Chemistry). (a) The crystal structures of (i) HEAs and (ii) high-entropy ceramics employed in energy-related sectors. *Source:* Reproduced from Ref. [5] / with permission of The Royal Society of Chemistry. (b) Crystal structure illustration of (i) FCC HEAs and (ii) HEOs with rock salt structures. *Source:* Reproduced from Ref. [7] / with permission of Elsevier. (c) The four main factors of HEMs are shown in a mechanism diagram. (i) Slow diffusion, (ii) high-entropy effects, and (iii) cocktail effect. *Source:* Reproduced from Ref. [5] / with permission of Elsevier.

Lattice Distortion Effect: In HEMs, there are several atoms with different sizes, atomic structures, and electronegativities. This inhomogeneity creates lattice distortions, thereby creating localized strain fields within the system. Interestingly, it has been shown that these lattice distortions and strain fields that emerge from them enhance the mechanical strength and improve the electrical properties and ionic conductivity, resulting in their exceptional overall properties [28].

Sluggish Diffusion Effect: This effect arises from complex atomic arrangements and reduces the atomic mobility by impeding phase separation and grain growth (Figure 15.3c). This improves the mechanical and high-temperature stability of HEMs; consequently, this works well for energy storage devices such as SIBs and LIBs [29].

Cocktail Effect: As there are multiple components in HEMs compared to conventional materials, synergistic interactions between properties of individual elements create novel properties (Figure 15.3c), which are highly important for catalytic applications where reaction kinetics can be tuned by adjusting surface and electrical properties [30, 31].

15.3 Characterization of HEMs

HEMs have gathered much attention because of their outstanding features and distinctive structural characteristics. They show exceptional performances because of their complex atomic configurations, which produce unique effects, including lattice distortion, cocktail effects, and slow diffusion [32]. The improved mechanical, thermal, and electrochemical properties make HEMs more appealing for energy storage applications. Accurate explanation of the structures of HEMs remains difficult because of their compositional and structural complexities. Several advanced characterization approaches, for instance X-ray diffraction (XRD), electron microscopy, and spectroscopy-based approaches, are used to provide accurate information about phase composition, atomic distribution, and electronic structure of HEMs, enabling a better understanding of their unique characteristics.

15.3.1 Diffraction Techniques

Crystalline materials are commonly analyzed to determine the phase composition and structure by diffraction techniques. Among these, the most popular technique is XRD, which yields both qualitative and quantitative structural information. XRD allows for the exact measurement of crystal properties such as lattice constants, crystallinity, grain size, strain, preferred orientation, and atomic occupancy within the lattice when combined with the Rietveld refinement technique. The XRD patterns were modified using this technique, confirming the development of a disordered rock salt (DRX) structure, with lattice parameters arising when more transition metal species are involved [6]. The expansion of crystal lattice is being controlled by various atomic radii and chemical interactions illustrated in high-entropy compositions, as shown in Figure 15.4a. However, the analysis of

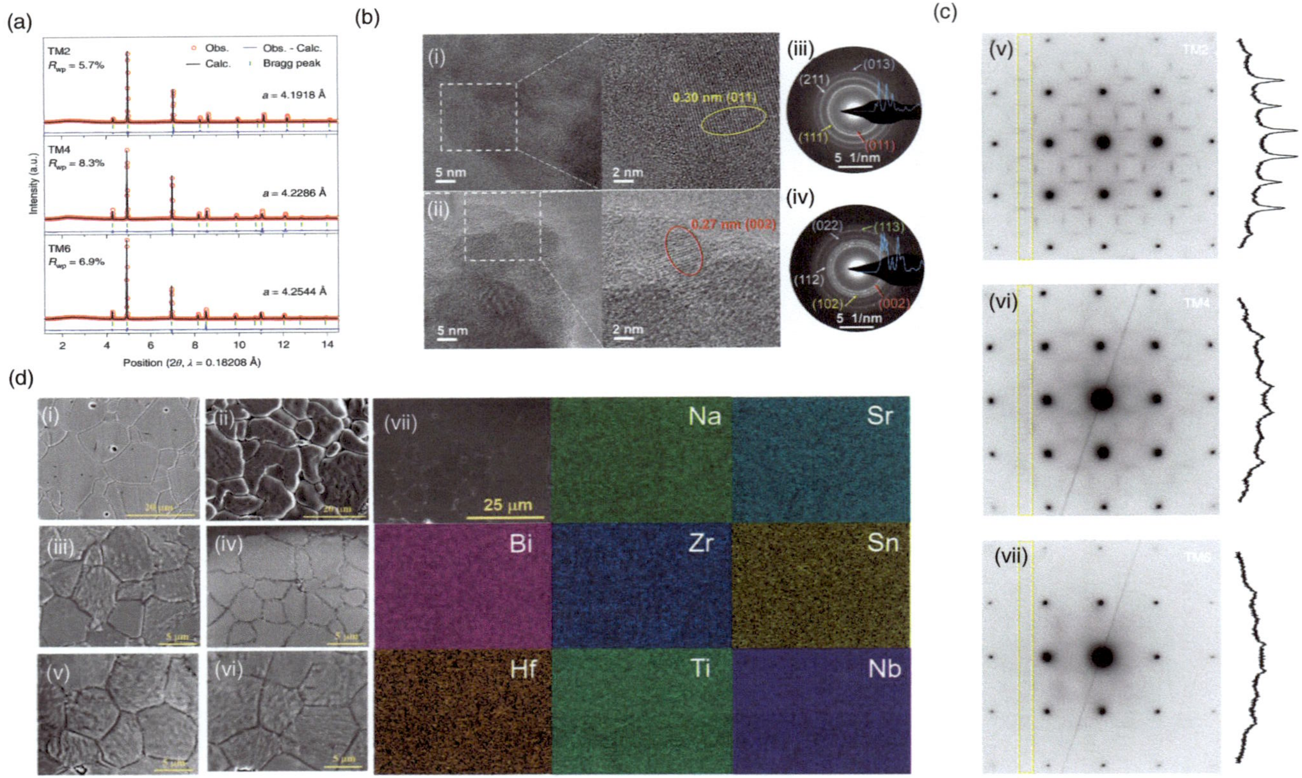

Figure 15.4 (a) The as-synthesized materials' synchrotron XRD spectra at wavelength $\lambda = 0.18208$ Å, where the refined lattice constant and Rwp is the refinement's *R*-factor. *Source:* Adapted from Ref. [33]. (b) HR-TEM picture with (i) MS and (ii) MS2 crystal lattices. SAED patterns of (iv) MS2 and (iii) MS. *Source:* Reproduced with permission from Ref. [33] / John Wiley & Sons / CC BY-NC 4.0. (c) TEM ED patterns obtained along the [100] zone axis on as-synthesized TM2 (v), TM4 (vi), and TM6 (vii). The materials' round spots are indicated as Fm-3m space group and come from LRO. The SRO is responsible for the diffuse scattering patterns that resemble squares. The SRO pattern intensity is shown in the ED patterns. *Source:* Reproduced with permission from Ref. [6] / Springer Nature. (d) Following thermal etching treatment, SEM graphs of $(1-x)$NBT–xSZSHTN ceramics show: (i) pure NBT, (ii) $x = 0.15$, (iii) $x = 0.2$, (iv) $x = 0.25$, (v) $x = 0.3$, and (vi) $x = 0.35$, and (vii) EDS mapping pictures of 0.8NBT–0.2SZSHTN ceramics. *Source:* Reproduced with permission from Ref. [34] / John Wiley & Sons.

HEMs using the traditional XRD method poses challenges due to weak signals from light elements, crystal lattice distortion, and nanocrystalline domains. To address these difficulties, high-intensity sources such as synchrotron XRD or neutron diffraction are utilized to provide detailed insights into lattice strain, atomic displacements, and defect formations.

15.3.2 Electron Microscopy Techniques

Electron microscopy techniques, which usually include scanning electron microscopy (SEM) [35], transmission electron microscopy (TEM), 3D atomic probe tomography (3D-APT), scanning transmission electron microscopy (STEM), energy-dispersive X-ray spectroscopy (EDS), and electron energy loss spectroscopy (EELS) equipped in instruments of SEM and TEM, are the most accessible methods for analyzing the structures, morphologies, and elemental distribution of samples. SEM is typically used to examine the direct three-dimensional morphologies of a greater area of samples, whereas TEM typically focuses on a compact region with better resolutions. The nanoscale crystal structure of HECs can be easily determined using TEM's selected area electronic diffraction (SAED).

Lin et al. also described the particle sizes of HESs [33]. The (011) diffraction plane in MS ((FeMnNiCoCr)S) with a lattice spacing of 0.30 nm is represented by the yellow circular region, while the (002) lattice planes of the pyrite structure for MS2 ((FeMnNiCoCr)S2) are represented by the red circled area. Figure 15.4b shows the SEM images of (1−x)NBT–xSZSHTN ($Sr(Zr_{0.2}Sn_{0.2}Hf_{0.2}Ti_{0.2}Nb_{0.2})O_3$) ($x$ = 0.15–0.35) and pure NBT ($Na_{0.5}Bi_{0.5}TiO_3$) ceramics. Each composition has a thick microstructure and is well sintered. Using SAED analysis, Lu et al. [6] investigated the changes in crystal structures in rock salt-type oxides as the number of cation component types increased from two to six. They observed that the formation of superlattice structures in binary oxides stopped as the number of metal cations increased to six (Figure 15.4c). Grain size reduces continuously as the concentration of SZSHTN increases, most likely due to the presence of the secondary phase, which limits grain expansion. Additionally, the EDS elemental mapping of 0.8NBT–0.2SZSHTN [34] (Figure 15.4d) confirms the consistent distribution of the elements and indicates the homogeneity of the material without any segregation.

15.3.3 Electrochemical Characterization

In an electrochemical energy storage (EES) system, electrical energy is first transformed into chemical energy during Li insertion, and then, it is reversed during Li extraction for the reproduction of electrical energy [36, 37]. These devices are essential for backup power, renewable energy integration, and electric car charging, all of which operate on the principle of reversible energy conversion. Multiple energy storage processes, including batteries [38] and supercapacitors, can be employed for HEMs depending on the electrochemical nature. The performance of HEMs in energy storage applications is determined by their electrochemical characterization.

Major techniques, such as cyclic voltammetry (CV) [39, 40], electrochemical impedance spectroscopy (EIS) [40, 41], and charge/discharge cycling [42, 43], are commonly used to elucidate the electrochemical behavior. CV was utilized to analyze the reversibility of electrode materials and Li^+-ion interactions and explain the voltage ranges corresponding to oxidation and reduction reactions. EIS is an effective method for examining processes in electrochemical systems. Its primary strength is its capacity to efficiently break down intricate electrochemical processes into a number of simpler ones depending on various relaxation times. A semicircular arc observed in the high-to-medium frequency range, which represents charge transfer resistance, and a straight line in the lower frequency range, which indicates Warburg impedance linked to lithium-ion diffusion, commonly appear on Nyquist plots derived from EIS measurements. Galvanostatic charge/discharge (GCD) tests are crucial for evaluating the capacity retention, cycling stability, and rate capability of HEM electrodes. Particularly, in GCD, the discharge curve indicates the Li-ion storage capacity in HEMs and vice versa.

15.4 Industrial Energy Storage Technologies Utilizing HEMs

EES systems operate by converting between electrical and chemical energy stored during the charging process. Redox-active materials undergo reversible chemical reactions that are useful in batteries. In supercapacitors, energy is stored by electrostatic charge separation at electrode–electrolyte interfaces.

15.4.1 HEMs in LIBs

The exceptional electrochemical stability and unique composition make HEMs suitable for energy storage systems, especially in LIBs. The structurally uniform solid solution structure with multiple metals can be superior to conventional electrodes in terms of electrochemical and mechanical durability. HEMs are used primarily as anode materials in LIBs. A Li^+-ion battery full cell representing a HEO (TM-HEO) anode based on transition metals, which is compiled with NCM (nickel–cobalt–Manganese) as a counterelectrode is shown in Figure 15.5. The visual representation showcases the initial electrochemical reactions occurring with the intercalation and deintercalation of Li^+ ions between two electrodes. Wang et al. [44] explored TM-HEOs as a prominent material for LIBs to be used as an anode, focusing on a $(Co_{0.2}Cu_{0.2}Mg_{0.2}Ni_{0.2}Zn_{0.2})O$ composition that has a rock salt structure. In full-cell configurations paired with a $LiNi_{1/3}Co_{1/3}Mn_{1/3}O_2$ cathode, after 50 and 100 cycles, respectively, the system's first cycle-specific discharge capacity of 446 mA h g^{-1} was maintained at 300 and 256 mA h g^{-1} (Figure 15.5b). Huang et al. [45] investigated the charge/discharge reaction mechanism in $(CrMnFeCoNi)_3O_4$ HEO anodes, which explains the complexity of electrochemical stability through a complex interplay of phase transformations and elemental redistribution. During Li^+-ion intercalation, the pristine spinel lattice undergoes

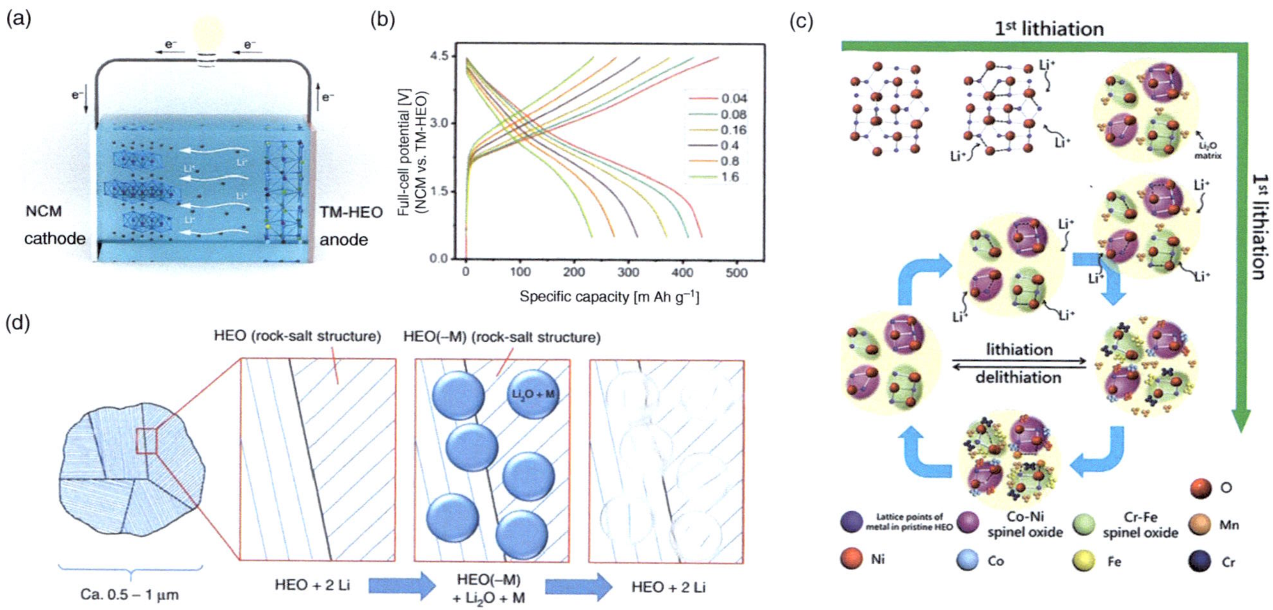

Figure 15.5 (a) Schematic of a full cell with TM-HEO and NCM111 electrodes, (b) Charge/discharge profiles of NCM111 and TM-HEO full cells at various specific currents at 25 °C. *Source:* Reproduced from Ref. [44] / with permission of Elsevier. (c) Diagrammatic representation of the evolution of HEO's atomic-scale microstructure during lithiation/delithiation cycling. *Source:* Reproduced from Ref. [45] / with permission of Elsevier. (d) Diagrams M in the figure solely represent the cations Co, Cu, Zn, and Ni. *Source:* Reference [46] / Springer Nature / CC BY 4.0.

decomposition into distinct domains: $Cr_xFe_{3-x}O_4$ spinel phases, $LiNi_xCo_{1-x}O_2$ layered oxides, and metallic Mn nanoparticles are distributed within a Li_2O matrix. In the deintercalation process, metallic nanoparticles are reincorporated into the oxide matrix. Mn species again reintegrate into the Cr-Fe spinel framework during charge recovery, a self-healing behavior enhanced by the multicationic environment's configurational entropy, as depicted in Figure 15.5c. These mechanisms overall reduce volume expansion (≤12% lattice parameter changes vs. ~300% in graphite) while sustaining capacities exceeding 600 mA h g^{-1} over 200 cycles. Figure 15.5d also explains the schematic representation of the intercalation and deintercalation mechanism of TM-HEO, where metal cations include Co^{2+}, Cu^{2+}, Zn^{2+}, and Ni^{2+}. During lithiation, transition metal cations undergo reduction, forming metallic nanoparticles (Co^0, Cu^0, Zn^0, Ni^0) in a Li_2O matrix [46]. Despite this change, the rock salt structure remains intact, allowing lithium ions to move through it. As charging progresses, internal stresses cause some lattice distortions, but the overall structure remains stable. During discharging, the metallic nanoparticles dissolve back into the structure, restoring the rock salt phase. This self-healing process, driven by high-entropy, ensures that the structure remains reversible and stable over many cycles. A recent study by An et al. [47] further demonstrated this effect in a nonequimolar six-component high-entropy spinel oxide, $(CoMnZnNiMg)_2CrO_4$, which exhibited exceptional cycling stability (608 mAh g^{-1} after 200 cycles) and rate capability (371 mAh g^{-1} at 2000 mA g^{-1}), attributed to entropy-induced phase stabilization and amorphous conversion during cycling. Several HEMs are used as LIB electrodes, which are listed in Table 15.2 with their respective electrochemical performances.

Table 15.2 List of HEMs used as LIB electrodes with performances.

	Composition	HEM structure	Method	Performance	Rate capability	References
HEO	(MgCo NiCuZn)O	Rock salt	Template-free salt-assisted strategy	424 mA h g^{-1} @250 cycles	280 mA hg^{-1} at 3 A g^{-1}	[48]
HEO	$Mg_{0.2}Co_{0.2}Ni_{0.2}Cu_{0.2}Zn_{0.2}O$	Rock salt	Electro-spinning	365 mA h g^{-1} @300 cycles	100 mA h g^{-1} at 3 A g^{-1}	[19]
HEO	$(FeCoNiCuZn)_3O_4$	Spinel	Hydrothermal method	540 mA h g^{-1} @500 cycles	306 mA h g^{-1} at 5 A g^{-1}	[49]
HEO	$(FeCoNiCrMn)_3O_4$	Spinel	Solid-state reaction	596.5 mA h g^{-1} @1,200 cycles	482 mA h g^{-1} at 3 A g^{-1}	[44]
HEO	$(CrFeMnCoMg)_3O_4$	Spinel	Solution combustion method	183.9 mA h g^{-1} @800 cycles	183 mA h g^{-1} at 2 A g^{-1}	[49]
HES	$Cu_4MnFeSnGeS_8$	Tetra-gonal	Solid-state method	430.4 mA h g^{-1} @1,200 cycles	459.2 mA h g^{-1} at 10 A g^{-1}	[50]
HEA	FeNiMn CrSiGeCo	BCC	Solid-state method	971 mA h g^{-1} @100 cycles	500 mA h g^{-1} at 1 A g^{-1}	[51]

15.4.2 HEMs in Supercapacitors

Several HEMs have emerged as prominent electrode materials because of their significant specific capacity and excellent cycling stability as compared to traditional materials. This is primarily due to their unique multielement composition, which leads to higher capacity and electrochemical stability. A study by Zhu et al. [52] explored a NiCoMoMnZn-LDH-based electrode, which showed a remarkable specific capacitance of 1,810.2 $F\,g^{-1}$ at 0.5 $A\,g^{-1}$ and an energy density of 62.1 $Wh\,kg^{-1}$ at 475 $W\,kg^{-1}$. This enhanced electrochemical performance can be expected because of the uniform elemental distribution, stable porous structure, and higher surface area. Jin and colleagues [53] introduced high-entropy metal nitrides (HEMNs) using a mechanochemical synthesis method. These HEMNs exhibited a notable specific capacitance of 230 $F\,g^{-1}$ when tested at 10 $mV\,s^{-1}$ in a 1 M KOH electrolyte, demonstrating robust electrochemical stability and efficient charge/discharge characteristics.

Figure 15.6 presents a description of nanoporous AlCoCrFeNi HEMs prepared through selective phase dissolution, as explored by Kong et al. [54] for enhanced supercapacitor applications. The original HEM ribbon featured integrated body-centered cubic (BCC) phases such as an Al-Ni-rich phase and a Cr-Fe-rich phase. The Al-Ni-rich phase is selectively removed due to its higher electrochemical reactivity, leaving behind a robust, interconnected nanoporous structure predominantly made up of the Cr-Fe-rich BCC phase after exposure to selective chemical dissolution. This nanoporous structure has a high surface area, resulting in enhanced active sites for ion diffusion and electron transport. As a result, this nanoporous HEM exhibits a binder-free electrode with excellent redox activity from multielectron composition. Additionally, it also exhibits common limitations, such as poor conductivity and sluggish ion transport like conventional electrode materials. Shen et al. [55] described HEA nanoparticles (NPs), synthesized on hyper-crosslinked polymer-derived carbon (HCPC) by the absorption of Fe^{2+}, Co^{2+}, Ni^{2+}, Cu^{2+}, and Sn^{2+} ions onto a polymer, following *in situ* reduction and carbonization. On the surface of HEA NPs, the oxide layer forms upon exposure to the air, which enhances the pseudocapacitive behavior resulting in electrochemical performance. This porous carbon matrix enhances the structural support and accelerates the electron transport, as shown in Figure 15.6b. Jiang et al. [56] explored a novel synthesis method in which high-entropy Prussian blue analogues (HEPBAs) are prepared at room temperature by combining of mechanochemistry with wet chemistry. They focused on the electrochemical performance of these HEPBAs, particularly their CV and specific capacitance. As shown in Figure 15.6c, the CV curves of the HEPBAs exhibited redox peaks, which indicate their strong pseudocapacitive behavior. The composition $K(MgMnCoNiCu)Fe(CN)_6$ of HEPBA demonstrates pronounced redox peaks, indicating efficient charge storage capabilities. The variation in peak positions and current responses between different HEPBAs is highly affected by the elemental composition. In Figure 15.6d, a comparative analysis of the specific capacitance values of the synthesized HEPBAs is presented in comparison with the traditional single-metal Prussian blue analogues (PBAs) [56]. $K(MgMnCoNiCu)Fe(CN)_6$ obtained a higher specific capacitance

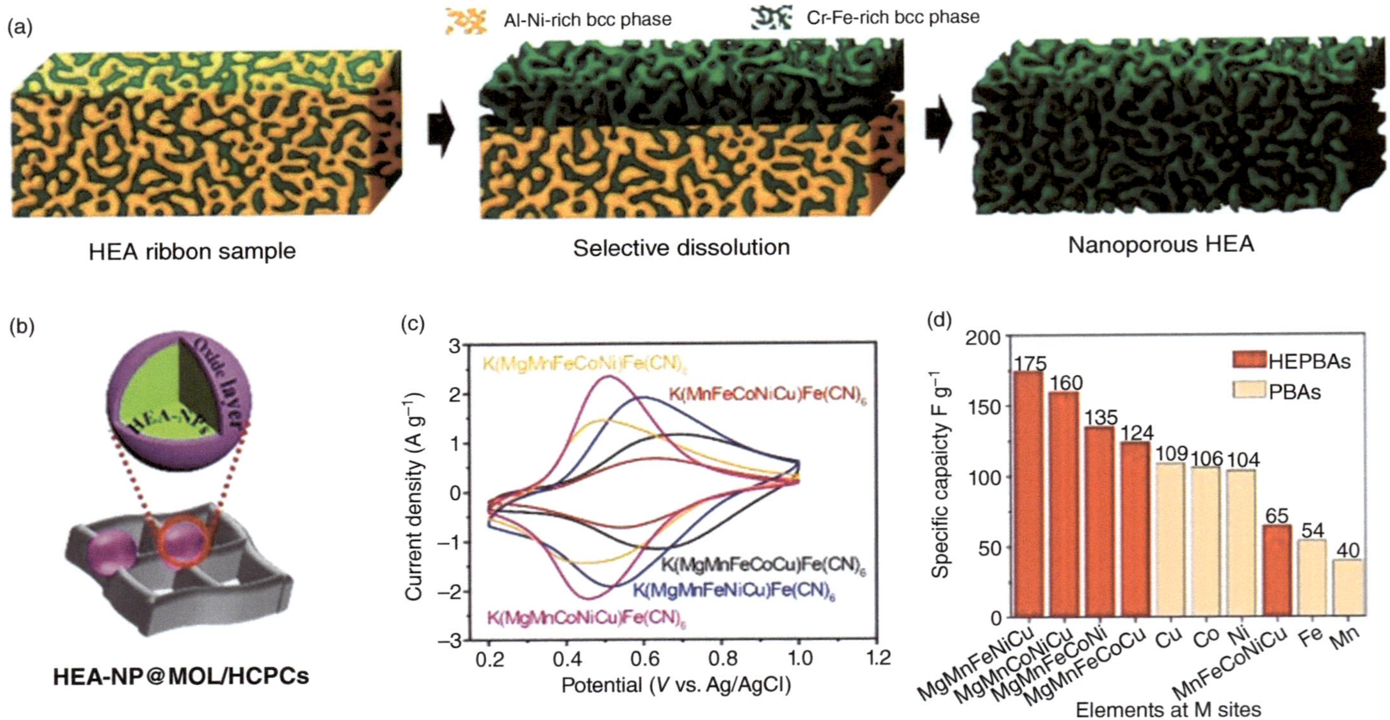

Figure 15.6 (a) Schematic of the phase-dissolving process in phase-separated HEA. *Source:* Reproduced from Ref. [54] / with permission of Elsevier. (b) HEA-NP@MOL/HCPC synthesis methods. *Source:* Reproduced from Ref. [55] / with permission of John Wiley & Sons. CV of $KMFe(CN)_6$ in 1 M Na_2SO_4 at 5 mV s^{-1}, and (d) summary of specific capacitances for HEPBAs and single component PBAs at 5 mV s^{-1}. *Source:* Reproduced from Ref. [56] / with permission of Elsevier.

Table 15.3 List of HEMs used as supercapacitor electrodes with their performance.

Material composition	Synthesis method	Specific capacitance ($F\,g^{-1}$)	Test condition	Cycling stability (%)	Retention cycles	References
HEA NP@ MOL/HCPC	*In situ* reduction and carbonization at 800°C	495.4	$10\,A\,g^{-1}$ current density	94.7	15,000	[55]
AlCoCrFeNi (nanoporous, Cr-Fe-rich)	Selective phase dissolution in 1 M H_2SO_4 at 313 K (Al-Ni phase removed)	700 F/cm^3	$1\,mA\,cm^{-2}$ (approx. $0.14\,A\,g^{-1}$) Current density	97	3,000	[54]
$BaTi_{0.97}Ca_{0.03}O_{2.97}$	Conventional solid-state reaction	$4.89\,J\,cm^{-3}$ (energy storage density)	Electric field of $240\,kV\,cm^{-1}$	91.2	—	[57]
HEA–CNT nano-composite	CVD with sol-gel precursors	286.6	$10\,mV\,s^{-1}$ scan rate	100	15,000	[9]
FeCoCr $MnNi_3O_4$ HEO	Sol-gel method	230	0.01 ($10\,mV\,s^{-1}$) scan rate	—	—	[58]
NiCoMo MnZn-LDH	Alkali etching	1,810.2	$0.5\,A\,g^{-1}$ current density	89.2	3,000	[52]
Fe-Co-Ni-Cu-Zn	Induction melting + ball milling + annealing	325.1	$1\,A\,g^{-1}$ current density	80	5,000	[59]
$(CrMnFeCoNi)_3O_4$	Copre-cipitation method	239	$0.5\,A\,g^{-1}$ current density	76	1,000	[60]

of $176\,F\,g^{-1}$, which is attributed to a synergistic effect of several metal ions in the high-entropy structure. These results highlight the superiority of HEMs as advanced electrode materials for supercapacitors with improved electrochemical performance. Table 15.3 presents another list of HEMs used in supercapacitor applications and their performance.

15.5 Manufacturing and Industrial Challenges in Energy Storage Applications

HEMs, due to their high-entropy, exceptional stability, and various applications in energy storage and related fields, have attracted the attention of industrial researchers. Despite having these qualities, limitations in synthesis, scalability, affordability, and performance tuning hinder its broad applications. In this section, we discuss the key challenges and possibilities related to associating HEMs with large-scale energy storage applications.

15.5.1 Industrial Challenges

15.5.1.1 Challenges in Production and Scalability

Although HEMs have shown their potential on a laboratory scale, the major challenge is enhancing the production volume. Almost all the preparation methods, including alloying, arc melting, and spark plasma sintering, are energy-consuming, less efficient, and less cost-effective [16]. For example, the arc melting method needs high-purity elements (>99.5%) under a controlled environment, which increases the production cost by 41% as compared to traditional alloys [61]. Additionally, in bulk production, maintaining phase stability and uniform composition is difficult due to the intrinsic multicomponent property. Thus, in most cases, phase separation and irregularity in the microstructure are observed [62, 63].

Potential Solutions Adopting cost-effective synthesis techniques: Sol-gel and highly effective chemical vapor deposition (CVD) methods can be used. With benefits including lower processing temperatures and higher homogeneity, the sol-gel process has emerged as an effective method for synthesizing nanocrystalline HEMs. This technique eliminates the limitations of high mixing entropy while enabling the formation of solid solution alloys. Moreover, HEOs with high-rate energy storage capacities have been prepared successfully using the sol-gel synthesis method [1, 13, 64].

Using additive manufacturing (3D printing): To improve compositional irregularity and process control, additive manufacturing (AM, 3D printing) can be helpful. HEAs with tuned microstructures and enhanced mechanical characteristics have been fabricated using AM such as laser powder bed fusion (LPBF) [65] and laser cladding. Defect-free, high-mechanical-strength, and wear-resistant HEMs can be prepared by the AM method due to its ability to precisely manipulate the microstructure and composition [66]. Tao et al. prepared $CoCrFeNiMo_{0.2}Nb_{0.2}/WC$, a composite made up of HEA and ceramic particles, using laser cladding, which provides a way to improve the high-temperature wear resistance of die-casting molds [63]. In another study, Lin et al. used LPBF to synthesize $FeCoCrNiMo_{0.5}$ HEA with exceptional strength and ductility at high temperatures [67].

15.5.1.2 Raw Material Cost and Availability

HEMs are challenging to implement in large-scale applications because of the high cost and low availability of exotic materials such as nickel, vanadium, and cobalt, which require carbon-intensive extraction (~6.97 GJ/metric ton energy consumption) [9, 61]. When compared with traditional materials, which have a good supply chain and are cost-effective, HEMs are difficult to prepare, thus making resource management very challenging [55] (Table 15.4).

Table 15.4 List of HEMs with various synthesis methods.

Material name	Synthesis method	Chemicals used	Temperature and duration	Morphology	Size (nm)	References
NiCoMo MnZn-LDH	Template re etching from MOF	Ni, Co, Mo, Mn, Zn MOF precursor	~120 °C (hydrothermal), 400–500 °C calcination	Layered nanoflowers	50–200	[52]
AlCo CrFeNi HEA	Selective phase dissolution	AlCoCr FeNi alloy in 1 M H2SO$_4$	313 K (~40 °C), 12 hours	Nanoporous (Cr–Fe-rich)	10–50 (pores)	[54]
$(CrMn FeCoNi)_3O_4$	Reverse coprecipitation + oxidation	Metal nitrates of Cr, Mn, Fe, Co, Ni	800 °C for 4 hours	Spinel nanoparticles	~30	[60]
HEA NP@ MOL/ HCPC	*In situ* reduction + carbonization	Fe, Co, Ni, Cu, Sn salts in polymer framework	Pyrolyzed at 800 °C	NPs on porous carbon	5–10	[55]
HEO–CNT composite	Sol-gel method + CVD	Metal precursors + CNT template	Calcined at 700 °C	Hybrid composite	20–50	[9]

Potential Solutions

Utilizing Earth-Abundant Elements: Expensive and rare materials can be replaced by elements such as iron, aluminum, manganese, and zinc [68]. For example, $MgFe_2O_4$, a high-entropy spinel oxide, was synthesized using Mg, Cu, and Zn that enhance stability and maintain structural integrity and electrochemical performance while reducing material cost by 65% [67]. In a recent study, Erdil et al. [69] demonstrated that high-entropy spinel ferrites synthesized from earth-abundant divalent cations (Mg, Cu, Zn) maintained identical crystal structures and electrochemical performance to traditional counterparts while offering improved sustainability and over 200 h of continuous battery operation. Spinel oxides made up of these earth-abundant elements show capacities of more than 600 mA h g^{-1} over 150 cycles, better than traditional spinel oxides such as Fe_3O_4 and $MgFe_2O_4$ [70].

Recycling Procedure: Improved techniques such as alloy mixing and catalytic upcycling [71] efficiently recover HEMs that are at the end of their productive lifespan [61]. For example, by melting alloy residuals such as 304 l stainless steel, electrical-grade copper, and nichrome 80 along with lower amounts of manganese and chromium, it is possible to synthesize near-equimolar CrCuFeMnNi HEA, as reported by Hariharan et al. [71]. With a 50% enhancement in yield strength and by reducing the cost of manufacturing, the resultant alloy exhibited the same microstructure as alloys made from pristine elements.

15.5.1.3 Challenges with Stability and Performance Optimization

Theoretically, HEMs show effectiveness in energy storage applications such as solid-state electrolytes, supercapacitors, LIBs, and SIBs. However, low corrosion resistance, low electrochemical stability, conductivity optimization, and cycling degradation limit their vast applications [72]. For example, under long-term cycling conditions, it has been found that HEOs undergo sluggish ionic diffusion and phase instability. These problems may reduce their cycle stability and energy storage capability. Recent research claims that complex multicomponent HEOs can lead to uneven ionic pathways and unstable phase transformations during long battery cycling; hence, the overall electrochemical performance is affected [73].

Potential Solutions

Material Doping Strategies: If HEMs are doped with several functional materials such as vanadium and tungsten, electrode stabilization and enhancement in electrical conductivity will be observed. Several techniques, such as elemental doping and nanostructuring, have been designed to enhance electrical conductivity and electrochemical performance as well, which has been demonstrated in several studies [12].

Entropy-Driven Ion Migration: This improves ionic conductivity by stabilizing disordered channels for ion diffusion in an effective way, lowering the energy barriers, and enhancing the configurational entropy, as discussed by Zeng et al. [9]. Zeng et al. showed that by decreasing reliance on certain chemistries and enhancing synthesizability, the ionic conductivity of a compound can be improved through the inclusion of various mixtures of metal cations. This method facilitates ion mobility to create a disordered structure through the high-entropy effect [74]. This development of high-entropy electrolytes (HEEs) has shown promise in improving the electrochemical performance of rechargeable batteries. According to the work by Ren et al. [75], HEEs can attain stability over a wide temperature range and enhance ionic conductivity by increasing the configurational entropy. Recent findings in high-entropy halide solid electrolytes have shown that a disordered lattice structure can promote lithium-ion mobility. Low activation energy and higher ionic conductivity resulting from this can be advantageous for all solid-state battery applications [76].

Nanoporous Architecture: The synergistic d-band electronic states and high surface area of dealloyed HEAs with 12+ metallic elements lead to higher catalytic current densities compared to traditional catalysts [77]. Recently, Zhang et al. [78] reviewed the emerging role of nanoporous HEAs and metallic glasses as advanced electrocatalysts for electrochemical water splitting, emphasizing that their three-dimensional interconnected structures and local chemical ordering significantly enhance catalytic activity and long-term stability. Nanoporous HEAs are highly effective as electrode materials, offering a higher number of active sites resulting from increased surface area. Besides, they possess remarkable stability and effective catalytic properties; hence, they are ideal for industrial applications requiring durable and potent catalytic materials [79]. Lanlan et al. [16] showed that novel synthesis techniques used in developing nanoporous HEAs promise stability and homogeneity and result in better catalytic activity and longer lifespans, so they can be extensively used in industrial applications.

Corrosion Resistance Strategies: This strategy becomes crucial for aqueous-based and marine energy storage systems, where galvanic corrosion may affect HEMs due to their multielemental nature. As per recent studies, HEAs with several elemental compositions can lead to differential corrosion potentials, which could speed up material degradation. In HEMs, by implementing surface treatments such as plasma electrolytic oxidation (PEO) and using protective coatings, their corrosion resistance can be greatly enhanced [80]. By forming dual-layer oxides, AlCoCrFeNi alloys reduce corrosion in NaCl solution [34], while Mo/Ti doping reduces chloride adsorption by increasing pitting resistance. In marine environments, TiZrHfNbTa coating by the magnetron-sputtering method exhibits exceptional wear resistance [7].

Surface Engineering: FeConiCrMn HEAs with denser and stronger surfaces are generated using laser fusion AM [79]. By using AI models and DFT studies, HEA phase stability and ideal doping can be predicted. As shown in a previous study [12], plasma oxidation recycling of HEAs generates solar-reflective coatings for energy-efficient buildings.

15.5.2 Future Perspectives and Opportunities for Industrial Deployment

Incorporation of HEAs Into Commercial Energy Storage Systems: HEAs, due to their intrinsic property, exceptional electrochemical performance, and structural robustness, can be used in next-generation batteries such as solid-state batteries, SIBs, and supercapacitor applications. When compared to traditional cathode materials used in LIBs, recent studies have shown that cation-disordered high-entropy cathodes exhibit enhanced cycling stability and capacity retention [6]. High-entropy spinel oxides can be used as electrodes, showing better electrochemical behavior, and have the potential to be used as large-scale commercial energy storage systems such as EV batteries and grid energy storage [8].

15.5.2.1 Potential Research Areas

There are several unexplored prospects for industrial energy storage applications of HEMs, some of which are explained below.

High-Entropy Solid Electrolytes: Solid-state electrolytes consisting of HEOs and ceramics possess tremendous ionic conductivities and thermodynamic stabilities. Further in-depth research is required in this area to produce safer and more efficient solid-state batteries [75].

Entropy-Driven Electrodes: Although there has been several studies on these types of electrodes, there are many possibilities to improve. Nanoporous high-entropy electrodes can show potential in catalytic applications, especially in electrolyzers and metal-air batteries. By surface engineering, such as higher surface area and synergistic electronic interactions, catalytic current densities can be enhanced and thus can be useful for storage capacity [9].

High-Entropy Spinel Ferrites: Using abundant earth elements such as magnesium, iron, copper, and zinc, more high-entropy spinel ferrites can be synthesized with better potential than the existing ones. These can show superior electrochemical performance than conventional electrode materials. Deeper research on this is required to produce cost-effective and suitable environmental-friendly electrodes that are highly beneficial for large-scale industrial battery applications and supercapacitor technologies [70].

Advanced Dielectric Capacitors: Perovskites, relaxor ferroelectrics, and other high-entropy ceramics can be used as potential capacitor dielectric materials. They possess increased dielectric constants, breakdown strengths, and higher thermal stability, and thus, they can be implemented in high-frequency power electric uses, modern EVs, and pulsed power technologies [57].

Use of Artificial Intelligence (AI): Implementation of AI and machine learning (ML) algorithms can be a highly effective technique in the optimization, discovery, and prediction of new HEM composite synthesis. It is an effective and rapid way to explore very large compositional spaces of HEMs suitable for any specific applications [15].

Safety Improvements in Battery Technologies: In several industries such as aerospace, automobile, nuclear, and defense, risks related to thermal runways, mechanical malfunctions, and fires are present. As high-entropy solid electrolytes have better thermal stability, they can be applied to enhance safety.

Scaling Up the Synthesis and Recyclability: Although HEMs exhibit tremendous applications in energy storage, large-scale synthesis for industrial uses is still limited because of higher complexity and cost. There should be more research on new cost-effective and productive synthesis methods such as microwave-assisted synthesis, continuous flow reactors, and many more advanced techniques that can be helpful. Furthermore, there should be more focus on HEMs and their composites derived from more abundant and recyclable resources so that the cost can be reduced.

15.6 Summary and Conclusions

In summary, HEAs and related HEMs provide significant benefits over conventional electrode materials, representing a transformative leap forward in industrial energy storage applications. The inherent multicomponent property and the resulting very high configurational entropy give rise to enhanced structural resilience, electrochemical stability, and ionic and electronic conductivity. Because of higher cycle stability, HEMs can be used in several energy storage systems such as LIBs and SIBs, supercapacitors, and solid-state batteries. Above all, as cost-effective and abundant elements are used in HEMs, they can be used in long-term industrial applications.

Despite significant promises, HEMs face several difficulties that need to be addressed to better understand their industrial potential. There are several conventional techniques, such as mechanical alloying, spark plasma sintering, and arc melting, that are economically challenging for large-scale production. Therefore, research on the development of scalable fabrication methods such as 3D printing and sol-gel techniques should be prioritized, which can provide better control over microstructures by minimizing energy consumption. The cost and availability of raw materials are another major challenge for elements such as cobalt, nickel, and vanadium. Hence, the substitution of rare and costly elements with abundant and eco-friendly alternatives such as magnesium, iron, zinc, and copper is needed for economic feasibility.

There are several challenges, such as improving corrosion resistance, electrochemical stability, and performance tuning. In addition, problems such as sluggish ionic diffusion, phase instability, and nonuniform ionic pathways that arise from the intrinsic multielemental complexity adversely affect the long-term performance of HEMs. Thus, to fix these issues, extensive research is needed into entropy-driven ion migration techniques, material doping, and surface engineering strategies. Besides, integration of recently evolving techniques such as AI into this research is capable of predicting new combinations of HEM composition and their optimization and synthesis methods, which can significantly accelerate material discovery and performance reliability.

To conclude, there are substantial opportunities for the large-scale production and commercialization of HEMs. The strategic implementation of HEMs into battery components, such as advanced cathodes, solid electrolytes, and thin film techniques, can potentially transform the battery industry with enhanced efficiency and reliability while improving safety concerns. There should be more in-depth research on high-entropy spinel ferrites to enhance their electrochemical and dielectric performance so that they can be extensively used in large-scale EV batteries, grid-level storage, pulsed-power technologies, and high-frequency power innovations. HEMs have the potential to emerge as a key component of sustainable energy storage solutions if focused efforts are made toward overcoming the challenges regarding synthesis techniques, cost, and performance.

Acknowledgement

J.H. acknowledges the financial support from the National Research Foundation of Korea (NRFK Grant No. RS-2024-00460248). This research was supported by Semiconductor-Secondary Battery Interfacing Platform Technology Development Project of National Nano Fab Center (NNFC).

References

1 Petrovičovà, B., Xu, W., Musolino, M.G. et al. (2022). High-entropy spinel oxides produced via sol-gel and electrospinning and their evaluation as anodes in Li-ion batteries. *Appl Sci* 12 (12).

2 Panda, P.K., Huang, H.Y., Dash, P. et al. (2025). Liquid-phase microwave synthesis of platinum-based high-entropy alloy catalysts on carbon supports for electrochemical hydrogen adsorption/desorption and oxygen evolution/reduction reactions. *Int J Hydrog Energy* 111: 536–545.

3 Liu, K.C., Kumar Panda, P., Chandra Mallick, B. et al. (2024). Solid-phase microwave synthesis of high-entropy graphene quantum dots as metal-free electrochemical catalysts. *Appl Surf Sci* 648: 159061.

4 Gu, S., Huang, H.Y., Lin, Y.C. et al. (2025). High-entropy alloys with low platinum content: novel catalysts for oxygen reduction and evolution reactions. *Emerg Mater* 8.

5 Li, C., Chen, X., Li, J. et al. (2024). High entropy anodes in batteries: From fundamentals to applications. *Energy Storage Mater* 71.

6 Lun, Z., Ouyang, B., Kwon, D.H. et al. (2021). Cation-disordered rocksalt-type high-entropy cathodes for Li-ion batteries. *Nat Mater* 20 (2): 214–221.

7 Liu, D., Guo, P., Pan, H., and Wu, R. (2024). Emerging high-entropy compounds for electrochemical energy storage and conversion. *Prog Mater Sci* 145.

8 Ma J, Huang C. High entropy energy storage materials: Synthesis and application: J Energy Storage. 2023;66.

9 Ma, Y., Ma, Y., Wang, Q. et al. (2021). High-entropy energy materials: challenges and new opportunities. *Energy Environ Sci* 14 (5): 2883–2905.

10 Yang, B., Zhang, Y., Pan, H. et al. (2022). High-entropy enhanced capacitive energy storage. *Nat Mater* 21 (9): 1074–1080.

11 Moskovskikh, D., Vorotilo, S., Buinevich, V. et al. (2020). Extremely hard and tough high entropy nitride ceramics. *Sci Rep* 10 (1).

12 Amiri, A. and Shahbazian-Yassar, R. (2021). Recent progress of high-entropy materials for energy storage and conversion. *J Mater Chem A* 9 (2): 782–823.

13 Torralba, J.M., Meza, A., Kumaran, S.V. et al. (2025). From high-entropy alloys to alloys with high entropy: a new paradigm in materials science and engineering for advancing sustainable metallurgy. *Curr Opinion Solid State Mater Sci* 36: 101221. http://arxiv.org/abs/2502.17279.

14 Jien-Wei, Y., Swee-Kai, C., Su-Jien, L. et al. (2004). Nanostructured high-entropy alloys with multiple principal elements: novel alloy design concepts and outcomes. *Adv Eng Mater* 6 (5): 299–303.

15 Lokhande, V., Malavekar, D., Kim, C. et al. (2024). Order within disorder: Unveiling the potential of high entropy materials in energy storage and electrocatalysis. *Energy Storage Mater* 72.

16 Fu, M., Ma, X., Zhao, K. et al. (2021). High-entropy materials for energy-related applications. *iScience* 24 (3).

17 Wei, Y., Zhao, Y., Chen, Y. et al. (2023). Lithium storage characteristic of nanoporous high-entropy alloy@high-entropy oxide with spin-dependent synergism of cations. Chem. *Engl J* 476.

18 Jianping, M. and Chengde, H. (2023). High entropy energy storage materials: synthesis and application. *J Energy Storage* 66.

19 Liu, C., Bi, J., Xie, L. et al. (2023). High entropy spinel oxides (CrFeMn NiCox)3O_4 ($x = 2, 3, 4$) nanoparticles as anode material towards electrochemical properties. *J Energy Storage* 71.

20 Cantor, B. (2021). Multicomponent high-entropy Cantor alloys. *Prog Mater Sci* 120.
21 Rost, C.M., Sachet, E., Borman, T. et al. (2015). Entropy-stabilized oxides. *Nat Commun* 6.
22 Miracle, D.B. and Senkov, O.N. (2017). A critical review of high entropy alloys and related concepts. *Acta Mater* 122: 448–511.
23 Berlijn, T., Alvarez, G., Parker, D.S. et al. (2021). Simulating spin waves in entropy stabilized oxides. *Phys Rev Res* 3 (3).
24 Zhang, Y., Bin, J.Z., Sun, S.K. et al. (2019). Microstructure and mechanical properties of high-entropy borides derived from boro/carbothermal reduction. *J Eur Ceram Soc* 39 (13): 3920–3924.
25 Wang, M., Dong, C.L., Huang, Y.C., and Shen, S. (2020). Operando spectral and electrochemical investigation into the heterophase stimulated active species transformation in transition-metal sulfides for efficient electrocatalytic oxygen evolution. *ACS Catal* 10 (3): 1855–1864.
26 Fracchia, M., Coduri, M., Ghigna, P., and Anselmi-Tamburini, U. (2024). Phase stability of high entropy oxides: a critical review. *J Eur Ceram Soc* 44 (2): 585–594.
27 Jiang, B., Yu, Y., Cui, J. et al. (2021). High-entropy-stabilized chalcogenides with high thermoelectric performance. *Science (80-)* 371 (6531): 830–834.
28 Song, H., Tian, F., Hu, Q.M. et al. (2017). Local lattice distortion in high-entropy alloys. *Phys Rev Mater* 1 (2).
29 Wan, Q., Zhang, F., and Xiong, Y. (2024). Facile synthesis, characterization, and mechanical properties of spinel-structured high-entropy oxides: lattice distortion and sluggish diffusion effects induced by aluminum cation. *J Eur Ceram Soc* 44 (11): 6629–6640.
30 Wang, L., Zhang, L., Lu, X. et al. (2023). Surprising cocktail effect in high entropy alloys on catalyzing magnesium hydride for solid-state hydrogen storage. Chem. *Engl J* 465.
31 Richardson, S.M., Bella, F., Mougel, V., and Milić, J.V. (2021). Scientific writing and publishing for early-career researchers from the perspective of young chemists. *J Mater Chem A* 9 (35): 18674–18680.
32 Wang, Q., Sarkar, A., Wang, D. et al. (2019). Multi-anionic and -cationic compounds: new high entropy materials for advanced Li-ion batteries. *Energy Environ Sci* 12 (8): 2433–2442.
33 Lin, L., Wang, K., Sarkar, A. et al. (2022). High-entropy sulfides as electrode materials for Li-ion batteries. *Adv Energy Mater* 12 (8).
34 Zhou, S., Pu, Y., Zhao, X. et al. (2022). Dielectric temperature stability and energy storage performance of NBT-based ceramics by introducing high-entropy oxide. *J Am Ceram Soc* 105 (7): 4796–4804.
35 Gonçalves, J.M., Santos, É.A., Martins, P.R. et al. (2023). Emerging medium- and high-entropy materials as catalysts for lithium-sulfur batteries. *Energy Storage Mater* 63.
36 Kim, J.H., Song, A., Park, J.M. et al. (2024). Analogous design of a microlayered silicon oxide-based electrode to the general electrode structure for thin-film lithium-ion batteries. *Adv Mater* 36 (14).

37 Xiao, C.F., Kim, J.H., Cho, S.H. et al. (2021). Ensemble design of electrode-electrolyte interfaces: toward high-performance thin-film all-solid-state Li-metal batteries. *ACS Nano* 15 (3): 4561–4575.

38 Park, J.S., Kim, J.H., Cho, S.H. et al. (2023). Silicon-protected, vertically grown MoS2 nanosheets for high-performance thin-film Li-ion batteries. *Ceram Int* 49 (3): 5538–5542.

39 Behera, S., Ippili, S., Jella, V. et al. (2024). Confluence of ZnO and PTFE binder for enhancing performance of thin-film lithium-ion batteries. *Energy Environ Mater* 7 (5).

40 Kim, J.H., Park, J.S., Cho, S.H. et al. (2022). Stabilizing the surface of Ni-rich cathodes via facing-target sputtering for high-performance lithium-ion batteries. *J Mater Chem A* 10 (47): 25009–25018.

41 Cho, S.H., Kim, J.H., Kim, I.G. et al. (2021). Reduced graphene-oxide-encapsulated mos2/carbon nanofiber composite electrode for high-performance na-ion batteries. *Nanomaterials* 11 (10).

42 Jella, V., Behera, S., Ippili, S. et al. (2024). Self-charging power system empowered by bismuth halide perovskite-based hybrid nanogenerator and lithium-ion battery. *Small* 20: 2400824.

43 Kim, J.H., Jung, J.W., Cho, S.H. et al. (2022). Investigation of ordering on oxygen-deficient $LiNi_{0.5}Mn_{1.5}O_{4-\delta}$ thin films for boosting electrochemical performance in all-solid-state thin-film batteries. *Small* 18 (24).

44 Wang, Q., Sarkar, A., Li, Z. et al. (2019). High entropy oxides as anode material for Li-ion battery applications: A practical approach. *Electrochem Commun* 100: 121–125.

45 Huang, C.Y., Huang, C.W., Wu, M.C. et al. (2021). Atomic-scale investigation of Lithiation/Delithiation mechanism in High-entropy spinel oxide with superior electrochemical performance. *Chem Eng J* 420.

46 Sarkar, A., Velasco, L., Wang, D. et al. (2018). High entropy oxides for reversible energy storage. *Nat Commun* 9 (1).

47 An, Q., Li, S., Zhou, J. et al. (2023). Novel spinel multicomponent high-entropy oxide as anode for lithium-ion batteries with excellent electrochemical performance. *Adv Eng Mater* 25 (20).

48 Liu, X., Wang, H., Dong, L. et al. (2023). Molten salt synthesis, morphology modulation, and lithiation mechanism of high entropy oxide for robust lithium storage. *J Energy Chem* 86: 536–545.

49 Liu, C., Bi, J., Xie, L. et al. (2023). Electrochemical properties of four novel high-entropy spinel oxides used as lithium-ion battery anodes synthesized by the glycine-nitrate SCS method. *J Mater Sci* 58 (19): 8005–8021.

50 Zhao, J., Zhang, Y., Chen, X. et al. (2022). Entropy-change driven highly reversible sodium storage for conversion-type sulfide. *Adv Funct Mater* 32 (45).

51 Cheng, W., Liu, J., Hu, J. et al. (2023). Pressure-stabilized high-entropy $(FeCoNiCuRu)S_2$ sulfide anode toward simultaneously fast and durable lithium/sodium ion storage. *Small* 19 (29).

52 Zhu, Z., Zhang, Y., Kong, D. et al. (2024). A novel high entropy hydroxide electrode material for promoting energy density of supercapacitors and its efficient synthesis strategy. *Small* 20 (20).

53 Jin, T., Sang, X., Unocic, R.R. et al. (2018). Mechanochemical-assisted synthesis of high-entropy metal nitride via a soft urea strategy. *Adv Mater* 30 (23).

54 Kong, K., Hyun, J., Kim, Y. et al. (2019). Nanoporous structure synthesized by selective phase dissolution of AlCoCrFeNi high entropy alloy and its electrochemical properties as supercapacitor electrode. *J Power Sources* 437.

55 Shen, E., Song, X., Chen, Q. et al. (2021). Spontaneously forming oxide layer of high entropy alloy nanoparticles deposited on porous carbons for supercapacitors. *ChemElectroChem* 8 (1): 260–269.

56 Jiang, W., Wang, T., Chen, H. et al. (2021). Room temperature synthesis of high-entropy Prussian blue analogues. *Nano Energy* 79.

57 Li, Z. and wei, Chen Z hui, Xu J jun. (2022). Enhanced energy storage performance of $BaTi_{0.97}Ca_{0.03}O_{2.97}$-based ceramics by doping high-entropy perovskite oxide. *J Alloys Compd* 922.

58 Yin, Y., Zhang, W.B., Zhang, X.L. et al. (2023). Future prospects of high-entropy alloys as next generation industrial electrode material. *Dalton Trans* 52: 9005–9016.

59 Mohanty, G.C., Gowda, C.C., Gakhad, P. et al. (2023). Iron-cobalt-nickel-copper-zinc (FeCoNiCuZn) high entropy alloy as positive electrode for high specific capacitance supercapacitor. *Electrochim Acta* 470: 143272.

60 Talluri, B., Aparna, M.L., Sreenivasulu, N. et al. (2021). High entropy spinel metal oxide $(CoCrFeMnNi)_3O_4$ nanoparticles as a high-performance supercapacitor electrode material. *J Energy Storage* 42: 103004.

61 Han, L., Mu, W., Wei, S. et al. (2024). Sustainable high-entropy materials? *Sci Adv* 10 (50): eads3926.

62 Bolar, S., Ito, Y., and Fujita, T. (2024). Future prospects of high-entropy alloys as next-generation industrial electrode materials. *Chem Sci* 15 (23): 8664–8722.

63 Singh, S., Wanderka, N., Murty, B.S. et al. (2011). Decomposition in multi-component AlCoCrCuFeNi high-entropy alloy. *Acta Mater* 59 (1): 182–190.

64 Zhang, X.L., Bin, Z.W., Yin, Y. et al. (2023). Sol-gel method preparation and high-rate energy storage of high-entropy ceramic (FeCoCrMnNi)C porous powder. *Ceram Int* 49 (17): 29327–29338.

65 Wu, T., Yang, C., Yu, L. et al. (2024). Microstructure characterization and high-temperature wear mechanism of high-entropy alloy matrix composite coating fabricated by laser cladding. *Appl Surf Sci* 677.

66 Mooraj, S., Kim, G., Fan, X. et al. (2024). Additive manufacturing of defect-free TiZrNbTa refractory high-entropy alloy with enhanced elastic isotropy via in-situ alloying of elemental powders. *Commun Mater* 5 (1).

67 Lin, D., Chen, Q., Xi, X. et al. (2024). Laser powder bed fusion to fabricate high-entropy alloy FeCoCrNiMo0.5 with excellent high-temperature strength and ductility. *Mater Sci Eng A* 900.

68 Ritter, T.G., Pappu, S., and Shahbazian-Yassar, R. (2024). High-entropy materials for lithium batteries. *Batteries* 10: 3.

69 Erdil, T., Ozgur, C., Geyikci, U. et al. (2024). Earth-abundant divalent cation high-entropy spinel ferrites as bifunctional electrocatalysts for oxygen evolution and reduction reactions. *ACS Appl Energy Mater* .

70 Nam, K.H., Wang, Z., Luo, J. et al. (2024). High-entropy spinel oxide ferrites for battery applications. *Chem Mater* 36 (9): 4481–4494.

71 Hariharan, K. and Sivaprasad, K. (2022). Sustainable low-cost method for production of high-entropy alloys from alloy scraps. *J Sustain Metall* 8 (2): 625–631.

72 Yao, Z., Chen, Y., Wang, X. et al. (2025). High-entropy alloys catalyzing polymeric transformation of water pollutants with remarkably improved electron utilization efficiency. *Nat Commun* 16 (1).

73 Zhao, J., Gao, S., Wei, H. et al. (2024). High entropy materials for reversible electrochemical energy storage systems. *ChemElectroChem* 11 (7).

74 Zeng, Y., Ouyang, B., Liu, J. et al. (2022). High-entropy mechanism to boost ionic conductivity. *Science (80-)* 378 (6626): 1320–1324.

75 Ren, K.F., Liu, H., Guo, J.X. et al. (2024). Working principles of high-entropy electrolytes in rechargeable batteries. *ACS Energy Lett* 9 (6): 2960–2980.

76 Ye, Y., Gu, Z., Geng, J. et al. (2025). Advanced high-entropy halide solid electrolytes enabling high-voltage, long-cycling all-solid-state batteries. *Nano Lett.*

77 Li, Y., Luo, Z., Skrabalak, S.E., and Xiong, Y. (2024). High-entropy materials in focus. *ACS Mater Lett* 6 (7): 2696–2697.

78 Zhang, Y., Cai, Y., and Ding, Y. (2025). Nanoporous high-entropy alloys and metallic glasses: advanced electrocatalytic materials for electrochemical water splitting. *Chem Commun* .

79 Li, X., Wang, Y.Q., Zhang, J.Y. et al. (2023). Research progress of high-entropy alloy coatings. *Surf Technol* 52 (1): 1–20.

80 Dehnavi, V., Shoesmith, D.W., Luan, B.L. et al. (2015). Corrosion properties of plasma electrolytic oxidation coatings on an aluminium alloy—the effect of the PEO process stage. *Mater Chem Phys* 161: 49–58.

16

Current Status, Challenges, and Prospects of High-Entropy Materials

Swagat Kumar Purohit[1], Abhaya Kumar Mishra[2], Deepak Kumar Pradhan[3], and Arpan Kumar Nayak[1]

[1] *Regional Institute of Education (NCERT), Department of Education in Science and Mathematics, Mysore, Karnataka, 570006, India*
[2] *Department of Materials Science and Engineering, Clemson, SC, 29634, USA*
[3] *Physics Section, Department of Education in Science and Mathematics, Regional Institute of Education (NCERT), Mysore, Karnataka, 570006, India*

16.1 Introduction

World energy consumption continues to increase and is expected to reach 30–50 terawatt (TW) in 2050, a staggering figure equivalent to 20 billion tons of oil per year, which poses a significant concern [1, 2]. This projection underscores rapid fossil fuel depletion, which contributes 85% of the total energy [3]. Moreover, burning of fossil fuels also causes environmental issues and global warming, leading to an adverse effect on future generations [4]. These challenges demand transitions toward an environment of sustainable, reliable, and renewable energy [5]. In this regard, the sun, wind, and tides are primary renewable power sources [6, 7]. However, their intimate nature, such as solar panels at night, necessitates the development of an advanced energy storage system [8]. Electrochemical devices, such as supercapacitors and batteries, have been employed as charge storage devices [9–12]. The so-called electrocatalysts are essential for the advanced applications of these devices [13]. Therefore, innovative strategies for electrocatalysts are strongly desired for better device performance.

High-entropy materials (HEMs) as electrocatalyst materials have been received tremendous attention in recent years due to their multielement composition and homogeneously mixed solid solution state [14–16]. Since the first report of crystalline high-entropy alloys (HEAs) in 2004, researchers have explored new classes of HEMs such as oxides, carbides [17–19], sulfides [20], high-entropy ceramics [21], halides [22], and perovskites [23, 24]. These materials have been applied to battery cathodes, anodes, solid electrolytes, and supercapacitors in recent years. HEMs have at least five elements, with the concentration of each element varying between 5 and 35 atomic percentages and mixed configuration entropy $S \geq 1.5R$ [25–27]. These intrinsic elements prefer single-phase solid solutions

High-Entropy Materials for Energy Storage Devices, First Edition.
Edited by Chien-Te Hsieh, Pradeep Kumar Panda, and Arpan Kumar Nayak.

with simple crystal structures due to the increased system entropy rather than multiphase compositions. Four different core effects are associated with HEMs, which include (i) high-entropy effect, (ii) severe lattice effect, (iii) sluggish diffusion effect, and (iv) cocktail effect. These effects provide the basis to understand the unique properties of HEMs. (i) High-entropy effect refers to the increase in the entropy of a system with an increase in the number of components. It promotes the formation of a single-phase solid solution rather than an intermetallic compound, along with the excellent stability of HEMs [28]. (ii) Lattice distortion effect arises due to the difference in the atomic size among the constituent atoms of HEMs [29]. The same probability of different atoms to occupy a lattice site causes lattice distortion, which can significantly vary the optical, electrical, thermal, and chemical behavior of the material. (iii) Sluggish diffusion effect is generated by the energy barrier from the lattice distortion, which contributes to nanocrystalline and amorphous structure by hindering atomic diffusion [30]. (iv) Cocktail effect refers to the synergistic responses from all constituent elements, resulting in exceptional properties [31]. All these factors collectively provide HEAs with numerous versatile unexpected properties and hence make them suitable under extreme conditions. Scheme 16.1 represents a schematic of various HEMs for the fabrication of energy storage devices.

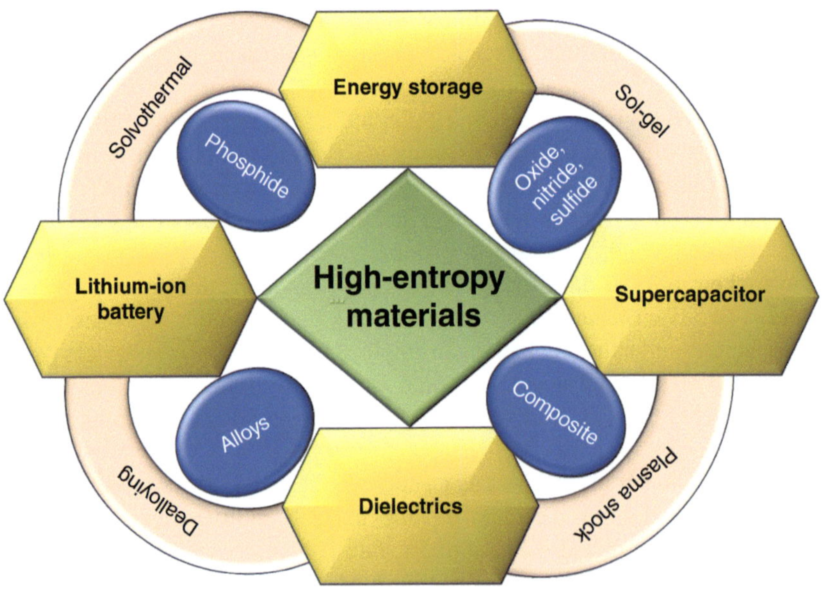

Scheme 16.1 Schematic of various HEMs used in energy storage devices.

16.2 Background

In order to study the current status of HEMs and energy storage devices, the PRISMA method was used for the selection of studies. Data from Scopus electronic database published between May 2025 and June 2025 were analyzed. The process comprised three phases: in the first phase, the inclusion and exclusion criteria were determined; in the second phase, the search strategy was selected; and in the third phase, the search was conducted and articles were chosen. The data were codified and analyzed quantitatively using VOS viewer version 1.6.20 software. Inclusion criteria for this bibliometric analysis were the following: English-language original research articles published between 2020 and 2025 were considered to capture recent trends. Documents were limited to Scopus subject areas relevant to the study, including Materials Science (MATE), Chemistry (CHEM), Engineering (ENGI), Energy (ENER), Physics (PHYS), Multidisciplinary (MULT), and Decision Sciences (DECI), reflecting the interdisciplinary nature of the research domain. Exclusion criteria included non-English publications and non-article studies such as reviews, editorials, conference papers, letters, and book chapters to ensure methodological consistency. Additionally, articles from irrelevant domains such as medicine, humanities, and social sciences were excluded to maintain disciplinary focus. Publications outside the 2020–2025 timeframe were also omitted to ensure the inclusion of contemporary and emerging research only. These criteria ensured a high-quality, focused dataset for effective bibliometric analysis. The search strategy considered two major fields: **HEMs** and **energy storage devices**, based on the following terms or key words: ("high-entropy material" OR "high-entropy alloy" OR "high-entropy oxide" OR "entropy-stabilized material*") AND ("supercapacitor" OR "lithium-ion battery" OR "Li-ion battery" OR "dielectric capacitor" OR "energy storage device" OR "electrochemical storage" OR "solid-state capacitor" OR "energy storage application" OR "electrode material"). In the Scopus database, the search was limited only to the title, abstract, and keywords, which returned 1,013 records. Then, abstracts of these articles were read, and 320 articles were accepted based on the inclusion criteria. Finally, 288 articles were considered for the analysis.

The bibliometric overview of country-level contributions to research on energy storage materials is presented in Figure 16.1. China significantly dominates the discipline, as evidenced by its 4,656 citations and 195 documents. Germany has contributed 27 documents and 1,107 citations, followed by the USA with 35 documents and 891 citations, indicating a high impact per paper. Taiwan and South Korea have also made substantial contributions, notably in terms of citations, with 1,042 and 273, respectively. India has shown active participation with 32 documents, despite a relatively modest citation count of 424. Countries such as the UK, Italy, and France have moderate productivity and impact. In general, these data emphasize a significant concentration of high-impact research in East Asia, North America, and certain regions of Europe, with China serving as the central hub for global collaboration and output in this field.

16.3 Synthesis Method for HEMs

Synthesis of single-phase HEMs with at least five elements as constituents remains a significant challenge. The inherent differences in physical and chemical properties among constituent elements commonly result in high chances of abnormal

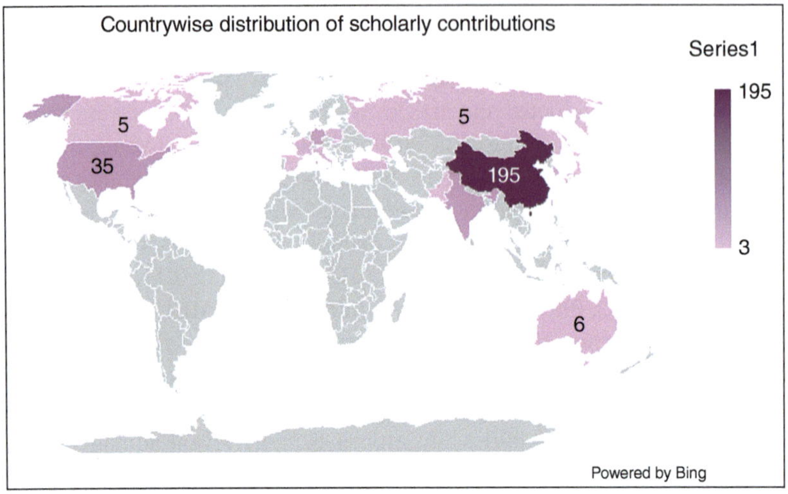

Figure 16.1 Countrywise distribution of scholarly contributions in energy storage materials.

nucleation and growth and secondary phase formation that drastically alter the desired properties [32]. Therefore, various synthesis methods have been explored to synthesize application-grade HEMs. For instance, carbon thermal shock synthesis of HEM materials involves a high temperature (~2,000 K) and a rapid cooling rate ($105\,\mathrm{K\,s^{-1}}$), facilitating atomic diffusion [33]. This process significantly influences the extent of the non-equilibrium state, while promoting structural ordering, ultimately leading to single-phase HEAs. Additionally, this method enables the synthesis of uniform nanostructures with optimized surface properties for advanced material applications, which is absent in the bulk form. Other kinetic mechanism-driven techniques based on carbon thermal shock, including sputtering, laser irradiation, and microwave heating, have been employed to synthesize high-entropy oxides (HEOs) [34–36]. However, the industrial application of these techniques is restricted by the complexity of instrumentation, limited scalability, material loss during synthesis, and the expensive nature of the process. The ball milling process is a simple and effective method to promote the formation of HEA powder. In this process, precursor powders are mixed and subjected to rotation in a chamber along with grinding media (steel or ceramic balls) [37]. The long-term high-speed friction continuously reduces the particle size while localized heating promotes the lattice defects, contributing to the formation of interatomic bonds. These techniques have been widely employed to synthesize various HEMs to address the need for energy storage devices. However, material contamination from milling media, particles with a relatively larger size, and high energy consumption are the major challenges associated with these techniques. Wet chemistry synthesis of HEMs is also a promising alternative to high-temperature synthesis methods, offering better control over material properties [38]. Synthesis parameters, such as precursors and their concentration, reaction medium, capping agent, reaction temperature, and duration, play a key role in designing and controlling the morphology of HEMs. Despite their advantages, solution-based chemical synthesis routes pose challenges such as decomposition of metal precursors and unwanted phase formation. However, optimizing nucleation and growth mechanisms, employing strong capping agents, and carefully managing drying and powder formation steps can potentially overcome the present challenges, paving the way for the synthesis of scalable and high-performance materials [39]. On the other hand, electrodeposition provides an effective alternative for synthesizing HEMs by adjusting nucleation and growth via depositing metal ions from an electrolyte by applying current or voltage. Recent literature has demonstrated its ability to produce nanomaterials as small as 10 nm with high-entropy while offering cost-effective processing, minimal equipment requirement, and high scalability[40]. Additionally, by controlling electrolyte concentration and applied current or voltage, the thickness of the deposited HEMs can be precisely controlled, facilitating optimized and tailored materials with optimal properties and performance in electrochemical devices. However, drifting of applied voltage and current can lead to the synthesis of nonuniform nanomaterials, affecting their functional properties.

Entropy catalyst	Synthesis methods	Particle size	Catalyst characteristics	Application	References
$(Co,Cr,Fe,Mn,Ni)_3O_4$	Coprecipitation	61±10 nm	Phase-pure spinel structure	Super-capacitor	[41]
$(Mn,Fe,Co,Ni,Cu)_3O_4$	Thermal plasma	16–32 nm	Phase-pure spinel structure	Super-capacitor	[42]
$(Na,Ni,Mn,Li,B)O_2$	Sol-gel	2–3 μm	Structural stability	Na-ion batteries	[43]
$Li_{5.4}La_3(Zr_{0.4}Hf_{0.4}Nb_{0.4}Ta_{0.4}W_{0.4})O_{12}$	Sol-gel		Higher Li-ion conductivity	Solid electrolyte	[44]
$Li_xMn_{0.6}Ni_{0.1}Co_{0.1}Fe_{0.1}Al_{0.1}O_y$	Coprecipitation	52 ± 14 nm	Oxygen vacancies	Li-ion batteries	[45]
$(Co_{0.2}Cr_{0.2}Fe_{0.2}Mn_{0.2}Ni_{0.2})_3O_4$	Polyacrylamide gel	52 nm	Good crystallinity	Super-capacitor	[46]
$Sb_{1.4}Bi_{0.2}Sn_{0.2}Co_{0.1}Mn_{0.1}Te_3$	Dealloying		Structural stability and volume variation	Potassium-ion batteries	[47]
$(Sr_{0.6}Bi_{0.2}Na_{0.2})(Ti_{1-x}Zr_{x/2}Al_{x/4}Nb_{x/4})O_3$	Solution combustion		High-entropy ceramics	Dielectric energy storage device	[48]

Entropy catalyst	Synthesis methods	Particle size	Catalyst characteristics	Application	References
$CoVMnFeZnPS_3$ ($HEPS_3$)	Electrochemical reconstruction	6–10 nm	Higher potassium-ion storage	Potassium-ion batteries	[49]
(Mn,Ni,Mo, Co,Fe)$_3$ O_4	Thermal plasma	13–29 nm	Quasi-spherical morphology	Super-capacitor	[50]

High-entropy nanomaterials have gained prominence in research, particularly for their applications in electrochemical energy storage devices. Ideally, these materials possess a single crystal structure and random uniform distribution of constituent elements. However, in reality, local variations in chemical composition regulated by entropy and diffusion barriers result in short-range order. Therefore, characterization of this multielement system, especially their localized bonding, coordination, and electronic properties, is highly challenging. Scanning electron microscopy (SEM) and transmission electron microscopy (TEM) can be used to investigate their morphology, microstructure, and size; X-ray diffraction (XRD) to determine their crystal phase structure; and X-ray photoelectron microscopy to identify their valence state [51]. However, the localized environment of HEMs cannot be effectively detected using the above conventional techniques due to resolution limitations. Synchrotron-based techniques with much shorter wavelengths provide high resolution for the precise detection of the localized environment. For instance, synchrotron XRD can be used to find phase structure, immiscible phases, and impurities in a HEM. X-ray adsorption spectroscopy (XAS) analyzes the local coordination environment of each element, which is crucial for understanding the possible short-range order in a multielement mixing[52, 53]. *In situ* TEM is used to directly visualize the formation and dispersion of the element during synthesis. Another advanced technique is four-dimensional scanning transmission electron microscopy (4D-STEM), which reveals local lattice distortion, structural heterogeneity, and short-range order. However, the use of these methods is limited due to high costs, extensive maintenance requirements, complex data analysis, and long measurement time.

16.4 Current Status of HEMs as the Electrode in Supercapacitors

In recent years, there has been an increase in the synthesis of HEM-derived supercapacitors [54]. The fascinating properties of HEMs such as HEAs [55] and HEOs [56] have received great attention. Researchers follow several processes such as the dealloying method [57], the liquid phase method at low temperatures [57], the arc and plasma shock method [50], and the carbon thermal shock method to synthesize HEAs. For the synthesis of HEOs, methods such as the solvothermal method [58], the solid-phase reaction method [59], the solution combustion method, and the sol-gel method [60, 61] are used. Various samples are also created by forming composites [62], nitrides [63], phosphides [64], oxides [65], carbides [66], and sulfides [67] of HEMs. The synthesized samples are used as electrode materials for supercapacitor devices. Supercapacitor electrodes derived from these materials show a long stability cycle, high energy and power density, and a desirable potential window [68]. The performance of the material does not depend on mechanical strength; instead, it is based on the morphology, type of material chosen, and entropy value. HEMs have enough strength to reduce the bandgap; thus, they have prominent applications in energy storage. Researchers have also reported that doping with a metal also reduces the bandgap up to a certain extent [69].

16.5 Current Status of HEMs as the Electrode in Lithium-Ion Batteries

In view of lithium-ion batteries, desirable progress can be tracked in recent years. Lithium-ion batteries are rechargeable and portable and have a great influence in the modern society. Hence, an appropriate material for its synthesis should be determined. HEMs are the most effective and suitable candidates when it comes to lithium-ion batteries [70]. In particular, HEOs [71] and HEA [72] have been used as efficient anode materials. Studies have found that these materials have a desirable stability and enhanced electrochemical performance. In a very few research works, HEMs were used as the cathode of the battery [73]. The procedures and materials for battery synthesis are the same as mentioned for supercapacitors. The promising potential of HEMs in battery synthesis can be understood by their ability to dope with various heteroatomic elements [74]. Various research studies have reported high ionic conductivity, high capacity, and enhanced battery life by applying HEMs [75]. The applications of HEMs are not just limited to electrode materials; instead, they can also be used as electrolytes in lithium-ion batteries. In recent research works, liquid-state HEMs are used in lithium-ion batteries for their desirable electrochemical output. These electrolytes are called high-entropy electrolytes [76].

16.6 Current Status of HEMs as the Electrode in Dielectric Materials

Dielectric materials have good insulating properties and energy storage capacity. To meet the demand due to advancements in energy storage devices, many researchers have started focusing on the synthesis of dielectric materials for energy storage. In recent studies, dielectric materials have been found to be superior over high-field and high-energy-density capacitors. In normal parallel plate capacitors, the dielectric is placed between them, but in dielectric capacitors, an external electric field is first applied [77]. Due to this, the dielectric material undergoes polarization, and just after removing the external field, the capacitor releases the stored energy. Thus, the dipole returns to its original state. Dielectric materials have also shown applications in bulk ceramics [78, 79] and multilayer ceramic capacitors. In bulk ceramics, a transition of phase change occurs with heating, and an irreversible phase transformation can be observed during the interaction of the external field [80]. In multilayer ceramics, there are multiple layers of ceramics and internal electrodes [81]. This type of ceramics can be fabricated by following certain procedures such as slurry preparation, tape casting, screen printing, lamination, cosintering, and termination in a systematic way. The effluent electrochemical characteristics can be enhanced by the disruption of long-range polar order, optimization of electrical properties, innovations in processing, and enhancements in high-entropy [82]. In most cases, high-entropy ceramics are used in the development of dielectric materials. They enhance the charge/discharge stability and temperature stability in the field of energy storage. Figure 16.2 shows the

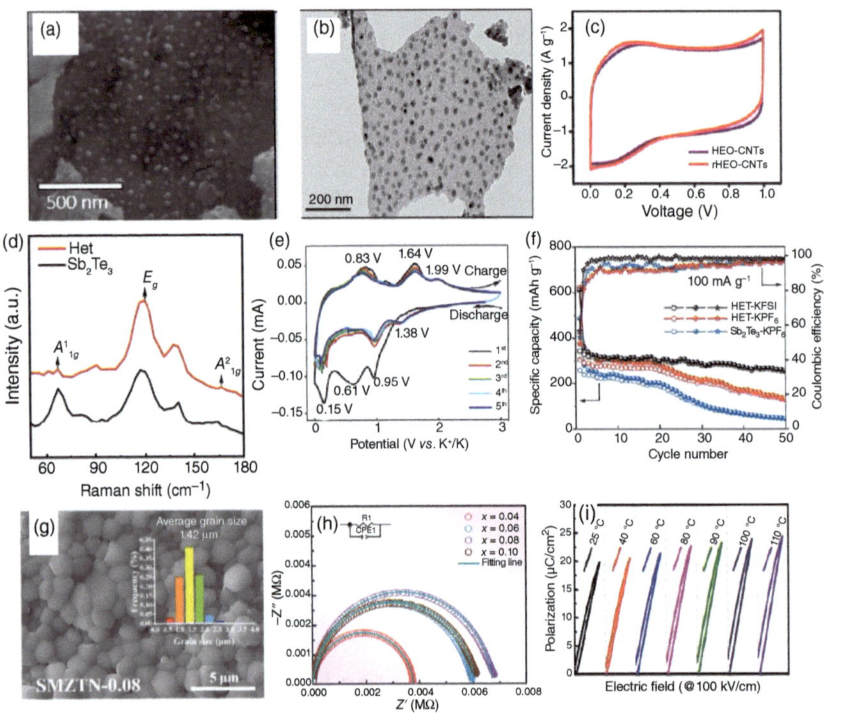

Figure 16.2 The current status of HEMs in energy storage devices. (a) SEM image of the HEO sample. (b) HRTEM image of the HEO nanoparticle. (c) CV curve for the fabricated supercapacitor electrode (HEO–CNT) at a voltage range of 0–1.0 V. *Source:* Reference [83] / American Chemical Society. (d) Raman spectrum plot of the high-entropy telluride as the electrode material for the potassium-ion battery. (e) CV curve of the fabricated battery electrode in the voltage range up to 3 V with a scan rate of 0.1 mV s^{-1} for the potassium-ion battery. (f) Charge discharge capacity of various high-entropy telluride samples at 100 mA g^{-1}. *Source:* Reproduced from Ref. [47] / with permission of American Chemical Society. (g) Morphological SEM image with grain size distribution of SMZTN-0.08 ceramics. (h) Impedance plotting of various samples synthesized of SMZTN-x at a heating temperature of 400 °C. (i) Temperature reliability in energy storage applications for SMZTN-0.08 ceramics. *Source:* Reproduced from Ref. [84] / with permission of American Chemical Society.

current progress of HEMs in energy storage applications. Figure 16.2a represents the SEM image of a HEO sample. Figure 16.2b represents a HRTEM image of the sample derived from the HEO. Figure 16.2c represents the cyclic voltammetry (CV) curve for the constructed electrode of the supercapacitor (HEO–CNT) [83]. Figure 16.2d presents the Raman graph of the high-entropy telluride-derived sample. Figure 16.2e presents the CV curve of the potassium-ion battery electrode in the voltage range of 0–3 V. Figure 16.2f presents the charge and discharge capacity of the high-entropy telluride-based battery [47]. Figure 16.2g presents the SEM image and grain size distribution of the as-synthesized SMZTN-0.08 ceramic. Figure 16.2h presents the impedance analysis of various synthesized samples of SMZTN-x at 400 °C. Figure 16.2i presents the energy storage applications of the SMZTN-0.08 ceramic based on temperature [84].

16.7 Challenges in Using HEMs as the Electrode in Supercapacitors

Till now, many studies have reported the synthesis of HEM-based supercapacitors. Numerous methodologies have been followed for material synthesis. Although there is extensive progress, there are some drawbacks. In many cases, it is quite difficult to determine the composition of material precursors [41]. In addition, the entropy and local structure of the molecules affecting the electrochemical performance are unclear. The materials are composed of multiple cations and an anion, making it difficult to predict the overall performance as it depends on the chemical and physical properties of the anions and cations used. The most common preparation techniques lack nucleation and growth of the synthesized material. As HEMs with complex structures are obtained sometimes, they become immiscible with the compounds that exist in more than one phase [85]. Thus, this makes it difficult to investigate their electrochemical performance. As materials are in a single phase, lattice distortion effect takes place, thus making it difficult to predict or change the morphological structure. The synthesis of HEMs requires highly pure compounds, making the process costly and energy-intensive. Commonly used characterization techniques, including TEM, XRD, EDS, and EXAFS, are efficient in determining the internal composition, but do not provide a good understanding of local structure [86]. Till now, many mechanisms remain undiscovered due to the complexity of the structure of HEMs.

16.8 Challenges in Using HEMs as the Electrode in Lithium-Ion Batteries

HEMs are mostly used as the anode of the battery. Thus, they have a direct impact on the stability, capacity, and charge/discharge cycles of the synthesized device. This is attributable to the complex structure of HEMs. This limits the overall electrochemical properties of the material. Although many studies have been reported, it is extremely challenging to fabricate a battery with an efficient output using HEMs [87]. The theories for the preparation of these materials are not clear, and there is no assurance of getting good outcomes. Furthermore, in many research works, a potential window of 0.1–3 V is obtained, which is inefficient when it comes to lithium-ion batteries [88]. Many studies lack deeper insights into HEMs and their enhanced performance. The lack of kinetics and thermodynamics theories also poses a challenge to in-depth studies of HEMs. In other precursors, the morphology can be modified to the desired shape, but in the case of HEMs, it is a huge challenge [89]. Furthermore, the relationship between the morphology of HEMs and their performance is yet to be determined.

16.9 Challenges in Using HEMs as the Raw Material for the Synthesis of Dielectric Materials

Although numerous research works have been conducted in this field, the mechanisms of HEMs still remain undiscovered, which has become a huge challenge for future researchers because of the complexities in the structure of HEMs. Existing characterization techniques do not provide adequate insights into HEMs. Hence, their internal properties are still a mystery. Synthesis of ceramic films is usually costly because of the use of the advanced physical vapor deposition method, which may cause explosion when directly integrated with the silicon technology [90]. The lack of *in situ* characterization methods makes it a great challenge to examine the final output and the proper internal structure of the synthesized materials [91]. The low efficiency of these materials makes us conclude that a huge amount of energy is consumed while charging. Theoretical research provides the basic idea about HEMs, but more detailed and intrinsic works are lacking [92]. Thus, insights into the synthesis of new and advanced materials with desirable productivity are limited.

16.10 Prospects of HEMs as the Electrode in Supercapacitors

Due to the rapid increase in energy demand, energy should be stored to avoid its loss. Thus, designing an efficient energy storage device is important. Supercapacitors are among those that have a serious impact in the field of energy storage. The materials used in designing supercapacitors for efficient energy storage also play a crucial role. HEMs that can enhance the energy density and power density are found to be suitable precursors for supercapacitors [93]. This has gained significant interest in the scientific community. The challenge in determining a well-structured process for the composition of HEMs can be addressed using data-driven machine learning technologies and artificial intelligence methods, which can lead to easy preparation of the samples with appropriate composition and enhanced properties. The presence of more than one phase can be tackled by developing procedures that can operate under low temperature and pressure conditions [94]. This makes HEMs superior to traditional processes, which have limitations in performing under high temperatures and pressures. To understand the overall performance of HEMs, the relationship between their composition and morphology should be evaluated. Kinetics and thermodynamics analyses can be used to understand the procedures involved in HEMs, which are lacking in the existing processes. To avoid the high cost and high energy consumption due to the use of pure materials, techniques that can help use raw precursors in the synthesis of HEMs are yet to be discovered. For the proper characterization of HEMs, their local structure should be investigated. In this regard, various available characterization techniques such as *in situ* TEM, *in situ* FTIR, *in situ* soft X-ray absorption spectroscopy, and surface-enhanced Raman scattering spectroscopy can be combined to develop a new advanced technique that can give an in-depth understanding of local structure [95]. Many theories regarding the complex structure of HEMs are yet to be validated.

16.11 Prospects of HEMs as the Electrode in Lithium-Ion Batteries

Batteries are another type of energy storage device that can be seen in objects ranging from small toys to huge instruments. Batteries can also be used to avoid the loss of the energy generated, and with proper storage mechanisms, the stored energy can be used in the future. In the fabrication of batteries, choosing appropriate cathode and anode materials is crucial. HEMs are one of these materials that ensure efficient electrochemical productivity and good battery life [96]. Thus, researchers have gained interest in using HEMs as the precursor in battery fabrication. To have a deeper understanding of the complex structure of HEMs, various available characterization techniques can be combined to obtain a new technique. This can address the limitations in synthesizing the desirable material; however, studies focusing on finding the relationship between the composition of HEMs and their performance need to propose ways to synthesize an efficient-energy storage material [97]. To date, many compositions of the materials have not been discovered. Researchers can attempt using other validated compositions through which better results may be obtained. Synthesis procedures that can perform well under low temperature and pressure conditions are yet to be explored to avoid phase transition during material synthesis [98]. There is also an urgent need for materials or methods through which the morphology of HEMs can be altered [99]. To understand the complex structure of HEMs, researchers need to explore data-driven machine learning and artificial intelligence-based methodologies, which predict the performance and properties of the as-synthesized material.

16.12 Prospects of HEMs in the Synthesis of Dielectric Materials

Dielectric materials act as insulators and can store electrical energy when an external electric field is introduced. Therefore, these materials are unique, thus fascinating

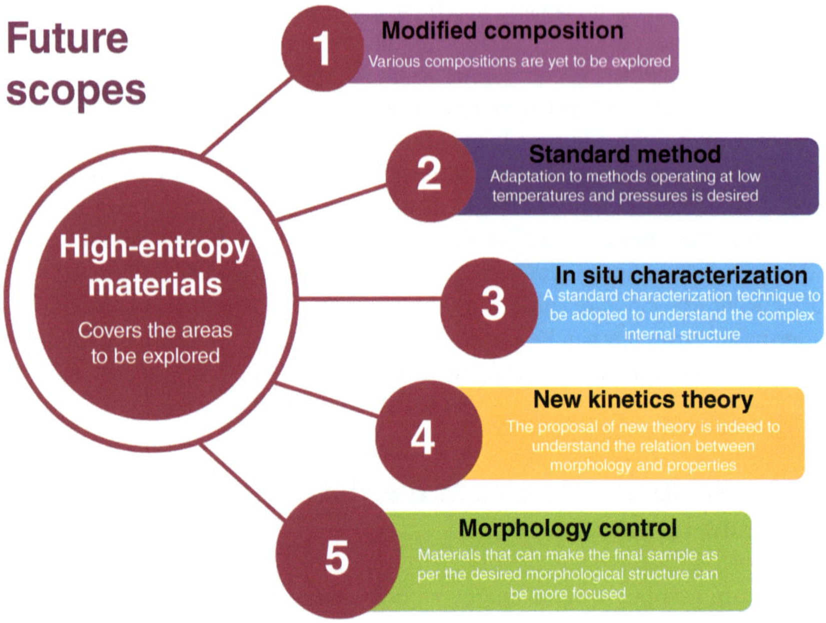

Figure 16.3 Schematic of future scopes for HEMs in various energy storage applications (supercapacitors, batteries, and dielectrics).

researchers worldwide [48]. The dielectric materials synthesized using HEMs are not enough to meet future demands. Thus, a significant increase in the number of research works is needed in this field. An efficient and specific method to understand the complexity in HEMs is yet to be discovered [100]. To gain deeper insights into HEMs, a modified *in situ* characterization technique needs to be developed. In the synthesis of ceramics, a new cost-effective method is desired that is not directly integrated with silicon technology. As energy consumption is higher, it leads to higher energy loss. Thus, metal doping or doping any other materials can be used with dielectrics to reduce energy consumption. Figure 16.3 presents a schematic of future scopes for HEMs in energy storage applications.

References

1 Zhu, J., Hu, L., Zhao, P. et al. (2020). Recent advances in electrocatalytic hydrogen evolution using nanoparticles. *Chem Rev* 120 (2): 851–918.

2 Fareza, A.R., Nugroho, F.A.A., Abdi, F.F., and Fauzia, V. (2022). Nanoscale metal oxides–2D materials heterostructures for photoelectrochemical water splitting – a review. *J Mater Chem A Mater* 10 (16): 8656–8686.

3 Megía, P.J., Vizcaíno, A.J., Calles, J.A., and Carrero, A. (2021). Hydrogen production technologies: from fossil fuels toward renewable sources. A mini Review. *Energy Fuel* 35 (20): 16403–16415.

4 Yue, M., Lambert, H., Pahon, E. et al. (2021). Hydrogen energy systems: a critical review of technologies, applications, trends and challenges. *Renew Sust Energ Rev* 146: 111180.

5 Gao, R., Zhu, J., and Yan, D. (2021). Transition metal-based layered double hydroxides for photo(electro)chemical water splitting: a mini review. *Nanoscale* 13 (32): 13593–13603.

6 Hassan, Q., Viktor, P.J., Al-Musawi, T. et al. (2024). The renewable energy role in the global energy transformations. *Renewable Energy Focus* 48: 100545.

7 Hereher, M. and El Kenawy, A.M. (2020). Exploring the potential of solar, tidal, and wind energy resources in Oman using an integrated climatic-socioeconomic approach. *Renew Energy* 161: 662–675.

8 Roger, I., Shipman, M.A., and Symes, M.D. (2017). Earth-abundant catalysts for electrochemical and photoelectrochemical water splitting. *Nat Rev Chem* 1 (1): 3.

9 Shruti, M.S., Khilari, S., Samuel, E.J.J. et al. (2023). Recent trends in graphene assisted vanadium based nanocomposites for supercapacitor applications. *J Energy Storage* 63: 107006.

10 Palm, R., Baena-Moncada, A.M., and Gonçalves, J.M. (2024). From medium- to high-entropy hydroxides for hybrid supercapacitors: a review. *J Mater Chem A Mater* 12 (43): 29402–29431.

11 Rudra, S., Nayak, A.K., Chakraborty, R. et al. (2018). Synthesis of Au-V_2O_5 composite nanowires through the shape transformation of a vanadium(iii) metal complex for high-performance solid-state supercapacitors. *Inorg Chem Front* 5 (8): 1836–1843.

12 Mishra, A.K., Nayak, A.K., Das, A.K., and Pradhan, D. (2018). Microwave-assisted solvothermal synthesis of cupric oxide nanostructures for high-performance supercapacitor. *J Phys Chem C* 122 (21): 11249–11261.

13 Masa, J., Andronescu, C., and Schuhmann, W. (2020). Electrocatalysis as the Nexus for sustainable renewable energy: the Gordian knot of activity, stability, and selectivity. *Angew Chem Int Ed* 59 (36): 15298–15312.

14 Guo, R., Yang, Y., Zhao, C. et al. (2024). The role of high-entropy materials in lithium-based rechargeable batteries. *Adv Funct Mater* 34 (18): 2313168.

15 Wu, J., Wang, H., Liu, N. et al. (2024). High-entropy materials in electrocatalysis: understanding, design, and development. *Small* 20 (43): 2403162.

16 Yan, X., Zhou, Y., and Wang, S. (2025). Nano-high entropy materials in electrocatalysis. *Adv Funct Mater* 35 (2): 2413115.

17 Dusza, J., Švec, P., Girman, V. et al. (2018). Microstructure of (Hf-Ta-Zr-Nb)C high-entropy carbide at micro and nano/atomic level. *J Eur Ceram Soc* 38 (12): 4303–4307.

18 Feng, L., Fahrenholtz, W.G., Hilmas, G.E., and Zhou, Y. (2019). Synthesis of single-phase high-entropy carbide powders. *Scr Mater* 162: 90–93.

19 Zhou, J., Zhang, J., Zhang, F. et al. (2018). High-entropy carbide: a novel class of multicomponent ceramics. *Ceram Int* 44 (17): 22014–22018.

20 Chen, K., Xie, H., Yang, H. et al. (2025). Construction of crystalline/amorphous heterostructures to enhance the supercapacitor performance of a high-entropy sulfide. *J Alloys Compd* 1015: 178921.

21 Wang, S., Wen, X., Huang, Z. et al. (2025). High-entropy strategy flattening lithium ion migration energy landscape to enhance the conductivity of garnet-type solid-state electrolytes. *Adv Funct Mater* 35 (9): 2416389.

22 Ye, Y., Gu, Z., Geng, J. et al. (2025). Advanced high-entropy halide solid electrolytes enabling high-voltage, long-cycling all-solid-state batteries. *Nano Lett* 25 (10): 3747–3755.

23 Guo, M., Liu, Y., Zhang, F. et al. (2022). Inactive Al3+-doped La(CoCrFeMnNiAlx)1/(5+x)O_3 high-entropy perovskite oxides as high performance supercapacitor electrodes. *J Adv Ceram* 11 (5): 742–753.

24 Wang, Y., Liu, J., Cao, H. et al. (2025). Facile synthesis of porous high-entropy perovskite nanoparticles through MOF gel method for solid-state supercapacitor application. *Chem Eng J* 509: 161246.

25 Yeh, J.-W. (2013). Alloy design strategies and future trends in high-entropy alloys. *JOM* 65 (12): 1759–1771.

26 Miracle, D.B. and Senkov, O.N. (2017). A critical review of high entropy alloys and related concepts. *Acta Mater* 122: 448–511.

27 George, E.P., Raabe, D., and Ritchie, R.O. (2019). High-entropy alloys. *Nat Rev Mater* 4 (8): 515–534.

28 Zhang, Y., Zuo, T.T., Tang, Z. et al. (2014). Microstructures and properties of high-entropy alloys. *Prog Mater Sci* 61: 1–93.

29 Ma, Y., Ma, Y., Wang, Q. et al. (2021). High-entropy energy materials: challenges and new opportunities. *Energy Environ Sci* 14 (5): 2883–2905.

30 Tsai, M.-H., Yeh, J.-W., and Gan, J.-Y. (2008). Diffusion barrier properties of AlMoNbSiTaTiVZr high-entropy alloy layer between copper and silicon. *Thin Solid Films* 516 (16): 5527–5530.

31 Xin, Y., Li, S., Qian, Y. et al. (2020). High-entropy alloys as a platform for catalysis: progress, challenges, and opportunities. *ACS Catal* 10 (19): 11280–11306.

32 Sun, L., Wen, K., Li, G. et al. (2024). High-entropy alloys in catalysis: progress, challenges, and prospects. *ACS Mater Au* 4 (6): 547–556.

33 Yao, Y., Huang, Z., Xie, P. et al. (2018). Carbothermal shock synthesis of high-entropy-alloy nanoparticles. *Science* 359 (6383): 1489–1494.

34 Lung, J.-K., Huang, J.-C., Tien, D.-C. et al. (2007). Preparation of gold nanoparticles by arc discharge in water. *J Alloys Compd* 434–435: 655–658.

35 Nair, R.B., Arora, H.S., Mandal, P. et al. (2018). High-performance microwave-derived multi-principal element alloy coatings for tribological application. *Adv Eng Mater* 20 (9): 1800163.

36 Lin, Z., Yue, J., Liang, L. et al. (2020). Rapid synthesis of metallic and alloy micro/nanoparticles by laser ablation towards water. *Appl Surf Sci* 504: 144461.

37 Feng, Y., Yang, L., Yan, Z. et al. (2023). Discovery of high entropy garnet solid-state electrolytes via ultrafast synthesis. *Energy Storage Mater* 63: 103053.

38 Feng, G., Ning, F., Song, J. et al. (2021). Sub-2 nm ultrasmall high-entropy alloy nanoparticles for extremely superior electrocatalytic hydrogen evolution. *J Am Chem Soc* 143 (41): 17117–17127.

39 Nayak, A.K., Lee, S., Sohn, Y., and Pradhan, D. (2014). Synthesis of In2S3 microspheres using a template-free and surfactant-less hydrothermal process and their visible light photocatalysis. *CrystEngComm* 16 (34): 8064–8072.

40 Yao, C.-Z., Zhang, P., Liu, M. et al. (2008). Electrochemical preparation and magnetic study of Bi–Fe–Co–Ni–Mn high entropy alloy. *Electrochim Acta* 53 (28): 8359–8365.

41 Talluri, B., Aparna, M.L., Sreenivasulu, N. et al. (2021). High entropy spinel metal oxide $(CoCrFeMnNi)_3O_4$ nanoparticles as a high-performance supercapacitor electrode material. *J Energy Storage* 42: 103004.

42 Pasupathi, A. and Subramaniam, Y. (2024). A novel strategy for rapid synthesis of nanostructured high-entropy metal oxides through thermal plasma for supercapacitor applications. *Energy Fuel* 38 (6): 5534–5544.

43 Nair, N., Nair, S.V., and Baskar, S. (2024). High Na-content P2 and P2-O_3 layered oxide cathodes for high performance Na-ion batteries. *Mater Lett* 354: 135397.

44 Fan, J., Wang, T., Bridges, C.A. et al. (2023). Entropy stabilized cubic $Li_7La_3Zr_2O_{12}$ with reduced lithium diffusion activation energy: studied using solid-state NMR spectroscopy. *RSC Adv* 13 (29): 19856–19861.

45 Huang, C., Luo, J., Mansley, Z.R. et al. (2024). Manganese-rich high entropy oxides for lithium-ion batteries:materials design approaches to address voltage fade. *J Mater Chem A Mater* 12 (38): 26253–26265.

46 Li, L., Ji, P., Geng, C. et al. (2023). Facile synthesis of high-entropy $(Co_{0.2}Cr_{0.2}Fe_{0.2}Mn_{0.2}Ni_{0.2})_3O_4$ nanopowders and their electrochemical properties as supercapacitor electrode. *J Energy Storage* 73: 109182.

47 Wang, Z., Qiao, S., Ma, M. et al. (2025). High-entropy conversion-alloying anode material for advanced potassium-ion batteries. *ACS Nano* 19: 15148–15160.

48 Zuo, C., Yang, S., Cao, Z. et al. (2023). Combustion synthesis of high-performance high-entropy dielectric ceramics for energy storage applications. *Ceram Int* 49: 25486–25494.

49 Chien, P.W., Chang, C.B., and Tuan, H.Y. (2023). High-entropy two-dimensional metal phosphorus trichalcogenides boost high-performance potassium ion storage devices via electrochemical reconstruction. *Energy Storage Mater* 61: 102853.

50 Pasupathi, A., Perumal, M., Narayanamoorthi, E. et al. (2024). Electrochemical charge storage performance of $(Mn,Ni,Mo,Co,Fe)_3O_4$ high entropy oxide nanoparticles produced via thermal plasma route. *Ceram Int* 50: 26740–26749.

51 Zhang, L., Ran, J., Qiao, S.-Z., and Jaroniec, M. (2019). Characterization of semiconductor photocatalysts. *Chem Soc Rev* 48 (20): 5184–5206.

52 Yao, Y., Huang, Z., Hughes, L.A. et al. (2021). Extreme mixing in nanoscale transition metal alloys. *Matter* 4 (7): 2340–2353.

53 Miao, J., Ercius, P., and Billinge, S.J.L. (2016). Atomic electron tomography: 3D structures without crystals. *Science* 353 (6306): aaf2157.

54 Vadivel, S., Dharmavarapu, V.S.H., Sujita, P. et al. (2025). A robust approach for designing a bismuth high entropy material $(BiOXCO_3)$ as a novel electrode material for supercapacitor applications. *Mater Res Bull* 184: 113225.

55 Mohanty, G.C., Chowde Gowda, C., Gakhad, P. et al. (2024). Enhanced energy density of high entropy alloy (Fe-Co-Ni-Cu-Mn) and green graphene hybrid supercapacitor. *Energy Storage* 6: e668.

56 Chillal, K., Halder, A., Jha, S.K. et al. (2024). Electrochemical investigation of synthesized $(Mg_{0.21}Cr_{0.21}Mn_{0.21}Fe_{0.21}Cu_{0.16})_3O_4$ high entropy oxide for supercapacitor electrode material. *Ceram Int* 51: 3890–3898.

57 Abid, T., Akram, M.A., Yaqub, T.B. et al. (2024). Design and development of porous CoCrFeNiMn high entropy alloy (Cantor alloy) with outstanding electrochemical properties. *J Alloys Compd* 970: 172633.

58 Hota, A., Das, J.K., Mohammed, A.A. et al. (2025). Temperature-mediated phase control in high entropy transition metal oxides for hybrid supercapacitors. *New J Chem* 49: 7824–7832.

59 Hu, H., Yang, C., Chen, F. et al. (2024). High-entropy engineering reinforced surface electronic states and structural defects of hierarchical metal oxides@graphene fibers toward high-performance wearable supercapacitors. *Adv Mater* 36: e2406483.

60 Gupta, A.K., Kumar, A., Marndi, M. et al. (2024). Studies on supercapacitor electrode performance of novel $(AlCuCoFeMnNi)_3O_4$ high entropy spinel oxide admixed with different carbon-based additives. *Energy Storage* 6: e70002.

61 Yin, Y., Zhang, W.B., Chen, B. et al. (2024). Porous powder synthesis of $(FeCoCrMnNi)_3O_4$ high entropy ceramic and its supercapacitor applications. *Ceram Int* 50: 10292–10304.

62 Guan, X., Fan, X., Zhu, E. et al. (2024). Controlled establishment of advanced local high-entropy NiCoMnFe-based layered double hydroxide for zinc batteries and low-temperature supercapacitors. *J Colloid Interface Sci* 658: 952–965.

63 Jiang, J., Xu, Y., Wang, Z. et al. (2025). High-entropy nitrides from dual entropic and enthalpic forces for high-efficiency oxygen evolution reaction. *Energy Mater* 5.

64 Feng, J., Zhang, W.-B., Batol, A. et al. (2025). Electrochemical capacitance rectification of FeCoCrMnNiP electrode in microcircuit rectification supercapacitor application. *Electrochim Acta* 533: 146513.

65 Mohanty, G.C., Ashish, A., Das, S., and Verma, A. (2024). Fabrication of aqueous asymmetric supercapacitor device by using spinel type $(FeCoNiCuZn)_3O_4$ high entropy oxide and green carbon derived from plastic wastes. *Ceram Int* 48938–48947.

66 Chen, B., Zhang, X.L., Zhang, W.B. et al. (2024). Electrochemical pseudocapacitance performance of high entropy carbide (FeCoCrMnNi)C film by magnetron sputtering technology. *Ceram Int* 50: 23302–23314.

67 Wang, S., Ji, Y., Zhang, B. et al. (2025). Walnut-like high-entropy sulfides via facile route for enhanced supercapacitor performance. *J Energy Storage* 110: 115336.

68 Shi, S.-J., Wei, L.-Z., Guo, M.-Y. et al. (2025). Self-supporting high-entropy-sulfide @ni electrode constructed by freeze casting and hydrothermal synthesis for supercapacitors. 136: 118460.

69 Wan, Z.Y., Lv, J.B., and Zhang, Y.X. (2025). Sr_{2+}-doped $La_{1-x}Sr_x(Co_{0.2}Mn_{0.2}Fe_{0.2}Ni_{0.2}Cu_{0.2})O_3$ high-entropy perovskite oxides as supercapacitor electrodes. *Ionics (Kiel)* 31: 3671–3682.

70 Lei, X., Wang, Y., Wang, J. et al. (2024). Si-based high-entropy anode for lithium-ion batteries. *Small Methods* 8: 2300754.

71 Wang, X.L., Kim, E.M., Senthamaraikannan, T.G. et al. (2024). Porous hollow high entropy metal oxides $(NiCoCuFeMg)_3O_4$ nanofiber anode for high-performance lithium-ion batteries. *Chem Eng J* 484: 149509.

72 Su, Y., Lei, X., Chen, W. et al. (2024). Si-based all-lithium-reactive high-entropy alloy for thin-film lithium-ion battery anode. *Chem Eng J* 500: 157197.

73 Kim, J., Yang, S., Zhong, Y. et al. (2025). High-entropy Li-rich layered oxide cathode for Li-ion batteries. *J Power Sources* 628: 235915.

74 Zhou, J., Hu, J., Zhou, X. et al. (2024). High-entropy doping for high-performance zero-cobalt high-nickel layered cathode materials. *Energy Environ Sci* 628: 235915.

75 Zeng, S., Zhu, Y., Si, J. et al. (2025). High-entropy doping for high-voltage $LiCoO_2$ with enhanced electrochemical performances. *J Power Sources* 626: 235726.

76 Wang, Q., Wang, J., Heringa, J.R. et al. (2024). High-entropy electrolytes for lithium-ion batteries. *ACS Energy Lett* 9: 3796–3806.

77 Zhou, Y., Zeng, Y., Wang, J. et al. (2024). Enhancement of the voltage output of droplet electricity generators using high dielectric high-entropy oxide composites. *Energy Environ Sci* 17: 3580–3593.

78 Zhang, J., Tian, J., Xing, B. et al. (2024). Preparation and thermal/dielectric properties of medium/high entropy perovskite titanate ceramics. *Ceram Int* 50: 29954–29965.

79 Zhao, H., Cao, W., Liang, C. et al. (2024). High-entropy design toward ultrahigh energy storage density under moderate electric field in bulk lead-free ceramics. *Adv Funct Mater* 35: 2411954.

80 Shih, H.-J., Shih, C.-T., and Lee, Y.-C. (2025). Microwave dielectric properties and microstructure of high entropy $(Mg_xCa_{0.6-x}Sr_{0.39}Ba_{0.01})ZrO_3$ ceramics. *Open Ceram* 22: 100793.

81 Liu, D., Wang, L., Wang, X. et al. (2025). Excellent energy storage performances for $BaTiO_3$-based multilayer capacitors through synergistic high-entropy and superparaelectric-relaxor strategy. *J Mater* 11: 100860.

82 Liu, H., Cao, L., Gu, J. et al. (2025). High-entropy processed high quality and low-temperature cofired $LiMgPO_4$-based dielectric ceramics for low-loss packaged millimeter-wave filters. *J Eur Ceram Soc* 45: 116957.

83 Lal, M.S. and Sundara, R. (2019). High entropy oxides – a cost-effective catalyst for the growth of high yield carbon nanotubes and their energy applications. *ACS Appl Mater Interfaces* 11: 30846–30857.

84 Chen, S., Wang, T., Xie, S. et al. (2024). High-entropy strategy for improved mechanical and energy storage properties in $BaTiO_3$-$BiFeO_3$-based ceramics. *ACS Appl Mater Interfaces* 16: 12521–12533.

85 Mohanty, G.C., Gowda, C.C., Gakhad, P. et al. (2023). High energy density liquid state asymmetric supercapacitor devices using Co-Cr-Ni-Fe-Mn high entropy alloy. *Mater Adv* 4: 3839–3852.

86 Yin, Y., Zhang, W.B., Zhang, X.L. et al. (2023). Low-dimensional high entropy oxide $(FeCoCrMnNi)_3O_4$ for supercapacitor applications. *Dalton Trans* 52: 9005–9016.

87 Lin, L., Wang, K., Sarkar, A. et al. (2022). High-entropy sulfides as electrode materials for Li-ion batteries. *Adv Energy Mater* 12: 2103090.

88 Xiao, B., Wu, G., Wang, T. et al. (2022). High-entropy oxides as advanced anode materials for long-life lithium-ion Batteries. *Nano Energy* 95: 106962.

89 Wang, Q., Sarkar, A., Li, Z. et al. (2019). High entropy oxides as anode material for Li-ion battery applications: a practical approach. *Electrochem Commun* 100: 121–125.

90 Salian, A., Pujar, P., Vardhan, R.V. et al. (2023). Evolution of high dielectric permittivity in low-temperature solution combustion-processed phase-pure high entropy oxide (CoMnNiFeCr)O for thin film transistors. *ACS Appl Electron Mater* 5: 2608–2623.

91 Schweidler, S., Botros, M., Strauss, F. et al. (2024). High-entropy materials for energy and electronic applications. *Nat Rev Mater* 9: 266–281.

92 Zhou, S., Pu, Y., Zhang, Q. et al. (2020). Microstructure and dielectric properties of high entropy $Ba(Zr_{0.2}Ti_{0.2}Sn_{0.2}Hf_{0.2}Me_{0.2})O_3$ perovskite oxides. *Ceram Int* 46: 7430–7437.

93 Zhu, Z., Zhang, Y., Kong, D. et al. (2024). A novel high entropy hydroxide electrode material for promoting energy density of supercapacitors and its efficient synthesis strategy. *Small* 20: 2307754.

94 Mohanty, G.C., Chowde Gowda, C., Gakhad, P. et al. (2023). Iron-cobalt-nickel-copper-zinc (FeCoNiCuZn) high entropy alloy as positive electrode for high specific capacitance supercapacitor. *Electrochim Acta* 470: 143272.

95 Shen, E., Song, X., Chen, Q. et al. (2021). Spontaneously forming oxide layer of high entropy alloy nanoparticles deposited on porous carbons for supercapacitors. *ChemElectroChem* 8: 260–269.

96 Duan, C.Q., Tian, K., Li, X. et al. (2021). New spinel high-entropy oxides $(FeCoNiCrMnXLi)_3O_4$ (X = Cu, Mg, Zn) as the anode material for lithium-ion batteries. *Ceram Int* 47: 32025–32032.

97 Zou, X., Zhang, Y.R., Huang, Z.P. et al. (2023). High-entropy oxides: an emerging anode material for lithium-ion batteries. *Chem Commun* 59: 13535–13550.

98 Xiao, B., Wu, G., Wang, T. et al. (2021). High entropy oxides $(FeNiCrMnX)_3O_4$ (X = Zn, Mg) as anode materials for lithium ion batteries. *Ceram Int* 47: 33972–33977.

99 Yang, X., Wang, H., Song, Y. et al. (2022). Low-temperature synthesis of a porous high-entropy transition-metal oxide as an anode for high-performance lithium-ion batteries. *ACS Appl Mater Interfaces* 14: 26873–26881.

100 Zhu, X., Xiong, S., Zhu, G. et al. (2024). Dielectric properties and excellent energy storage density under low electric fields for high entropy relaxor ferroelectric $(Li_{0.2}Ca_{0.2}Sr_{0.2}Ba_{0.2}La_{0.2})TiO_3$ ceramic. *J Alloys Compd* 984: 173987.

Index

High-Entropy Materials for Energy Storage Devices, First Edition.
Edited by Chien-Te Hsieh, Pradeep Kumar Panda, and Arpan Kumar Nayak.

d

t